PROGRESS IN DIFFERENTIATION RESEARCH

# PROGRESS IN DIFFERENTIATION RESEARCH

Proceedings of the Second International Conference on Differentiation Copenhagen, Denmark, 8-12 September, 1975

*Editor:*

N. Müller-Bérat

*Co-editors:*

C. Rosenfeld
D. Tarin
D. Viza

1976

NORTH-HOLLAND PUBLISHING COMPANY - AMSTERDAM · OXFORD
AMERICAN ELSEVIER PUBLISHING COMPANY, INC. - NEW YORK

North-Holland ISBN : 0 7204 0580 7
American Elsevier ISBN: 0 444 11200 6

Published by:

North-Holland Publishing Company – Amsterdam
North-Holland Publishing Company, Ltd. – Oxford

Sole distributors for the U.S.A. and Canada:

American Elsevier Publishing Company, Inc.
52 Vanderbilt Avenue
New York, N.Y. 10017

**Library of Congress Cataloging in Publication Data**

International Conference on Differentiation, 2d,
Copenhagen, 1975.
Progress in differentiation research.

Includes index.
1. Cell differentiation. I. Müller-Bérat, N.
II. Title.
QH607.148 1975 574.8'761 76-7349
ISBN 0-444-11200-6 (American Elsevier)

# Preface

This book does not pretend to provide ready made solutions nor stunning slogans applicable to any of the intriguing problems in the field of "Differentiation". It hardly even provides a definition of the word Differentiation. This difficulty in approaching a satisfying definition of the general phenomenon of Differentiation occurs because technology and experimental models have developed tremendously since embryologists founded this discipline which has extended in numerous directions including studies on adult tissues and disease processes. Therefore at this time even a partial synthesis of information and views is useful. This was the purpose of the Second International Conference on Differentiation and is also the purpose of this book which records many of the contributions to that meeting.

At certain periods of scientific activity, workers in a field may feel that they have reached a plateau where only details can be added to their models. We hope that the assembly of data from various workers presented here in the form of new experimental approaches, speculations and partial conclusions, can open up new avenues for progress in the study of Differentiation. The subdivisions of this volume are obviously artificial since, as stated above, the objective was to provide a synthesis of available information to the elucidation of the general principles involved. For instance, the data on cell surface physiology studied on normal tissue and its neoplastic counterpart could as well have fitted in the section on cell surfaces as in the section on neoplasia. General considerations on hemopoietic stem cells could as as well have fitted in the section on cell proliferation as in that on hemopoiesis. Some systematic scheme of presentation is essential, however, and we rely on the desired synthesis taking place in the mind of the reader. We are most indebted to the President of the Society of Differentiation, Professor Brachet, and to the members of the Steering Committee as well as to many of the Guest Lecturers for their advices in selecting the program of these Proceedings.

N. Müller-Bérat

December, 1975

# Contents

SECTION 6. NORMAL AND MALIGNANT HEMOPOIESIS AS A MODEL OF DIFFERENTIATION

# Section 1.
# CELL PROLIFERATION, GROWTH, AND EXPRESSION OF DIFFERENTIATION POTENTIAL BY PROLIFERATING CELLS

*Progress in Differentiation Research, ed. N. Müller-Bérat et al.*



# LINEAGES, QUANTAL CELL CYCLES AND CELL DIVERSIFICATION.

H. Holtzer and S. Holtzer
Department of Anatomy, School of Medicine
University of Pennsylvania

## 1. Introduction

Much work over the past two decades has attempted to demonstrate a primary role for exogenous molecules and cell-cell interactions in diversifying "undifferentiated" or "multipotential" cells [1-3]. In our opinion this vast literature in fact demonstrates that exogenous molecules merely permit already diversified cells to express metabolic options that otherwise might remain unexpressed. Cell-cell interactions, nerve growth factor[4], colony stimulating factors[5,6], etc. at most function as hormones[7,8]. They cue already diversified target cells that have been differentiated by earlier endogenous events. Hydrocortisone, for example, induces three very different metabolic reactions in chondroblasts, liver cells and oocytes respectively. Clearly even before being exposed to the hormone the chondroblast, the liver cell, or the oocyte was primed to respond, each in its unique way, by prior diversifying events. Changes in a cell's microenvironment cannot be a primary "cause" of cell diversification any more than a hormone can "cause" the phenotype of its target cell. Though exogenous molecules may be required for the expression of a particular differentiated state, or influence a particular "binary decision"[9,10], in general they must be of secondary importance regarding the basic mechanisms of cell diversification.

Recently we have reviewed the old idea that lineages are central to cell diversification[10,11]. A scheme which postulates the passage of cells through an obligatory sequence of compartments within a lineage is incompatible with any scheme based on the diversification of virginal, undifferentiated cells by means of alterations in the cells microenvironment. The concept of lineage excludes the existence of undifferentiated cells. The concept of lineage requires, however, a mechanism that promotes transit of cells through successive compartments within the lineage. It must also account for the way in which cells within one compartment, endowed with one set of circumscribed metabolic options, can yield daughter cells with predictably different metabolic options. We have proposed, accordingly that transition from one compartment to the next requires reprogramming of the genome, and that this type of reprogramming - in contrast to changes induced by hormone line agents - requires DNA synthesis and passage through a "quantal cell cycle". By definition, a quantal cell cycle is one that allows a region of the genome that was not available for transcription in the mother cell to become available for transcription in the daughter cell[8,10,11]. Quantal cell cycles diversify by "moving" cells from one compartment in the lineage into the next. This type of cell cycle differs from the more commonly studied cell cycle, the proliferative cell cycle, in which the daughter's options for transcription do not differ from those of the mother. Proliferative cell cycles increase the numbers of cells within a compartment. The mosaic type of developmental system is characterized by a succession of quantal cell cycles with no intervening proliferative cell cycles, whereas regulative systems permit variable numbers of proliferative cycles between the fixed number of quantal cell cycles.

## 2. Erythrogenesis

During erythrogenesis there may be no more than 5 quantal cell cycles separating the zygote from the 1st generation erythroblast [8,11,12]. Here we will focus on that quantal cell cycle effecting transit from the erythrogenic hematocytoblast compartment.

The primitive line of chick erythroblasts develop as a relatively homogenous cohort of cells and the Hb can be measured microspectrophotometrically in individual cells [13,14]. The erythrogenic homatocytoblasts do not synthesize Hb. However by the time their daughter cells, the 1st generation erythroblasts, are only a few hours old they synthesize detectable quantities of Hb. Experiments based on uptake of $^{3}$H-TdR demonstrated that a single erythrogenic hematocytoblast divides symmetrically, yielding two 1st generation erythroblasts[14,15]. These 1st generation erythroblasts serve as progenitors for six subsequent generations of erythroblasts. Each class of erythroblasts synthesizes characteristic amounts of Hb and displays characteristic cytologies and characteristic G1, S and G2 periods. The hematocytoblast and at least two different classes of erythroblasts co-exist in the same circulation. This finding means that microenvironmental influences alone cannot influence the cell divisions that separate one class of erythroblast from another, nor induce that cell division which separates erythrogenic hematocytoblast from 1st generation erythroblast. These and other observations [11,12] have lead us to conclude that the hematocytoblast is endogenously programmed in such a way that its daughter erythroblasts serve as progenitors for only six generations, or 128 Hb-producing cells. This limited potential for replication of 1st generation erythroblasts is also displayed by other Hb-producing cells. Once normal erythrogenic cells or Friend leukemic cells induced with DMSO initiate the synthesis of Hb they function under stringent restraints with regard to DNA synthesis. There are no tumors in which each neoplastic cell synthesizes large quantities of Hb. Cells synthesizing significant quantities of Hb do not have the option to yield the large progeny that characterize "stem" cells earlier in the lineage. A 1st generation erythroblast yields 128 descendents, whereas a 5th generation erythroblast yields two postmitotic daughters.

Current evidence favors the notion that a given erythrogenic hematocytoblast divides symmetrically, yielding two 1st generation erythroblasts. This means that with respect to the broader hematopoetic lineages, the erythrogenic hematocytoblast must be past the binary decision point [9,10,11] where a given stem cell had the option to divide assymetrically and establish white and red blood cell lineages respectively.

With regard to Hb synthesis there are significant differences between the erythrogenic hematocytoblasts and the erythroblasts in response to BudR and to the DNA inhibitors, Ara-C and FudR[16] Incorporation of BudR into the DNA of replicating erythroblasts does not block these cells from continuing to synthesize Hb. In contrast, incorporation of BudR into the DNA of replicating erythrogenic hematocytoblasts blocks the initiation of Hb synthesis in cells that otherwise would have begun to synthesize this molecule[17]. Blocking DNA synthesis with either Ara-C or FudR does not block any class of erythroblast from continuing to synthesize Hb. Indeed Ara-C blocked 3rd generation erythroblasts continue to make Hb until their content exceeds that of 4th generation erythroblasts. These blocked cells enlarge greatly and eventually lyse. In contrast, Ara-C treated hematocytoblasts do not initiate the synthesis of Hb. If embryos rich in erythrogenic hematocytoblasts are blocked from synthesizing DNA before erythroblasts have emerged they never synthesize Hb. Taken together, these experiments show that movement from the penultimate compartment into the erythroblast compartment is not simply dependent upon "maturation" of the erythrogenic hematocytoblast. The hematocytoblast itself cannot become an erythroblast. Only if the erythrogenic hematocytoblast synthesizes DNA and completes its nuclear division will cells emerge that have the option to synthesize Hb. Work to be published elsewhere has demonstrated that the relationship between the absence of Hb in the hematocytoblasts and its presence in their daughter erythroblasts also extends to the appearance of spectrin and carbonic anhydrase.

To determine whether the failure of erythrogenic hematocytoblasts to synthesize Hb was due to a failure in transcription or translation, the following experiments were performed: $^{3}$H-cDNA against adult Hb-mRNA was prepared using E. coli reverse transcriptase[18]. Hybridization experiments with this labelled probe revealed that if globin message was transcribed in the erythrogenic hematocytoblast it was less than $10^{-5}$ times that found in erythroblasts. From this we concluded that : (1) erythrogenic hematocytoblasts did not transcribe active or inactive globin messages, and (2) one consequence of the quantal cell cycle was to make the globin genes that were not available for transcription in the hematocytoblast, available for transcription in their daughter erythroblasts. Needless to say it would be of interest to know whether BudR and the Friend leukimic virus act on the same "master locus" and that it is this locus that is altered during passage through the quantal cell cycle which separates the hematocytoblast from the erythroblasts.

3. Myogenesis

Since the crucial role of quantal cell cycles during myogenesis has been reviewed recently [10,11] only a brief account will be sketched here. If it is assumed that the same mesenchyme cell (Ms cell) is the common progenitor to the myogenic, chondrogenic, and fibrogenic lineages, then at a minimum three quantal cell cycles separate the Ms cell from the postmitotic myoblast. One quantal cell cycle separates the Ms cell from the PMbFb compartment and the PCbFb compartment. The PMbFb cell is a bipotential cell that yields a presumptive myoblast and/or definitive fibroblast, whereas the PCBFb cell is a bipotential cell that yields a chondroblast and/or fibroblast [9,10,11].

The abrupt switch in metabolic options which accompany a quantal cell cycle is best documented by observing the changes in synthesis of the myosin heavy and light chains by presumptive myoblasts on the one hand and their daughter postmitotic myoblasts on the other. It has been known for years that replicating presumptive myoblasts do not bind labelled antibody against skeletal myosin, whereas their postmitotic daughters, the myoblasts, do [19,20]. These old findings have recently been confirmed by collecting large numbers of mononucleated, postmitotic myoblasts with Cytochalasin-B and EGTA [21,22].

By culturing replicating presumptive myoblasts, fibroblasts, chondroblasts and BudR-suppressed myogenic cells in $^{14}C$-leucine it can be shown that all of these kinds of cells synthesize a 200,000 myosin heavy chain. The myosins from these cells in SDS are indistinguishable from the myosins isolated from myoblasts, myotubes and mature muscle. Nevertheless, in Ouchterlony diffusion plates antibody against skeletal heavy chains does not even cross-react with the myosins from presumptive myoblasts or from non-myogenic cells. These findings confirm the suggestion[23] that the myosin heavy chain synthesized in the myoblast is the product of a different structural gene from that active in its mother, the replicating presumptive myoblast. These results are contrary to the many reports by Yaffe and co-workers [24,25] and by Strohman and co-workers [26,27]. Work in progress is designed to determine whether the myosin in replicating presumptive myoblasts is the product of the same structural gene active in such non-myogenic cells as fibroblasts, chondroblasts and nerve cells. If this should prove to be the case this myosin might be considered a "constitutive" contractile protein.

Given current interest in contractile proteins in non-muscle cells, it is worth mentioning that the myosin molecules assembled into thick filaments in myoblasts are different from those that may be associated with a) the outside of the plasma membrane[28], or (2) the microfilaments sub-tending the plasma membrane[29]. Needless to say the issue of polymorphism of a molecule such as myosin gives a new dimension to the problem of cell diversification[30].

Chi et al.[31,32] have found comparable differences in the myosin light chains synthesized by myoblasts versus those synthesized by presumptive myoblasts and non-myogenic cells. Myoblasts synthesize the definitive light chains found in mature skeletal muscle. They have molecular weights of 25,000 and 18,000 respectively. Presumptive myoblasts and non-myogenic cells (eg. fibroblasts, chondroblasts, nerve cells and BudR-suppressed myogenic cells) synthesize two light chains that are indistinguishable from one another in SDS gels; their molecular weights are 20,000 and 16,000. Again the issue is open as to whether these are "constitutive" light chains common to all cells, whereas the definitive myoblasts synthesize their characteristic light chains only after passing through their terminal quantal cell cycle.

The mother of the postmitotic myoblast, the replicating presumptive myoblast, does not have the option to fuse to form myotubes[10,20,33] or to initiate the synthesis of the definitive myosin heavy and light chains. These are the unique properties of postmitotic myoblasts. This transition between replicating presumptive myoblast and myoblast is not dependent upon collagen[34,35] or other mysterious factors in conditioned medium. Transition, however is blocked by the incorporation of BudR into the DNA of the presumptive myoblast[20,36,37]. BudR-suppressed myogenic cells do not withdraw from the cell cycle. Operationally the analog appears to promote proliferative cell cycles and thus preclude the quantal cell cycle required for transit into the myoblast compartment. In this regard BudR in the DNA of presumptive myoblasts mimics the effect of RSV on transformed myogenic cells[38].

If presumptive myoblasts are treated with Ara-C or FudR they do not acquire the cell surface required for fusion nor do they initiate the synthesis of the definitive myosin heavy and light chains[23,39].

Clearly there are striking differences in the metabolic options open to the myoblast as compared to those available to their mother cells, the presumptive myoblast. Currently we are performing experiments based on the assumptions that the differences between the two types of cells is due to activation of a "master switch" that in turn makes available for transcription those myosin genes that are uniquely active in myoblast: it is assumed that the genes for such myosins are not available for transcription in the mother presumptive myoblast.

## 4. Summary

Much confusion in the literature could be avoided if the distinction between the genetic regulation of a given cell's on-going metabolic activity was not confused for the regulatory mechanisms responsible for cell diversification. With this distinction in mind there is no reason to postulate different mechanisms between "determination" and "differentiation", or between regulative and mosaic systems. Fortunately there is a simple operational test for this distinction: if mother and daughter cells respond in a similar fashion to all exogenous molecules then they are separated by a proliferative cell cycle and then one is studying cell physiology. If there is a difference, then mother and daughter were separated by a quantal cell cycle and they have diversified.

It is obvious that the role of the quantal cell cycle in generating metabolic options in daughters that were not available in the mother cell, is not synonymous with simply altering the synthetic activity of a given cell. The striking changes in response to exogenous influences in Protozoan and Metazoan cells without an intervening cell cycle, proliferative or quantal, are too well known to enlarge on here. What is being stressed is that there is one regulatory mechanism that (1) determines which of the available metabolic options a cell will use, and another that (2) determines which set of available options a cell will be endowed with. The former involves problems in cell physiology, the latter the unique problem of cell diversification. Much confusion and irrelevant semantics could be avoided by recognizing these two very different aspects of a given cell's metabolic behavior.

References

1. Fleischmajer, R. and Billingham, R. 1968 Epithelial-Mesenchymal Interactions. Williams and Wilkins, Baltimore.
2. Rutter, W., Pictet, R. and Monis, P. 1973 Ann. Rev. Biochem., 42, 601.
3. Slavkin, H. and Greulich, R. 1975 Extracellular matrix influences on gene expression. Academic Press, New York.
4. Levi-Montalcini, R. 1974 In: Dynamics of degeneration and growth in neurons (eds. K. Fuxe, L. Olson and Y. Zotterman). Pergamon Press, Oxford.
5. Medcalf, D. 1974 In: Control of Proliferation in animal cells. Eds., B. Clarkson and R. Baserga. Cold Spring Harbor Symp. Quant. Biol.p. 887.
6. Till, J., Messner, H., Price, G., Aye, M. and McCulloch, E. 1974) In: Control of Proliferation in Animal Cells. Eds., B. Clarkson and R. Baserga. Cold Spring Harbor Symp. Quant. Biol. p. 907.
7. H. Holtzer, 1968 In: Epithelial-Mesenchymal Interactions (eds. Fleischmajer and Billingham). Williams and Wilkins, Baltimore.
8. Holtzer, H., Weintraub, and Mayne, R. 1972 Current Topics in Developmental Biology (eds. Moscona and Monroy). Academic Press, New York.
9. Abbott, J., Schiltz, J., Dienstman, S. and Holtzer, H. 1974 Proc. Nat. Acad. Sci., 71, 1506.
10. Dienstman, S. and Holtzer, H. 1975 In: Cell Cycle and Cell Differentiation (eds. Reinert and Holtzer). Springer-Verlag, Heidelberg.

11. Holtzer, H., Rubinstein, N., Fellini, S., Chi, J. and Okayama, M. 1975 Quart. Rev. of Biophysics. Nov. Issue.

12. Weintraub, H. 1975 In: Cell Cycle and Cell Differentiation (eds. Reinert and Holtzer). Springer-Verlag, Heidelberg.

13. Weintraub, H., Campbell, G., Mayall, B. and Holtzer, H. 1971 Cell Biol. 50, 652.

14. Campbell, G., Weintraub, H. Mayall, B., and Holtzer, H. 1971 J. Cell Biol. 50, 669.

15. Hagopian, H., Lippke, J., and Ingram, V. 1972 J. Cell Biol., 54, 98.

16. Weintraub, H., Campbell, G. and Holtzer, H. 1972 J. Mol. Biol., 70,337.

17. Campbell, G., Weintraub, H. and Holtzer, H. 1974 J. Cell Physiol., 83, 11.

18. Groudine, M., Holtzer, H., Scherrer, K. and Therwath, A. 1974 Cell, 3, 243.

19. Holtzer, H., Marshall, J. and Finck, H. 1957 J. Cell Biol., 3, 705.

20. Okazaki, K. and Holtzer, H. 1965 J. Hist. Cytochem., 13, 726.

21. Holtzer, H., Croop, J. Dienstman, S., Ishikawa, H., and Somlyo, A. 1975 Proc. Nat. Acad. Sci. USA 72, 513-517.

22. Holtzer, H., Strahs, K., Biehl, J., Somlyo, A., and Ishikawa, H. 1975 Science, 188, 943-945.

23. Holtzer, H., Sanger, J., Ishikawa, H., and Strahs, K. 1973 Cold Spring Harbour Symp. Quant. Biol. 37, 549-566.

24. Yaffe, D. 1969 Curr. Top. Dev. Biol. 4, 37-77.

25. Yaffe, D. and Dym, H. 1973 Cold Spring Harbour Symp. Quant. Biol. 37, 543-547.

26. Paterson, B., and Strohman, R. 1972 Develop. Biol. 29, 112-138.

27. Przybyla, A. and Strohman, R. 1974 Proc. Nat. Acad. Sci. USA 71, 662-666.

28. Willingham, M., Ostlund, R., and Pastan, I. 1974 Proc. Nat. Acad. Sci. USA 71, 4144-4148.

29. Weber, K. and Groeschel-Stewart, U. 1974 Proc. Nat. Acad. Sci. USA 71, 4561-4564.

30. Holtzer, H. 1976 In: Non-muscle motility. Cold Spring Harbour Symp. Quant. Biol. (eds. Goldman, R., Pollard, T. and Rosenbaum, J. In press.

31. Chi, J., Fellini, S. and Holtzer, H. 1976 Proc. Nat. Acad. Sci. In press.

32 Chi, J., Rubinstein, N., Strahs, K. and Holtzer, H. 1975 J. Cell Biol. In press.

33. Holtzer, H., Rubinstein, N., Dienstman, S., Chi, J., Biehl, J., and Somlyo, A. 1975 Biochimie, 56, 1575-1580.

34. Hauschka, S. and Konigsberg, I. 1966 Proc. Nat. Acad. Sci, 55,119-126.

35. Konigsberg, I. 1972 In: Chemistry and molecular biology of the intercellular matrix. Ed. E. Balazs. 3, 1779. Academic Press, New York and London.

36. Stockdale, F., Nameroff, M., Okazaki, K. and Holtzer, H. 1964 Science, 146, 533.

37. Bischoff, R. and Holtzer, H. 1969 J. Cell Biol. 41, 188.

38. Holtzer, H., Biehl, J., Yeoh, G., Maganathan, R. and Kaji, A. 1975 Proc. Nat. Acad. Sci. In press.

39. Yeoh, G. and Holtzer, H. 1976 Exp. Cell Res. In press.

This work was supported by NIH grants CA-18194, GM-20138, and HL15835 to the Pennsylvania Muscle Institute, and by grants from the Muscular Dystrophy Association and the National Science Foundation.

*Progress in Differentiation Research, ed. N. Müller-Bérat et al.*



# CELL LINEAGE AND CELL INTERACTIONS IN NEURONAL DIFFERENTIATION

R. Kevin Hunt

Jenkins Biophysical Laboratories
The Johns Hopkins University
Baltimore, MD 21218 U.S.A.

## 1. Introduction

Retinal ganglion cells of Xenopus undergo true position-dependent (patterned) differentiations in the developing eye. These differentiations lead to the deployment of unique affinity/recognition properties called locus specificities[1,2] which enable each cell's axon to reach its appropriate synaptic site in the topographic retinotectal map[3,4]. My colleagues and I have been studying the early programming events which specify these positional differentiations and their relation to the primary processes of cell replication, retinal histogenesis and growth, and the general cytologic differentiation of the ganglion cell phenotype.

## 2. Cell lineages and retinal histogenesis

The eye bud originates as an epithelial bulge off the prosencephalic region of the neural tube and reaches the periphery of the head region by stage 22 in *Xenopus*[5]. At stages 23-25 it pinches off its anterior aspect to form the optic vesicle and then (at stages 26-28) invaginates to form the optic cup. The first neural cells of the presumptive retina cease to incorporate 3H-thymidine into DNA at stage 29; by stage 31/32, most of the cells in the central retina have withdrawn from the mitotic cycle, and this majority embraces neuroblasts destined to mature into all major classes of retinal neurons: photoreceptors, retinal interneurons (bipolars, Amocrines, horizontal cells), as well as ganglion cells[6-9]. By late thirties stages, replication in the central retina has ceased, intraretinal circuits have assembled (and an electroretinogram can be recorded), and ganglion cell axons have begun to grow out from the eye and innervate the brain[7,8]. Still this early larval retina contains only about 500 ganglion cells and only 2000 neurons overall. The remaining 70,000 ganglion cells (and perhaps $10^6$ photoreceptors and interneurons) are gradually added to the periphery of the growing retina over the entire larval and early juvenile period, from the continued division of neurogenic stem cells which reside in a 'ring' at the ciliary margin of the retina[9,10]. In this radial pattern of retinal growth, each generation of stem cells divides to produce a ring of neurons (added to the edge of the morphologically differentiated central retina) and a new generation of stem cells (passively displaced outward to remain at the retinal margin). To a first approximation, then, the clonal boundaries[11] which divide the adult ganglion cell population are defined by <u>arcs</u> at some angle (i.e., the clones are shaped like 'pie slices'); and the sibship relations between adjacent cells as one moves centrifugally along a given radius are 'aunt -- niece -- great niece'.

Several lines of evidence converge to suggest that cell lineage and the cell cycle play a causal role in the <u>phenotypic</u> diversification of retinal neurons, during the early phases of retinal histogenesis[11]. Cumulative labelling with 3H-thymidine from stage 26 or stage 28 produces a linear increase in labelling index to 100% in 10 hours (at 18°C); similar protocols from stage 29 produce a linear increase to a plateau of 78% in 11 hours, and all 22% of the unlabelled nuclei were ultimately tracked to the ganglion cell layer[6]. This suggests that two populations of neuroepithelial cells are cycling, out of phase, in the stage 26-28 optic cup, one of which is in <u>S</u> at stage 28-29 and is specifically <u>gangliogenic</u>. This inference was confirmed in perturbation experiments using the thymidine analog 5-Bromodeoxyuridine (BUdR) which selectively blocks differentiation when embryonic precursor cells are allowed to replicate in its presence[12,13]. Treatment with BUdR from optic vesicle stage 24 completely suppresses the production and differentiation of <u>all</u> major neuronal phenotypes, leading to giant neuroepithelia at early larval stages 37-41; treatment from stage 28/29 (when neuron birth is imminent) has no effect; treatment from the intermediate stage (27) permits the emergence and morphologic differentiation of the ganglion cells while completely blocking the emergence of photoreceptors and sensory interneurons[14] (Figures 2-5).

In addition, the precursor cells of Xenopus retina show a remarkable refractoriness to experimental 'telescoping' of the neural cell lineages. A remarkably normal sequence of cell production and phenotypic diversification occurs in eye buds explanted into saline solution at stage 22/23 -- a stage at which the cells are at least two cycles away from generating frank neuroblasts -- and these explants go on to assemble functional neural circuits in vitro[15,16]. Stage 22 eye buds also generate the same number of each neuronal phenotype, with roughly normal timecourse, when grafted to stage 28 or stage 38 host orbits or to ectopic body sites as when grafted to the orbit of stage 22 hosts[17]. Half-eye primordia prepared at stage 25/26 generate about half the normal number of neurons of each phenotype, ruling out control by critical mass[18]. Finally, when fluorodeoxyuridine (FUdR) or cytosine arabinoside or KCl are used to block or slow the passage of cells through the cell cycle, the precursor cells acquire the morphological properties (axons, dendrites, etc.) of frank neurons only upon release from the block and completion of the ensuing cycles[11]. It appears thus far that early precursor cells, which do not normally generate frank neurons, are reluctant to do so, even when their cell cycle is prolonged or their microenvironment is manipulated; nor have such conditions yet enticed last-generation precursor cells to display definitive neuronal properties within their own lifetimes.

3. Developmental information for locus specificities

Concurrent with these processes of histogenesis and phenotypic diversification of retinal cell types, a developmental program is set in motion for the position-dependent differentiation of the individual ganglion cells. During the earliest phases of optic cup formation, the stage 22-28 eye bud already possesses full information about its anteroposterior (AP) and dorsoventral (DV) orientation on the embryo, and the capacity to use this information to institute a complete program for patterned differentiation of the ganglion cells. An eye-bud cultured in saline for several days from stage 22/23 form functional larval eyes *in vitro* and, upon reintroduction into a host orbit, will assemble an orderly retinotectal map based on its original axial plan from the donor embryo[6]. However, the orientational information at stages 22-28, if sufficient and complete, is also 'replaceable': if the orientation of a stage 22-28 eye bud is acutely changed by rotation or transplantation *in vivo*, the anlage interacts with the axial signals of the embryo and acquires new information appropriate to the new orientation after surgery. Thus, even though the eyeball itself develops upside-down following an 180°-rotation at stage 28 -- the ganglion cells undergo their position-dependent differentiations in accordance with their new locations and assemble a normally-oriented retinotectal map.

At stages 28-31, control mechanisms within the retina stabilize the existing information in the anlage, and establish a permanent developmental program for ganglion cell patterning. The stabilization process is called *axial specification* and takes place in two steps (AP before DV) over about five hours. After specification, the retinal field is refractory to the axial signals of the embryo; and even when rotated into a pre-stage 28 host, or serially passed through a sequence of four or more hosts, the retina follows its original program for position-dependent differentiation; and the orientation of the retinotectal map assembled in the final host deviates from the normal by the same angle as the retina, in its final disposition, deviates from what its orientation was at stages 28 to 31 (reviewed in Ref. 7). Most importantly, the entire retinotectal map is consistent, indicating that the $10^5$ ganglion cells added to the retinal periphery after stage 31 underwent *their* position-dependent differentiation in accordance with the specified developmental program.

## 4. The cell cycle and orientational information

There is evidence that cell lineage plays little or no role in either the memory functions or the signalling functions of *unspecified* stage 22-28 eye buds[11]. The tentative orientational information present in explanted stage 22 eye buds survives treatment with

Figure 1. To show that stage 22 eye buds already contain stable information about retinal orientation on the embryo. Stage 22 right eye bud was cultured in saline for four days (from stage 22) and reimplanted in 90°-rotated orientation in the right orbit of a stage 39/40 host embryo (Top). At metamorphosis, the retinotectal map from the right eye to the left tectum was rotated by 90°; the map from the normal left eye to the right tectum is normally oriented. Each number in the visual field shows the position of the stimulus that optimally evoked potentials recorded when a microelectrode impaled the tectum at the correspondingly numbered position. Distance between tectal electrode positions is shown by the bar; the visual field spans 100° from center to periphery.

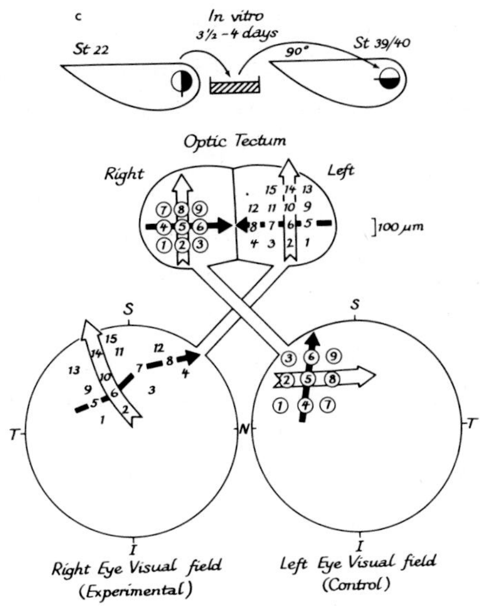


---

BUdR from stage 24 (suppression of cytodifferentiation of all neurons) or stage 27 (suppression of photoreceptors and interneurons); when allowed to recover in normal medium and reimplanted into the orbit of a normal host, these retinae finally differentiate and assemble retinotectal maps based on their original orientation in the donor embryo. Likewise, eye buds treated with BUdR and acutely rotated in vivo are able to interact with the axial signals of the embryo and acquire new information appropriate to the new (rotated) orientation; after recovery during prolonged culture (allowing time for the cells to finally differentiate), these eye buds assemble retinotectal maps based on information in the rotated orientation and acquired while under treatment with BUdR. Finally, both the survival of tentative information in vitro and the ability to undergo axial replacement are independent of passage of the cells through the cell cycle: as with BUdR, both functions persist in eye buds treated with FUdR or cytosine arabinoside to block or greatly slow DNA synthesis. (Hunt, Berman & Holtzer, in prep.)

Axial specification, on the other hand -- stabilization of the information and setting up of the definitive program for position-dependent differentiation of the ganglion cells -- does appear to be lineage dependent. Axial specification can occur in vitro[15], and its timing is intrinsically controlled within the eye primordium. Thus the time of specification always correlates with the stage of the optic cup and, in eyes grafted between embryos, is neither accelerated nor delayed by altering the stage of the host[19]. Bergey, Holtzer and I exploited the effects of BrUdR on the developing optic cup, to show that the time of specification was a programmed component of ganglion cell differentiation. When BrUdR (from stage 24) blocks the onset of differentiation of all retinal neurons (Figure 4), the eye bud is still able to acquire new orientational information at stage 32. Neither the prolonged time needed for these eyes to reach stage 32, nor the attainment of unusually large size, nor the appearance of terminally differentiated non-neural cell types is sufficient to trigger axial specification. However, when some of the ganglion cells are allowed to commence their terminal cytodifferentiation (after BrUdR from stage 27, see Figure 5), axial specification occurs on schedule. These observations thus eliminate an obligatory role for the interneurons and photoreceptors in the control of axial specification; they further suggest that the trigger for axial specification is localized to the gangliogenic precursor cells in the pre-stage 29 optic cup, and is activated at a predetermined point in their differentiation.

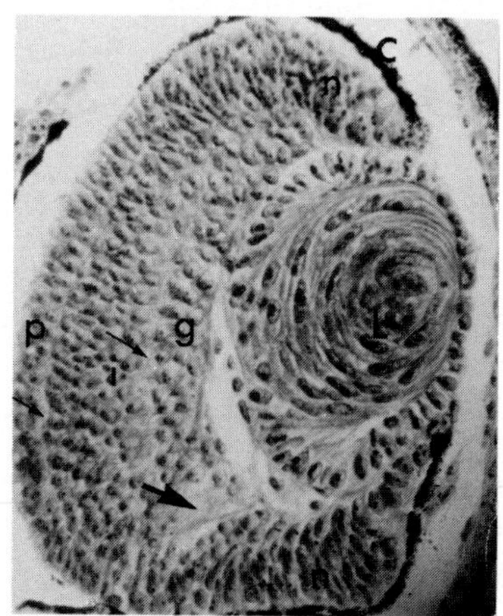


Figure 2. To show a normal stage 39 retina, including ganglion (g), interneuron (i), photoreceptor (p), and plexiform (light arrows) layers, lens (L) and optic nerve (dark arrow).

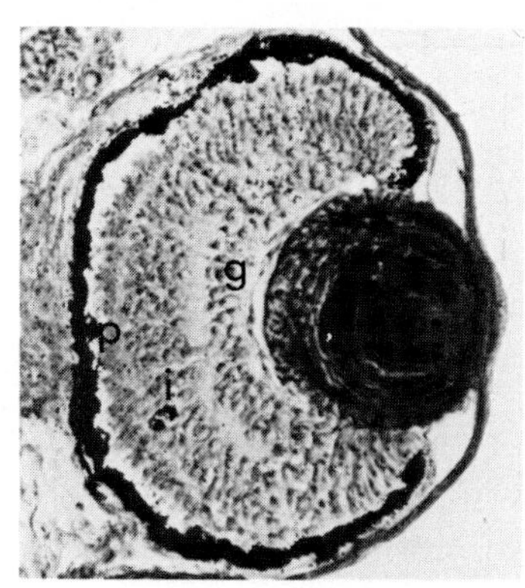


Figure 3. To show a nearly normal retina following BUdR treatment from stage 28/29; fixed stage 39, labels as in Figure 2.

The importance of the precursor cell cycle is further underscored by observations on axial specification after treatment with FUdR (Hunt et al., in preparation). Although FUdR blocks or greatly slows DNA synthesis and will eventually kill 100% of the retinal cells, thymidine rescue (within 18 hours, by injection of the nucleoside into the normal host embryo) circumvents the FUdR block long enough for the eye bud to replenish its thymidylate synthetase -- permitting recovery of functional eyes and histologically normal retinae. Thus, such eye buds can be challenged to acquire new orientational information (by grafting them, in 180°-rotated orientation, into a normal stage 27/28 host), and then 'rescued' after the transplantation. When such a block is imposed from early stage 28 (arresting the last generation of gangliogenic precursors in $G_1$ or S), AP-specification always occurs and DV-specification almost always occurs by stage 32. Following the axial replacement challenge and rescue with thymidine, these eyes go on to generate completely rotated, or AP-rotated retinotectal maps. In contrast, early FUdR

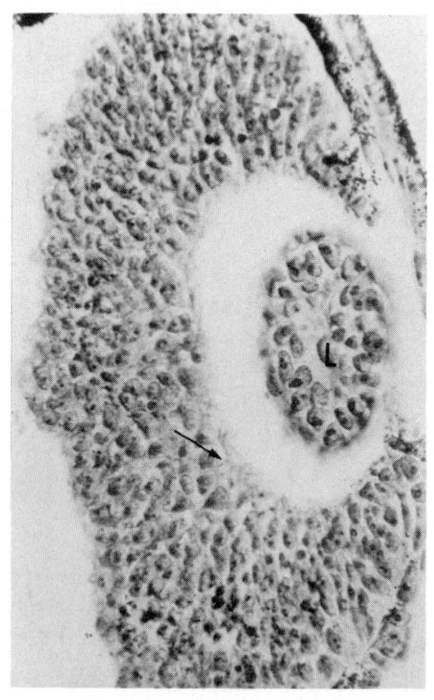


Figure 4. To show completely suppressed cytodifferentiation following BUdR treatment from stage 24; fixed stage 39. There is one overtly differentiated neuron in this retina (arrow).

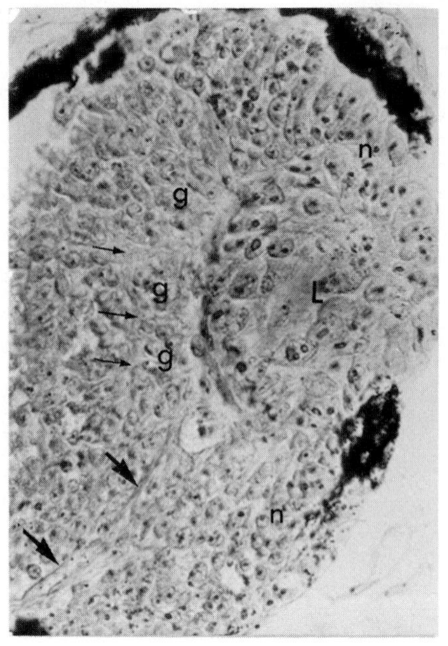


Figure 5. To show the selective suppression of the non-ganglion cell phenotypes following BUdR treatment from stage 27; fixed stage 39. Note islands of ganglion cells (g), their dendrites (light arrows), and their axons joining to form an optic nerve head (heavy arrow).

treatment (from stage 20-24, when cells are two generations away from generating frank neurons) does delay the time of specification, by intervals similar to the duration of the block.

These experiments permit two important inferences: First, the actual cessation of DNA synthesis and birth of definitive ganglion cells may not be necessary for axial specification (a possibility also raised by the findings of Jacobson et al., 1974, on Rana pipiens). Second, the trigger for axial specification may be activated several hours in advance of the time at which the optic cup actually loses the capacity to undergo axial replacement -- early in the lifetime of the immediate gangliogenic precursor cell or, perhaps, late in the lifetime of the penultimate precursor. Once activated, the events of specification can proceed to completion even if the final phase of ganglion cell cytodifferentiation is postponed.

These preliminary observations only begin to define the role of the cell cycle and cell lineage in the programming phase of ganglion cell patterning. The exact cellular events involved in axial signalling and in the fixation of orientational information in the eye are totally unknown, though an abrupt disappearance of gap-like intercellular junctions from the central stage 30 optic cup has been emphasized by some authors[20,21]. Whatever the ultimate mechanisms, it appears that the optic cup of Xenopus has used lineage-dependence to solve two of its major pattern-formation problems, namely, coordination of spatial differentiation with cytodifferentiation and provision of a permanent blueprint for cellular positional information in advance of the cells' needs to express position-dependent properties. This is accomplished by consigning control of axial specification to the precursor cells of those neurons whose need for positional information is most pressing and immediate: the ganglion cells of the central retina.

5. Positional information in the growing larval eye

The foregoing experiments address only the earliest phases of ganglion cell patterning. And though these programming events specify the position-dependent differentiation of all the ganglion cells the retina will ever contain, some $10^5$ ganglion cells remain to be integrated into the program as they are added to the retina during its larval growth. As yet we have no clear picture of how this integration is achieved; but several rather disconnected data, recently acquired, suggest a model in which cell lineage and cell interactions interdigitate in conveying positional information to the individual ganglion cell.

(a) Signalling continues in the growing retina. The specified state is exceedingly stable in whole stage 31 or older eyes, surviving serial transplantation into four or more younger hosts, chronic tissue culture, or more than thirty days deprivation of tectal connections[1,22]. Surgical rearrangements of the retinal field, on the other hand, can produce dramatic perturbations in this program. Figure 6c shows a mirror-image redundant pattern of ganglion cell differentiations, which arose in a surgical recombinant (left-posterior/right-anterior) stage 32 eye[23]. By fusing such eye fragments for 14 or 30 hours before separating them (and allowing the left-temporal fragment to round up to form a complete eye and map into the tectum alone) Frank and I could show that the departure of the left-temporal fragment from its original program (Figure 6b) involves a rapid, stepwise reprogramming of first its AP axis (Figure 6d) and then its DV axis (Figure 6e). Following the interaction, all cells added to the peripheral retina undergo

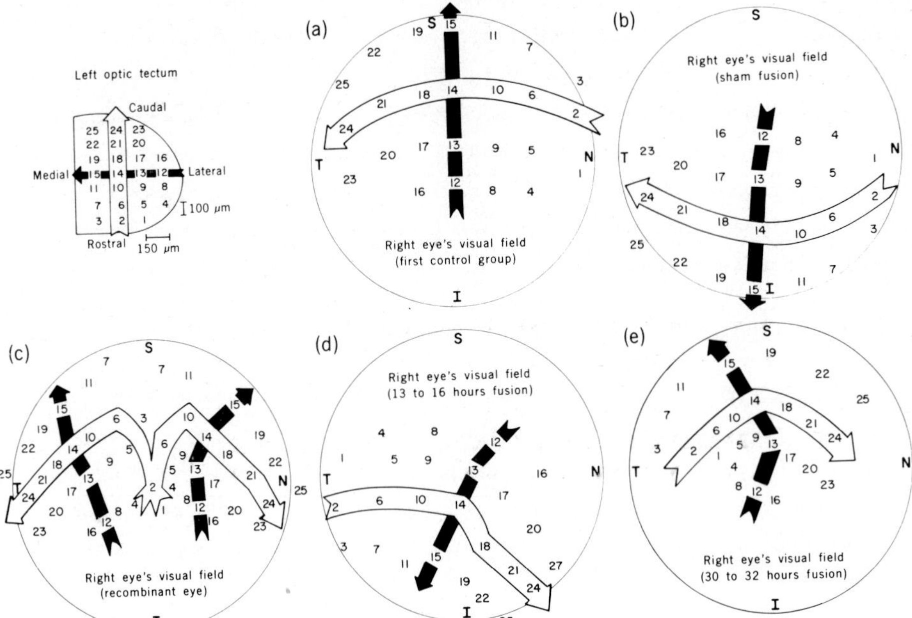


Figure 6. To show retinotectal maps from a control right eye (a), a rounded-up left-posterior eye fragment grafted to the right orbit (b), a permanent left-posterior/right-anterior recombinant eye (c), a rounded-up left-posterior fragment which had been fused with a right-anterior fragment for 14 hours and then isolated (d), and a left-posterior fragment fused for 30 hours and isolated. All fragments prepared at stage 32.

---

position-dependent differentiation with respect to this new program; and other experiments[18,24] indicate that the rapid reprogramming involves signals within the retinal fragments and not extraretinal tissues or the extraocular microenvironment.

(b) Propagation of the signal is radial. Occasionally reprogramming is incomplete, or for one reason or another, a small patch of retina either fails to undergo orderly position-dependent differentiations or differentiates according to an autonomous alternative plan (Figure 7). Whenever this occurs, the involved region of the adult retina is 'pie-slice' in shape, extending from near the center to the extreme periphery of the retina, and varying in size from

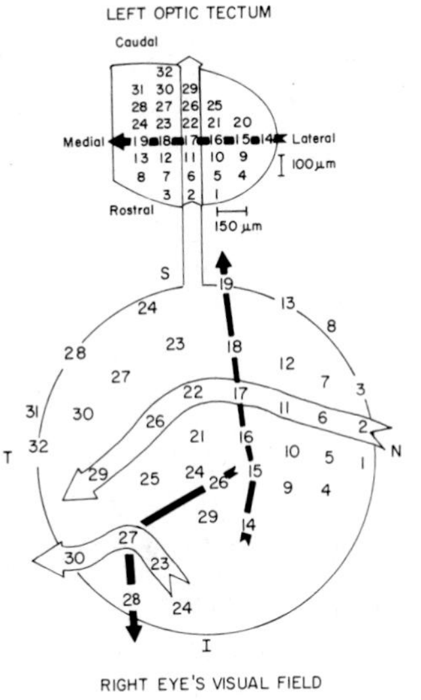

Figure 7. To show a retinotectal map from a retina containing an aberrant patch of ganglion cells (shaped like a 'pie-slice') whose position-dependent differentiation followed an alternative program.

180° of arc (indeed, one could consider all complete reprogrammings, as in Figure 6c, to fall into this category) down to 10° or less -- the latter mapping as the descendants of 1-3 cells on the original peripheral ring.

(c) Positional information is Cartesian in organization. If the localization of the aberrant cells suggests that the positional information is conveyed radially and more or less autonomously to a 'polyclone'[25] of retinal cells, the nature of the information that is deranged or lost is Cartesian. That is, the positional code appears not to involve angular (θ) and radial (r) parameters -- even though the geometry by which the information is handed down to new ganglion cells is radial. Thus, the aberrant 'pie-slice' of retina shows specific perturbations in the in the AP or DV axis -- either a reversal of one )Figure 7) or a complete deletion of AP or DV information from the system (Figure 8). Indeed, by systematically 'jamming' the AP or DV axial signal during the replacement of orientational informa-

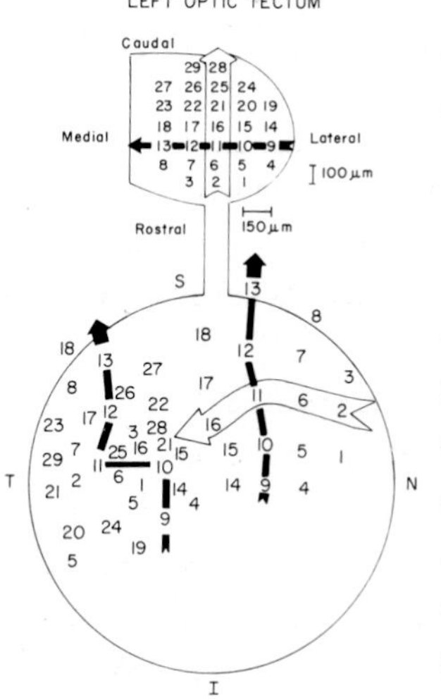


Figure 8. To show a retinotectal map from a retina, in the posterior half of which, ganglion cell patterning is based on incomplete information; the defect, disorganization in the AP direction, suggests a Cartesian organization for the positional information code.

tion in young unspecified eyes, we have succeeded in knocking out only the AP component (or only the DV component) of positional information for the entire retina. The ganglion cells accurately discriminate mediolateral level in the tectum but not rostrocaudal level (or vice versa). Detailed physiologic analysis on such animals[26] indicates that there are multiple effector sites with different functions on the tip of the individual ganglion cell: one of them discriminates AP tectal level and derives from information about the cell's position along the AP axis of the retina; another discriminates mediolateral tectal level and derives from information about the cell's position in the DV axis of the retina.

(d) Individual precursor cells are dispensible. When early larval eyes are treated with FUdR without thymidine rescue, up to 75% of the precursor (stem) cells at the ciliary margin are killed, and growth of the eye abruptly ceases[11]. The survivors undergo 'circumferential' divisions and repopulate this stem cell 'ring', before resuming the normal pattern of radial growth. Thus, the clonal boundaries of the adult peripheral retina are severely altered from the normal; yet, a normal pattern of position-dependent differentiations occurs in the entire ganglion cell population, based on the original program instituted at stages 28-31. These results suggest that positional information is not programmed into the individual precursor cell, to be handed down in sequence each time a new ganglion cell daughter is born. More likely, the nascent ganglion cell interacts with the older antecedent generation(s) of neurons to acquire its positional information -- an hypothesis supported (though not proved) by the observation that ultraviolet microbeaming of these older more central cells has led to scrambling of the map[11].

Synthesizing these lines of evidence suggests a model of integration of new ganglion cells in which the nascent ganglion cell interacts locally with the few neurons (or one neuron) immediately adjacent to it along a radius. Based on the positional values of that older cell (coded as $X_m$, $Y_n$), and a pair of transform operations ($F_x$ and $F_y$) which are a function of the cell's angle ($\theta$) on the ring, the nascent cell adopts positional values of $X_{m'}$ and $Y_{n'}$.

Only the transform operations ($F_x$, $F_y$) are dependent on stable cellular relationships around the eye, and can be reset for a given $\theta$ by disruption and internal rearrangement of the retinal field.

## 6. Summary

The role of cell cycle, cell lineage and cell interactions has been considered in the three phases of position-dependent differentiation of retinal ganglion cells in Xenopus. The stability and exchange of tentative orientational information in the optic vesicle is independent of cell lineage and passage of cells through the cell cycle; the stabilization of that information and evolution of a definitive program of patterning is completely dependent on the cell cycle in the gangliogenic lineage. The integration of new ganglion cells into the program may involve cell interactions, between daughter cell and 'aunt' neuron, and so propagated without cell heredity through clones descended from a single cell.

## Acknowledgements

Thanks are extended to Dr. H. Holtzer for his suggestions and continued encouragement, Ms. Pat Burck for preparation of the manuscript, and the National Institutes of Health (NS-12606) and National Science Foundation (BMS-75-18998) for support.

## References

1. Hunt, R.K. and Jacobson, M., 1972, PNAS, US 69:2860.
2. ----, 1974, Curr. Tops. Develop. Biol. 8, 202.
3. Sperry, R.W., 1951, Growth Symp. 10, 63.
4. ------, 1963, PNAS, US 50, 703.
5. Nieuwkoop, P.D. and Faber, J., 1956, Normal Table of Xenopus laevis (Daudin) (North Holland, Amsterdam).
6. Jacobson, M., 1968, Develop. Biol. 17, 219.
7. Bergey, G.K., Hunt, R.K. and Holtzer, H., 1973, Anat. Rec. 175, 271.
8. Fisher, S. and Jacobson, M., 1971, Z. Zellforsch. 104, 165.
9. Straznicky, K. and Gaze, R.M. 1971, J. Embryol. Exp. Morph. 26, 67.

10. Chung, S.-H., Stirling, R.V., and Gaze, R.M., 1975, *ibid.*, 33, 915.

11. Hunt, R.K., 1975, Res. Probs. Cell Differentiation 7, 43.

12. Holtzer, H., 1970, Symp. Soc. Cell Biol. 9, 69.

13. Weintraub, H., 1975, Res. Probs. Cell Differentiation 7, 27.

14. Hunt, R.K., Bergey, G.K., and Holtzer, H., 1975, Develop. Biol., In press.

15. Hunt, R.K. and Jacogson, M., 1973, PNAS, US 70, 503.

16. Frank, E., Hunt, R.K., and Holtzer, H., 1974, Unpublished data.

17. Hunt, R.K., 1976, In: R. Llinas & W. Precht, eds., *Handbook of Frog Neurobiology* (Springer, Heidleberg), In press.

18. Berman, N. and Hunt, R.K., 1975, J. Comp. Neurol. 162, 23.

19. Hunt, R.K. and Jacobson, M., 1974, PNAS, US 71, 3616.

20. Dixon, J. and Cronly-Dillon, J.R., 1972, J. Embryol. Exp. Morph. 28, 659.

21. Jacobson, M. & Hunt, R.K., 1973, Sci. Amer. 228 (2), 27.

22. Hunt, R.K., 1975, Ciba Found. Symp. 29, 131.

23. Hunt, R.K. & Frank, E., 1975, Science 189, 563.

24. Hunt, R.K. and Berman, N., 1975, J. Comp. Neurol. 162, 43.

25. Garcia-Bellido, A., 1972, Res. Probs. Cell Diff. 5, 59.

26. Hunt, R.K., 1975, In: D. McMahon and C.F. Fox, eds., *Developmental Biology: Proceedings of the ICN-UCLA Winter Conference* (W.A. Benjamin, San Francisco).

*Progress in Differentiation Research, ed. N. Müller-Bérat et al.*



# STUDIES ON THE NATURE OF THE LAST CELL DIVISION DURING THE DIFFERENTIATION OF RAT SKELETAL MUSCLE CELLS

Gad Lavie and David Yaffe

Department of Cell Biology, The Weizmann Institute of Science, Rehovot, Israel

## ABSTRACT

Myoblasts triggered to fuse in mass cultures and then isolated during mitosis and plated in cloning density are able to divide once, and form binucleated fibers by fusion of the two offspring of that division. This was used to investigate the question of whether cells become committed to fuse after a specific cell cycle. Rat skeletal muscle mass cultures were grown in a nutritional medium which promotes proliferation without cell fusion, and then transferrred to a fusion-permissive medium. This treatement induced the initiation of a phase of rapid cell fusion which starts 18 h after the change of medium. At various time intervals following the change of medium, mitotic cells were isolated and grown in cloning conditions. It was found that mitotic cells collected during the first 3-4 h after the change to the fusion-stimulating medium did not form binucleated fibers when plated at cloning density. However, increasing numbers of binucleated fibers were found as the collection of mitotic cells was delayed, reaching its maximum when collected about 14 h after the change of medium. If, however, such cells were put back in a medium that promotes proliferation, formation of binucleated cells was strongly suppressed. A follow-up of the fate of individually marked cells strongly suggests that cells having already completed a mitotic cycle which in mass cultures is followed by cell fusion, could be induced after that mitosis to resume proliferation, if put in the appropriate culture conditions. The results indicate that no irreversible loss of the capacity of the cells to proliferate takes place until very close to their fusion, which occurs about 6-10 h after their last mitosis.

## INTRODUCTION

The role of DNA synthesis and cell replication in the differentiation of muscle cells, as well as of several other cell types, has been a matter of great interest[1-5]. Most myoblasts obtained from developing muscle divide at least once in cell culture before fusing into multinucleated fibers. Application of inhibitors of DNA synthesis soon after plating the cells inteferes with the formation of multinucleated fibers. Exposure of the cells to halogenated deoxynucleotides such as BUdR or IUdR, at concentrations which do not prevent cell division, strongly inhibits the formation of multinucleated fibers[6,7]. Such observations led to the notion that major changes in the programming of gene activity are linked to a specific cell division, sometimes termed "quantal mitosis". After this division, the cells withdraw from the proliferative cycle and enter a stage of postmitotic differentiation[1,8].

An important question with regard to the possible causal relation between cell division and terminal differentiation is whether the cells lose their capacity to proliferate after a specific cell cycle. It was shown that DNA synthesis and karyokinesis can be induced in multinucleated muscle fibers by infection with oncogenic viruses[9-11]. However, since introduction of the viral genome alters cellular control mechanisms, the implication of these observations regarding the normal course of differentiation is very limited. The notion of the existence of a specific cell division, after which the cells are committed to fuse, has been challenged recently by experiments with normal chick muscle cultures. These experiments showed that cell density and fresh medium can determine whether the cells will form fibers or continue to replicate and that growing the cultures in conditioned medium collected from older cultures increases the rate of myoblasts which could fuse without division in culture[12,13].

In primary rat skeletal muscle cultures, it is possible to control the time of initiation of the phase during which most of the cells make the transition from proliferation to postmitotic differentiation solely by changing the type of nutritional medium[14]. Cells obtained from newborn rat thigh muscle and plated in standard mass culture conditions (S medium) multiply during the first 2 days, but no significant fusion takes place at that time regardless of whether plating is done at high or at low cell density. At about 52 h after plating, the cultures enter a period of rapid cell fusion. Growing the cells in a medium enriched with fetal calf serum and chick embryo extract (FE medium) promotes proliferation without cell fusion. When cultures maintained in a proliferative state by FE medium are transferred to S medium, a period of very rapid cell fusion starts after a lag of 18-20 h. This phase of intense cell fusion proceeds for about 12-15 h and then slows down and eventually levels off. 60-70% of the cells are fused during this period. This experimental system was used to study whether the cells become irreversibly committed to fuse after a specific round of cell replication, or whether they can be induced to resume proliferation after this division has taken place.

## MATERIALS AND METHODS

### Cultures

Primary skeletal muscle cultures were prepared from newborn Wistar rat thigh muscle, as described previously[15, 16].

Nutritional media

Two types of media were employed in the present study: (1) Standard medium (S medium) composed of 90 parts of a 1:4 (v/v) mixture of M 199 (Gibco, Cat No. E11) and Dulbecco's modified Eagle's medium (commercially prepared), 10 parts of tested horse serum (Gibco) and 1 part of chick embryo extract prepared as described previously[16]. (2) Fetal calf serum-chick embryo extract-rich medium (FE medium), composed of 72 parts of a 1:4 (v/v) mixture of M 199 and Dulbecco's modified Eagle's medium, supplemented with 20 parts of tested fetal calf serum (Gibco) and 8 parts of chick embryo extract. Embryo extract was clarified by spinning at 15,000 x g for 30 min, just prior to use.

Cells are plated in S medium at a concentration of $1.5 \times 10^6$ cells per 60-mm Falcon tissue culture plate, pre-coated with gelatin[16]. After about 20 h, the medium was replaced by FE medium, and at 40 h after plating, or at the time indicated, the medium was changed back to S medium. This treatment resulted in the onset of a phase of very rapid cell fusion at 18-20 h after the change from FE medium to S medium.

Conditioned S medium was produced by transferring 40-h old cultures from FE medium to S medium and incubating them for a period of 14 or 24 h ( as indicated later). This medium was applied immediately (after centrifugation for 10 min at 1,000 x g) to the experimental groups in place of fresh S medium.

Inhibition of DNA synthesis

This was done by applying cytosine β- D-arabinofuranoside (cytosine arabinoside) to the cultures at a concentration of 10 μg/ml. This dose of cytosine arabinoside inhibited 98% of the incorporation of $^3$H-thymidine into the acid-insoluble fraction of rat muscle cell cultures.

Autoradiography

Autoradiography of cells labelled with $^3$H-thymidine (TRK 120, Amersham) was performed as described previously[15], using Ilford $K_5$ emulsion mixture, Kodak DA-19B developer and Kodak acid fixer.

Cell fusion was measured as described previously[14].

Isolation of cells in mitosis

Cells grown in tissue culture detach easily from the surface of the plate during mitosis[17]. To obtain cells in mitosis, plates were shaken gently and the medium containing floating cells in mitosis was collected. The cells were then transferred in their original medium to gelatin-coated plates at a concentration of 500-1,000 per plate. The degree of synchronization, i.e., the percentage of cells in mitosis during the transfer, was determined by fixing samples of plates 3 h after the transfer and counting the bicellular colonies formed. In most of the experiments, about 80% of the cells which settled were found to produce bicellular clones.

The efficiency of plating of collected mitotic cells was determined by counting the number of trypan blue-excluding cells which were found in the medium collected after shaking, and comparing that value to the total number of cells which settled on the surface of the gelatin-coated tissue culture plates after these cells were replated (fixation after 3h). Efficiency of plating was about 98%.

RESULTS

DNA synthesis and cell fusion in mass cultures

When cultures grown in FE medium, which promotes proliferation without fusion, are transferred to the fusion-permissive S mdeium, they start to fuse about 18 h later. To check the relation between the period of DNA synthesis and the time of

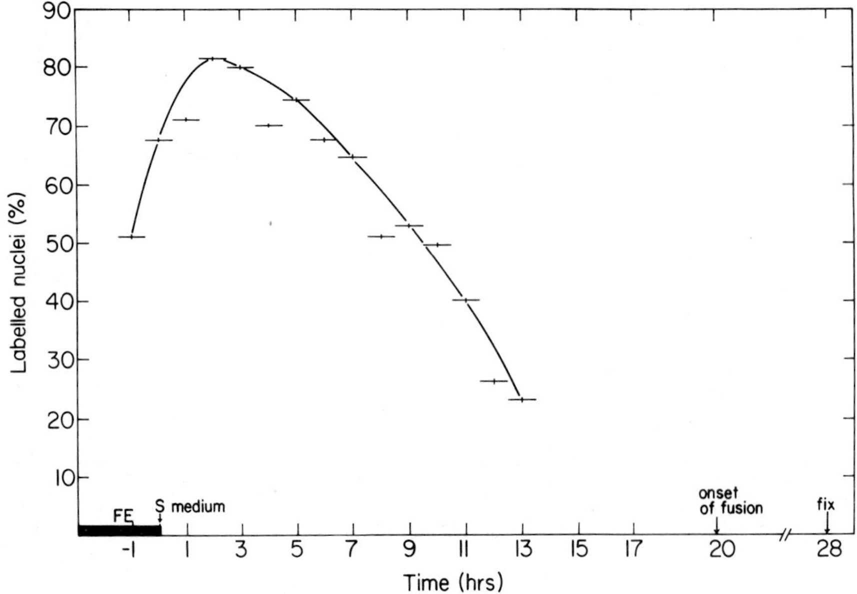


Fig. 1. : Effect of time of exposure to a pulse of $^3$H-thymidine on the percentage of labelled nuclei within fibers in cultures fixed 8 h after onset of fusion. Cultures grown in FE medium were transferred to S medium 40 h after plating. 1 h pulses of $^3$H-thymidine (0.5 $\mu$ Ci/ml) were given to groups of cultures at 1 h intervals, beginning 1 h prior to change of medium. 8 h after onset of fusion, all cultures were fixed and processed for autoradiography. The percentage of labelled nuclei within fibers was determined.

fusion, groups of cultures were exposed to 1 h pulses of $^{3}H$-thymidine at different times and fixed for autoradiography 8 h after the onset of cell fusion. It can be seen from Fig. 1 that most of the cells which participated in fiber formation during this period did synthesize DNA during the early part of the lag between the change of medium and cell fusion.

In another experiment, the effects of inhibition of DNA synthesis by cytosine arabinoside on cell fusion were tested. It was found that application of the inhibitor during the first few hours following the change from FE to S medium resulted in an effective inhibition of cell fusion; however, as the cultures approach the phase of cell fusion the effect of application of cytosine arabinoside gradually diminishes (Table 1).

These experiments indicate that most of the cells which fuse complete one round of replication in S medium and then enter a nonproliferative stage.

TABLE I

Inhibition of fiber formation by cytosine arabinoside administered at different times after change to S medium .

Cultures grown in FE medium were transferred to S medium 40 h after plating. Groups of 4 plates received 10 µg/ml cytosine arabinoside (CA) at 2 h intervals, starting at time of change to S medium and were further incubated. Fixations were made 34 h after the change of medium. The plates were stained and scored for the average number of nuclei within fibers, as described in Materials and Methods. Inhibition of fusion of each group was calculated relative to the fusion in plates which had not received the drug.

| Time of application of CA (h after change to S medium) | Average No. of nuclei within fibers/field | % of fusion relative to control (No CA) |
|---|---|---|
| 0 | 0.5 | 0.5 |
| 2 | 2.6 | 2.6 |
| 4 | 3.6 | 3.6 |
| 8 | 9.4 | 9.5 |
| 12 | 29.3 | 29.9 |
| 16 | 57.2 | 58.1 |
| 24 | 85.5 | 86.0 |
| No CA | 98.4 | 100.0 |

Effect of conditioned medium on the lag between change of medium and cell fusion

Investigations dealing with chick myogenesis have indicated an enhancing effect of conditioned medium (CM) on cell fusion and a delaying effect of fresh medium[12,13,18]. This raised the possibility that the 18 h lag between the change from FE to S medium and the onset of fusion reflects the requirement for conditioning the fresh S medium before it becomes favorable to fusion. To test this, primary muscle cultures were grown in FE medium as described in Methods. At 40 h after plating, the cultures were divided into four groups which received the following media: (1) fresh S medium; (2) fresh S medium plus 10 μg/ml cytosine arabinoside (CA); (3) S medium collected from cultures prepared one day earlier, which were already in the phase of rapid fusion at that time (conditioned medium); and (4) conditioned medium plus cytosine arabinoside. The results are shown in Fig. 2. It can be seen that (a) conditioned

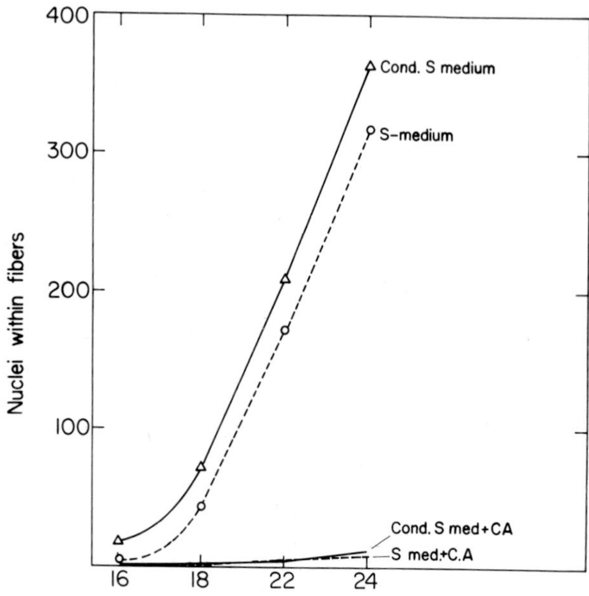


Fig. 2. : Effect of conditioned S medium on cell fusion. Primary cultures plated at a density of 1.5 x $10^6$ were transferred to FE medium 24 h after plating. At 40 h, the cultures were changed to the following media: (a) Freshly prepared S medium; (b) S medium which was conditioned for 24 h in a previous set of cultures which underwent fusion; (c) fresh S medium plus 10 μg/ml cytosine arabinoside; (d) conditioned S medium plus cytosine arabinoside. Starting 16 h later, 4 fixations were performed at 2 h intervals (2 plates of each group); the plates were stained and nuclei within fibers were counted in 10 randomly selected fields in each plate at magnification x 400. The results are expressed as the average number of nuclei within fibers per field.

medium did not affect very much the duration of the lag period between the change to the permissive medium and onset of cell fusion; (b) cytosine arabinoside given during the change of medium had a highly inhibitory effect on cell fusion, whether the cultures were exposed to S medium or to conditioned medium. These results indicate that the 18 h lag reflects intrinsic changes in the cells[14] and is not due to the need for the conditioning of the cell environment.

## Formation of binucleated fibers by single cells isolated during mitosis

When very dilute populations of myoblasts are plated in cultures so that the cells stay separated from each other, they start to proliferate and colonies, each originating from a single cell , are formed. Under such conditions, fusion is delayed till the population in the clones becomes sufficiently dense[19,20]. This phenomenon was utilized in the following experiments to examine the capacity of single cells that are placed in cloning conditions to differentiate or resume proliferation after having been exposed to the fusion-permissive S medium during a defined phase of their mitotic cycle. It was anticipated that if cells previously grown in FE medium become triggered to fuse following exposure to S medium, they will not form colonies when plated as single cells. Instead, the cells will complete their division and the two isolated descendant cells will fuse with each other and form a binucleated fiber (Fig. 3).

The first question which was examined was : How soon after the change to S medium do mitotic cells become triggered to fuse and form binucleated fibers? Primary rat myoblasts were plated at a density of 1.5 x $10^6$ cells/plate and grown in FE medium. 40 h after plating, the cultures were transferred to S medium. From then on, plate duplicates were shaken and cells in mitosis were collected at intervals of 2 h and replated in gelatin-coated plates with the original S medium at a denstiy of 800 cells/ plate. Each collection was carried out using a separate culture plate, to avoid any possible effects of repeated change of medium. The collection of cells was continued for a period of 16 h following the change to S medium (cell fusion started in the primary mass cultures 18 h following the change to S medium). 2 h after replating, sets of 2 cultures were fixed and the paired cells counted. This was used later to calculate the percentage of mitotic cells that formed binucleated fibers. The remainder of the plates containing the collected mitotic cells were incubated for a period of 24h. Then they were fixed and scored for the number of binucleated fibers formed. It was found (Fig. 4) that mitotic cells collected during the first 4 h after change of medium did not produce binucleated fibers when plated under cloning conditions, whether they were maintained in S medium or in FE medium. However, mitotic cells collected at later times and left in S medium produced increasing numbers of binucleated cells, the maximum occurring when the cells were collected 14 h after the change to S medium.

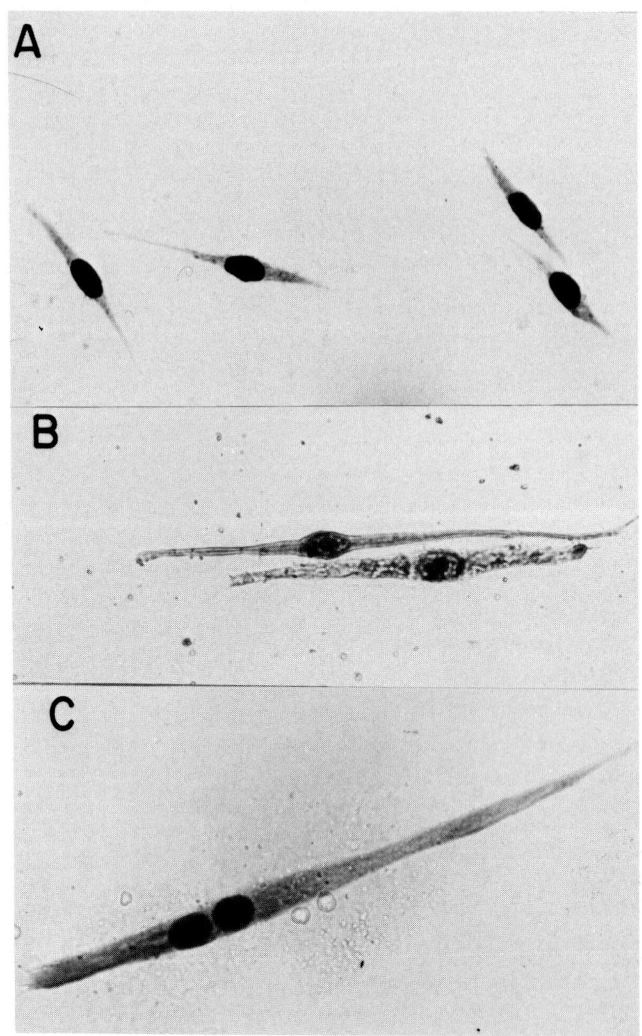


Fig. 3. : Cells which are collected during mitosis after having been exposed to S medium in mass culture and which are plated as clones, divide and develop into one of the following forms: (a) A proliferating clone of myoblastic cells; (b) a non-dividing pair of markedly elongated cells, and (c) a binucleated fiber resulting from the fusion of the two daughter cells formed after mitosis. (Giemsa. Magnification x 300).

Figure 4 shows also that significant numbers of binucleated cells were formed only when the collected cells were plated in S medium. However, when cells that had been collected during mitosis were subsequently grown under cloning conditions in FE medium formation of binucleated cells was very low, irrespective of when collection of mitotic cells was made.

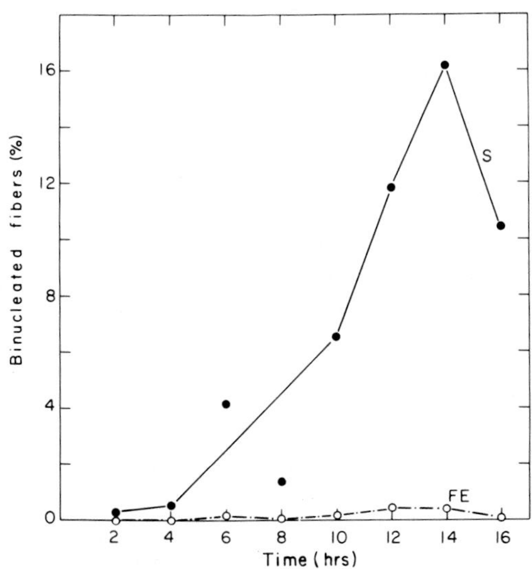


Fig.4.: Dependence of binuclear fiber formation on the time after change to S medium at which the mitotic cells are collected. Cells plated at a density of 1.5 x $10^6$ per plate were transferred 40 h later from FE to S medium. At 2 h intervals, cells in mitosis were collected and subcultured as clones in gelatin-coated plates, in groups of 4 plates. Two plates of each group were maintained in the original S medium (●——●) and the other two were transferred to FE 2 h after subculture (o---o). Each group was fixed 24 h after the collection of mitotic cells, stained, and scored for the number of binucleated fibers formed. The percentage of cells in mitosis at the time of collection was determined by fixing samples 2 h after collection, and counting the percentage of clones composed of paired cells. Abscissa: Time (in h) of collection of the mitotic cells from mass cultures (0 h is the time of change from FE to S medium). Ordinate: Percentage of cells collected during mitosis, which formed the binucleated cell fibers (from Yaffe et al., see Ref. 26).

The kinetics of formation of binucleated fibers in the secondary cultures plated with small numbers of cells isolated during mitosis in S medium is shown in Fig. 5. It can be seen that most binucleated cells are formed between 6 and 10 h after plating (i.e., 6-10 h after mitosis).

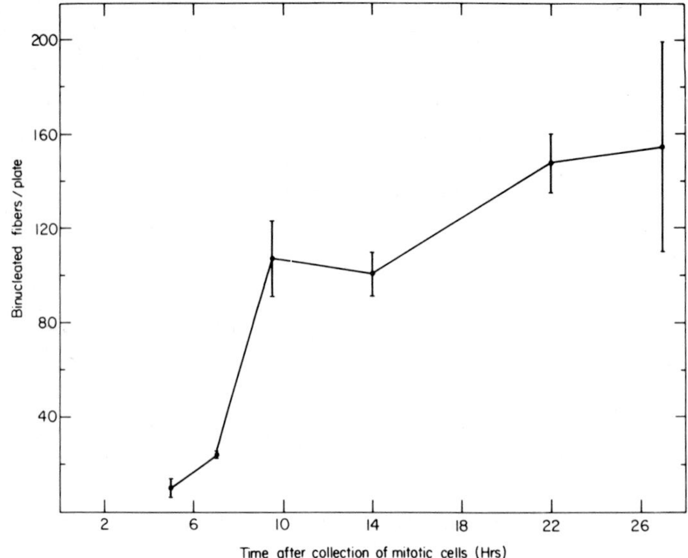


Fig. 5. : Kinetics of formation of binucleated fibers by cells collected during mitosis from mass cultures and plated at low density in S medium. Cultures grown in FE medium were exposed to S medium at 40 h after plating. 14 h later, the plates were shaken gently and mitotic cells were collected and plated in gelatin-coated plates, at a concentration of 1,500 cells per plate. Fixations were performed at the indicated times. The stained plates were scored for the number of binucleated fibers formed. Each point indicates the average of 2 plates. The vertical bars indicate the variation between the actual values of the plates. Abscissa: Time after collection of mitotic cells, at which fixations were made.

## Can myoblasts be induced to resume proliferation after the mitotic cycle which is followed by cell fusion ?

The great reduction in number of binucleated fibers in groups in which collected mitotic cells were returned to FE medium (Fig. 4) suggested that when cells in mass culture that have passed their last mitosis and are triggered to proceed to cell fusion are transferred to cloning conditions in FE medium, they become capable of resuming proliferation. However, the absence of binucleated fibers in cultures returned to FE medium did not exclude the possibility that cells, although inhibited from fusing, did not resume proliferation. Therefore, an experiment was designed to observe the fate of cells collected in mitosis and marked individually after settling in their new environment.

Mass cultures were transferred from FE to S medium. 14 h later, mitotic cells were collected and plated under cloning density in S medium. 2 to 4 h after plating, i.e., after the cells had settled and completed division, the cultures were screened on an inverted microscope and the location of well-isolated pairs of cells was marked on the underside of the plates, with a needle. Half of the cultures were then changed to FE medium. All cultures were incubated for an additional 32 h, then fixed and stained. The number of labelled pairs which had either formed binucleated fibers or proliferated were counted, and the results of four independent experiments are shown in Table II. It can be seen that a change to FE medium resulted in a great increase in the proportion of marked cells which proliferated and a reduction in the number of marked cells which formed binucleated fibers or remained as pairs.

TABLE II

Stimulation of proliferation by a shift back to FE of cells collected from mass cultures and plated at low density.

Cells grown in FE medium in mass cultures were exposed to S medium 40 h after plating. 14 h after change of medium, the plates were gently shaken and cells in mitosis were collected, diluted in S medium collected from mass cultures, and plated in gelatin-coated plates at a concentration of approximately 500 cells per plate. As soon as the cells adhered to the surface of the plates, the location of well-isolated pairs of cells formed after cytokinesis was marked under the microscope on the underside of the plates, with the aid of a needle. The marked plates were divided into two groups: in one, the medium was changed back to FE 4 h after collection of mitotic cells and the second was maintained in the original S medium. The two groups were incubated for 32 h then fixed and stained for observation of the formation of binucleated fibers, nonproliferating bicelluar clones and proliferating clones in the marked areas. The table summarizes 4 independent experiments.
A significant shift toward proliferation in the groups grown in FE medium was established in all four experiments. The overall significance (combined $\chi$ square statistics) is extremely high.

| Exp. No. | Medium* | No. of marked colonies | Missing** | Nonproliferating cells(%) Bi-nucleated | 2-cell colonies | Proliferating col. (%) |
|---|---|---|---|---|---|---|
| 1 | S | 76 | 1 | 18.4 | 26.3 | 55.3 |
| | FE | 133 | 4 | 2.2 | 6.7 | 91.0 |
| 2 | S | 150 | 5 | 16.6 | 43.3 | 40.1 |
| | FE | 118 | 7 | 0.8 | 21.2 | 77.9 |
| 3 | S | 67 | 1 | 25.3 | 17.9 | 56.7 |
| | FE | 42 | 2 | 0 | 19 | 80.9 |
| 4 | S | 75 | 1 | 26.7 | 32.3 | 41.0 |
| | FE | 68 | 2 | 4.4 | 23.6 | 72.0 |

* Nutritional medium in secondary cultures.

** Number of marked pairs which could not be recovered after 32 h.

The close follow-up of the behavior of the marked pairs of cells grown in FE medium suggests that some of them resume proliferation only after a lag of more than 24 h (data not shown).

How long after mitosis can triggered cells be affected by change back to FE medium?

The results descrived above strongly support the conclusion that cells past the mitotic cycle which is followed by fusion can be induced to resume proliferation. As indicated in Fig. 5, most binucleated cells are formed between 6 and 9 h after mitosis. Experiments were performed to see how late during this period cells can still be prevented from fusing, by changing to FE medium. Mitotic cells were isolated from mass cultures 14 h after the change from FE to S medium. The cells were plated at a density of 500 per plate in S medium. At indicated time intervals (Fig.6), three cultures were changed to FE medium. At each time point, two additional plates were fixed for counting the number of binucleated fibers already present in the cultures when the medium was changed (group A, dotted lines).

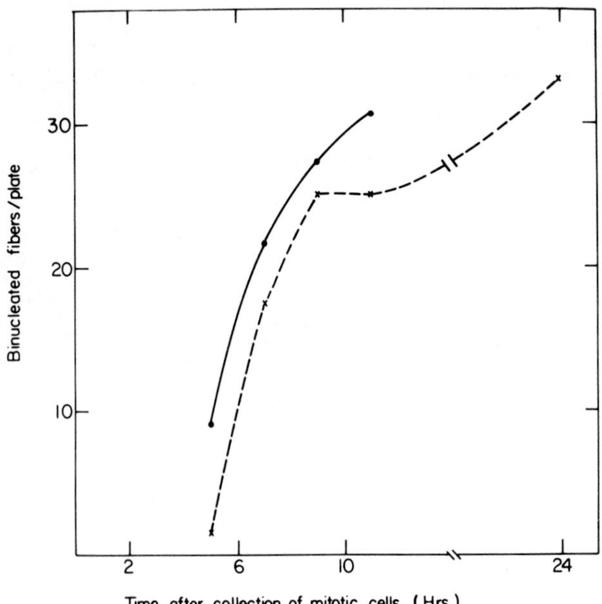


Fig. 6. : Formation of binucleated fibers after change back to FE medium, at different times following mitosis. Mitotic cells were isolated from mass cultures 14 h after the change from FE to S medium. The cells were plated at a density of 500 cells/plate in the original medium. At 5,7,9 and 11 h after plating, cultures were fixed and stained for counting the binucleated fibers already present in the cultures (group A; x---x). In 3 other cultures, the medium was change to FE. These cultures were incubated and fixed 24 h after plating, then stained and scored for the number of binucleated fibers (group B; •——•). Abscissa: Time after collection of mitotic cells, at which fixation (group A) or change to FE medium (group B) were made.

The cultures which were transferred to FE medium were incubated, and all of them were fixed 24 h after plating. They were stained and the binucleated fibers were counted (group B). It can be seen that the number of muscle fibers found after further incubation in FE medium is roughly equal to the number of binucleated cells which were expected to form if the cultures had been grown in S medium for one additional hour (compare groups A and B). These results imply that in triggered cells, processes leading to fusion can be reversed by changing to FE medium up until very shortly before fusion takes place.

Effect of fresh medium on formation of binucleated fibers

Fresh medium has a stimulatory effect on the proliferation of several cell types[21,22]. Therefore, it was of interest to test the effect of fresh S medium on the formation of binucleated fibers. Mitotic cells were collected from mass culture 14 h after the change to S medium and replated at low density in the original medium. One group was left in the original medium until the end of the experiment. At 2 h after plating, groups of cultures were changed to either fresh S medium or FE medium. All cultures were incubated, fixed 24 h after plating, stained, and the number of binucleated cells was counted. It was found (Table III) that the change to fresh S medium, although much less effective than to FE medium, nevertheless reduced by 70% the formation of binucleated fibers.

TABLE III

The effect of changing to fresh medium on formation of binucleated fibers in cultures of cells isolated during mitosis

Mitotic cells were collected from mass cultures 14 h after the change from FE to S medium and were plated, in the original medium, at a density of 500 cells per plate. 2 h after plating, the medium in Group 1 was replaced by FE medium and that of Group 2 by fresh S medium. Group 3 was left in the original S medium throughout. All plates were fixed and stained 24 h after plating and the number of formed binucleated fibers was counted.

| Group | Type of medium | Binucleated fibers formed | % of control (Group 3) |
|---|---|---|---|
| 1 | FE | 0 ; 2 | 2.9 |
| 2 | Fresh S medium | 8 ; 12 | 29.4 |
| 3 | Original S medium not replaced | 34 | 100.0 |

Effect of conditioned medium on formation of binucleated fibers

The inhibitory effect of fresh S medium on the formation of binucleated fibers (Table III) made it necessary to check whether the difference between the number of binucleated fibers formed by mitotic cells collected from mass cultures during the first 4 h after the change to S medium and the number of binucleated fibers formed by mitotic cells collected at later times is not due to the conditioning of the medium by the mass cultures, in the groups which were collected several hours later. Cultures grown in FE medium were transferred, 40 h after plating, to S medium or to S medium which had been conditioned in a previous set of cultures for 14 h (which is the length of time found to yield the largest number of binucleated fibers (Fig. 4)). At 2 h and 14 h after the change of medium, cells in mitosis were collected from these groups, as described previously. The collected cells were replated in the same medium, incubated for 24 h, and then fixed, stained and assayed for total binucleated fiber formation. The results are presented in Table IV. It can be seen that exposing the cells to conditioned S medium rather than fresh medium for 2 h in the mass cultures, and subsequently growing the cells in that medium, did not result in a significant increase in binucleated fiber formation.

TABLE IV

Effect of conditioned medium on formation of binucleated fibers by single cells collected during mitosis from mass cultures at different times after the change to S medium.

A. Mass cultures grown in FE medium were changed to S medium or conditioned S medium as described in text. 2 h and 14 h after change of medium, groups of cultures were shaken and detached cells were collected, replated and grown in the same medium. 24 h after plating, the secondary cultures were fixed and stained and binucleated fibers were counted. The figures are the average of 3 plates.

B. Similar to A, except that conditioned medium was added to the appropriate group of cultures only after isolation of mitotic cells.

| Time of collection of cells* | Type of medium | Average number of binucleated fibers/plate |
|---|---|---|
| A. | | |
| 3 h | S medium | 1.3 |
| 3 h | Conditoned S medium | 2.6 |
| 14 h | S medium | 60.0 |
| B. | | |
| 2 h | S medium | 3.5 |
| 2 h | Conditoned S medium | 3.0 |
| 14 h | S medium | 41.5 |

* Hours after change from FE to S medium or conditioned S medium.

## DISCUSSION

Muscle cultures afford one of the best examples of an apparent relation between cell division and the expression of differentiation. There is a clear-cut transition from a proliferative state to postmitotic differentiation; this is associated with a major morphological change, i.e., cell fusion and extensive changes in the rate of synthesis of many proteins. Data presented here and elsewhere[14] has shown that when myoblasts which were kept in a proliferative state in FE medium were stimulated to fuse by changing to S medium, the great majority went through a period of DNA synthesis and mitosis before fusing. Inhibition of DNA synthesis after the change to S medium strongly inhibited formation of fibers.

The follow-up of the behavior of single cells isolated from mass cultures during mitosis at different times after the change from FE to S medium showed that formation of binucleated fibers by the descendants of these cells plated in low density in S medium occurred only if the cells were exposed to S medium in mass culture for at least 4-5 h before mitosis. Since the single isolated cells were then maintained in the same medium, one could argue that the requirement for exposure of the cells to S medium in mass cultures for several hours prior to their transfer to cloning conditions in order for the cells to form binucleated fibers is due to the necessity of conditioning this medium by the mass cultures. However, this does not seem to be the case, since virtually the same results were obtained when the collected cells were subsequently grown in S medium which was conditoned by exposing it for 14 h to differentiating mass cultures. This shows that the triggering of the process leading to formation of binucleated fibers takes place in mass cultures.

The generation time of these cells is about 12 h. Autoradiography of cells exposed to short pulses of $^3$H-thymidine has shown that the S period is about 8 h. G2 is about 1.5 h (Lavie, unpublished). Therefore, cells in mitosis that were collected during the first 4-5 h following the change to S medium were exposed to this medium in mass culture after having completed most or all of their DNA synthesis period. Cells collected later were exposed to the S medium during a substantial part of their DNA synthesis period, and some of them proceeded to differentiate under cloning conditons. These experiments may suggest that the cells are most responsive to the stimulatory effect on fusion produced by the change to S medium during a specific phase of the mitotic cycle (i.e., late S). However, the design of the experiment does not exclude the possibility that the duration of the exposure to S medium in mass culture per se is the determining factor.

The question of whether cells which already passed the last mitosis can resume proliferation was examined by the follow-up of single cells isolated during mitosis, 14 h after the change from FE to S medium. These cells were exposed for a complete generation to S medium and if left in mass cultures the majority of them would have fused without further division. When such cells were plated without change of medium as single cells, between 16% and 27% of them completed their mitosis and the pairs of daughter cells fused into binucleated fibers. However, when the medium was changed to FE, very few binucleated cells were found. The way the experiments were performed rules out the possibility of some kind of selection of different cell populations caused by the change back to FE medium. All cultures were plated in the original medium, pairs of cells obtained after completion of mitosis were marked 2-4 h later and only then was the medium changed. Almost all marked clones were recovered, and their proliferation in FE medium was confirmed. The decrease observed in the formation of binucleated fibers by just changing to fresh S medium is an additional indication that the results are not due to some unidentified effect unique to FE medium.

It was shown earlier that inhibition of RNA synthesis by application of actinomycin D during the last 6-8 h preceding cell fusion does not prevent the initiation of the phase of rapid cell fusion nor the increase in the synthesis of muscle-specific proteins. These experiments suggest that mRNA molecules which code for the proteins which are required for cell fusion are present in the cells at least several hours prior to fusion[23-25]. Examination of the behavior of individual mitotic cells which were previously triggered to fuse in mass culture and then returned to FE medium under cloning conditions has shown that these cells can resume proliferation for at least 4 h after their otherwise last mitosis. Comparison between the curve describing formation of binucleated fibers by isolated cells grown in S medium and the effect on fiber formation of the time at which the medium is changed back to FE (Fig. 6) shows that cells were responsive to the effect of FE medium until very close to the time of actual fusion. This suggests that the cells can be affected by this change of medium after they contain mRNA molecules which specify proteins involved in the process of fusion and related synthetic activities.

The experiments described in the present communication do not answer the question whether DNA synthesis or cell replication are required for a change in the pattern of gene expression. However, they do indicate that, whatever the nature of the relationship between the last period of DNA synthesis and the programming of cells to proceed to differentiation, no irreversible loss of the capacity of the cells to proliferate occurs until very close to the time of actual fusion. These results are thus incompatible with the idea of the existence of a unique mitosis, after which the cells are committed to fuse. Rather, they show that in this cell system, after the cells have replicated for some time in culture, any round or replication can be made to be the last one, by exposing the cells to the appropriate culture conditions during that cycle. Following such a division, the cells gain the option to fuse, but they are also able to revert and resume proliferation. It seems that many environmental factors may be working together to determine the pathway taken by these cells. The presence of conditioned S medium or fresh S medium might direct the cells to continue to proliferate. Experiments in which avian muscle cells were plated at different cell densitites or subjected to a different schedule of medium changes also led to similar conclusions[12, 13, 18].

ACKNOWLEDGEMENTS

Thanks are due to Dr. M. Feldman, Dr. S. Penman and Ms. Peggy Arps for their comments on the manuscript, and to Ms. Ora Saxel for her skillful technical assistance. This research was supported by the U.S.-Israel Binational Science Foundation and the Muscular Dystrophy Association of America.

REFERENCES

1. Okazaki, K. and Holtzer, H.: *Proc. Nat. Acad. Sci., U.S.*, 56, 1484, 1966.
2. Weintraub, H., Campbell, G.L.M. and Holtzer, H. :*J. Cell Biol.*, 50, 652, 1971.
3. Campbell, G.L.M., Weintraub, H., Mayall, B.H. and Holtzer, H.: *J. Cell. Biol.*, 50, 669, 1971.
4. Lockwood, D.M., Stockdale, F.E. and Topper,Y.J.: *Science*, 156, 945, 1967.
5. Nakamura, I., Segal, S. Globerson, A. and Feldman, M.: *Cell Immunol.*, 4, 351, 1972.
6. Coleman, J.R., Coleman, A.W. and Hartline, E.J.H. : *Develop. Biol.*, 19, 527, 1969.
7. Okazaki, K., Nameroff, M. and Holtzer, H. :*Science*, 146, 533, 1964.
8. Bischoff, R.:In *Regeneration of Striated Muscle and Myogenesis.*, Eds. A.A. Mauro, S. Shafiq and A. Milhorat, pp. 218- 231. Excerpta Medica, Amsterdam, 1970.
9. Yaffe, D. and Gershon, D.: *Nature (Lond.)*, 215, 421, 1967.
10. Yaffe, D. :*Curr. Top. Dev. Biol.*, 4., 37, 1969.
11. Graessmann, A., Graessmann, M. and Fogel M. :*Develop. Biol.*, 35, 180, 1973.
12. O'Neill, M.C. and Stockdale, F.E. : *Develop. Biol.*, 29,410, 1972.
13. Doering, J. L. and Fischman, D.A. : *Develop. Biol.*, 36, 225, 1974.
14. Yaffe, D. : *Exp. Cell Res.*, 66, 33, 1971.
15. Yaffe, D. and Fuchs, S.: *Develop. Biol.*, 15, 33, 1967.
16. Yaffe, D. : In *Tissue Culture: Methods and Application*. Eds. P.F. Kruse and M.K. Patterson, pp. 106-114, Academic Press, New York, 1973.
17. Terasima, T. and Tolmach, L.J.: *Exp. Cell Res.*, 30,344, 1963.
18. Buckley, P.A. and Konigsberg, I.R.: *Develop. Biol.*, 37, 186, 1974.
19. Hauschka, S.D. and Konigsberg, I.R. :*Proc. Nat. Acad. Sci., U.S.*, 55, 119, 1966.
20. Richler, D. and Yaffe, D.: *Develop. Biol.*, 23, p. 18, 1970.

21. Balk, S. D.: *Proc. Nat. Acad. Sci., U.S.*, 68, 1689, 1971.

22. Rovera, G., Farber, J. and Baserga, R.: *Proc. Nat. Acad. Sci., U.S.*, 68, 1725, 1971.

23. Shainberg, A., Yagil, G. and Yaffe, D.: *Develop. Biol.*, 25, 1, 1971.

24. Yaffe, D. and Dym, H.: *Cold Spring Harbor Symp.Quant. Biol.*, 37, 534, 1972.

25. Yaffe, D., Yablonka, Z., Kessler, G. and Dym, H. : In *Proceedings 10th FEBS Meeting, Paris, 1975*, Vol. 38, Eds. G. Bernardi and F. Gros, pp.313-323, North-Holland, Amsterdam, 1975.

26. Yaffe, D., Lavie, G. and Dym, H. : In *Colloquium on Normal and Pathological Protein Synthesis in Higher Organisms*, INSERM Paris, 1973, p. 451.

Progress in Differentiation Research, ed. N. Müller-Bérat et al.



# MYOGENESIS: THE ROLE OF EXOGENOUS FACTORS VS. ENDOGENOUS PROGRAMMING

G.C.T. YEOH

Department of Physiology,
University of Western Australia,
Nedlands, Western Australia, 6009.

## 1. Introduction

It is possible to group investigators in the field of myogenesis into two factions; those who contend that the role of the environment is trivial and thus of little consequence, and those who believe that exogenous cues can dictate the manner in which myogenic cultures differentiate.

Advocates of a major role for the microenvironment in promoting myogenesis propose that all the myogenic cells in a culture are equivalent and capable of fusion _or_ replication. In the appropriate environment cells fuse, then as a consequence irreversibly withdraw from the cell cycle. If they do not fuse, they re-enter the cell cycle and divide. These views are shared by Konigsberg and co-workers[1,2], Stockdale and O'Neill[3], Doering and Fischman[4] and Yaffe and Dym[5,6]. In contrast, Holtzer and co-workers[7-9] have proposed that the myogenic population is not homogeneous but consists of cells in different compartments of the myogenic lineage. At a minimum, myogenic cultures consist of replicating presumptive myoblasts and their daughters, the postmitotic myoblasts. By definition, replicating presumptive myoblasts do not have the option to transcribe and translate the definitive muscle myosin mRNAs _or_ to fuse. These unique properties are restricted to the postmitotic myoblast population. The reprogramming required of a replicating presumptive myoblast to generate a myoblast has been attributed to passage of that cell through a critical cell cycle, termed a 'quantal cell cycle'

Evidence supporting a primary role of the microenvironment in myogenesis stems from two types of experiments: (1) rearing myogenic cells on a collagen substrate and (2) growing myogenic cells in conditioned medium. The contention that collagen as a molecule has a unique function in the event of fusion[10] is not supported by the recent findings of Dientsman and Holtzer[8]. It is much more likely that the collagen substrate by promoting adhesion of cells, and by permitting cell replication and cell migration only <u>indirectly</u> allows the accumulation of myogenic cells with the capacity to fuse. Further reasons for being dubious of the contention that exogenous collagen supplied a unique metabolite that directly mediated the event of fusion, is the finding that myogenic cultures themselves synthesize and secrete collagen[8].

In this report, evidence will be presented demonstrating that while conditioned medium does enhance fusion, it does so not by inducing cells that would otherwise have replicated to fuse, but by allowing more fusion-competent cells of the original inoculum to adhere to the substrate. In brief, conditioned medium, like exogenous collagen, does not stimulate the processes involved in the melding of cell surfaces;it allows more myogenic cells to survive and thus only indirectly allows more fusible cells to accumulate.

2. Materials and Methods.

Primary myogenic cell cultures were prepared from embryonic chicks as described by Bischoff and Holtzer[11]. Secondary myogenic cultures were prepared from seven-day old primary cultures as described by Yeoh and Holtzer[12]. All cultures were grown in 8:1:1 medium[11] and fed on alternate days. Unless otherwise specified, cultures were inoculated at a density of $5x10^5$ cells/ml using volumes of 1 ml, 3 ml and 9 ml for 35 mm, 60 mm and 100 mm culture dishes (Falcon plastic) respectively.

Accurate measurements of fusion in cultures of high or low density, particularly shortly after the initiation of fusion, are impossible with the light microscope. Even using the electron-microscope it is difficult, in the early stages, to determine whether two cells have actually fused. Alternatively, many investigators have demonstrated a close correlation between the accumulation of creatine phosphokinase (CPK) activity and the recruitment of nuclei into myotubes[12-14]. This correlation between fusion as judged conservatively under the light microscope and CPK activity has been confirmed under the conditions of culturing used in these experiments[15]. Monitoring fusion by following CPK accumulation was especially useful in high density cultures where cells overlie each other and the myotubes making microscopic nuclear scores grossly unreliable. The limitations of using CPK levels for monitoring fusion have been discussed by Yeoh and Holtzer[15].

3. Results.

Effect of conditioned medium: Several investigators have shown that there is an increase in both the absolute number of fused cells and the percentage of fused cells in cultures reared in conditioned medium[1,2,4]. However, what is not clear in these experiments is whether fusion occurred <u>earlier</u> in cultures grown in conditioned medium To determine whether fusion actually occurs precociously, it is necessary to demonstrate that fusion is initiated in conditioned medium proir to its occurrence in standard medium. In the experiments of Doering and Fischman[4] for instance, at the times selected for study, fusion had occurred in both sets of cultures. Therefore the investigators could not justifiably have concluded that fusion was initiated earlier in conditioned medium cultures[4]. In view of this, myogenic cultures in conditioned medium and standard medium were set up. They were monitored microscopically, and CPK levels were assayed every 24 hours for evidence of an earlier initiation of fusion in conditioned medium. As shown in Fig.1, there is no difference in the two sets of cultures regarding the time when levels of CPK begin to rise. The rise in CPK activity coincides with the first appearance of myotubes as observed with the microscope.

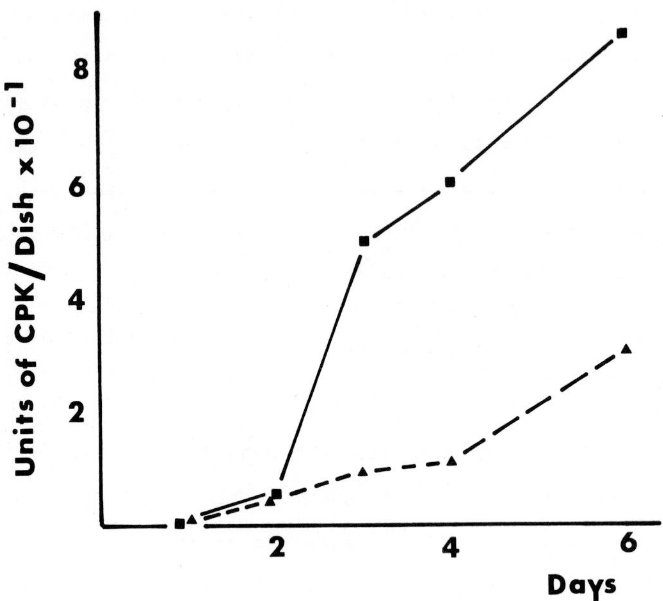


Fig.1 The accumulation of CPK in myogenic cultures grown in conditioned medium (▲---▲) and standard medium (■—■). Cells were harvested from culture dishes daily by gentle scraping, then washed with saline and homogenized. Aliquots (25μl) of the supernatent were assayed for CPK activity as outlined by Yeoh and Holtzer[15].

In brief, conditioned medium does not induce precocious fusion. A minimal amount of time is necessary to prepare cells for the event of fusion and this period is unaltered by conditioned medium.

It is worth noting that in studies involving myogenic clones, White et al.[16] also concluded that conditioned medium does not promote precocious initiation of fusion.

Fig.2a and 2b compare myogenic cultures grown in standard medium and conditioned medium. The myotubes in conditioned medium are broader and flatter. They form swirling patterns in conditioned medium as opposed to the more parallel orientation seen in standard medium. Cultures in conditioned medium at 3 days do contain more myotubes than controls. Nuclear counts reveal that there is an increase in both percentage fusion and absolute numbers of nuclei in myotubes. In this respect our findings confirm the observations of Konigsberg[1] and Doering and Fischman[4]. This enhancement of fusion is only temporary and observed on days 2 and 3, for subsequently, more myotubes are seen in control cultures. This is likely to be due to the depressed rates of replication in conditioned medium, and is consistent with the notion that with the exception of the fusible myoblasts in the original inoculum, presumptive myoblasts must divide before yielding cells with the option to fuse[8,9].

Question: Does the finding of greater numbers of myotubes in cultures grown in conditioned medium for 3 days mean that cells which otherwise could have replicated, were instead induced to fuse?

To answer this question, it is necessary to consider the various ways the yield of myotubes in a given culture can be increased. This could be achieved by (i) increasing the original plating efficiency or by (ii) increasing rates of replication. Both events would lead to an increase in the absolute numbers of fused cells. However, the percentage fusion index would remain constant if there were no differential selection for myogenic cells over the non-myogenic cells in the culture.

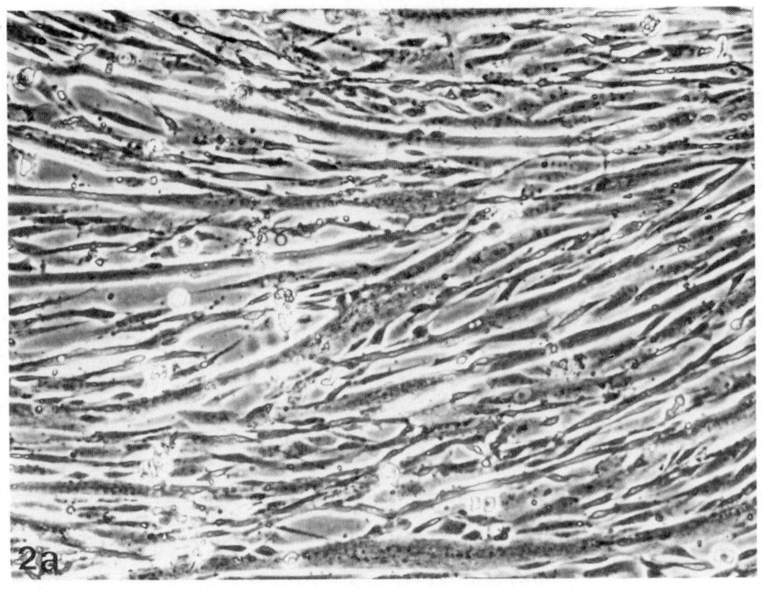


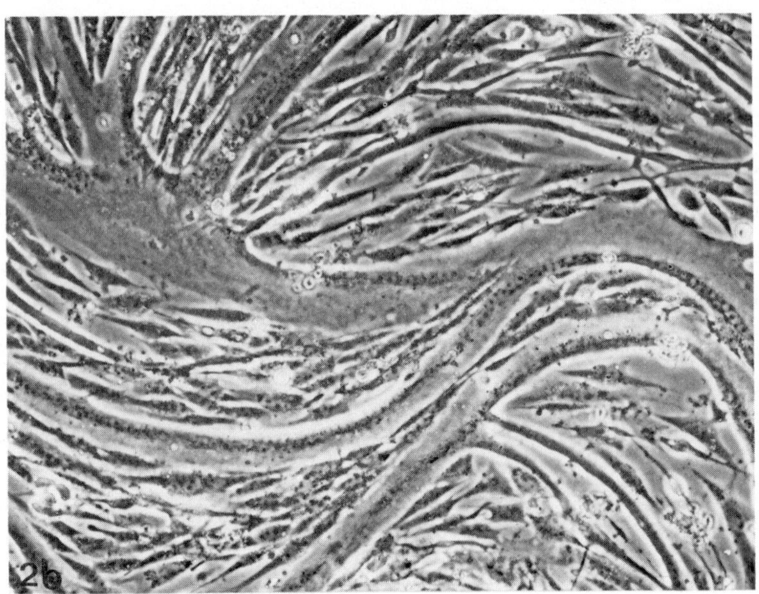


Fig.2. Primary myogenic cultures grown in standard 8:1:1 medium (a) and conditioned medium (b) after 3 days of culture. Magnification x160

Selection for myogenic cells could take the form of i) enhancing the proliferation of myogenic cells, ii) selective survival of myogenic cells or iii) promoting presumptive myoblasts to undergo the quantal cell cycle that yields postmitotic myoblasts rather than a proliferative cell cycle that yields more presumptive myoblasts. Obviously, the observation that conditioned medium increases the yield of myotubes on day 3 does not distinguish among the above possibilities.

The following experiments demonstrate that conditioned medium does selectively enhance the survival of postmitotic myoblasts in the initial inoculum used to set up primary myogenic cultures. This alone can explain the enhancement of fusion by conditioned medium that has been reported in the literature[1,4].

Experiment 1: If myogenic cultures are maintained for 1 day in conditioned medium, then transferred to standard medium for 2 days, they are indistinguishable from cultures which had been grown continuously in conditioned medium for 3 days. Alternatively, cultures kept in standard medium for 1 day, then transferred to conditioned medium for 2 days, do not display enhanced fusion. Conclusion: Conditioned medium exerts its effect during the first 24 hours of culture. Since fusion occurs late on day 2 and on day 3, the effects of conditioned medium are exerted considerably earlier than the event of fusion.

Experiment 2: Cultures maintained in medium containing EGTA for 4 days accumulate large numbers of postmitotic myoblasts[8]. Such a culture is shown in Fig.3a. On day 4 these cultures are fed with either standard medium or conditioned medium. Fusion occurs in both cultures over the next 24 hours (Fig.3b and 3c). Fusion is not greater in cultures treated with conditioned medium. Conclusion: Conditioned medium does not promote the fusion process per se, and it does not act on cells that have acquired the option to fuse i.e. the myoblasts.

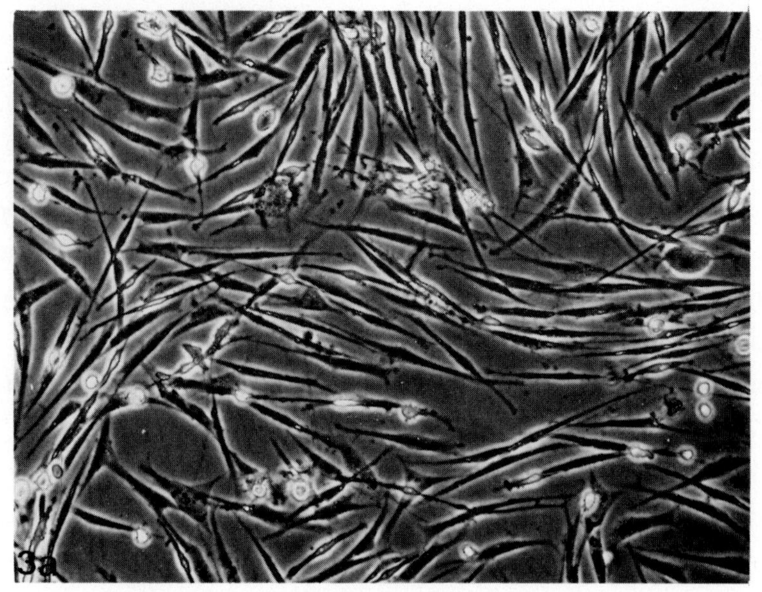
3a

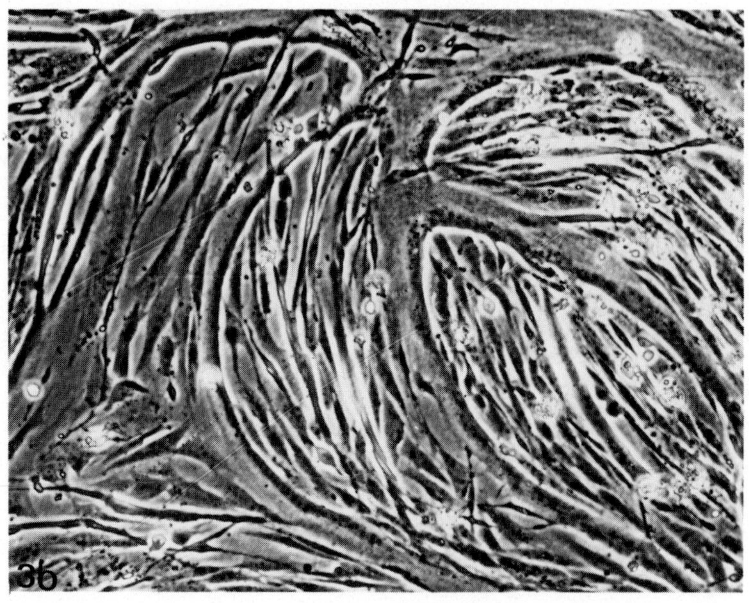
3b

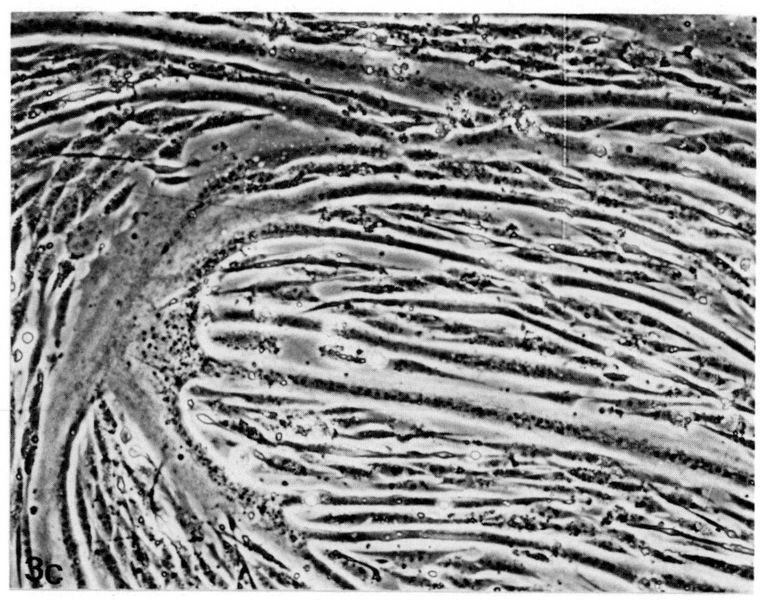


Fig.3. Myogenic cultures grown in EGTA for 4 days (a). Note elongated mononucleated cells. On day 4 these cultures are transferred to standard medium (b) or conditioned medium (c). These cultures display similar morphology to conditioned medium cultures. Mag.x160

Experiment 3: Myogenic cultures are maintained in EGTA for 4 days. One group is not sub-cultured but simply fed with standard medium. The other group is trypsinized and sub-cultured into standard medium. The extent of fusion is assessed after 24 hours by i) phase microscopy and ii) CPK activity. As shown in Fig.4a and 4b, fusion is substantially greater in the series that did not undergo sub-culturing. The level of CPK in the respective cultures (Table 1) confirm the microscopic observation that more fusion occurred in cultures not subjected to the sub-culturing procedure. Conclusion:

A significant proportion of postmitotic myoblasts do not survive sub-culturing. The difference between the CPK/mg DNA values for non-sub-cultured vs. sub-cultured gives an indication of the degree of selection against myoblasts.

Table 1

Recovery of creatine phosphokinase (CPK) activity after re-feeding and sub-culturing EGTA-derived myoblasts

| Procedure | Units CPK per dish | DNA per dish | Units CPK per mg DNAx$10^{-3}$ |
|---|---|---|---|
| Refeeding | 266 | 24 | 11.1 |
| | 220 | 23 | 9.6 |
| Sub-culturing | 40 | 26 | 1.6 |
| | 41 | 23 | 1.8 |

Cultures were initially maintained in EGTA medium for 4 days and then either refed with standard 8:1:1 medium or entire dishes sub-cultured into identical dishes in 8:1:1. After 24 hours, dishes were rinsed with BSS and cells were removed and pelleted and then homogenized in 0.9% NaCl. Aliquots were taken for CPK assay and DNA determination.[11] Results are means of duplicate determinations.

Experiment 4: Large numbers of postmitotic myoblasts accumulate in EGTA-treated cultures during days 2 and 3[8]. When 3 day EGTA-treated cultures are exposed to $^{3}$H-thymidine for the next 24 hours, only the replicating cells i.e. fibroblasts and presumptive myoblasts incorporate the label. Most of the myogenic cells which are labelled prove to be presumptive myoblasts, a few are myoblasts. When such a culture is trypsinized and replated in standard or conditioned medium any selective effects against the survival of the unlabelled myoblasts will show up as an increase in the specific activity of the culture. If more postmitotic myoblasts (i.e. unlabelled cells) survive and adhere to the dish in conditioned medium in contrast to

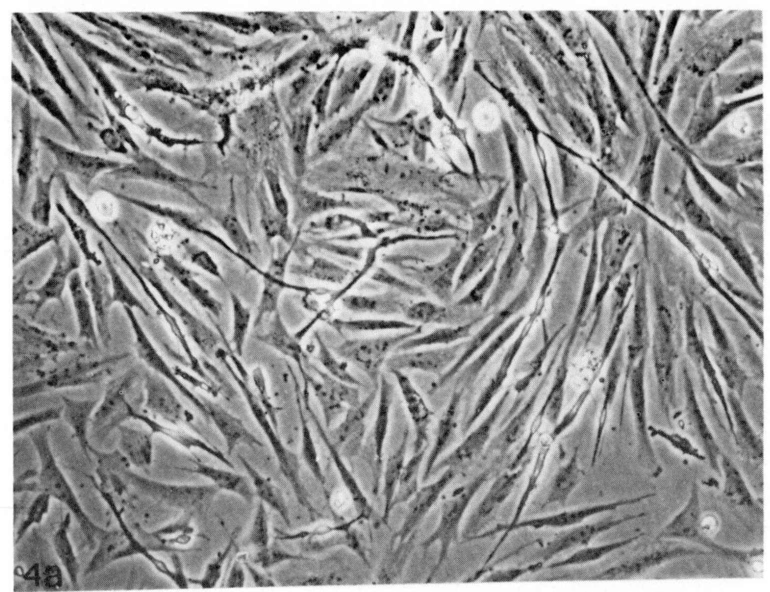


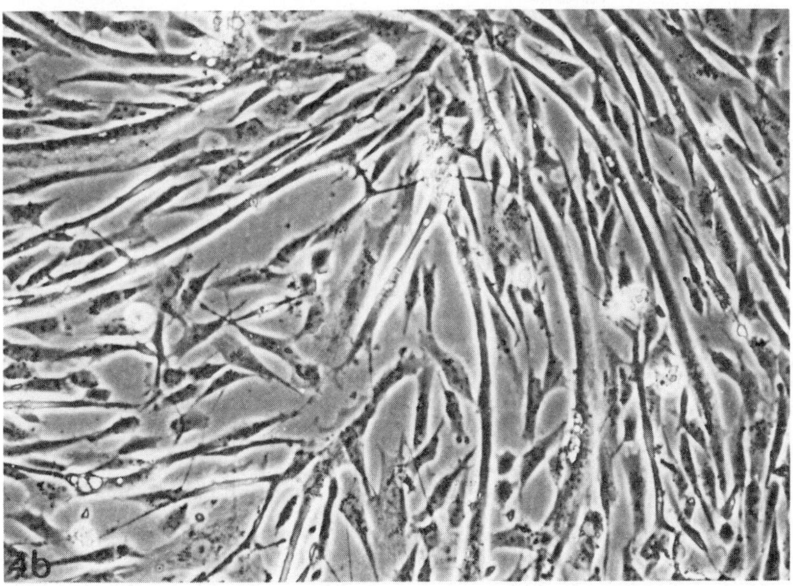


Fig.4. Photomicrograph comparing the extent of fusion 24 hours after 3 day-old EGTA treated myogenic cultures are fed with standard 8:1:1 medium without sub-culturing (a) and with sub-culturing (b). Magnification x160. Sub-culturing was performed as described by Yeoh and Holtzer[15].

standard medium, then the specific activity of cells in such cultures would be lower than in controls. The reverse would be true for the cells which remain in the medium. The results of such an experiment are shown in Table 2.

Table 2

A comparison of the specific activity of pre-labelled cells which attach to culture dishes and remain in the medium in standard cultures and cultures grown in conditioned medium.

| | | µg DNA per dish | CPM per µg DNA |
|---|---|---|---|
| Cells on the dish | Standard medium | 8.8 | 7090 |
| | | 7.6 | 6122 |
| | | 7.6 | 6863 |
| | Conditioned medium | 10 | 4406 |
| | | 10.4 | 4472 |
| | | 12 | 3861 |
| | Standard medium | 4.3 | 735 |
| | | 3.7 | 878 |
| | | 4.1 | 766 |
| | Conditioned medium | 1.3 | 1052 |
| | | 1.3 | 1258 |
| | | 1.1 | 1277 |

Myogenic cultures were grown in EGTA for 3 days and then exposed to $^{3}$H-thymidine (0.25µCi) for the next 24 hours. Cultures were then trypsinized and re-platted in standard 8:1:1 medium or conditioned medium and kept under normal culture conditions for 8 hours. Cells were harvested from the bottom of the dish and from the medium, washed twice with 0.9% NaCl, pelleted and solubilized in 0.5N NaOH. Aliquots were taken for DNA determination[17] and liquid scintillation counting.

The lower specific activity of cells which had attached to the dish in conditioned medium demonstrates that conditioned medium favours the survival of postmitotic myoblasts when compared with standard medium. Conclusion: Condition medium allows more postmitotic cells of an inoculum to survive than standard medium.

Experiment 5: When primary myogenic cultures are grown continuously in the presence of $^{3}$H-thymidine, a fraction of the nuclei which eventually enter myotubes are unlabelled. These cells entered the terminal compartment of the myogenic lineage in vivo and consttitute cells in the initial inoculum which are committed to fuse. Cultures grown in standard and conditioned medium continuously in the presence of $^{3}$H-thymidine are radioautographed after 3 days. The myotubes are examined and scored for the % nuclei which are unlabelled. Table 3 shows that whereas only 38% of standard culture myotube nuclei are unlabelled, 49% of nuclei in myotubes are unlabelled in cultures grown in conditioned medium. Conclusion: More postmitotic myoblasts in the initial inoculum survive in primary cultures grown in conditioned medium.

Table 3

Postmitotic myoblasts in conditioned medium and standard medium

| | Unlabelled nuclei | Labelled nuclei | % unlabelled nuclei |
|---|---|---|---|
| Standard medium | 287<br>251 | 423<br>477 | 40<br>35 |
| Conditioned medium | 369<br>393 | 398<br>412 | 48<br>49 |

Cultures were set-up in conditioned medium and standard medium in the presence of $^{3}$H-thymidine. After 3 days of culture, cells were fixed and prepared for radioautography. Twelve fields were examined for each culture dish.

The above experiments suggest that conditioned medium does not promote the event of fusion per se, but that it enhances the survival of postmitotic myoblasts. These observations do not preclude the possibility of other effects of conditioned medium on myogenic cultures. For example, it may induce replicating presumptive myoblasts to favour quantal rather than proliferative cell cycles, thus thus increasing the yield of fusible cells in a culture.

Analysis of fusion in secondary cultures: A detailed analysis of fusion in secondary cultures has revealed that these differ from primary cultures in the following respects; i) fusion is delayed by at least 48 hours, ii) fusion is less synchronous and occurs over many more days, iii) myotubes contain fewer nuclei i.e. they are hyponucleated and iv) there are fewer myotubes[15].

Little fusion is detectable in secondary cultures on day 4 (Fig. 5a), whereas fusion is almost maximal by day 4 in primary cultures. Fusion occurs after day 4 and multinucleated myotubes are present by day 7 (Fig.5b). The delay in fusion is of interest and there are several possible explanations which might account for it. The density of myoblasts could be too low to permit earlier fusion; thus the delay is the time required to establish a prerequisite density of myoblasts for fusion to occur. The delay could also be due to a low myogenic/fibrogenic cell ratio. The presence of relatively more fibroblasts could impede the movement of myoblasts so decreasing the collison frequency of the fusion competent cells. Alternatively, the delay could be the time that is required to generate fusible cells by division, suggesting that fusible myoblasts are simply not present in the inoculum used to set-up secondary cultures. In order to investigate these alternatives, the following experiments were performed:

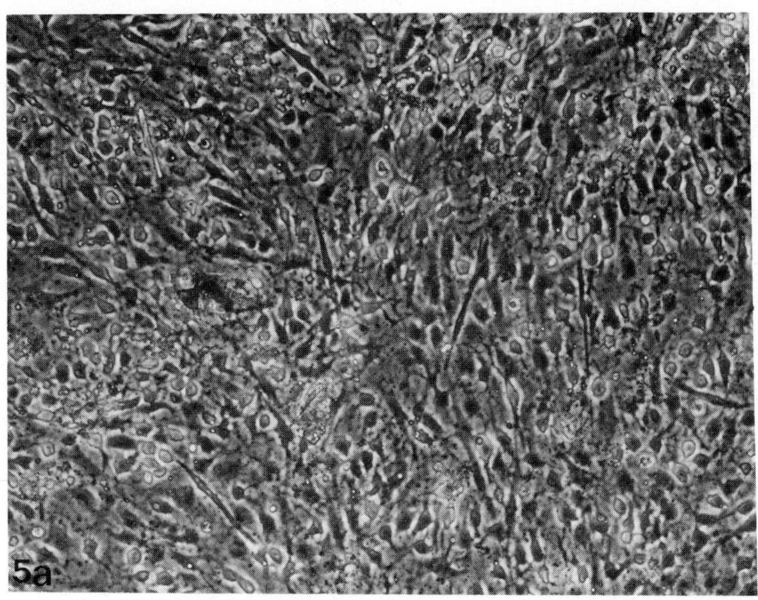


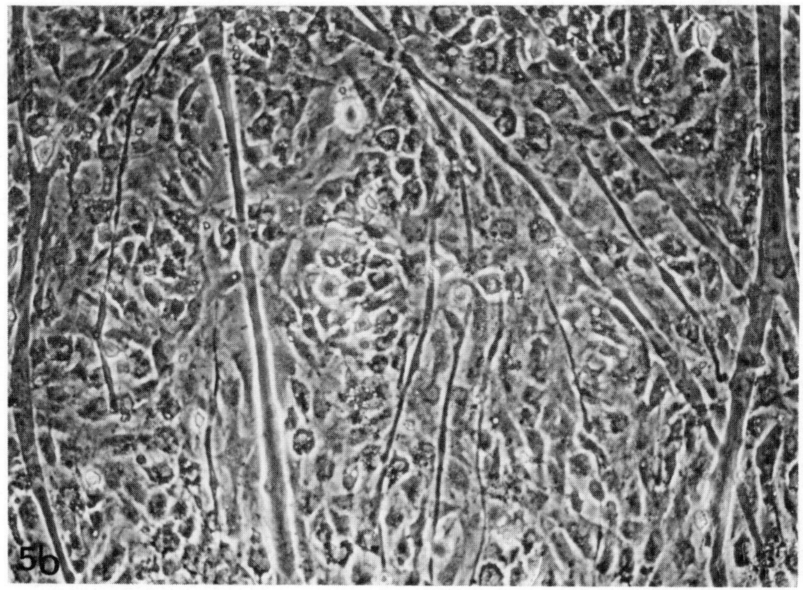


Fig.5. Secondary myogenic cultures prepared from 7 day-old primary myogenic cultures contain elongated bipolar mononucleated cells but no myotubes after 4 days of culture (a). After 7 days of culture myotubes are clearly evident although present at a lower density than primary cultures (b) (See Fig.1a) Magnification x160.

Experiment 1: That the delay in fusion is not due to a density effect is shown by initiating secondary cultures using different sized inoculums. The progress of fusion is followed by the accumulation of CPK and Fig.6 shows that this is the same for the different cultures. Microscopic observations reveal that fusion is initiated in these cultures after the same delay irrespective of the initial densities.

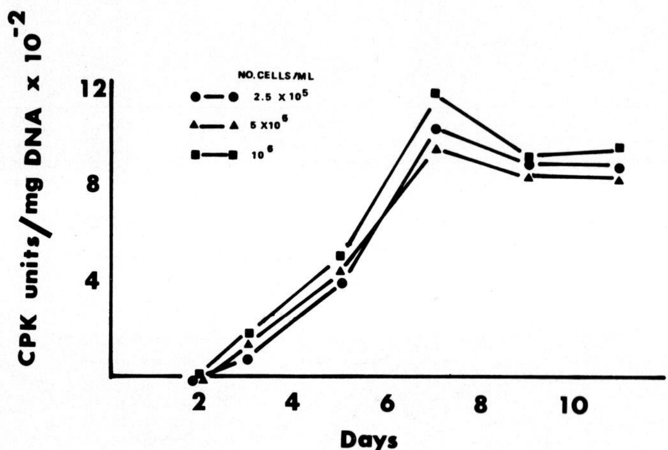


Fig.6. Fusion as monitored by the accumulation of CPK in secondary myogenic cultures set-up at different initial densities.

Experiment 2: That a low myogenic/fibrogenic cell ratio does not delay fusion is shown by using an inoculum of $2.5 \times 10^5$ myogenic cells/ml mixed with a one-fold, two-fold and four-fold excess of fibroblasts. As shown in Fig.7, the kinetics of fusion in these cultures which were deliberately contaminated with fibroblasts[18] is indistinguishable from controls.

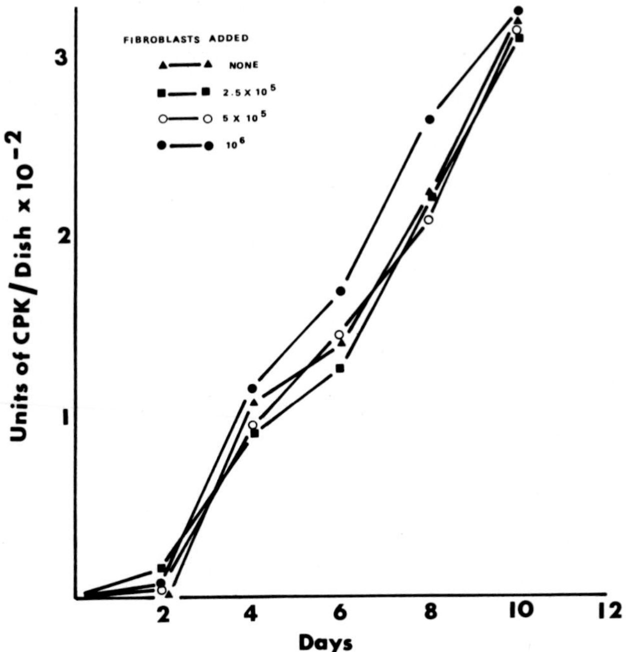


Fig.7. Thekinetics of fusion as monitored by the accumulation of CPK in primary myogenic cultures mixed with differing amounts of fibroblasts as indicated. In all experiments, 2.5 x $10^5$ primary myogenic cells were used.

Experiment 3: That cell division is obligatory for myotube formation in secondary cultures is shown by using cytosine arabinoside (Ara-C). Secondary cultures grown continuously in the presence of Ara-C show no fusion. In contrast, in primary cultures, the postmitotic myoblasts present in the original inoculum will fuse in the presence of Ara-C. Failure of myotubes to form in secondary cultures exposed to Ara-C argues that no myoblasts were present inthe inoculum used to set-up these cultures. These results are incompatible with the proposals of Doering and Fischman[4] that myogenic cells can be induced to fuse in the presence of Ara-C.

Experiment 4: To confirm that cells that do fuse in secondary cultures were obligated to first replicate, such cultures were grown in $^3$H-thymidine for the first 18 hours of culture and prepared for radioautography on day 8. As shown in Fig.8, the nuclei in the myotubes were all labelled. Therefore it is concluded that all cells which acquire the capacity to fuse in secondary cultures were derived from replicating presumptive myoblasts and that these cells themselves do not have the option to fuse. Only their daughters, the postmitotic myoblasts have this option.

In conclusion, these series of experiments suggest that altering the environment might permit a myoblast that already has the option, to excercise that option and fuse. However, the microenvironment is incapable of confering on a presumptive myoblast the option to fuse. The option to fuse is a unique property of the myoblast. It is a property that is acquired by that cell by virtue of its passage through a quantal cell cycle (see Holtzer, this volume)

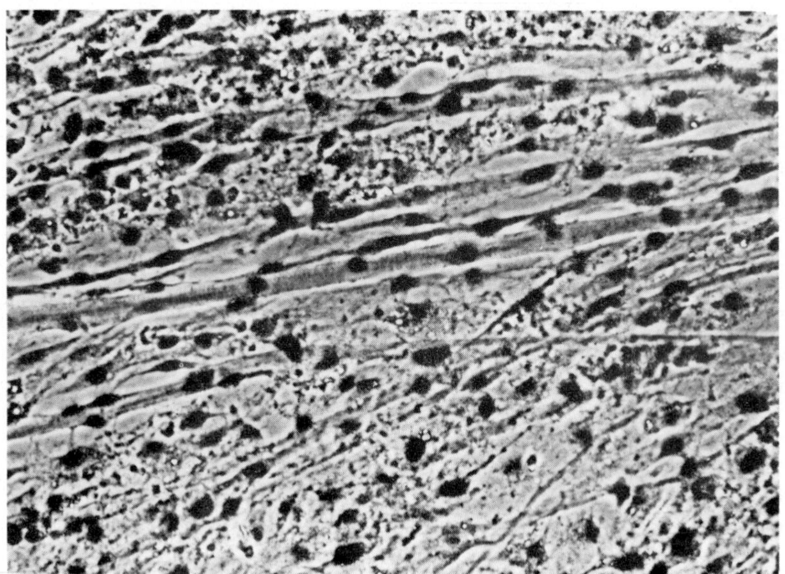

Fig.8. Radioautograph of a secondary myogenic culture exposed to $^3$H-thymidine during the first 18 hours of culture. Culture was subsequently maintained for 8 days. All nuclei in myotubes are labelled. Note 'hyponucleated' condition of myotubes. Magnification x 480

As such, this is not an immediate response to exogenous cues. This view of myogenesis differs profoundly from the views of other workers in the field[1-6].

References


1. Konigsberg, I.R. (1971) Develop. Biol., 26, 133-152

2. Buckley, P.A. and Konigsberg, I.R. (1974) Develop. Biol., 37, 193-212

3. Stockdale, F. and O'Neill, M.C. (1972) In vitro, 8, 212-227

4. Doering, J.L. and Fischman, D.A. (1974) Develop. Biol., 36, 225-235

5. Yaffe, D. and Dym, H. (1973) Cold Spring Harbor Symp. Quant. Biol., 37, 543-547

6. Yaffe, D. and Dym, H. (1975) Abstract, EMBO workshop on 'Muscle cell culture in the study of cell differentiation' Israel

7. Holtzer, H. (1970) in 'Cell differentiation' (eds. O. Schjeide & J. de Vellis) 476-503 Nostrand-Reinhold N.Y.

8. Dientsman, S. and Holtzer, H. (1975) In 'The cell cycle and cell differentiation' (eds. J. Reinert & H. Holtzer) 7, 1-25 Springer Verlag, Heidelberg

9. Holtzer, H., Mayne, R., Weintraub, H. and Campbell, G. (1973) In 'Biochemistry of gene expression in higher organisms' (eds. J. Pollack & J. Wilson-Lee) New Zealand Book co., Sydney.

10. Hauschka, S.D. and Konigsberg, I.R. (1966) Proc. Nat. Acad. Sci. U.S.A., 55, 119-126

11. Bischoff, R. and Holtzer, H. (1968) J. Cell. Biol., 36, 111-128

12. Coleman, J.R. and Coleman, A.W. (1968) J. Cell Physiol., 72 supp 1, 19-34

13. Shainberg, A., Yagil, G. and Yaffe, D. (1971) Develop. Biol., 25, 1-29

14. Turner, D.C., Maier, V. and Eppinberger, H.M. (1974) Develop. Biol., 37, 63-89

15. Yeoh, G.C.T. and Holtzer, H. submitted for publication

16. White, N.K., Bonner, P.H., Rae Nelson, D. and Hauschka, S.D. (1975) Develop. Biol., 44, 346-361

17. Hinegardner, R.T. (1971) Anal. Biochem., 39, 197-201

18. Abbott, J., Schiltz, J., Dientsman, S. and Holtzer, H. (1974) Proc. Nat. Acad. Sci. U.S.A., 71, 1506-1510

Acknowledgement

I am especially grateful to Dr. Holtzer and members of his laboratory for their helpful suggestions and stimulating discussion during my stay in Philadelphia. Special thanks to Mrs. Biehl for her expert technical instruction. This work was made possible by NIH grants HD-00189 and HL 15835 to the Pennsylvania Muscle Inst., Muscular Dystrophy Association, National Science Foundation and American Cancer Society. Travel and support was made possible by a grant from the National Health & Medical Research Council of Australia in the form of The C.J.Martin Overseas Travelling Fellowship.

*Progress in Differentiation Research, ed. N. Müller-Bérat et al.*



# CELL PROLIFERATION AND CYCLIC NUCLEOTIDES

R. Kram, W. Moens and A. Vokaer

Département de biologie moléculaire
Université Libre de Bruxelles

## 1. Introduction

Recent work from various laboratories has implicated 3':5'-cyclic AMP (cAMP) and 3':5'-cyclic GMP (cGMP) in the regulation of proliferation in fibroblast cultures. As described below, these cell lines provide a useful in vitro system to study the mechanisms of growth regulation in normal cells and their alterations in malignant cells.

A regulatory role for cAMP in cell proliferation was originally suggested on the basis of observations relating a defect in cAMP metabolism with the expression of malignant phenotype in cultured fibroblasts. Bürk[1], working with BHK hamster fibroblasts, found reduced adenylate cyclase activity in polyoma-transformed BHK cells. On the other hand, methylxanthines such as caffeine and theophylline, which inhibit cAMP degradation, slowed the growth rate of both untransformed and transformed BHK fibroblasts. Further work in different laboratories[2–4] indicated that several of the abnormal properties of transformed cells (e.g. abnormal morphology, decreased adhesiveness, rapid growth rate, increased agglutinability by plant lectins) could be reverted toward that of normal cells by treatment with dibutyryl cAMP or agents which raise intracellular cAMP levels. Transformed fibroblasts were then shown to have lower cAMP concentrations than their normal counterparts[5,6]. Furthermore, comparison of several cultured cell lines indicated an inverse relationship between cAMP content and generation time[5,7].

In non transformed fibroblasts, low levels of cAMP were found in exponentially growing cultures and high levels in cells that were reversibly arrested in a quiescent state of growth[6,8,9]. Early decreases in intracellular cAMP concentrations follow exposure of resting fibroblasts to a variety of mitogenic agents, e.g. serum, insulin, proteases[9–12]. Conversely, addition of dibutyryl cAMP to the culture medium partially prevents the induction of cell proliferation[9,11–13]. It was further shown that addition of cGMP counteracts the inhibitory effects of dibutyryl cAMP on serum-induced stimulation of some transport systems activated early in the mitogenic response[14] (fig. 1).

This report describes experiments undertaken in order to investigate the physiological relevance of the antagonism observed between high concentrations of exogenously added dibutyryl cAMP ($10^{-4}$ – $10^{-3}$ M) and cGMP ($10^{-3}$ M). Therefore, we have studied the effects of cell density and serum concentration on intracellu-

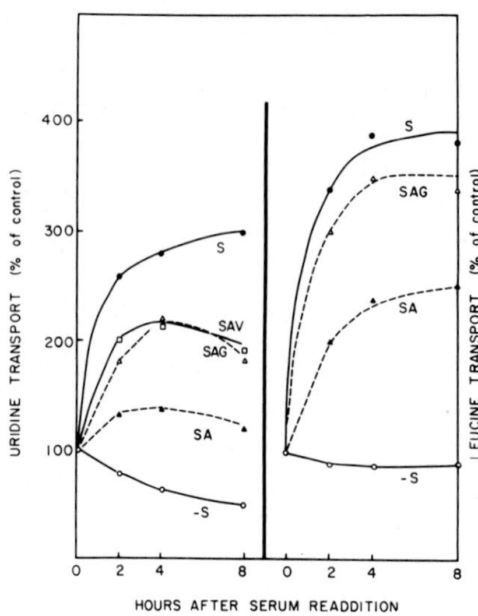


Fig. 1 : Antagonistic effects of cyclic AMP and cyclic GMP on the serum-stimulated rates of uridine and leucine transport in 3T3 fibroblasts. After serum deprivation for 16 hr, the cells were, from time 0, further incubated in either medium without serum (-S), or medium plus 10 % calf serum (S), with or without the following additions symbolized as follows : A = dibutyryl cAMP (0.2 mM) and theophylline (1 mM) ; G = cGMP (1 mM). V refers to the addition of vinblastine at 0.01 μg/ml : note that microtubules inhibitors also antagonize the transport inhibition caused by cAMP [14].

lar cyclic nucleotide concentrations in several fibroblast lines, derived from the mouse 3T3 cell line. This system has also allowed us to investigate whether the cyclic nucleotide levels respond specifically to these regulatory influences. Indeed, data from different laboratories agree that cAMP levels increase in serum-restricted cultures[8-10,15]. In contrast, the mediation by cAMP of density-dependent inhibition of growth remains controversial, since some[5,7,16], but not other[6,15] investigators have observed a rise in cAMP when cells approach confluence.

## 2. Cell lines

The growth of nontransformed fibroblasts, such as the mouse 3T3 line, is regulated by serum factors and cell-to-cell contact. Transition from a growing state to a resting state occurs upon depletion or deprivation of serum and upon cell confluence. Simian virus 40 (SV40) virus-transformed 3T3 clones often lose sensitivity to these environmental factors. Using negative selection pressures on SV40-transformed 3T3 fibroblasts, one can isolate revertant cell lines.

The cell lines used in this study are Swiss 3T3, clone SV101 of SV-transformed 3T3, and four revertant lines derived from SV101 by Pollack and coworkers [17-19]. The density-dependent revertant clones FlSV101 and BuSV2 were obtained by negative selection with fluorodeoxyuridine and bromodeoxyuridine, respectively. The serum-dependent clones (LsSV2 and AγSV5) were selected for their inability to grow at low serum concentrations (1 %) and in gamma-depleted (= agamma) 10 % calf serum, respectively. The density revertants have regained density-dependent inhibition of cell growth comparable to that of normal 3T3, whereas the serum revertants are in addition sensitive to serum restriction.

## 3. Results

Transformed SV3T3 fibroblasts are characterized by cAMP and cGMP levels about half and twice, respectively, those found in growing, untransformed 3T3 cells. Either cAMP or cGMP concentrations in the range of normal cells are found in the revertant lines (Table 1, columnsa).

Table 1 : Cyclic nucleotide concentrations in growing* and serum-restricted* fibroblast cultures.

| Class | Cell line | cAMP * (a) | cAMP * (b) | cGMP * (a) | cGMP * (b) |
|---|---|---|---|---|---|
| normal | 3T3 | 23 | 59 | 0.44 | 0.06 |
| transformed | SV3T3-101 | 10 | 20 | 0.82 | 0.32 |
| density revertants | FlSV101 | 22 | 24 | 0.58 | 0.05 |
| | BuSV2 | 20 | 35 | 0.71 | 0.11 |
| serum revertants | LsSV2 | 20 | 42 | 0.45 | 0.04 |
| | AγSV5 | 10 | 52 | 0.40 | 0.05 |

* All assays were performed on sparse cultures (about $1.5 \times 10^6$ cells per 75 $cm^2$ plate, at time of harvesting) either (a) growing in medium supplemented with 10 % serum or (b) after 1 day of serum-restriction in medium + 0.5 % serum. Procedures for purification and assays of cyclic nucleotides are described in reference 20. Results are expressed in picomoles of cyclic nucleotide per mg of cell protein.

Upon serum restriction (Table 1, columns b), the serum-dependent cell lines showed a greater increase in intracellular cAMP content than serum-insensitive cell lines. This difference is emphasized by the response to serum deprivation of line AγSV5, despite its low cAMP content during logarithmical growth in medium supplemented with 10 % serum. The levels of cGMP were greatly reduced in serum restricted cultures of normal 3T3 fibroblasts, whereas, in SV3T3, they decrease only to a value close to that found in exponentially growing 3T3 cells. However, greatly reduced cGMP concentrations after 1 day in medium + 0.5 % serum were found in the serum-insensitive density-revertants as well as in the serum-revertants. Thus, this response does not appear to be specific for serum restriction, although this discrepancy could be accounted for on the following basis.

The density-revertant lines grow as well as SV3T3 in 1 % calf serum, but reach a lower saturation density[18,19], unfortunately, in the range of cell density at which the cultures were harvested. Although this interpretation remains to be tested, it is possible that the density-revertant lines were already arrested in a quiescent state of growth, hence the significant decrease in cGMP found under our experimental conditions.

Figure 2 illustrates the changes in cyclic nucleotide concentrations following serum readdition to sparse cultures of 3T3 fibroblasts maintained in medium + 0.5 % serum. Reinitiation of growth in these quiescent cultures is correlated with early, transient increases in cGMP and decreases in cAMP levels, reaching maximum and minimum values within 20 min. after serum readdition (fig. 2).

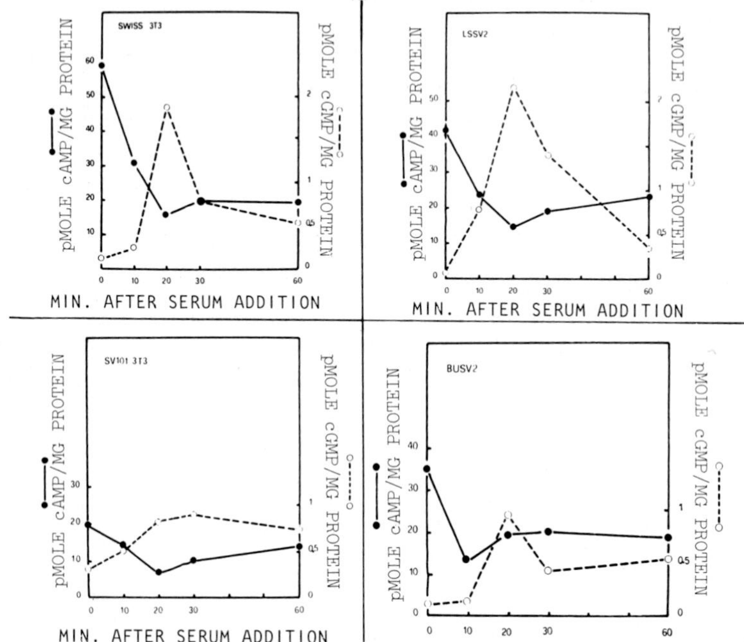


Fig. 2 : Kinetics of cyclic nucleotide responses to serum readdition to sparse cultures of 3T3, SV3T3 and revertant lines. The cultures had been maintained for 24 hr in 0.5 % serum before shifting up, at zero time, serum concentration to 20 %.

Both serum revertant lines[20] responded similarly to 3T3 cells with parallel but reciprocal changes in the cyclic nucleotide concentrations, whereas these responses to serum readdition were absent or much less pronounced in SV3T3 and in the serum-independent revertant lines (fig. 2).

To test whether cell density influences cyclic nucleotide concentrations, cultures were inoculated and grown to confluence in medium + 10 % serum. Fresh medium was replaced every other day to prevent interference from depletion of nutrients and serum factors. Under these conditions, the cAMP levels are relatively unaffected by density restriction. In agreement with Oey et al.[15], we did not observe an increase, but rather a slight decrease of cAMP at confluence (fig. 3). On the contrary, density-dependent inhibition of growth is correlated with greatly reduced cGMP concentrations in all contact-inhibited cell lines. In 3T3, cGMP decreases continuously down to values, in confluent monolayers, about one fifth of those found at a density of $10^6$ cells per 75 $cm^2$ plate. As the revertant cell lines entered a state of density inhibition, their cGMP content dropped abruptly (fig. 3).

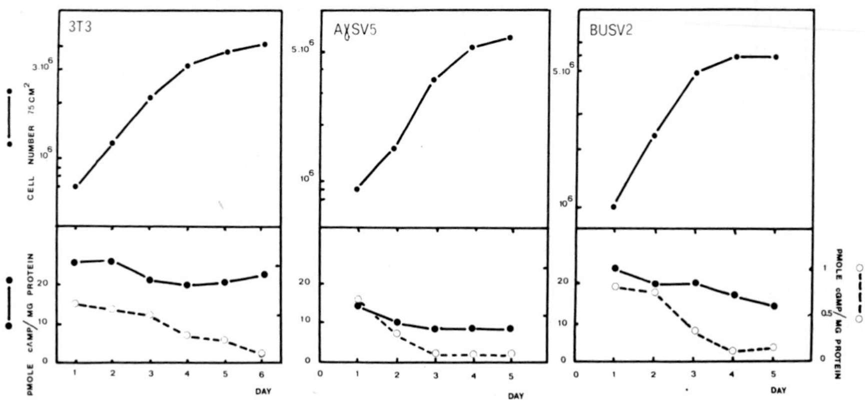


Fig. 3 : Response of cyclic nucleotides to density restriction in 3T3 and revertant lines.

In contrast, the cGMP concentration in SV3T3 fibroblasts decreased only slightly over the same range of cell population (fig. 4). The transformed cells grew exponentially up to cell densities higher than 20 to 30 x $10^6$ cells, where the cultures began to peel off. Only then, they showed a drop in cGMP but with a concomittant decrease of cAMP levels, as reported previously[5,15].

## 4. Discussion

The data presented here indicate that cGMP concentrations are correlated with the growth state of untransformed mouse 3T3 fibroblasts. Greatly reduced cGMP levels are found in quiescent cells, whether at confluence or upon limiting serum concentration in sparse cultures. Serum-restricted cultures also show

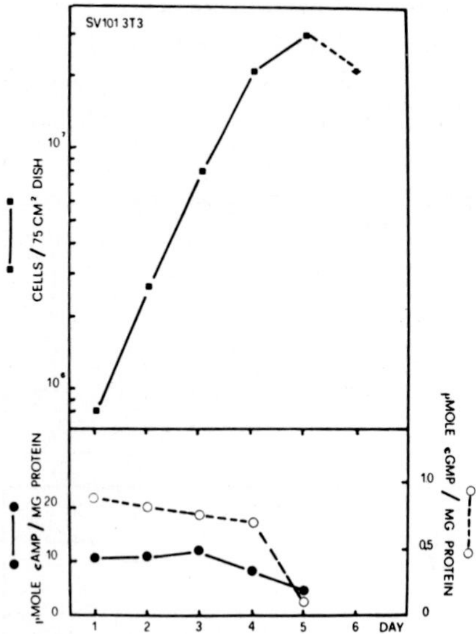


Fig. 4 : Effects of cell density on cyclic nucleotide concentrations in SV3T3 fibroblasts.

increased cAMP content. These changes in cyclic nucleotide concentration are largely lost in transformed SV3T3 fibroblasts which are relatively insensitive to growth regulation by serum restriction and cell density. Rudland et al.[21] have reported similar differences between untransformed lines of mouse, hamster and monkey fibroblasts, on the one hand, and SV3T3 or polyoma-transformed BHK hamster cells, on the other, when growing and stationary cultures are compared.

The use of revertant lines, selected from SV3T3, led Oey et al.[15] to conclude that intracellular cAMP concentrations respond specifically to growth regulation by serum. Our data do not allow a definitive conclusion to be drawn with respect to the specificity of cGMP response to serum restriction. Its levels are greatly reduced,not only in 3T3 and serum revertants, but also in serum-insensitive density revertants, although not in SV3T3. A plausible interpretation, suggested by the properties of the density revertants[18,19], is that these lines, unlike SV3T3, were harvested at a density above their saturation density in 0.5 % serum.

Our results suggest that decreased cGMP concentrations might specifically mediate sensitivity to density-dependent inhibition of growth. Since the cAMP levels do not increase concomitantly, it is unlikely that these reduced cGMP levels reflect a lower ability of dense cultures to utilize components of the medium, especially serum[22,23]. Taken together with the possibility of dissociating serum restriction from density restriction in some revertant lines[18,19], the data suggest that these two modes of regulation of fibroblast proliferation are distinct.

Early transient decreases in the intracellular levels of cAMP[9,10,12,15] and increases in cyclic GMP[20,21] are observed after growth reinitiation by serum. A drop of cAMP is also produced by other mitogenic stimuli, such as treatment with proteases[11] and insulin[6,10]. Conversely, addition of $PGE_1$ with serum increases cAMP levels[6,10,21], prevents the rise in cGMP[21] and partially inhibits the induction of DNA synthesis in quiescent fibroblasts. Other observations relating transient increases in cGMP, but little change in cAMP, with stimulation of cell proliferation include the response of fibroblasts to phorbol myristate acetate[24] (a tumor promoter isolated from croton oil), a purified fibroblast growth factor[25], prostaglandin $F_{2\alpha}$[26] and the activation of human peripheral blood lymphocytes by phytohaemagglutinin[27]. Cyclic GMP has been implicated in the stimulation of purine and pyrimidine biosynthesis induced by concanavalin A in mouse spleen lymphocytes[28].

These data are consistent with the hypothesis that cAMP and cGMP act respectively as negative and positive intracellular signals for controlling the growth of cultured cells[14,27]. However, mitogenic stimuli trigger complex membrane changes which result in increased rates of transport of phosphate, $Ca^{++}$ and $K^+$ ions, and essential nutrients. These early events associated with the induction of fibroblast proliferation by serum have also been implicated as regulatory signals[29-31]. A great deal of further investigation will be required to answer a number of important questions : a) whether or not all these biochemical changes are essential for the initiation of cell division; b) whether these events are ordered in a causal sequence, triggered by a single primary event or regulated independently. Although cyclic nucleotides might play an important role in the changes of some transport systems[9,14] and in the stimulation of precursor biosynthesis for DNA synthesis[28], other early events appear to be elicited by mechanisms not involving cyclic nucleotides[31,32].

## 5. References

1. Bürk, R.R. (1968) Nature 219, 1272-1275.
2. Hsie, A.W. and Puck. T.T. (1971) Proc. Nat. Acad. Sci. USA 68, 358-361.
3. Johnson, G.S., Friedman, R.M. and Pastan, I. (1971) Proc. Nat. Acad. Sci. USA 68, 425-429.
4. Sheppard, J.R. (1971) Proc. Nat. Acad. Sci. USA 68, 1316-1320.
5. Otten, J., Johnson, G.S. and Pastan, I. (1971) Biochem. Biophys. Res. Commun. 44, 1192-1198.
6. Sheppard, J.R. (1972) Nature 236, 14-16.
7. Heidrick, M.L. and Ryan, W.L. (1971) Cancer Res. 31, 1313-1315.
8. Seifert, W. and Paul, D. (1972) Nature New Biol. 240, 281-283.
9. Kram, R., Mamont, P. and Tomkins, G.M. (1973) Proc. Nat. Acad. Sci. USA 70, 1432-1436.

10. Otten, J., Johnson, G.S. and Pastan, I. (1972) J. Biol. Chem. 247, 7082-7087.

11. Burger, M., Bombik, B., Breckenridge, B. and Sheppard, J. (1972) Nature 239, 161-163.

12. Froehlich, J. and Rachmeler, M. (1972) J. Cell. Biol. 55, 19-31.

13. Frank, W. (1972) Exptl. Cell Res. 71, 238-241.

14. Kram, R. and Tomkins, G.M. (1973) Proc. Nat. Acad. Sci. USA 70, 1659-1663.

15. Oey, J., Vogel, A. and Pollack, R. (1974) Proc. Nat. Acad. Sci. USA 71, 694-698.

16. Anderson, W.B., Russel, T.R., Carchman, R.A. and Pastan, I. (1973) Proc. Nat. Acad. Sci. USA 70, 3802-3805.

17. Pollack, R., Green, H. and Todaro, G. (1968) Proc. Nat. Acad. Sci USA 60, 126-133.

18. Vogel, A., Risser, R. and Pollack, R. (1973) J. Cell. Physiol. 82, 181-188.

19. Vogel, A. and Pollack, R. (1973) J. Cell. Physiol. 82, 189-198.

20. Moens, W., Vokaer, A. and Kram, R. (1975) Proc. Nat. Acad. Sci. USA 72, 1063-1067.

21. Rudland, P.S., Seeley, M. and Seifert, W. (1974) Nature 251, 417-419.

22. Dulbecco, R. and Elkington, J. (1973) Nature 246, 197-199.

23. Stoker, M.G.P. (1973) Nature 246, 200-203.

24. Estensen, R.D., Hadden, J.W., Hadden, E.M., Touraine, F., Touraine, J.L., Haddox, M.K. and Goldberg, N.V. (1974) in : Control of proliferation in animal cells, ed. by Baserga, R. and Clarkson, B., 627-634 (Cold Spring Harbor Laboratory).

25. Rudland, P.S., Gospodarowicz, D. and Seifert, W. (1974) Nature 250, 741-774.

26. Jimenez,de Asva, L., Clingan, D. and Rudland, P.S. (1975) Proc. Nat. Acad. Sci. USA 72, 2724-2728.

27. Hadden, J.W., Hadden, E.M., Haddox, M.K. and Goldberg, N.S. (1972) Proc. Nat. Acad. Sci. USA 69, 3024-3027.

28. Chambers, D.A., Martin, D.W. and Weinstein, I. (1974) Cell 3, 375-380.

29. Holley, R.W. (1974) in : Control of proliferation in animal cells, ed. by Baserga, R. and Clarkson, B., 13-18 (Cold Spring Harbor Laboratory).

30. Dulbecco, R. and Elkington, J. (1975) Proc. Nat. Acad. Sci. USA 72, 1584-1588.

31. Jimenez de Asua, L., Rozengurt, E. and Dulbecco, R. (1974) Proc. Nat. Acad. Sci. 71, 96-98.

32. Jimenez de Asua, L. and Rozengurt, E. (1974) Nature 251, 624-626.

Acknowledgement

This work was supported by a grant from the Caisse Générale d'Epargne et de Retraite (Fonds Cancer). We thank Michèle Rivière for valuable technical assistance.

*Progress in Differentiation Research, ed. N. Müller-Bérat et al.*



# STUDIES ON THE MOLECULAR BASIS FOR INHIBITION OF CELL PROLIFERATION BY 5-BROMODEOXYURIDINE.

Anne E. Lykkesfeldt and H. A. Andersen
The Biological Institute of the Carlsberg Foundation,
16 Tagensvej, DK-2200 Copenhagen N, Denmark.

The thymidine analogue 5-bromodeoxyuridine (BUdR) is a valuable tool in the study of mechanisms underlying cellular differentiation. Incorporation of the analogue into DNA inhibits differentiation at some stage of development in all differentiating systems studied so far, while exerting little influence on the proliferating rate or the general viability of the cells (for review see ref. 1). How incorporation of BUdR into DNA influences cell metabolism with the consequences reported by several workers is still a puzzle. However, we think that the study of effects of BUdR incorporation into DNA of the ciliated protozoan *Tetrahymena pyriformis* might contribute to the understanding of these effects.

Under most growth conditions *Tetrahymena* cells tolerate incorporation of BUdR into DNA without measurable damaging effects. Under special growth conditions, however, the presence of BUdR in nuclear DNA may reach a level where the cell proliferation ceases and macromolecular metabolism is severely altered (2,3). In the following we want to give a brief description of our findings, which lead to the conclusion, that the presence of BUdR in the DNA may inhibit transcription of specific genes. Inhibition of specific enzyme activities has been reported several times (4,5,6), but *Tetrahymena* is the first cell in which a specific transcriptional inhibition has been observed.

The amount of BUdR substitution into nuclear DNA of *Tetrahymena* is highly dependent on the growth conditions of the cells. The thymidine necessary for the DNA synthesis may be synthesized via uridine or cytidine using tetrahydrofolic acid as a cofactor, or the cells may take up thymidine from the growth medium directly. The concentrations of these compounds in the medium are therefore of the utmost importance for the amount of BUdR incorporated. To be able to control various concentrations, the cells were grown on a fully chemically defined medium lacking thymidine (7), where the known concentration of tetrahydrofolic acid in turn controlled the endogenous synthesis of thymidine. With a fixed concentration of

BUdR in the medium (0.8 mM) the amount of BUdR incorporation into DNA was measured as a function of the tetrahydrofolic acid concentration in the medium. The results are shown in fig. 1. With a fixed concentration of tetrahydrofolic acid in the medium the amount of substitution could then be controlled by varying the ratio between BUdR and thymidine in the medium, as it may appear from the results shown in fig. 2.

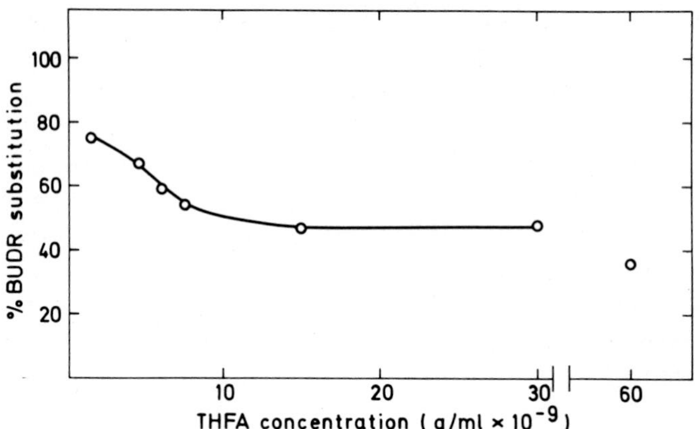


Fig. 1.

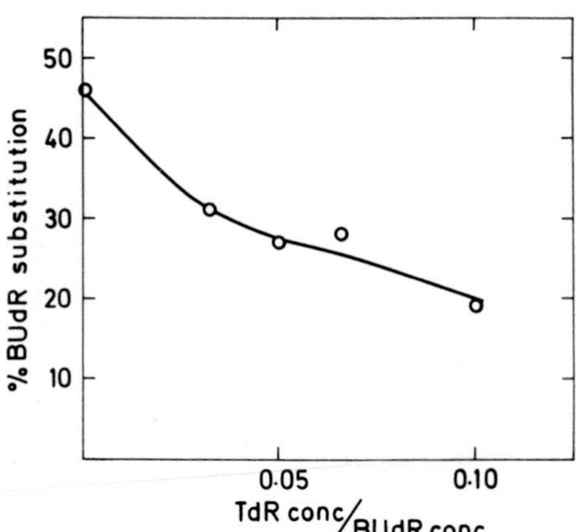


Fig. 2.

At the BUdR concentration of 0.8 mM and a tetrahydrofolic acid concentration of $7.5 \times 10^{-9}$g/ml the cell proliferation ceased after a little more than a doubling in the cell number. As it may be seen in fig. 1. these growth conditions give a BUdR substitution of about 60% of the thymidine sites in DNA. When the increase in amount of DNA, RNA and protein was measured in such a culture, it was found, that the RNA synthesis was strongly inhibited already within the first cell generation, wheras the growth in DNA strictly followed the increase in cell number, and the protein synthesis was only slightly reduced. These conclusions appear from the results shown in fig. 3.

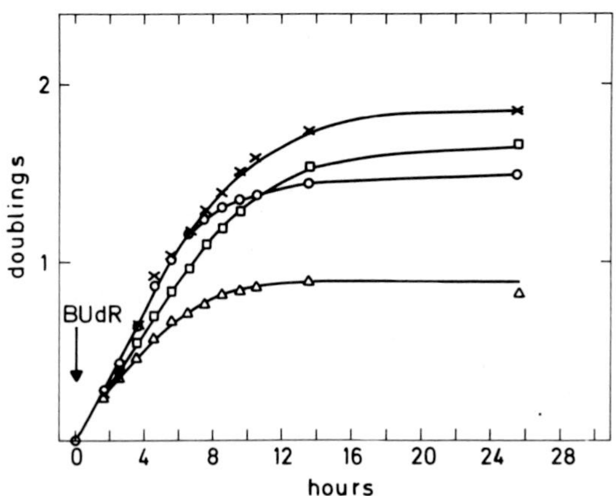


Fig. 3. The increase in: Cell number (o), DNA (x), RNA (Δ) and protein (□) after addition of BUdR (0.8 mM) to an exponentially growing population.

In spite of this rapid and preferential effect observed on the RNA synthesis, the effect is probably indirect and mediated through substituted DNA. This has been demonstrated by studying populations in synchronous growth, where the DNA synthesis takes place in distinct periods, and where the effect of BUdR on RNA synthesis without contemporary DNA synthesis could be studied.

Synchronization of cell division and DNA replication in <u>Tetrahymena</u> may be obtained by several different treatments (8,9). We have used two of these, namely the so called heat shock procedure described by Zeuthen (10) and the starvation/refeeding synchronization described by Cameron and Jeter (11). In both of these synchronized cell cycles DNA replication has been studied in some details (12,13, 14). The two synchronized cell cycles are indicated on figs. 4 and 5. In fig. 4 is shown the heat synchronized cell cycle. The cells have a short G1-period followed by a 2,5 hr S-period and a 2 hr G2-period. Fig. 5 shows the cell "cycle" after starvation and refeeding. After refeeding the cells have a 4 - 5 hr G1-period before the DNA replication is initiated. The synchrony obtained in the latter procedure is not very prominent, but the major advantage of the system in this context, is the long G1-period in which the BUdR effect on RNA synthesis can be followed. It has been shown in both of the synchronized S-periods that the genes coding for ribosomal RNA are replicated in a discrete time interval at the beginning of the nuclear S-period (15,16). The results presented in figs. 4 and 5 show, that the presence of BUdR has no effect on the RNA synthesis as long as no BUdR has been incorporated into nuclear DNA. However, as soon as nuclear DNA replication is initiated and the genes coding for ribosomal RNA have incorporated BUdR, a rapid reduction in the rate of RNA synthesis is observed.

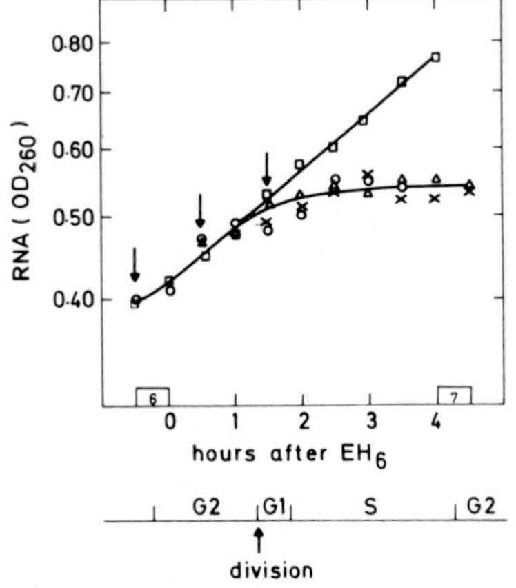


Fig. 4. Relative amount of RNA as a function of time. The arrows indicate the addition of BUdR (0.8 mM) to subcultures (o,Δ,x). The control (□) was treated with 0.8 mM TdR. The cell cycle is indicated at the lower part of the fig.

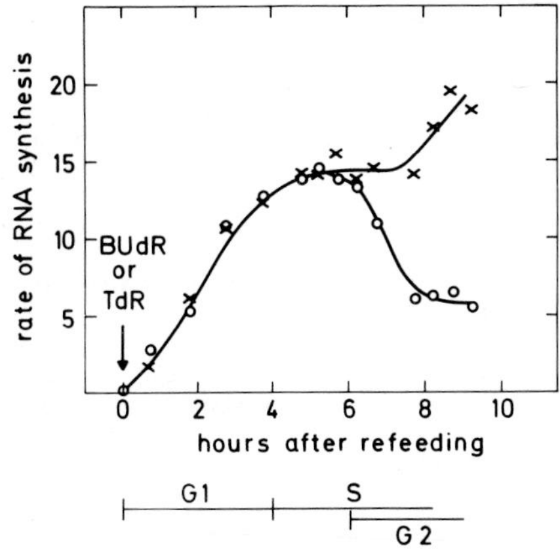


Fig. 5.

Rate of RNA synthesis measured as C14-uridine incorporation during a 30 min. pulse (cpm x $10^{-2}$) as a function of time after refeeding. BUdR culture: (o), TdR culture: (x). The cell cycle is indicated at the lower part of the fig.

Based on these and previous (2,3) results we have suggested, that the presence of high amounts of BUdR in the genes coding for ribosomal RNA causes a structural change of the DNA and inhibits the transcription of these genes. Ribosomal RNA amounts to about 90% of the total RNA present in Tetrahymena (17) and a change in the amount of this single gene product could therefore be measured as a change in total RNA. The results shown here do not rule out the possibility that transcription of other genes might also be inhibited. However, preliminary experiments have shown, that there is a preferential inhibition of the synthesis of ribosomal RNA.

REFERENCES.

1. Rutter, W.J., Pictet, R.L. and Morris, P.W. (1973) Ann Rev. Biochem. 42, 601.

2. Lykkesfeldt, A.E. and Andersen, H.A. (1974) J.Cell Biol. 62,316.

3. Lykkesfeldt, A.E. and Andersen, H.A. (1975) J.Cell Sci. 17, 495.

4. Tomida, M.,Koyama, H. and Ono, T. (1974) Biochim. Biophys. Acta. 338, 352.

5. Marzullo, G. (1972) Dev. Biol. 27, 2o.

6. Stellwagen, R.H. and Tomkins, G.M. (1971) Proc. Natl. Acad. Sci. 68, 1147.

7. Rasmussen, L. and Modeweg-Hansen, L. (1973) J. Cell Sci. 12, 275

8. Zeuthen, E. and Rasmussen, L. (1972) Research in Protozoology, 4 Oxford, Pergamon Press.

9. Andersen, H.A., Rasmussen, L. and Zeuthen, E. (1976) Current Topics in Microbiology and Immunology, 72, Springer-Verlag.

10. Zeuthen, E. (1971) Exptl. Cell Res. 68, 49.

11. Cameron, I.L. and Jeter, J.R. (1970) J. Protozool. 17, 429.

12. Andersen, H.A. (1972) Exptl. Cell Res. 75, 89.

13. Andersen, H.A. and Zeuthen, E. (1971) Exptl. Cell Res. 68, 309.

14. Mowat, D., Pearlman, R.E. and Engberg, J. (1974) Exptl. Cell Res. 84, 282.

15. Andersen, H.A. and Engberg, J. (1975) Exptl. Cell Res. 92, 159.

16. Engberg, J., Mowat, D. and Pearlman, R.E. (1972) Biochim. Biophys. Acta, 272, 312.

17. Leick, V. (1967) Compt. Rend. Trav. Lab. Carlsberg, 36, 113.

*Progress in Differentiation Research, ed. N. Müller-Bérat et al.*
*© 1976, North-Holland Publishing Company - Amsterdam, The Netherlands.*


# NUCLEAR PORES AND NUCLEAR ENVELOPE IN SYNCHRONOUS YEAST

E. G. Jordan, N. J. Severs and D. H. Williamson*
Biology Department, Queen Elizabeth College,
University of London, Campden Hill,
London W8

## 1. Introduction

The relationship of nuclear pore complexes to the cell-cycle has been studied in higher eukaryotic cells in culture and has shown certain changes related to the phases of the cell-cycle[1,2]. The number of pores has been shown to increase rapidly early in $G_1$ and rise again at the time of DNA synthesis[1]. The latter effect together with other correlations has raised the possibility that nuclear pores are involved in DNA replication. Although high numbers of pores are sometimes found in cells which are inactive, their increase is normally associated with an increase in cell activity[1-4]. It is hoped that a careful correlation of changes in nuclear pore number with changes in different metabolic process could give a better understanding of nuclear pore function. To investigate these relationships we have utilized synchronous cultures of the yeast Saccharomyces cerevisiae; a lower eukaryote. The cells were studied during the first cycle of synchronous growth. The synchronisation procedure depends upon the selection of the larger cells from a stationary phase population induced by starvation[5]. Nutrient depletion is thought to operate a block very early in $G_1$ and the growth which follows release from this block can be thought of as a process of differentiation from a stationary phase cell to one following the normal cell division cycle[6].

*National Institute for Medical Research, Mill Hill, London NW7.

## 2. Materials and methods.

Yeast cells were synchronised according to the technique of Williamson[5] and prepared for examination by freeze-fracturing without any chemical fixation or cryoprotection[7].

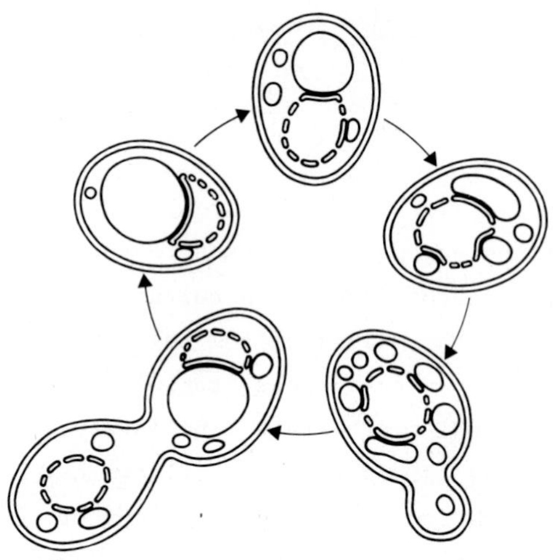

Fig. 1. Diagrammatic representation of the stages in vacuole fragmentation and influence on the distribution of pore-free areas on the nucleus during the first cycle of synchronous growth.

---

Fig. 2. Freeze-fractured yeast cell at the start of synchronous growth showing a large depression, devoid of nuclear pores, in the surface of the nuclear membrane (Nm) caused by close apposition of the vacuole membrane (Vm). Nu, nucleoplasm; arrow indicates direction of shadow; bar represents 1 μm.

Fig. 3. Freeze-fractured yeast cell 30 minutes after the start of synchronous growth. Close association of the nucleus with a vacuole (V) is still shown. Numerous nuclear pores are present on the nuclear membrane (Nm) where no vacuole association occurs. Arrow indicates shadow direction; bar represents 1 μm.

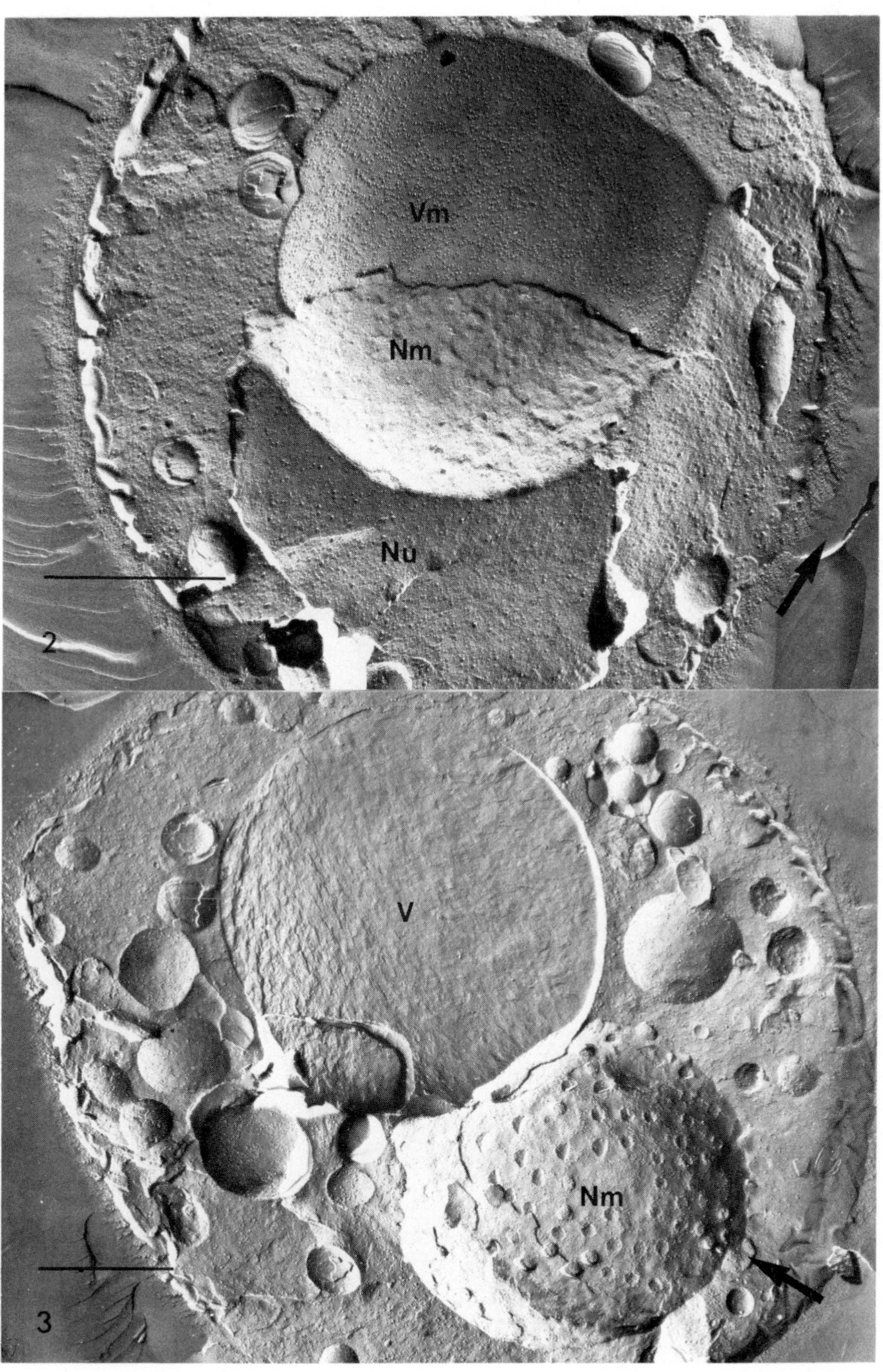
Vm
Nm
Nu
2
V
Nm
3

3. Results

Changes in both arrangement and number of nuclear pore complexes accompany the return of the stationary phase cells to normal growth and division. The change in arrangement is illustrated in figs. 1-3 and concerns the changing distribution of pore-free areas on the nuclear envelope. The cells at the start of the experiment were found to have a large pore-free area on one side of the nucleus, fig. 2. This pore-free area occured at a site on the nucleus where a close association between the large vacuole and the nucleus of these cells could be seen. Pore-free areas in association with vacuoles were found at all stages but their distribution changed markedly in the early stages as the cells left the stationary phase. The change in

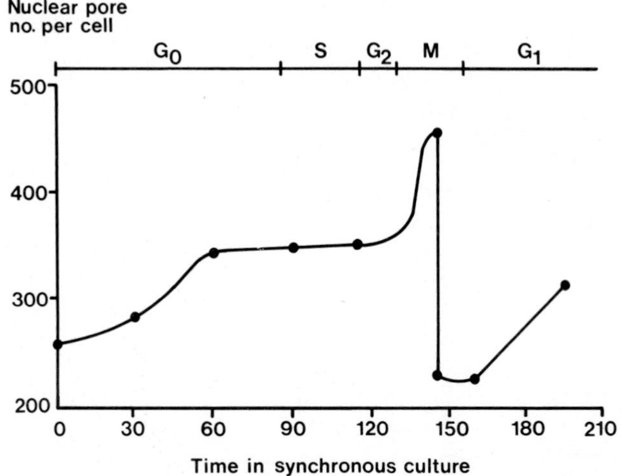


Fig. 4. Number of nuclear pores per nucleus during the first cycle of synchronous growth in Saccharomyces cerevisiae. The number during mitosis has been plotted twice to indicate the halving which occurs. The value calculated is the lower one but since fractures only represent half a dumbell shaped nucleus at this time a point showing double the number is also plotted. The position of the cell-cycle markers were established in relation to budding and cell division which were monitored for this purpose.

distribution relates to the changing sizes and distributions of vacuoles and their influences upon the nucleus as shown diagrammatically in fig. 1.

The number of nuclear pores per nucleus was calculated from measurements of their frequency observed in face fractured nuclei; and multiplied by the surface area of the nucleus obtained from measurements of nuclear profiles in cross fractures. The number of nuclear pores for the different phases of the cell-cycle is shown in fig. 4.

The number of pores per nucleus rose from 255 to 343 during the first hour of synchronous growth and then remained steady at about 350 per nucleus until a second increase which occurred around the time of nuclear division. The S phase occurs between the two periods of pore increase at the same time that the number of pores remains steady. The next cell-cycle begins with a $G_1$ pore increase similar to the $G_o$ increase at the start of the experiment.

## 4. Discussion

A full explanation of the changes in distribution of the pore-free areas must await information on the rate of nuclear pore and nuclear membrane turnover. However, if the nuclear pore is a long lived entity then the observations could be understood in terms of nuclear pore mobility[8]. The observation that the number of nuclear pores remains constant throughout the S phase is clearly different from that reported in higher eucaryotes[1], showing that a doubling in nuclear pore number is not essential for the DNA replication process.

The increase in pore number which follows the S phase may be a parallel of the early $G_1$ increase in higher eucaryotes and an earlier part of the yeast $G_1$ increase; the onset of nuclear pore synthesis being possible before division in yeast because the nuclear envelope does not break down during mitosis.

The increase which occurs early in $G_o$, however, may not be a direct parallel of the $G_1$ increase. In $G_o$ the increase in pore number occurs at a time when there is little net synthesis of RNA or protein[9]. This change in pore number coincides with the transition from a stationary phase to a growth phase[6,9]. Perhaps this change from one type of growth to another may require a change at the level of the nuclear pores. It is not perhaps too speculative to ask whether the increase in pores following mitosis at the time of nuclear swelling might also have some significance in terms of differentiation. It has been suggested by Gurdon and Woodland that a reprogramming event accompanies each cell division and may be evidenced by the nuclear swelling observed[10]. Nuclear pores have been shown to increase at this time[1,2]. All these findings and suggestions may indicate a role for the nuclear pore in RNA metabolism, with increase in nuclear pore numbers at times of mRNA transport.

Acknowledgements

An award from the SRC is gratefully acknowledged by N. J. Severs.

References

1. Maul, G. G., Maul, H. M., Scogna, J. E., Lieberman, M. W., Stein, G. S., Hsu, B. Y. and Borun, T. W., 1972, J. Cell Biol. 55, 433.
2. Scott, R. E., Carter, R. L. and Kidwell, W. R., 1971, Nature New Biology. 233, 219.
3. Jordan, E. G. and Chapman, J. M., 1973, J. Exp. Bot. 24, 197.
4. Severs, N. J. and Jordan, E. G., Experentia. In Press.
5. Williamson, D. H., 1964, in: Synchrony in cell division and growth, ed. E. Zeuthen (John Wiley and Sons, Inc., New York) p. 351.
6. Hartwell, L. H., 1974, Bacteriological Revs. 38, 164.
7. Severs, N. J. and Jordan, E. G., 1975, J. Ultrastruct. Res. 52, 85.
8. Severs, N. J., Jordan, E. G. and Williamson, D. H., J. Ultrastruct. Res. In Press.
9. Williamson, D. H., 1966, in: Cell synchrony, eds. I. L. Cameron and G. M. Padilla (Academic Press Inc., New York) p. 81.
10. Gurdon, J. B. and Woodland, H. R., 1971, Current Topics in Dev. Biol., 11, 39.

*Progress in Differentiation Research, ed. N. Müller-Bérat et al.*



# CELL CULTURE OF HYDRA ATTENUATA: AN APPROACH TO THE STUDY OF CNIDARIAN CELL DETERMINATION, DIFFERENTIATION AND INTERACTION IN VITRO.

P. Pierobon, G. Quagliarotti and S. Aurisicchio

Laboratorio di Cibernetica del CNR
Arco Felice, Naples, Italy.

## 1. Introduction

Because of its limited number of cell types and simple morphology Hydra represents a useful system for investigating the mechanisms that control cell proliferation and differentiation in a multicellular animal. The mature animal has the form of a cylinder made of two epithelial cell layers, surrounding a central gastric cavity, and it reproduces asexually by budding.

Epithelial and interstitial cells (I cells) are proliferating cell types, I cells representing the multipotent stem population, while nerve and gland cells and nematocytes are formed primarily by differentiation from I cells[1,2]; this question, however, is still controversial, some authors raising doubts as to the origin of gland cells [3,4]. In the normal animal the percentages and ratios of the different cell types are fairly stable[5], and growth, proliferation and differentiation occur at constant rates in an unchanging tissue.[6]

The control of these developmental phenomena and the stability of the differentiated state are attributed to the existence in the animal of morphogenetic gradients [7], involving the rôle of the nervous system [8].

Tissue culture techniques would greatly enhance the study of these problems at the cellular level, but most attempts have so far been unsuccessful, owing to the difficulty of developing adequate media and of overcoming contamination problems. Recently Trenkner et al. [9] and Pierobon et al. [10] separately have reported two different methods for the culture of Hydra cells and explants in a synthetic medium, obtaining cell survival, proliferation and macromolecular synthesis.

In this work we present autoradiographic evidence of cell differentiation in vitro and a quantitative analysis of the variations of cell type ratios and labelling index at different times in vitro. The hypothesis that I cells are not an homogeneous cell population but constitute a complex cell system is discussed.

## 2. Materials and methods

The animals (Hydra attenuata) were obtained from Prof. P. Tardent (Zurich) and grown asexually in our Laboratory by the method of Loomis and Lenhoff [11].

Before preparation of the cultures, animals without buds and of approximately uniform size were selected; they were starved for at least 7 days and during this period repeatedly washed in sterile culture solution; on the last day they were exposed to rifampicin (50μg/ml) for 40', then centrifuged at 900 g for 10', washed in BSS to remove rifampicin and resuspended in culture or disaggregation medium (see Table 1).

Table 1

| Substance | mM/liter | Substance | mM/liter |
|---|---|---|---|
| L-Arginine | $5 \cdot 10^{-1}$ | Uracil | $2 \cdot 10^{-2}$ |
| L-Histidine | $3 \cdot 10^{-1}$ | Cytosine | $2 \cdot 10^{-2}$ |
| L-Lysine | $9 \cdot 10^{-1}$ | Guanine | $2 \cdot 10^{-2}$ |
| L-Tyrosine | $14 \cdot 10^{-1}$ | Thymidine | $2 \cdot 10^{-2}$ |
| L-Tryptophan | $4 \cdot 10^{-1}$ | Adenine | $20 \cdot 10^{-2}$ |
| L-Phenylalanine | $6 \cdot 10^{-1}$ | ATP | $4 \cdot 10^{-2}$ |
| L-Cystine | $2 \cdot 10^{-1}$ | Xanthine | $2 \cdot 10^{-2}$ |
| L-Methionine | $10 \cdot 10^{-1}$ | Hypoxanthine | $2 \cdot 10^{-2}$ |
| L-Serine | $24 \cdot 10^{-1}$ | Thiamine | $3 \cdot 10^{-3}$ |
| L-Threonine | $8 \cdot 10^{-1}$ | Riboflavin | $3 \cdot 10^{-3}$ |
| L-Isoleucine | $6 \cdot 10^{-1}$ | Pyridoxine | $5 \cdot 10^{-3}$ |
| L-Leucine | $14 \cdot 10^{-1}$ | Biotin | $2 \cdot 10^{-3}$ |
| L-Valine | $8 \cdot 10^{-1}$ | Nicotinamide | $8 \cdot 10^{-3}$ |
| L-Glutamic Acid | $20 \cdot 10^{-1}$ | p-Aminobenzoic acid | $7 \cdot 10^{-3}$ |
| L-Aspartic acid | $8 \cdot 10^{-1}$ | Vitamin A | $3 \cdot 10^{-3}$ |
| L-Glutamine | $10 \cdot 10^{-1}$ | Inositol | $5 \cdot 10^{-3}$ |
| L-Asparagine | $12 \cdot 10^{-1}$ | Calcium pantothenate | $2 \cdot 10^{-3}$ |
| L-Alanine | $12 \cdot 10^{-1}$ | Folic Acid | $2 \cdot 10^{-3}$ |
| L-Proline | $4 \cdot 10^{-1}$ | Choline | $7 \cdot 10^{-2}$ |
| Glycine | $14 \cdot 10^{-1}$ | $CaCl_2$ | 1 |
| Glucose | 25 | $MgCl_2$ | 1 |
| Ribose | $2 \cdot 10^{-2}$ | $Na_2HPO_4$ | 20 |
| Deoxyribose | $2 \cdot 10^{-2}$ | $KH_2PO_4$ | 10 |

Culture medium composition. Balanced salt solution (BSS) contains: $CaCl_2$, $MgCl_2$, $Na_2HPO_4$, $KH_2PO_4$, and glucose at the same concentrations of culture medium. Disaggregation medium contains: 6 mM KCl, 2 mM $CaCl_2$, 9 mM $Na_3$ citrate, 9 mM Na pyruvate, 20 mM glycine. All the media are adjusted to pH=6.8.

For explant cultures two animals with their tentacles removed were placed on a round, collagen-coated coverslip and cut in 8 to 10 pieces; the coverslip was then covered with a strip of dializing membrane and mounted on the lower half of a Rose chamber [12] of 1 ml volume. The chamber was closed and filled with 0.8 ml of culture medium, to which penicillin G was added (100 I.U./ml).

For cell cultures cell suspensions were prepared by mechanical disaggregation: 50 animals were collected in 2 ml of disaggregation medium and homogenized by pipetting until a suspension was obtained. The cells were then washed by repeated centrifugation at 900 g for 20', resuspended in culture medium and diluted to a final concentration of $2\cdot10^3$ cells/ml, in order to obtain sufficiently dispersed cultures; 1.5 ml of the final suspension was then seeded in 35 mm Falcon culture dishes. Cultures of higher cell density, up to $5\cdot10^5$ cells/ml, were also prepared by the same technique.

In both types of culture the preparation time was kept as short as possible and most of the steps were carried out in an ice bath. The cultures were kept at $18\pm0.5\ ^{\circ}C$ and the medium changed weekly.

Cell proliferation was measured by $^{14}C$-thymidine incorporation and also by direct cell counts in the low density cell cultures. $^{14}C$-thymidine (Amersham) was added to the medium immediately after preparation of the cultures, with a final thymidine concentration of $2\cdot10^{-2}$ mM/l and a final specific activity of $10^{-1}$ C/mM thymidine. At the appropriate time the cells were collected in pools of 5 cultures, centrifuged at 900 g for 10', heated to 100°C for 5' in TCA 5%, and centrifuged; the insoluble fraction was washed in 80% ethanol, dissolved in 0.2 ml of Nuclear Chicago Solvent and counted in a Nuclear Chicago liquid scintillation counter.

Autoradiographs were prepared from explants grown in the presence of $^{3}H$-thymidine (Amersham) added to the medium immediately after preparation of the cultures, with a final thymidine concentration of $2\cdot10^{-2}$mM/l and a final specific activity of 3 C/mM thymidine. The explants were washed three times in BSS while still in the culture, then collected in pools of 5 cultures and washed again by centrifugation in order to reduce background activity; they were then resuspended in 0.5 ml of a maceration solution and disaggregated by David's technique[13]. After fixation, slides were prepared, covered with NT2B film (Kodak) and exposed for 15 days.

Cell counts were carried out in phase contrast optics on slides prepared from cultures of 1, 4, 7 days and also of 0 days, that is, from cultures freshly prepared; this precaution was taken in order to correct possible systematic errors due to selective cell loss during centrifugation, for instance. Cell counts were also carried out on whole normal animals using David's maceration procedure. For cell classification David's criteria were used [13]; for each culture time the data were obtained from three independent counts and referred to about 15.000 cells each.

## 3. Results

The curve of $^{14}C$-thymidine incorporation in explant cultures shows that maximum proliferation occurs between the second and fourth day in vitro and is followed by a plateau (fig. 1). The levels, measured up to 14 days, give a maximum incorporation of about $6\cdot10^{-4}$ $\mu g$ of thymidine per culture; this value, assuming a theoretical (A+T) ratio of 60% in the DNA of Hydra, and on the basis of an average content of 3.5 pg of DNA per cell [6], gives an increment of about $8-9\cdot10^3$ cells per culture at the end of the proliferating phase, that is an increment of 15% on an initial population of $5-6\cdot10^4$ cells.

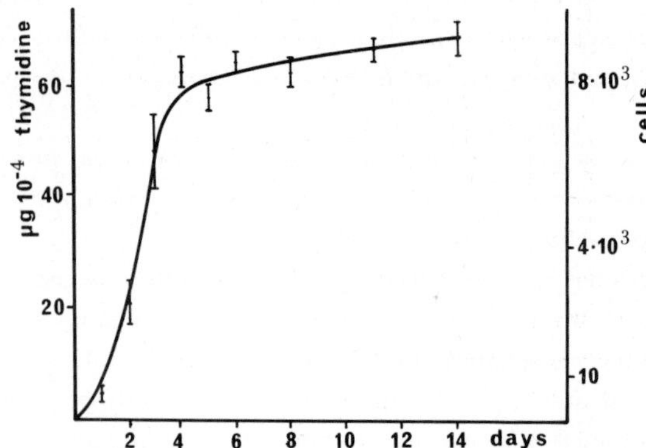


Fig. 1. Curve of $^{14}$C-thymidine incorporation in explant cultures at different times in vitro.

Two distinct stages of development are also recognizable in the living culture by optical microscope observations:in the first two or three days an outgrowth zone forms on the borders of the explant; the cells are rounded, freely floating and of different sizes (plate 1). In the next stage, until 15 days,the cells start to adhere to the supporting collagen and rearrange in groups, partly resuming the typical morphology. Intracellular bridges or junctions do not become evident.

The following degenerative phase has a quite variable duration; it is characterized by a gradual cell reduction and by the appearance of anomalous, unidentifiable elements. Nematocytes and nematoblasts are still clearly identifiable, and in the final stages they represent more than 30% of the cells.

Cell proliferation is not observed, on the other hand, in cell cultures: neither the measurements of thymidine incorporation nor direct counts carried out on the low density cultures give significant increment values. The cultures with a higher cell density maintain for relatively long periods the ability to form new cultures by dilution of the primary suspensions; in this case a small increment of about 2% is observed.

Autoradiography of explant cultures shows that already at day 1 in vitro labelled nuclei appear in all cell types, with the exception of battery cells . In the slides prepared from cultures of 1 and 4 days immature forms, i.e. nematoblasts or I cells in the first differentiation stages, are prevailing; at 7 days the first active nematocytes appear. Labelled nerve cells, apparently mature within the limits of optical microscope observations, are already visible at 4 days. Finally, after the fourth day, labelled anomalous

---

Plate 1. Photographs of the living culture at day 2 (photo 1,2,3) and day 5 (photo 4,5). Cell growth from the explant (photo 1) reveals at higher magnification a nest of I cells (photo 2) and gland and nerve cells (photo 3) between free, rounded cells. At day 5, nests of nematoblasts (photo 4) and epithelial cells (photo 5) can be identified.

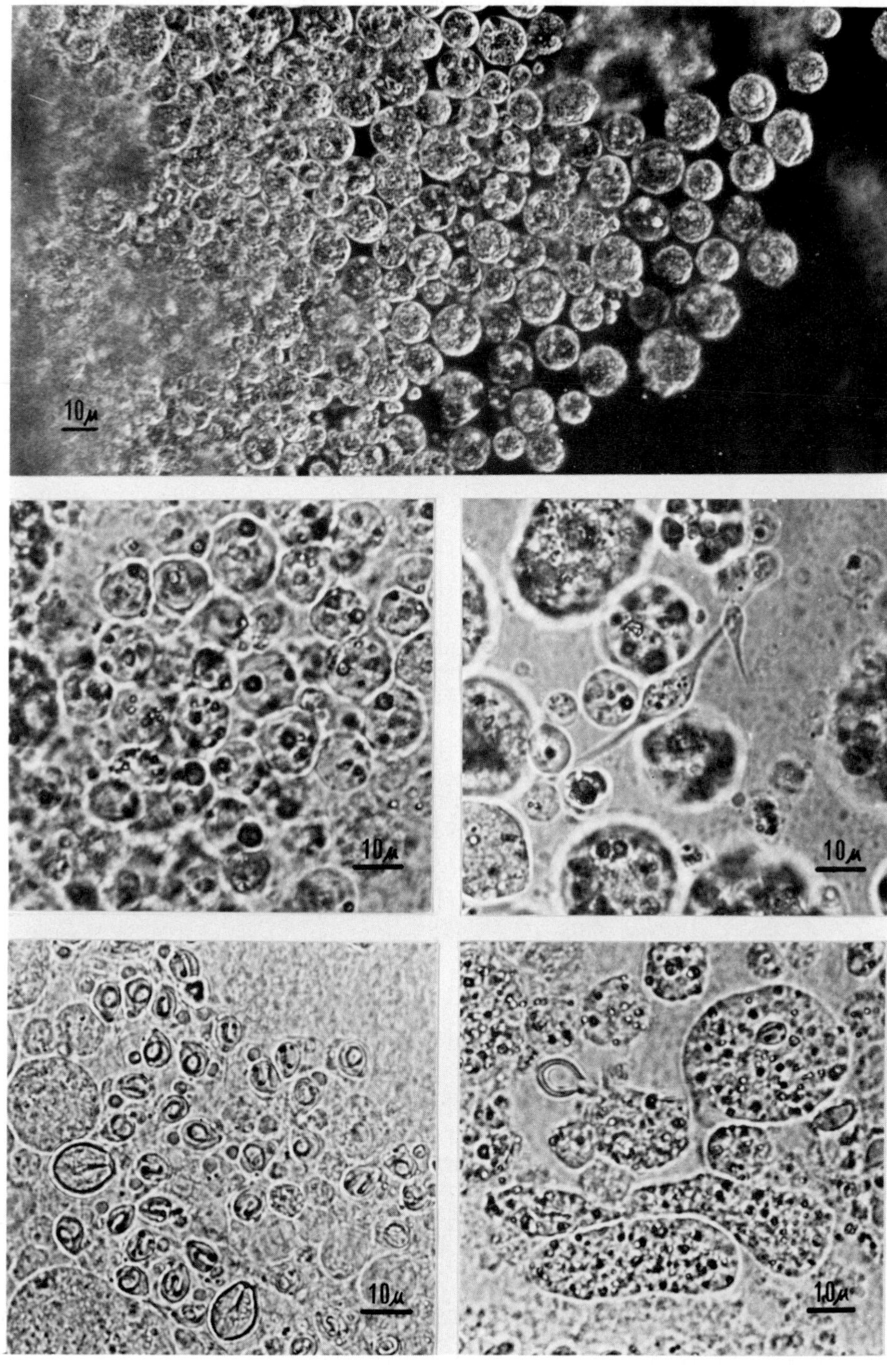
10µ
10µ
10µ
10µ
10µ

cells appear: they are no longer identifiable on the basis of shape and cytoplasmic inclusions and have quite different sizes, apparently representing degenerating/dedifferentiating cells from all the various cell types (plate 2).

Since the nuclei have different label intensity, only those cells whose nuclei contained more than 15 grains were considered. The label intensity is not dependent on the conditions of treatment with the thymidine which is present in the medium for the whole length of the experiment. Besides, since it is administered immediately after preparation of the cultures, the labelling proves that the active cells have undergone at least one mitotic cycle in vitro and therefore, in the case of the cells that are formed by differentiation, at least the last mitosis of precursor cells before the start of morphological maturation has taken place in vitro.

The quantitative analysis of the variations of cell type ratios and of the labelling index (L.I.) of each cell type, carried out on the autoradiographs of cultures at different times, is referred to time 0 (see Table 2). Comparison of this value with that obtained for whole, normal animals shows that they are practically coincident, even in the case of nematoblasts + nematocytes; this could probably be due to the fact that the tentacle cell number is small and therefore the partial ablation of tentacles does not significantly affect the nematoblast percentage.

Table 2

Percentage of the various cell types at different days in vitro.

| days | battery cells | epithelial cells | large I-cells | small I-cells | nematoblast +cytes | gland+ mucous | nerve | unidentified |
|---|---|---|---|---|---|---|---|---|
| Hy | 4.7 | 31.5 | 17.5 | 18.1 | 14.4 | 9.3 | 4.5 | 0 |
| 0 | 4.8 | 32.2 | 17.0 | 17.5 | 14.3 | 9.9 | 4.3 | 0 |
| 1 | 7.2 | 32.2 | 13.0 | 20.8 | 14.5 | 6.3 | 4.8 | 1.2 |
| 4 | 6.5 | 38.8 | 11.6 | 15.2 | 15.1 | 5.8 | 3.4 | 3.6 |
| 7 | 6.7 | 43.4 | 9.7 | 11.0 | 18.7 | 4.1 | 2.6 | 3.8 |

The total L.I. in the cultures increases from an initial level of 10.9% at day 1 to 14.8% at day 4 and stabilizes at 14.3% at day 7. Maximum proliferation occurs between the first and fourth day and its final value is in good agreement with the calculations based on the results of 14 C-thymidine incorporation.

The analysis of the L.I. of the single cell types (see Table 3) shows that they have different time courses: the small I cells have a high L.I. already at day 1 (24.5%), which increases slightly further during the next days and at day 7 is stabilized at 27.9%.

---

Plate 2. Autoradiographs of epithelial cells (photo 1), I cells (photo 2), differentiating I cells (photo 3) nerve (photo 4), gland (photo 5) mucous (photo 6) cells and nematoblasts (photo 7). Phase contrast optics, 4 days in vitro.

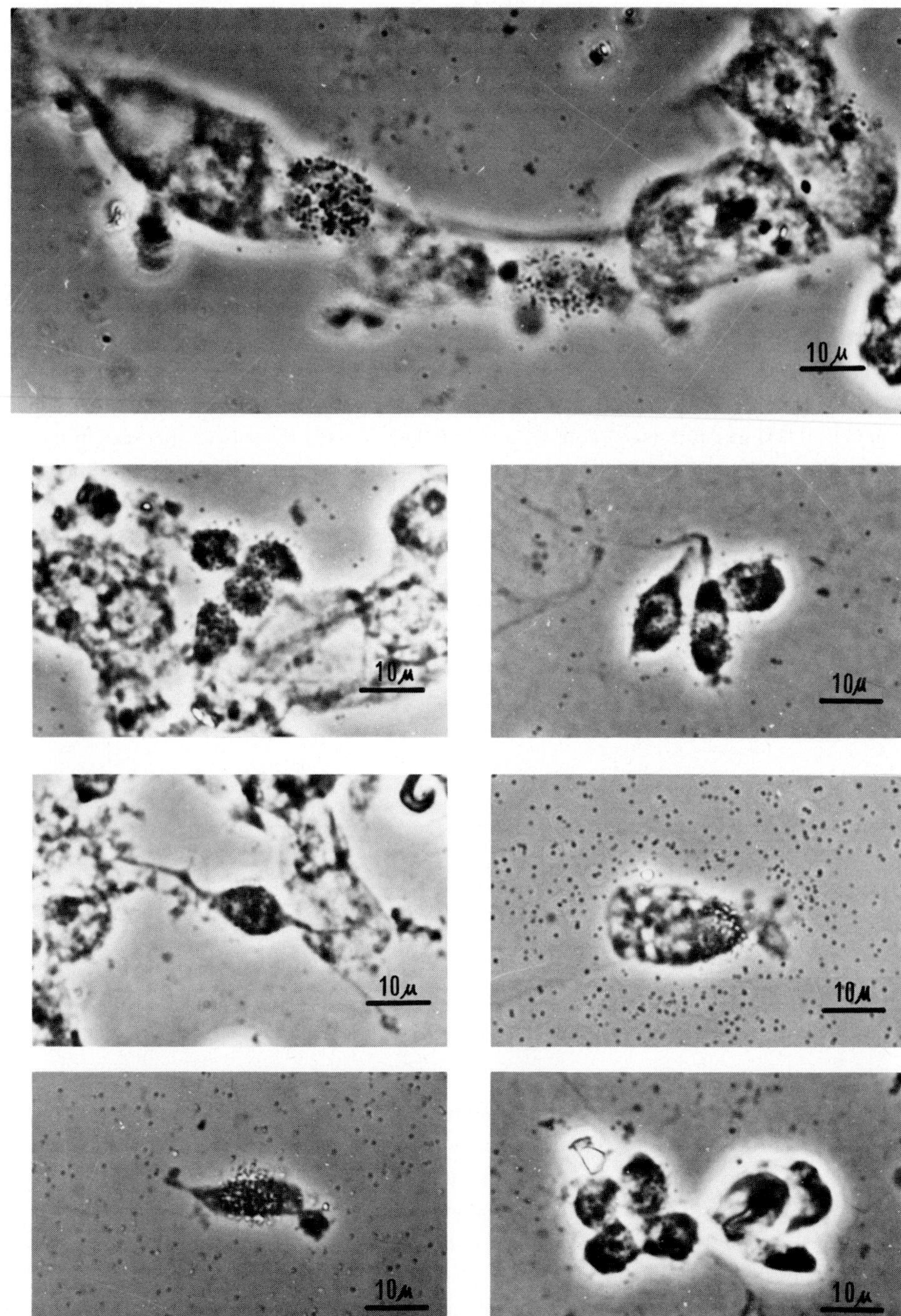
10μ
10μ
10μ
10μ
10μ
10μ
10μ

Table 3

Labelling index of the various cell types at different days in vitro.

| days | battery cells | epithelial cells | large I-cells | small I-cells | nematoblast +cytes | gland+ mucous | nerve | unidentified |
|---|---|---|---|---|---|---|---|---|
| 1 | 0 | 4.5 | 25.6 | 24.5 | 3.9 | 6.0 | 3.3 | 1.2 |
| 4 | 0 | 8.9 | 39.5 | 27.8 | 10.2 | 13.5 | 8.3 | 5.1 |
| 7 | 0 | 8.8 | 39.6 | 27.9 | 11.9 | 15.3 | 12.6 | 7.7 |

In the large I cells, instead, the L.I. at 24 hours is comparable to that of the small ones (25.6%), then it increases significantly and at day 7 it stabilizes at 39.6%. The L.I. of epithelial cells follows a similar pattern, but it reaches much lower levels (8.8% at 7 days). The L.I. of nematoblasts, nerve and gland cells increases linearly and at 7 days is not yet stabilized.

The active cell population then consists mainly of I cells (about 80%) and epithelial cells (13%) at day 1 (see Table 4); in the following days there is an increase in the ratio of cells that are formed by differentiation (mainly of nematoblasts) and of epithelial cells, while I cells sharply decrease.

The analysis of total cell ratios shows that epithelial cells and nematoblasts follow the same time course, while nerve and gland cells decrease sharply. Large I cells start decreasing already at day 1, and at day 4 both large and small I cells are decreasing, even though their L.I. is still increasing.

Finally a progressive increase is observed both of the L.I. and of the total percentage of unidentified cells, that at day 7 reach a level of 3.8%, with a L.I. of 7.7%.

Table 4

Labelled cell types ratios (percentage values) at different days in vitro.

| days | battery cells | epithelial cells | large I-cells | small I-cells | nematoblast +cytes | gland+ mucous | nerve | unidentified |
|---|---|---|---|---|---|---|---|---|
| 1 | 0 | 12.8 | 30.7 | 46.8 | 4.4 | 3.5 | 1.4 | 0.4 |
| 4 | 0 | 23.8 | 30.8 | 28.5 | 8.6 | 5.3 | 1.9 | 1.2 |
| 7 | 0 | 26.9 | 27.0 | 21.6 | 15.6 | 4.5 | 2.3 | 2.1 |

## 4. Discussion

The method we have developed permits the culture of disaggregated cells and of small tissue fragments. Its main features are: the use of a limited number of animals to overcome contamination problems; the use of the Rose chamber; the use of a synthetic medium in order to avoid introducing into the system unknown and unmeasured morphogens possibly contained in biological fluids such as sera or extracts.

In these conditions proliferation and differentiation of gland and nerve cells and nematoblasts observed in explant cultures indicate that the system still retains such elementary yet complex functions as cell growth and differentiation.

The different results obtained in the two types of cultures could be tentatively explained by the maintenance in the explants of cell adhesion and of a certain degree of cell organization and contiguity which would permit synthesis and release of morphogenetic factors from cell to cell at least in part, without loss to the surrounding medium.

In order to carry out a finer analysis of the pattern of development in the culture, cell type ratios and their variations in time have been measured. To interpret these data some caution is required, since they are relative values by which selective cell loss is not measured directly; however they point out clearly different variation patterns of the single cell types.

Epithelial cells are quite stable in culture; their increase is accompanied by the increase of their L.I. The fact that the L.I. is rather low is probably due to the length and variability of $G_2$ phase, in which as much as 80% of the epithelial cell population is resting, even under conditions of rapid growth[6]. It is therefore reasonable to assume that part of the newly formed cells may not be labelled.

Nematoblasts and nematocytes are also quite stable in culture; their increase is accompanied by an increase of the L.I. and by the appearance of mature active cells.

Nerve and gland cells decrease in culture; at 7 days their levels are practically halved, even though their L.I. is increasing. This result suggests a rather high rate of cell death which is not compensated for even by the high rate of mobilitation of precursor cells.

The different results obtained for small and large I cells suggest the possibility that they do not constitute an homogeneous cell population, either because they are predetermined at an early stage to give specific cell types or because part of the population is made by stem cells, that represent its reserve pool. In the first case the arbitrary division into large and small I cells would result in pointing out different rates of differentiation and accordingly different mobilitation rates of precursor I cells, due either to different times of morphological and functional maturation or to different sizes of the single cell pools. In this regard it is interesting to observe that the appearance of active mature nematocytes under our conditions is delayed with respect to nerve cells, in agreement with similar observations reported by the literature[4]. In the second instance, on the other hand, the two I cell groups would approximately coincide with the two different subpopulations, the large I cells representing the stem cells and the small ones the group of determined cell precursors, according to the model proposed by Campbell et al.[14]. In this case the L.I. variations agree with Campbell's results on I cell cycle kinetics: the small, i.e. the determined ones, that have a short $G_2$ phase of fixed duration and a short total cycle, undergo an initial increase at day 1 both in percentage and in L.I.; the latter remains practically constant, while the total percent ratio decreases concurrantly with the beginning of differentiation. The large I cells, i.e. the stem cells, whose $G_2$ phase is longer and variable, undergo instead a progressive increase of the L.I., that stabilizes at day 7, showing the same time course as epithelial cells. In both cases the L.I. increase is much higher than that observed in epithelial cells, in agreement with the shorter duration of the $G_2$ phase in these cells.

The decrease of I cells, associated with the increase of their L.I. and the increase of the L.I. of nematoblasts, nerve and gland cells (that under our conditions represents an index of differentiation from precursor I cells rather than a proliferation index) is the hardest result to be analyzed, each value resulting at any given time from the difference between cell growth and death rates. Therefore even though the total cell loss is low, if it is restricted to a specific cell type, as seems to happen in our case, it will affect the other data, the evaluation of the true magnitude of the changes becoming then impossible.

The reduction of I cells could then depend on the fact that: 1) the rate of exit of I cells from the pool, either by differentiation or degeneration/dedifferentiation is higher than that of growth, which in these conditions is decreased with respect to normal for lack of specific growth factors, for instance; 2) the rate of growth is normal but still lower than that of cell loss because of the high rate of nerve and gland cell death, which would lead to extinction of the precursor cells.

It is interesting to note that the higher and earlier death rates are shown by nerve and gland cells, that is by the Hydra cells specialized for secretory functions and, in the case of nerve cells, for production of morphogenetic factors[8]. Under our conditions this results in the inadequacy of the system to keep sufficiently high proliferation rates and to keep intact the mechanisms responsible for the maintenance of equilibrium at the tissue organization level rather than in the loss of differentiation capabilities, which seem to be maintained.

It is our purpose to further investigate these aspects of differentiation and growth in connection with the rôle of specific morphogenetic factors.

References

1. Lentz, T.L., 1966, The cell biology of Hydra (North-Holland, Amsterdam)
2. Burnett, A.L., 1968, in: The stability of the differentiated state; ed. H. Ursprung (Springer-Verlag, Berlin), p. 109.
3. Davis, L.E., 1970, Z. Zellforsch. 105, 526.
4. David, C.N. and Gierer, A., 1974, J. Cell Sci. 16, 359.
5. Bode, H. et al., 1973, Roux' Archiv 171, 269.
6. David, C.N. and Campbell, R.D., 1972, J. Cell Sci. 11, 557.
7. Webster, G., 1971, Biol. Rev. 46, 1.
8. Schaller, H. and Gierer, A., 1973, J. Embryol. exp. Morph. 29, 39.
9. Trenkner, E. et al., 1973, J. Exp. Zool. 185, 317.
10. Pierobon, P. et al., 1973, Atti I Congr. BPA, Camogli, 205.
11. Loomis, W.F. and Lenhoff, H.M., 1956, J. Exp. Zool. 132, 555.
12. Rose, G., 1954, Tex. Rep. Biol. Med. 12, 1074.
13. David, C.N., 1973, Roux' Archiv 171, 259.
14. Campbell, R.D. and David, C.N., 1974, J. Cell Sci. 16, 349.

*Progress in Differentiation Research, ed. N. Müller-Bérat et al.*



# CELL DEGENERATION IN THE MOUSE EMBRYO: A PREREQUISITE FOR NORMAL DEVELOPMENT

Rob E. Poelmann and Christl Vermeij-Keers
Department of Anatomy and Embryology, University of Leiden
Wassenaarseweg 62, Leiden
The Netherlands

## INTRODUCTION

Cell degeneration followed by cell death during the normal development of vertebrates was first described by Ernst[1]. Since then, cell degeneration has been described by some[2], but denied by others. Andersen and Matthiessen[3], for example, interpreted degenerating cells as peripherally sectioned mitotic figures, which would require a basic change in the theory of embryonic development[4]. Menkes[5] and Silver and Hughes[6], on the contrary, postulated that an abnormal number of degenerating cells can result in severe abnormalities such as myeloschisis and cardiac interventricular communications[5], and anophthalmia[6].

Both Ernst[1] and Glücksmann[2] listed a number of organs and tissues in which cell degeneration takes place. In the present report two more examples are given and discussed in detail. The first occurs in the embryonic ectoderm during gastrulation, in which the relationship between cell degeneration, locally abundant mitoses, and trunk elongation was investigated. The second concerns the transformations occurring in the facial region, e.g. during the formation of the primitive palate via the epithelial plate of Hochstetter. Special attention is given to two questions: Is disappearance of the plate due to differentiation of epithelial cells into mesenchyme-like cells and/or to degeneration of these epithelial cells? Does the plate disrupt at one place or at several places simultaneously, as suggested by Barry[7]?

In addition, the relationship between lysosomal activity and cell degeneration was investigated with cytochemical methods.

## MATERIAL AND METHODS

Age of embryos: The investigations were performed in mouse embryos of the CPB-S strain, the developmental stages of which can be calculated[8]. The age of the gastrulating embryos was estimated to be 5.7-6.9 days post coitum (p.c.). Twelve embryos were used for light microscopy and nine for ultrastructural investigation; about forty embryos were processed for enzyme cytochemistry. The epithelial plate of Hochstetter was investigated light microscopically in seventeen embryos aged 9.8-10.7 days p.c., and sixty five and thirty embryos were used for electron microscopy and enzyme cytochemistry, respectively.

Light microscopy (LM): The intact embryos were fixed in Carnoy or Bouin, and cut routinely into 8 or 10 μm serial sections, which were stained with hematoxylin-eosin or toluidin blue.

Electron microscopy (EM): The 5.7-to 6.9-day-old embryos were perfusion-fixed with paraformaldehyde-glutaraldehyde[9] via the maternal circulation and subsequently immersed in the same fixative. The 9.8-to-10.7-day-old embryos were prepared free of the uterus, after which the nasal parts were immersed in the same fixative. The total fixation time ranged between 1 and 24 hours. Before being embedded in Epon, tissues were postfixed in cacodylate-buffered 1% $OsO_4$ and before dehydration in some cases also with uranyl acetate.

Enzyme cytochemistry: For LM enzyme cytochemistry, unfixed frozen sections were used, and for EM prefixed material processed as described above. Incubation for the demonstration of acid phosphatase activity was performed according to Gomori[10] for LM and Barka and Anderson[11] for LM and EM. Aryl sulphatase activity was demonstrated according to Hopsu Havu et al.[12] for LM and EM and according to Hanker et al.[13] for EM only. Controls were performed in media without substrate; no "enzyme inhibitors" were used. Controls were always negative. After postfixation, the tissues were routinely processed.

RESULTS

Stages of cell degeneration: Glücksmann[2,14] and Silver and Hughes[15] gave a light microscopical description of stages of degeneration determined solely by nuclear changes. We distinguish the following stages on the basis of our ultrastructural findings. Stage Ia: the first sign of degeneration is an increased electron-density of the cytoplasm; the nuclear chromatin becomes coagulated and concentrated immediately beneath the nuclear envelope. The polygonal cell outline becomes irregular, probably due to a relative shrinkage of the cell (see Fig. 2). Stage Ib: the degenerating cell then loses its specialized cell contacts (such as desmosomes) with the surrounding "normal" cells, after which the cell becomes globular by retraction of the irregular protrusions (see Fig. 3). These altera-

---

Fig. 1. Electron micrograph of a para-sagittal section of a 6½-day-old embryo. The mid-ventral region of the ectoderm shows and abundance of degenerating and phagocytosed cells (▲).
Fig. 2. Degenerating ectodermal cell representing stage la. Cell contacts (▲) with neighboring cells are still visible. Note the increased electron density of the cytoplasm, the swollen ER, and the irregular outline of the cell.
Fig. 3. Detail of Fig. 1, showing degeneration of cells in stages Ib, II, and III.

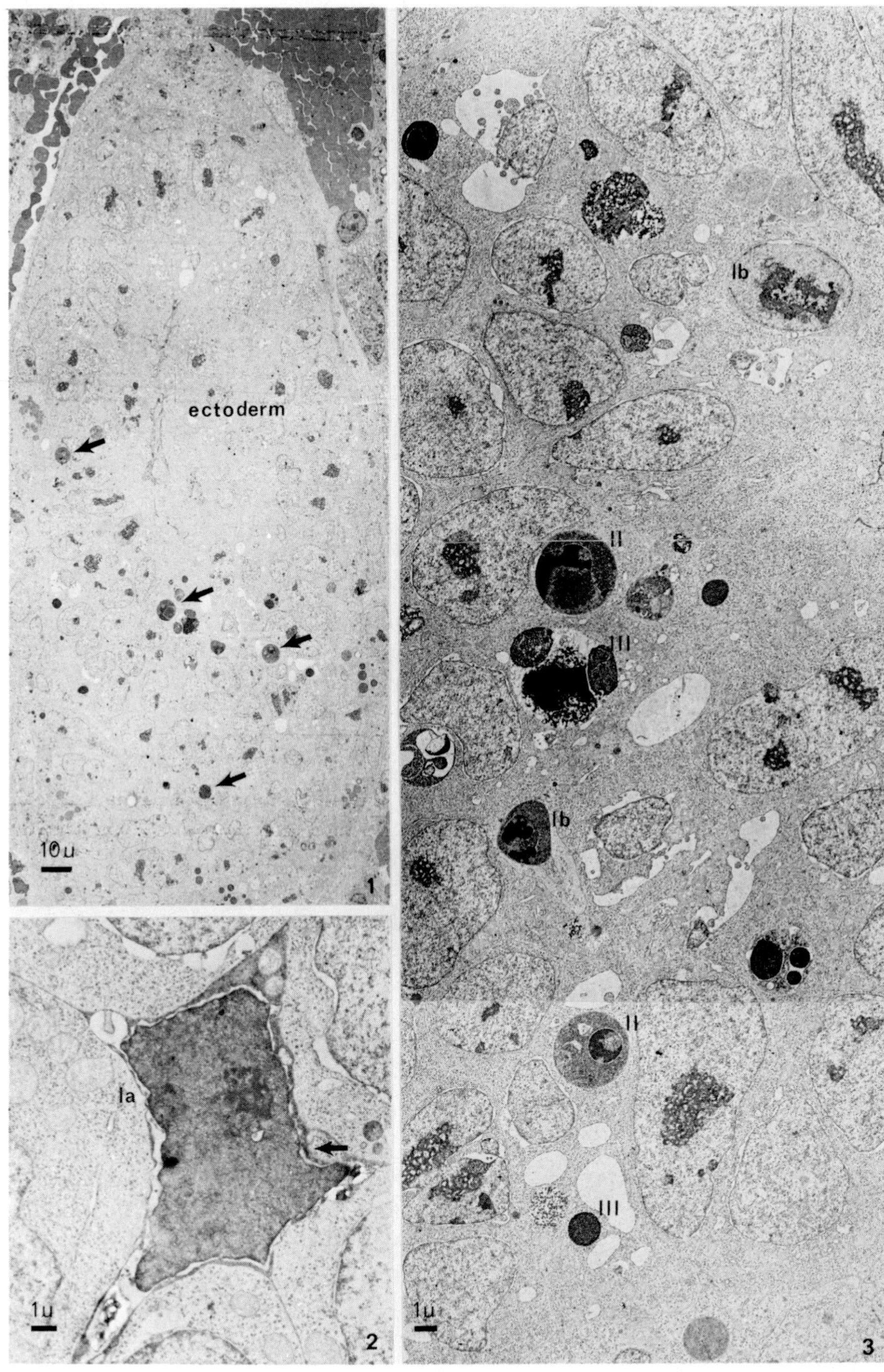
ectoderm
10u
1
Ia
1u
2
Ib
II
III
Ib
II
III
1u
3

tions are accompanied by shrinkage of the nucleus and increasing coalescence of the chromatin, forming the typical half moon at one side of the nucleus. Other organelles, such as the mitochondria and ER, take on a swollen appearance. Stage II: the degenerating cell is phagocytosed by neighboring cells, which can be called macrophages or phagocytes (see Fig. 3). Stage III: the content of the phagosomes is gradually broken down into unrecognizable structures (see Fig. 3). These stages of cell degeneration are identical for both the primary ectoderm and the epithelial plate.

Embryos aged 5.7-6.9 days p.c.: At approximately six days p.c., shortly after the formation of the proamniotic cavity[16], cell degeneration is restricted almost

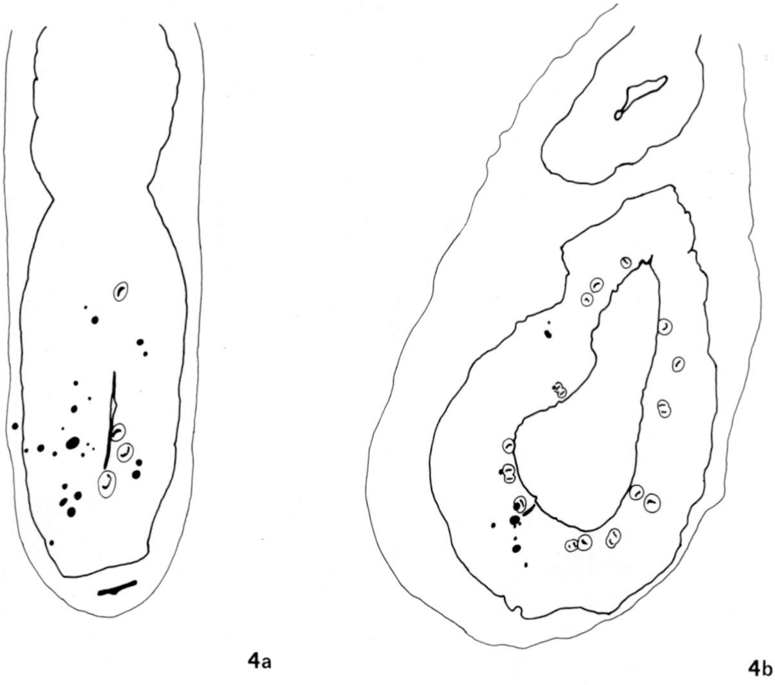


Figs. 4a and b. Drawings after electron micrographs of approximately longitudinal sections of a 6- and 7-day-old embryo, respectively. Degenerating (●) and dividing (◉) cells are indicated. Note the relative increase in the number of mitoses in the 7-day-old embryo. The total number of mitoses is even higher than indicated here, because only those cells are shown in which chromosomes were clearly visible.

exclusively to the mid-ventral region of the embryonic ectoderm (Fig. 1) of the mouse, but can occasionally be seen in the region where the third germ layer, the mesoderm, will be formed. Sometimes, degenerated cells are found between the cells of the newly formed mesoderm. In the ectoderm some mitotic figures are observed in the vicinity of degenerating cells. Cell degeneration was seen in the endoderm in some embryos but never at similar places. In 7-day-old embryos phagocytosed cells are still located in the same ectodermal area as in the 6-day-old embryos, but now cell mitosis exceeds cell degeneration by far.

The increasing number of mitoses in the mid-ventral region of the ectoderm results in a tremendous elongation of the trunk in a very short time; in the space of four hours on the 7th day (7.3 to 7.5 days p.c.) the distance between the anterior and posterior ectodermal wall of the proamniotic cavity increases three-fold, i.e., from 22 to 70 μm in embryos processed for LM.

The number of cells involved in cell degeneration in 6-day-old embryos is difficult to estimate, but in relation to the number of apparently healthy cells they account for at least 15 to 20% of the total number of cells in mid-sagittal sections of the mouse embryo (Fig. 1). Schematic drawings of a 6- and 7-day-old embryo, made after electron micrographs, are given in Figs. 4a and b.

Embryos aged 9.8-10.7 days p.c.: The transformation of the nasal placode via the nasal groove into the nasal tube with interposition of the epithelial plate of Hochstetter has been studied in human and mouse embryos. In the latter species this process takes place in embryos aged from 9.8 to 10.7 days p.c. In both species the isthmus maxillae (a mesenchymal contact between the medial nasal, lateral nasal, and maxillary processes) develops in this period[4,17,18,20]. The formation of the isthmus involves the disappearance of the epithelial plate, which is brought about by cell degeneration[19].

Because we now have a sufficient number of embryos in the relevant developmental stages at our disposal, we are able to describe the exact mechanism underlying the disappearance of the epithelial plate. Shortly before 10.5 days the plate has reached its maximum extension. In this stage the following features can be seen microscopically. An area showing a high frequency of cell degeneration and macrophages as well as enlarged intercellular spaces, is located in the middle of the plate, immediately above the palatine groove. At this place the epithelial basal membrane seems to be interrupted. Degenerating cells are found not only in the plate itself, but also in the mesenchyme of the medial nasal process bordering the occipital part of the plate. More frontally, degenerating cells and macrophages are present in the lateral nasal process near the middle of the plate. In the next developmental stage an 8 μm interruption is present in the basal membrane near the middle of the plate, showing ingrowth of mesenchyme accompanied by a capillary from the medial nasal process. This is the first indication of the isthmus maxillae. At 10.5 days the width of the interruption has reached 40 μm in the right plate and 20 μm in the left plate (Figs. 5a and b). In older stages this isthmus enlarges in the cranial, caudal, and frontal directions along with a gradual reduction of the plate. During this reduction, cell degeneration continues in the remnants of the epithelial plate (at a higher rate in the nasal than in the oral part) until the plate remnants disappear completely at 10.7 days, except for the part which will give rise to the bucconasal membrane.

Lysosomal enzyme activity: Studies done to determine whether lysosomal enzymes are involved in the process of cell degeneration, indeed demonstrated the presence of hydrolytic enzymes. Acid phosphatase and arylsulphatase activity was detected light microscopically in the ectoderm of 5.7-to 6.9-day-old embryos at places where cell degeneration is to be expected. In the epithelial plate of Hochstetter and in the mesenchyme near the plate, acid phosphatase activity was found (LM and EM) within macrophages (Figs. 6 and 7). Ultrastructurally, aryl sulphatase and acid phosphatase activity are restricted to membrane-bound vesicles also containing more or less recognizable material from apparently degenerated phagocytosed cells. However, the results of the aryl sulphatase experiments are less convincing than those of the acid phosphatase experiments.

## DISCUSSION

Glücksmann[2] distinguished three forms of cell degeneration, i.e.,morphogenetic, histiogenetic, and phylogenetic, on the basis of a "classification of cell degeneration according to their developmental functions" (p. 75). He himself stated, however, that the properties of the three classes are vague, and it is indeed not clear from his examples which criteria are conclusive. Furthermore, we have found unclassifiable instances of cell degeneration scattered in areas of the ectoderm, endoderm, and mesoderm. The view that phylogenetic cell degeneration occurs during ontogeny is erroneous, because it involves the assumption that onto-

---

Figs. 5a and b. The interruption in the left epithelial plate of Hochstetter is seen in de middle of the plate immediately above the palatine groove. n.r.=nasal remnant.

Fig. 6. A transversely cut frozen section of a left epithelial plate. Acid phosphatase activity inside the macrophages (▲) demonstrated according to Gomori[10].

Fig. 7. Electron micrograph of a macrophage situated in the mesenchyme between the remnants of the epithelial plate. The macrophage contains several phagolysosomes in different stages of lysis. Acid phosphatase activity (▲) demonstrated according to Barka and Anderson[11].

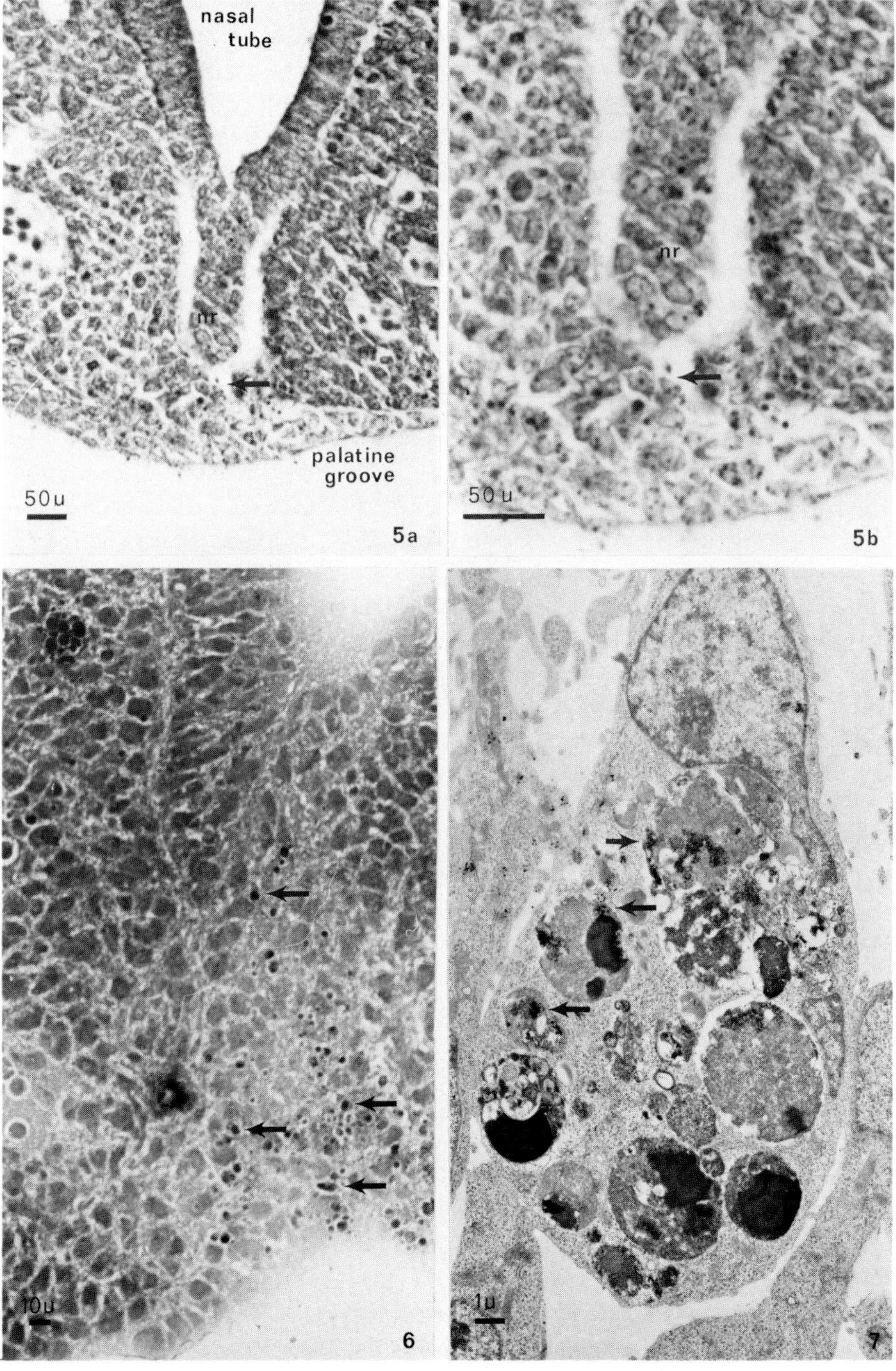
nasal
tube
nr
palatine
groove
50u
5a
nr
50 u
5b
10u
6
1u
7

geny is at least partially a recapitulation of phylogeny. This more or less Haeckelian philosophy was abandoned long ago. On practical and theoretical grounds, therefore, we do not accept Glücksmann's distinction, and propose that only the term cell degeneration be used. In the literature the term cell death is used as well, but cell degeneration is a process having several aspects that can be analysed morphologically and cytochemically, whereas cell death is only the final stage of this process.

The occurrence of cell degeneration can be observed very distinctly in the early mouse embryo (5.7-6.9 days) and in the epithelial plate of Hochstetter (9.8-10.7 days) of the same species. In these cases the morphology and enzyme patterns are comparable in most respects, e.g. the nuclear and cellular changes, phagocytosis by neighboring cells, and digestion by hydrolytic enzymes. It is clear that lysosomal enzymes play a role in these phenomena. The results obtained so far indicate that these enzymes are present only in cells showing phagocytotic activity. This means that the hydrolases are not involved in the earliest stages of the degeneration process per se[21] and do not make their appearance before stage II. Consequently, other factors related to cell degeneration must be sought.

Cell degeneration in the ectoderm of 5.7-to 6.9-day-old embryos seems to play a role either in tissue growth or in differentiative events, e.g. mesoderm formation. In the ectoderm cell degeneration takes place in a region which shortly afterward shows an abundance of mitoses, which indicates that cell degeneration precedes tissue growth. Maruyama and D'Agostino described something similar in the telencephalic roof-plate before ventricle formation. Abnormal mitoses[23,24] were not, however, observed in our material and hence cannot account for the massive cell degeneration.

Cell degeneration may also be involved in the differentiation of parts of the ectoderm into mesoderm. Kärner and Leikola[25] showed cell degeneration and acid phosphatase activity in Hensen's node of the chicken. Thus, in both the mouse and the chicken cell degeneration and lysosomal enzymes may be involved in the formation of the third germ layer, the mesoderm. El-Shershaby and Hinchliffe[26] reported cell degeneration in the inner cell mass of the pre-implantation mouse blastocyst at 95 hours p.c. They concluded that cells that fail to differentiate normally may be the source of the dead cells. In our opinion, the phenomenon they describe may be related to the differentiation of a part of the inner cell mass into the second germ layer, the endoderm, whereas in our material cell degeneration seems to be involved into mesoderm formation.

In the epithelial plate of Hochstetter, cell degeneration does not precede mitosis, nor is it involved in differentiation per se. Starting with the onset of the formation of the plate, we have observed cell degeneration on an enormous scale but only a single case of mitosis. After disruption at one place (in the middle, immediately above the palatine groove) the plate shows only regression of its remnants accompanied by spatial replacement by mesenchymal structures forming the isthmus maxillae. Cell degeneration occurs even before the plate is actually formed[19]. Thus, it is evident that the degeneration is not evoked by the formation of the plate itself, but coincides with the outgrowth of the mesenchymal nasal and maxillary processes. The disappearance of the plate has been reviewed by Vermeij-Keers[4].

In our opinion, an abnormal number of degenerating cells[5,6] can result in congenital malformations; for example, when Hochstetter's epithelial plate does not disappear, leaving the nasal and maxillary processes separated, the result is a cleft lip[20]. Cell degeneration seems to occur in almost all tissues of normally developing embryos. In some tissues, as illustrated by the examples given above, the phenomenon of cell degeneration is more pronounced and results in spatial alterations of the tissues concerned. Some of the relationships between cell degeneration and other phenomena are already known, particularly those involving extrinsic factors such as hormones[27,28], or genetic control[29,30], but the great majority are still obscure.

SUMMARY

Cell degeneration was studied in CPB-S mouse embryos by light and electron microscopical techniques. In 5.7-to 6.9-day-old embryos cell degeneration followed by phagocytosis by neighboring cells predominates in the embryonic ectoderm bordering the pro-amniotic cavity, the total number of degenerating cells being estimated at about 15% of the embryonic ectoderm cells. Cell degeneration precedes tissue growth, because in a later developmental stage a large number of cell mitoses are observed in the same area. The relationship between these two processes is not yet quite clear. It is also conceivable that cell degeneration is linked to differentiation of the mesoderm.

In embryos aged 9.8-10.7 days cell degeneration occurs during the transformation of the nasal placode via the nasal groove into the nasal tube. During this transformation the epithelia of the nasal and maxillary processes make contact to form the epithelial plate of Hochstetter. This plate becomes disrupted at one place (in the middle of the plate, immediately above the palatine groove) shortly before 10.5 days. Cell degeneration coincides with the disappearance of Hochstetter's plate. Next, the mesenchyme of the nasal and maxillary processes meet to give rise to the isthmus maxillae. When cell degeneration does not occur in a normal frequency the consequence may be a cleft lip.

Lysosomal enzymes, e.g. aryl sulphatase and acid phosphatase, were found in the ectoderm and in the epithelial plate. Lysosomes do not play a role in the onset of cell degeneration, but are active in the removal and breakdown of degenerated cells by neighboring cells, some of which are transformed into macrophages.

## REFERENCES


1. Ernst, M. (1926) Zeitschr. Anat. Entwickl.-Gesch. 79, 228-262.
2. Glücksmann, A. (1951) Biol. Rev. 26, 59-86.
3. Andersen, H. and Matthiessen, M.E. (1967) Acta Anat. 68, 473-508.
4. Vermeij-Keers, C. (1972) Erg. Anat. Entwickl. 46,5, 1-30.
5. Menkes, B. (1968) Rev. Roum. d'Embryol. Cytol.: sér. d'Embryol. 5, 139-149.
6. Silver, J. and Hughes, A.F.W. (1974) J. Comp. Neurol. 157, 281-302.
7. Barry, A. (1961) in Congenital anomalies of the face and associated structures (Pruzansky, S., ed.) pp. 46-62 Thomas, Springfield Ill.
8. Goedbloed, J.F. (1972) Acta Anat. 82, 305-336.
9. Karnovsky, M.J. (1965) J. Cell Biol. 27, 137A.
10. Gomori, G. (1952) Microscopic histochemistry: principles and practice, The Univ. of Chicago Press, Chicago.
11. Barka, T. and Anderson, P.J. (1962) J. Histochem. Cytochem. 10, 741-753.
12. Hopsu Havu, V.K., Arstilla, A.U., Helminen, H.J. and Kalimo, H.O. (1967) Histochem. 8, 54-64.
13. Hanker, J.S., Thornburg, L.P., Yates, P.E. and Romanovicz, D.K. (1975) Histochem. 41, 207-225.
14. Glücksmann, A. (1930) Zeitschr. Anat. Entwickl.-Gesch. 93, 35-93.
15. Silver, J. and Hughes, A.F.W. (1973) J. Morph. 140, 159-170.
16. Poelmann, R.E. (1975) J. Anat. 119, 421-434.
17. Vermeij-Keers, C. (1967) De facialis musculatuur en transformaties in het kopgebied. Thesis, Leiden.
18. Vermeij-Keers, C. (1971) Acta Morphol. Neerl.-Scand. 9, 127-128.
19. Vermeij-Keers, C. (1972) Acta Morphol. Neerl.-Scand. 9, 386-387.
20. Vermeij-Keers, C. (1975) Acta Morphol. Neerl.-Scand. 13, 126-127.
21. Saunders, J.W. (1966) Science 154, 604-612.
22. Maruyama, S. and D'Agostino, A.N. (1967) Neurology 17, 550-558.
23. Forsberg, J.G. (1967) Experientia 23, 841.
24. Forsberg, J.G. and Källén, B. (1968) Rev. Roum. d'Embryol. Cytol.: sér. d'Embryol. 5, 91-102.
25. Kärner, J. and Leikola, A. (1975) presented at the Second Int. Conf. on Differentiation, Copenhagen.
26. El-Shershaby, E.M. and Hinchliffe, J.R. (1975) J. Embryol. exp. Morph. 33, 1067-1080.
27. Forsberg, J.G. and Åbro, A. (1973) Acta Anat. 85, 353-368.
28. Helminen, H.J. and Ericsson, J.L.E. (1971) J. Ultrastruct. Res. 36, 708-724.
29. Grüneberg, H. (1956) Britt. Med. Bull. 12, 153-157.
30. Tansley, K. (1954) J. Hered. 45, 123-127.

# Section 2.
# REGULATION OF GENE ACTIVITY AND CHROMATIN ACTIVITY DURING CELL DIFFERENTIATION

*Progress in Differentiation Research, ed. N. Müller-Bérat et al.*



# REGULATION OF TRANSCRIPTION OF RIBOSOMAL RNA-GENES DURING AMPHIBIAN OOGENESIS

U. Scheer, M.F. Trendelenburg, and W.W. Franke

Division of Membrane Biology and Biochemistry,
Institute of Experimental Pathology,
German Cancer Research Center
D-69 Heidelberg, Federal Republic of Germany

## 1. Introduction

The mechanisms that govern the regulation of the synthetic rate of a gene product are still poorly understood. In principle, such regulative mechanisms may act at two different levels of RNA biogenesis. The rate of synthesis could be modulated by either (i) direct coupling of the transcription rate to the requirement of the specific cell for a certain amount of specific RNA molecules or (ii) post-transcriptional events, for example degradation, which select the appropriate number of functional RNA molecules.

The ribosomal RNA (rRNA) synthesis is especially well suited for elucidating such control mechanisms since under different physiological conditions the ribosome production of a given cell may show pronounced variations [1-11]. From biochemical experiments alone, however, an unequivocal differentiation between the alternative regulative mechanisms of ribosome biogenesis at the transcriptional or posttranscriptional levels is often impossible. A decrease in the amount of the pre-rRNA species or in the incorporation of nucleotides into acid-precipitable material does not necessarily mean that the transcriptional rate of the rRNA-genes is concomitantly reduced but may likewise indicate a higher instability and turnover rate of the primary transcription product. To avoid these interpretative difficulties, we have directly analyzed the transcriptional activity of rRNA-genes in the electron microscope using the spreading technique of Miller and coworkers [12,13]. The amphibian oocyte has been chosen for this study because marked variations in the rate of rRNA synthesis occur during the prolonged diplotene stage of oogenesis [14-18]. Moreover, the spreading technique for visualization the activity of rRNA-genes is particularly successful with this cell system, due to the amplification of rDNA which takes place very early in oogenesis, in the pachytene stage [19,20]. Using this technique we have recently shown that structural alterations of rDNA transcription complexes can be correlated with stages of experimentally induced reduction of transcriptional activity [21]. In this article we report on naturally occurring processes involved in activation and inactivation of rRNA-genes.

## 2. Results

Alpine newts (*Triturus alpestris*) were collected during their breeding season. Gel electrophoresis of oocyte RNA labelled *in vivo* has revealed that radioactivity present in the 28S and 18S rRNA peaks was diminished in mature oocytes and drastically reduced in the previtellogenic oocytes, compared to the lampbrush chromosome stage [22]. The main product synthesized in the previtellogenic oocytes migrated in the 4-5S RNA region. From our biochemical data we calculated that the rate of rRNA synthesis was less than 1% in previtellogenic oocytes and about 30% in mature oocytes relative to that

---

Fig. 1 a,b. Electron micrographs of spread and positively stained nucleolar material from previtellogenic oocytes of *Triturus alpestris*. The typical appearance of the amplified nucleoli of this oogenic stage is demonstrated in an ultrathin section (inset of Fig. 1b), showing the firm attachment of the nucleolus to the nuclear envelope and its rather uniform composition of densely packed fibrils. When this nucleolar material is spread it becomes evident that the whole nucleolar body (diameter of about 10 μm) consists of aggregated "naked" fibrillar material. In the preparation shown in Fig. 1a no matrix units are seen. Only occasionally are individual, isolated lateral fibrils noted (inset) which probably contain transcription products. Note, however, that dense 80-120 A particles, similar to those interpreted as polymerase molecules in matrix units, are rather evenly distributed along the axes (Fig. 1a). Fig. 1b shows a large aggregate of "naked" axial fibrils with one identifiable matrix unit (arrow). Note that this single matrix unit already reveals the maximal packing density of lateral fibrils, thus suggesting full gene activity. Note also the many small aggregates of densely stained fibrillar material which perhaps represent components from chromosomes or the nuclear sap. Scales indicate 2 μm (Fig. 1 a,b); 0.2 μm (inset of Fig. 1a); 1 μm (inset of Fig. 1b).

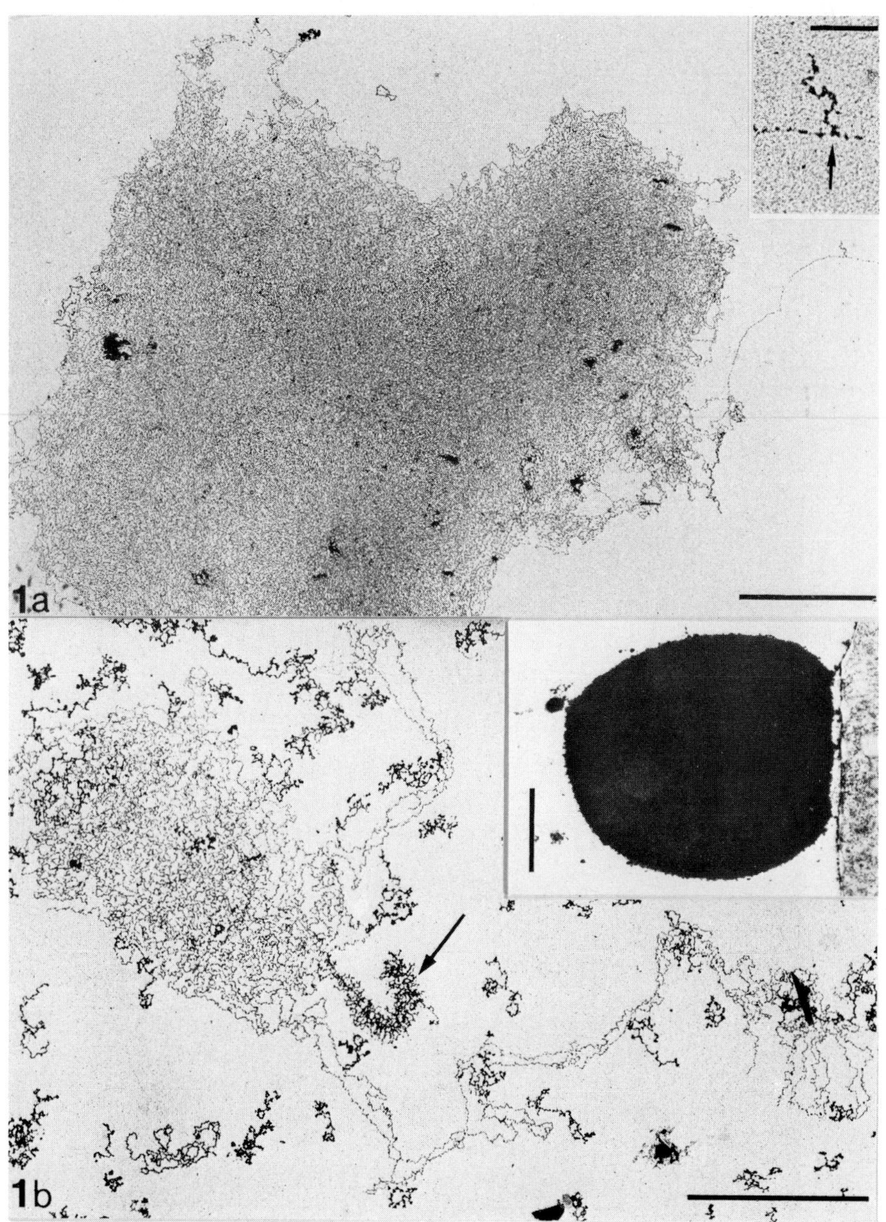
1a
1b

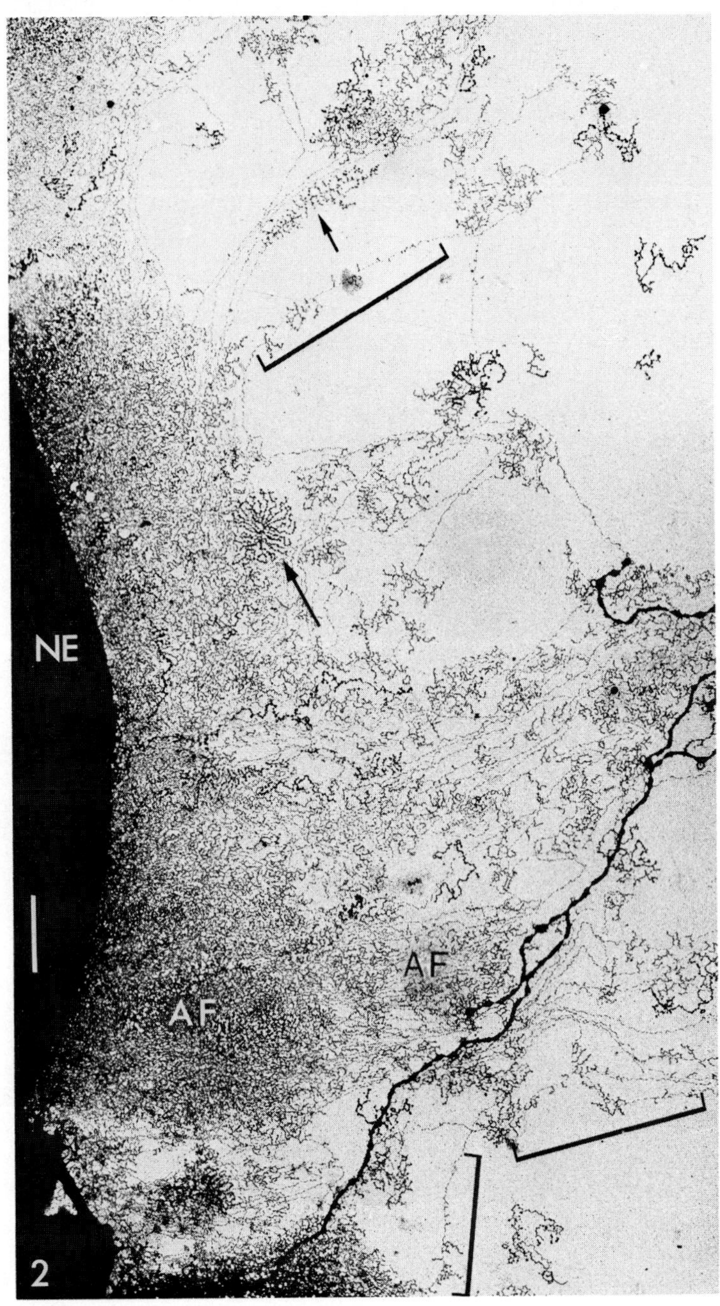


Fig. 2. Spread preparation of a nuclear envelope (NE) together with the attached nucleolar material, isolated from previtellogenic oocytes. Most of the nucleolar fibrils occur in aggregated "naked" fibrils (AF) but some groups of lateral fibrils, either densely clustered in typical matrix units (arrows) or with larger intervals (e.g. as indicated by the brackets) are seen. Scale indicates 1 $\mu$m.

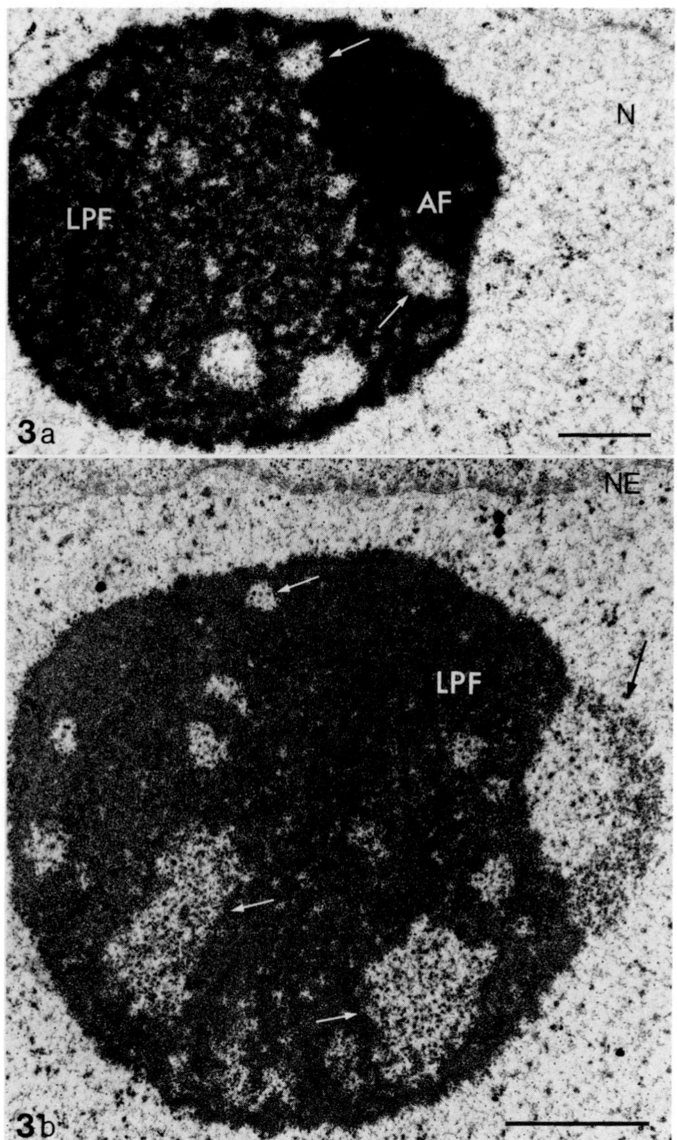


Fig. 3 a,b. Typical ultrastructure of peripheral nucleoli as revealed in ultrathin sections at early lampbrush oocyte stages of T. alpestris. Note the separation of zones with densely packed fibrils (AF in Fig. 3a) from zones with loosely packed fibrils (LPF), and vacuolized regions in which some granular particles are identified (arrows). Fig. 3b represents the characteristic appearance of a more advanced state of nucleolar transcriptional activity in which most of the fibrils are relatively dispersed. N, nucleus; NE, nuclear envelope; scales indicate 1 μm.

in lampbrush chromosome stage oocytes [22]. This suggested that ribosome biogenesis is not a continuous process throughout oogenesis but that at the onset of yolk platelet deposition the synthesis of the high molecular weight rRNAs is dramatically increased until the oocyte has reached its final size, whereupon rRNA production is again reduced (for similar data obtained in Xenopus laevis oocytes see [14-18]). It should be emphasized, however, that rRNA synthesis is never completely suppressed but that in all oocyte stages examined a significant amount of synthesis takes place (see also [23]).

The amplified nucleoli of previtellogenic oocytes of Triturus alpestris are closely associated with the inner nuclear membrane and consist of a dense fibrillar mass without a typical pars granulosa (Fig. 1b, inset). In spread and positively stained preparations (for methodology see [13,24]) these fibrillar aggregates became partially unravelled and were shown to consist of long strands of nucleolar chromatin (Fig. 1a). Typical "matrix units", i.e. transcriptional units characterized by the attachment of ribonucleoprotein fibrils that contain the growing precursor molecules for the rRNAs, could be found at a few sites within these aggregates, most of which were composed of "naked", i.e. transcriptionally inactive, rDNA-containing axes (for nomenclature see [12,13,24]). Most of the lateral fibrils, however, were separated from each other by relatively large distances compared to the maximal possible packing (see below), resulting in incomplete, "diluted" matrix units which contained 10 to 80 lateral fibrils per 2.8 µm of axial length, i.e. the average length of a "normal" matrix unit [12,24]. The relatively low number of lateral fibrils per matrix unit resulted either in homogeneously diluted matrix units or in the formation of local groups of nascent fibrils alternating with fibril-free, intramatrical intervals (Fig. 2).

---

Fig. 4 a,b. Electron micrographs showing the appearance of a fully transcribed nucleolus isolated from a lampbrush stage T. alpestris oocyte as revealed after spreading and staining. The abundance of more than 100 complete matrix units in this nucleolar body is demonstrated in the survey of Fig. 4a. Almost all genes seem to be transcribed. Fig. 4b presents details of fully transcribed rRNA-genes of nucleolar material from the same stage. Note the close packing of the lateral fibrils within the matrix units and the occurrence of "knobs" at the free ends of the lateral fibrils attached to the more terminal regions of matrix units. Scales indicate 5 µm (Fig. 4a) and 1 µm (Fig. 4b).

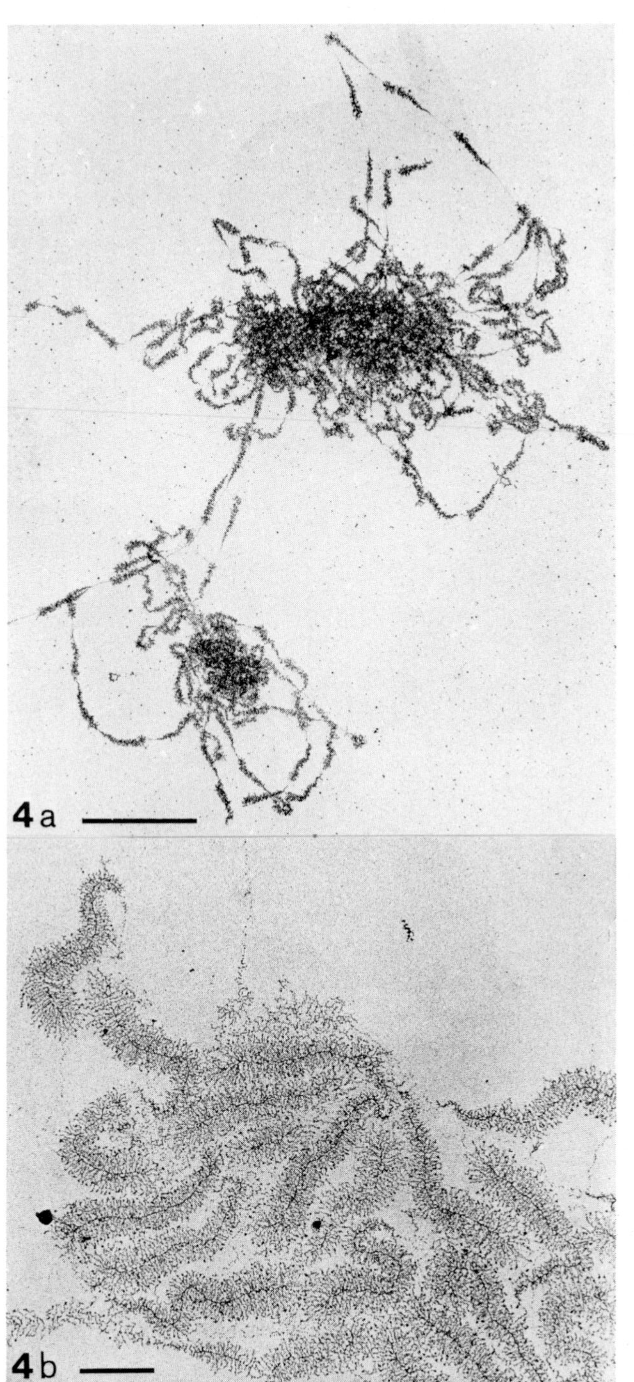
4a
4b

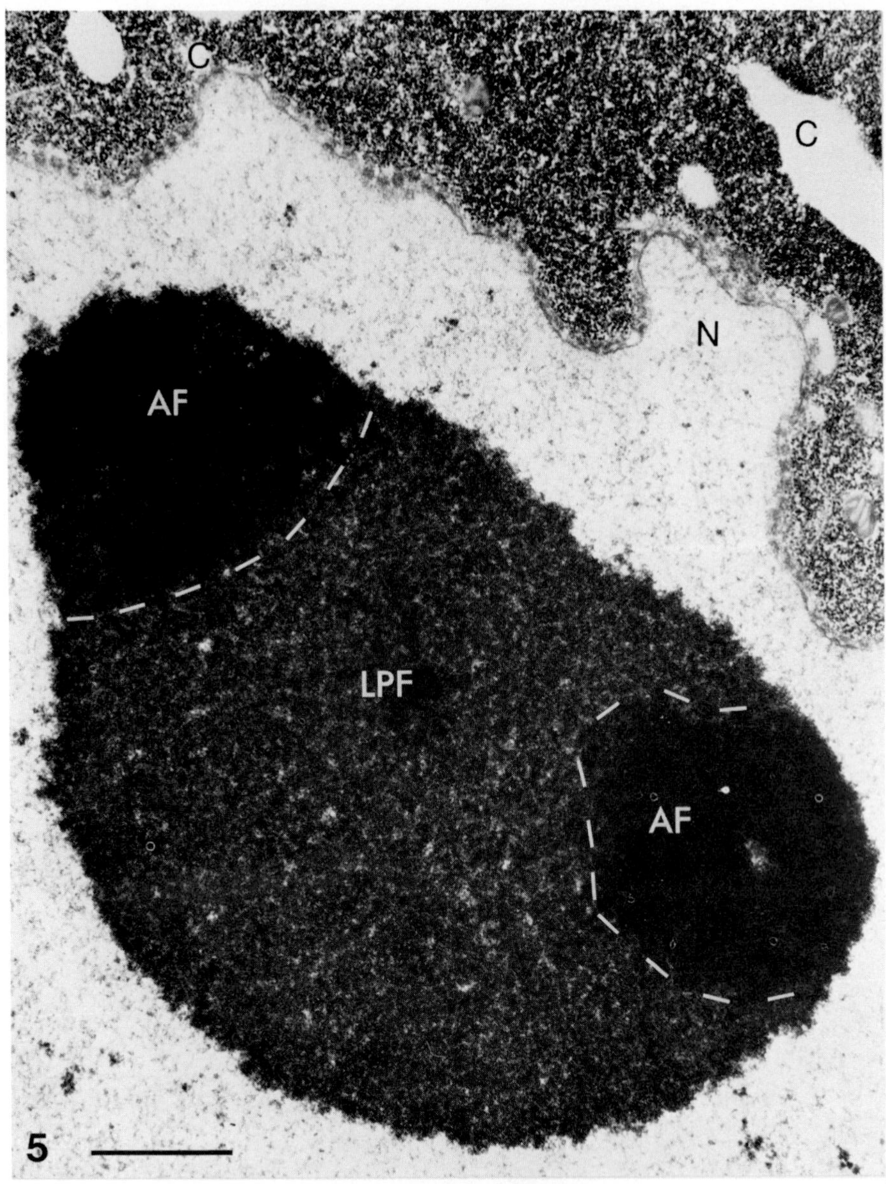


Fig. 5. Electron micrograph of an ultrathin section through the nuclear (N) periphery of a mature T. alpestris oocyte. The nucleoli show a "segregation" into zones with densely packed fibrillar aggregates (AF), usually somewhat excentrically located, and a central zone with looser packed fibrils (LPF). Note the high ribosome density in the cytoplasm (C). Scale indicates 1 μm.

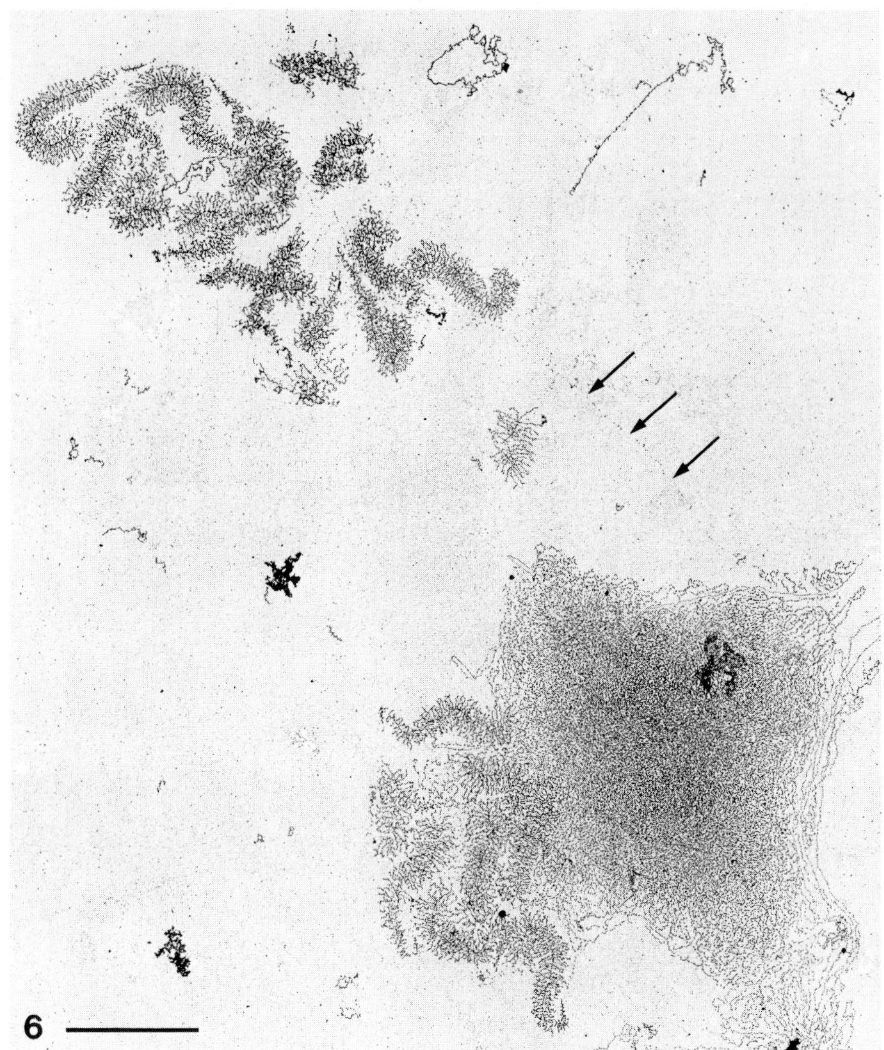


Fig. 6. Spread preparation of a nucleolus from a mature oocyte of T. alpestris. A large portion of the rDNA is not at all transcribed and appears in the form of large aggregates of "naked" axial fibrils (lower right). Fully transcribed rRNA-genes are found adjacent to this transcriptionally inactive nucleolar chromatin and are connected to it by extended axial fibrils without visible transcription complexes (arrows). Scale indicates 2 μm.

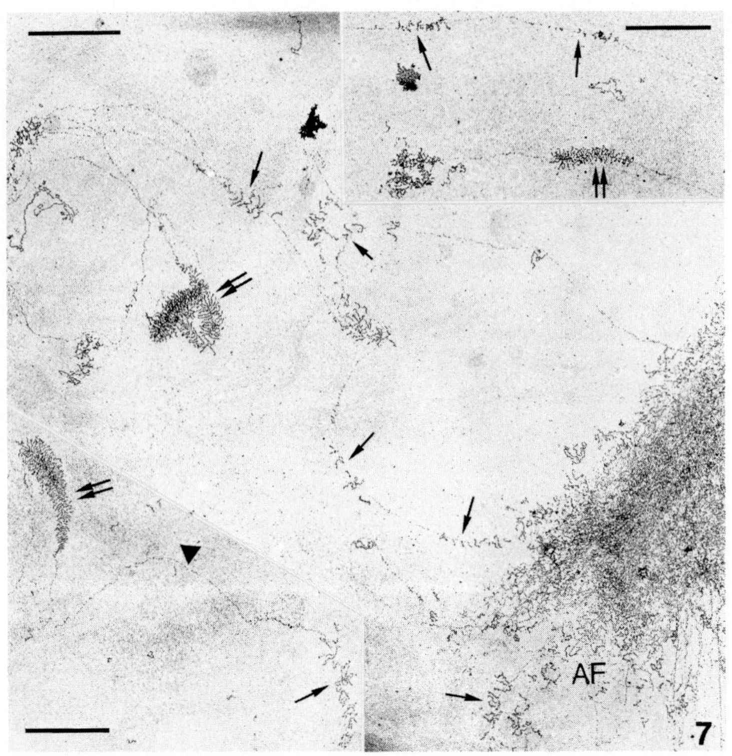


Fig. 7. Spread rDNA-containing axes of nucleoli from mature oocytes of T. alpestris, showing the coincident occurrence of different degrees and forms of reduction in lateral fibril density in adjacent matrix units. Fully transcribed or almost fully transcribed matrix units (double arrows) alternate with intercepts that show a much larger spacing (arrows) or a complete absence of lateral fibrils (e.g. at the triangle in the lower inset). Note that non-transcribed regions are much less readily spread and tend to appear as fibrillar aggregates (AF). Scales indicate 2 μm.

Fig. 8 a,b. Sequence of 15 successive rRNA-genes, revealing the typical appearance of reduced nucleolar transcription activity in the mature T. alpestris oocyte. The individual cistrons are numbered in sequence. Note the wide range of morphological forms, from almost complete matrix units (e.g. No. 15; see also the lower left of Fig. 8a) to totally untranscribed intercepts (e.g. Nos. 1,13,14) and the whole spectrum of cistrons with intermediate forms of coverage with lateral fibrils. The arrows in the upper part of Fig. 8a and the bottom part of Fig. 8b denote the same marker structure and allow the identification of this axis. Scale indicates 2 μm.

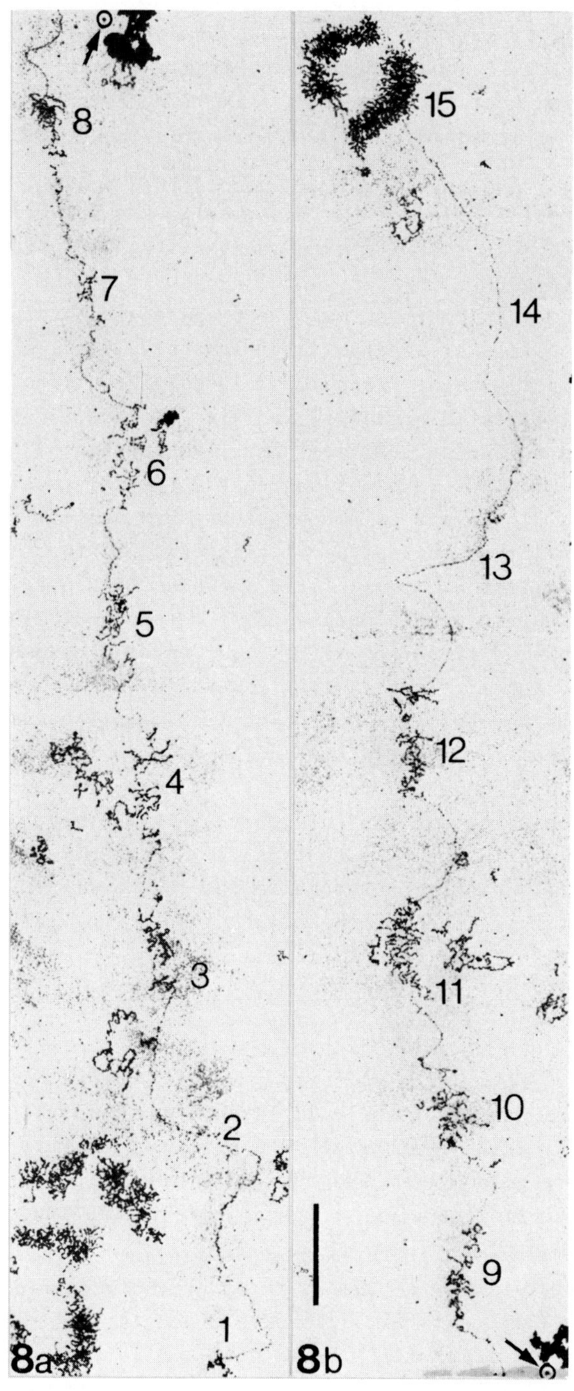
8
7
6
5
4
3
2
1
8a
15
14
13
12
11
10
9
8b

The nucleoli of lampbrush chromosome stage (vitellogenic) oocytes showed a progressive loosening and appearance of nucleolar granules (Fig. 3). Spread nucleoli from this stage were characterized by the closest possible arrangement of matrix units along the rDNA-containing axes (Fig. 4a), almost all of which showed maximal packing of the lateral fibrils, i.e. of the functioning RNA-polymerase A molecules (Fig. 4b). This indicates maximal transcriptional activity of the pre-rRNA-genes.

In mature oocytes most nucleoli were aggregated in the center of the germinal vesicle around the chromosome bivalents, which at this stage had retracted their "lampbrush" loops. The ultrastructure of these centrally located nucleoli, and also that of the few nucleoli which remained in the nuclear periphery, was dominated by the appearance of distinct zones of aggregated, densely stained fibrillar material (Fig. 5). In spread preparations such nucleoli showed an appearance similar to that of previtellogenic oocytes: aggregates of "naked" fibrillar axes were found in which only a low number of complete or almost complete matrix units were noted, often in a regionally enhanced frequency (Figs. 6,7). Again, however, the reduction of transcriptional activity resulted in a widely heterogeneous pattern of the distribution of transcription complexes along the nucleolar chromatin axes. From Fig. 7 it is evident that along a given axis complete matrix units alternate with dilute units or even cistronic regions that are completely "dormant". Fig. 8, which shows a continuous axis containing 15 tandemly arranged repeating units, demonstrates that in the same nucleolus some genes can be fully transcribed whereas adjacent ones show either no activity at all or only reduced fibril density.

## 3. Discussion

Spread preparations of nucleoli taken from previtellogenic or mature oocytes, which synthesize rRNA at a reduced rate as compared to lampbrush stage oocytes, showed significant morphological changes of the transcription complexes. In contrast to the lampbrush stage oocytes where nearly all pre-rRNA-genes appeared to be transcribed with maximal efficiency as demonstrated by the occurrence of tandemly arranged matrix units with closely packed lateral fibrils [12,24], the total number of lateral fibrils attached to the rDNA-containing axes was significantly lower in previtellogenic and mature oocytes.

The resulting arrangement of the lateral fibrils in nucleolar chromatin strands revealed a marked morphological heterogeneity. On the same nucleolar axis, individual or grouped complete matrix units alternate with fibril-free cistronic segments and matrix units with a variable, reduced number of lateral fibrils ("diluted matrix units").

In lampbrush stage oocytes a total of about 23,000 lateral fibrils were attached per millimeter traced nucleolar chromatin axis, whereas this value was about 640 in previtellogenic oocytes and 3,350 in mature oocytes. Assuming that the lateral fibril density present in nucleoli from lampbrush stage oocytes represents a transcriptional activity of 100%, then the relative transcriptional activity is about 3% and 15% in previtellogenic and mature oocytes respectively. From the basic agreement between the number of lateral fibrils containing the nascent pre-rRNA and the rate of rRNA synthesis as determined by biochemical methods [22], it can be concluded that during amphibian oogenesis the rate of formation of 28S and 18S rRNA is primarily regulated at the transcriptional level. The alternative possibility, namely that in stages of reduced ribosome formation pre-rRNA synthesis proceeds at a normal rate and posttranscriptional events are exclusively responsible for the reduced rRNA production, is apparently excluded.

The activation or inactivation of the rRNA-genes apparently is individually controlled by the frequency of initiation events by the RNA-polymerases. Mechanisms controlling exclusively the elongation rate of the nascent pre-rRNA chains should lead to the appearance of the whole set of complete matrix units in stages of reduced rRNA production. Furthermore, our data do not support the concept that a common promotor-like region controls a whole set of genes [25] but rather suggest that each pre-rRNA-gene is regulated individually (see also [26]). From the investigations of Roeder [27], it is obvious that the modulation of the rate of rRNA synthesis is not due to different levels of the activity of the nucleolar RNA-polymerase, since this enzyme is present in excess during all stages of amphibian oogenesis. Thus the quantitative regulation of transcription of pre-rRNA-genes seems to be controlled by regulatory factors [28] which interact either with the RNA-polymerases or with the template.

References


1. Cooper, H.L., 1970, Nature 227, 1105.
2. Emerson, P.C., 1971, Nature New Biol. 232, 101.
3. Fantoni, A., and Bordin, S., 1971, Biochim. Biophys. Acta 238, 245.
4. Hill, J.M., 1975, J. Cell Biol. 64, 260.
5. Papaconstantinou, J., and Julku, E.M., 1968, J. Cell. Physiol. 72, Suppl. 1, 161.
6. Perry, R.P., 1973, Biochem. Soc. Symp. 37, 105.
7. Rizzo, A.J., and Webb, T.E., 1972, Eur. J. Biochem. 27, 136.
8. Stephenson, J.R., and Dimmock, N.J., 1974, Biochem. Biophys. Acta 361, 198.
9. Tsai, R.L., and Green, H., 1973, Nature New Biol. 243, 168.
10. Vaughan, M.H., 1972, Exp. Cell Res. 75, 23.
11. Weber, M.J., 1972, Nature New Biol. 235, 58.
12. Miller, O.L., and Beatty, B.R., 1969, Science 164, 955.
13. Miller, O.L., and Bakken, A.H., 1972, Acta Endocrinol. Suppl. 168, 155.
14. Brown, D.D., and Littna, E., 1964, J. Mol. Biol. 8, 688.
15. Davidson, E.H., 1968, Gene activity in early development (Academic Press, New York).
16. Ford, P.J., 1972, in: Oogenesis, eds. J.D. Biggers and A.W. Schuetz (University Park Press, Baltimore; Butterworth, London) p. 168.
17. Mairy, M., and Denis, H., 1971, Develop. Biol. 24, 143.
18. Thomas, C., 1974, Develop. Biol. 39, 191.
19. Bird, A.P., and Birnstiel, M.L., 1971, Chromosoma 35, 300.
20. Watson Coggins, L., and Gall, J.G., 1972, J. Cell Biol. 52, 569.
21. Scheer, U., Trendelenburg, M.F., and Franke, W.W., 1975, J. Cell Biol. 65, 163.
22. Scheer, U., Trendelenburg, M.F., and Franke, W.W., 1976. J. Cell Biol., in press.
23. Leonard, D.A., and La Marca, M.J., 1975, Develop. Biol. 45, 199.
24. Scheer, U., Trendelenburg, M.F., and Franke, W.W., 1973, Exp. Cell Res. 80, 175.
25. Perry, R., and Kelley, D.E., 1970, J. Cell. Physiol. 76, 127.
26. Hackett, P.B., and Sauerbier, W., 1975, J. Mol. Biol. 91, 235.
27. Roeder, R.G., 1974, J. Biol. Chem. 249, 249.
28. Crippa, M., 1970, Nature 227, 1138.

*Progress in Differentiation Research, ed. N. Müller-Bérat et al.*



# ON THE EXISTENCE OF ARRESTED TRANSCRIPTIONAL MACHINERY IN LATE STAGES OF AVIAN ERYTHROPOIESIS

Hanswalter Zentgraf, Ulrich Scheer, and Werner W. Franke

Division of Membrane Biology and Biochemistry,
Institute of Experimental Pathology,
German Cancer Research Center,
D-69 Heidelberg, Federal Republic of Germany

A large number of cell differentiation processes is characterized by the predominance of the synthesis of one or few special proteins, and such a "specialization" is often concomitant with the switching off of most other genes. This progressive inactivation of transcriptional activity may then proceed even further and finally result in a highly differentiated cell which shows little, if any, transcription and translation. The erythropoiesis of animals with nucleated erythrocytes such as amphibia, reptiles and birds is a particularly well defined differentiation process which shows dramatic changes in cellular and nuclear function. During the formation of the red blood cell the RNA synthesis decreases at the same time as the pattern of transcription and protein biosynthesis changes[1]. In the end product of this differentiation, the mature avian erythrocyte, nearly all cytoplasmic structures, including ribosomes and mitochondria, have disappeared and most, if not all, chromatin is in a highly condensed form[2,3]. The disappearance of polyribosomes and nucleoli and the vast decrease in the synthesis of RNA led to the general concept that the fully mature avian erythrocyte is completely inactive in RNA synthesis[4-7]. Recently, however, there have been indications in the literature[8-10] of some continued RNA synthesis even in fully mature erythrocytes. Therefore, we have examined the question as to the presence and formation of some, perhaps specific, RNA in the hen erythrocyte in more detail. The results can be summarized as follows (for details and experimental procedures see[11]).

(1) In order to estimate the minimal amount of RNA in the mature hen erythrocyte the total nucleic acids were extracted from large numbers ($10^{10}$-$10^{11}$) of purified red blood cells and isolated

erythrocyte nuclei[12-15], and RNA and DNA was determined by various procedures. The minimal amount of RNA in the mature hen erythrocyte is 0.02 to 0.04 pg, that is about 1 to 2% compared to the 1.7 pg of DNA present in the nuclei. As demonstrable by cytochemistry[16] most of this RNA is located in the interchromatinic regions of the nucleus.

(2) As previous authors[8,9,17] we found that radioactive precursors are incorporated into TCA-precipitable RNA of mature erythrocytes. Incorporation of $^3$H-uridine was noted *in vivo* after intravenous injection of the precursor as well as *in vitro* after the addition of the precursor to the freshly prepared blood or after incubating the purified erythrocytes in artificial medium[11]. The incorporated radioactivity was stable against cold TCA and pronase but was hydrolyzed by ribonucleases and with 0.4 N alkali.

(3) When the labelled nucleic acids were extracted from erythrocytes and centrifuged in cesium sulfate gradients, with or without dimethylsulfoxide, DNA and RNA (ca. 1% of the total UV light absorbing material) were clearly separated. The incorporated radioactivity was exclusively recovered in the position of the RNA, demonstrating that the RNA synthesized in the mature hen erythrocyte is not stably associated with DNA.

---

Fig. 1. Localization of $^3$H-uridine incorporated *in vivo* (a-c) and *in vitro* (d) in hen erythrocytes as revealed by light (a-c) and electron (d) microscopic autoradiography. Figs. a-c. Light microscopic autoradiographs of 1 μm thick sections through mature hen erythrocytes taken from blood that has been collected at 1 hour after injection of 3 mCi $^3$H-uridine (45 Ci/mmol) into the wing vein. Most of the cells are labelled; the majority of the silver grains is located over the nuclei. Scales indicate 10 μm. Fig. d. Electron microscopic autoradiograph of an ultrathin section of tightly pelleted mature hen erythrocytes. The cells had been prepared as described[11] and incubated for three hours *in vitro* with 0.25 mCi/ml $^3$H-uridine contained in minimal essential medium containing Hanks salts. The majority of the silver grains is located over nuclear areas. In some nuclei the labelling seems to be restricted to interchromatinic regions. Cells with different degrees of labelling occur side by side. Cytoplasmic labelling is hardly above background, indicating that considerable amounts of labelled substance have not been transported to the cytoplasm during the incubation time. Scale indicates 5 μm.

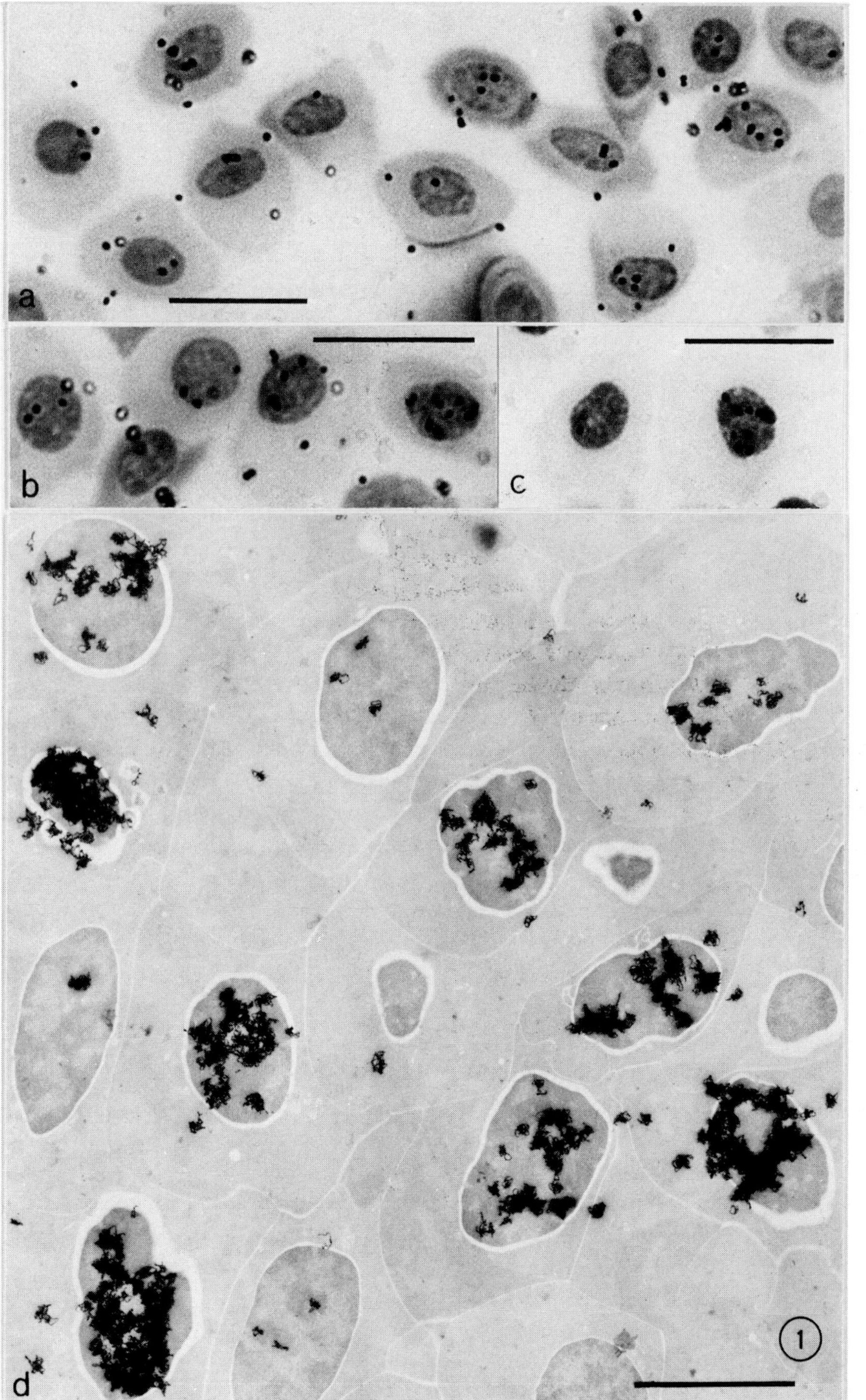
a
b
c
d
1

(4) When labelled and unlabelled RNA was applied to gel electrophoresis[11] under non-denaturing conditions it appeared very heterogeneous. Most of the radioactivity migrated in a position corresponding to molecular weights higher than $10^6$ daltons. Distinct peaks could not be recognized. In gel electrophoresis under denaturing conditions[11] about three quarters of the RNA appeared with significantly decreased apparent molecular weight. Some peaks appeared at positions corresponding to molecular weights of 1.8, 1.0 and 0.5 million daltons. No peak did appear at a position corresponding to the molecular weight determined for the globin chain messenger RNAs[18]. When the molecular weights of the chick erythrocyte RNA were determined in spread preparations (95% formamide[19]) in the electron microscope a heterogeneity in size was also noted; the largest molecules found were between two and three million daltons.

(5) Chromatography of the labelled RNA on poly(U)sepharose showed a variable proportion (17-33%) of the binding most of which, however, was not significantly specific, compared to poly-A-containing mRNAs, as demonstrable by selective elution experiments.

(6) The sensitivity of the uridine incorporation to Actinomycin D and $\alpha$-Amanitin revealed charcteristics close to that of mRNA formation but was clearly different from the sensitivity of the synthesis of the ribosomal RNA precursors on the one hand and those for 5S RNA and transfer RNA on the other hand[11,2o].

(7) The radioactivity incorporated in vivo and in vitro can also be demonstrated by light and electron microscopic autoradiography. It is mostly localized over the nuclei (Fig. 1) and remains in the nucleus even several hours after incubation and injection, respectively. The percentage of labelled cells, however, was somewhat variable (70-95%) and cells with different degrees of labelling occured side by side. We could not see an inverse correlation of the stage in erythropoiesis and the degree of incorporation.

These findings, which correspond to reports of other authors on the continuing presence of RNA-polymerase B, but not of polymerase A, in the fully mature avian erythrocyte[21-23] are perplexing since they point to the existence of RNA and RNA synthesis in nuclei even in the complete absence of translational machinery. The question whether this nuclear RNA and this RNA synthesis in the mature hen erythrocyte nucleus reflects the presence of conserved, though reduced, transcriptional activity left over from that present in earlier stages of erythropoiesis or whether it reflects synthesis of a specific RNA of yet unknown function remains open.

References

1. Ringertz, N.R., and Bolund, L., 1974, in: The cell nucleus, ed. H. Busch (Academic Press, New York) Vol. III, p. 417.

2. Kartenbeck, J., Zentgraf, H., Scheer, U., Franke, W.W., 1971, Adv. Anat. Embryol. Cell Biol. 45, 1.

3. Grasso, J.A., 1973, J. Cell Sci. 12, 491.

4. Cameron, J.L., and Prescott, D.M., 1963, Exptl. Cell Res. 30, 6o9.

5. Scherrer, K., Marcaud, L., Zajdela, F., London, J.M., and Gros, F., 1966, Proc. Nat. Acad. Sci. 56, 1571.

6. Kolodny, G.M., and Rosenthal, L.J., 1973, Exptl. Cell Res. 83, 442.

7. Sanders, L.A. Schechter, N.M., McCarthy, K.S., 1973, Biochem., 12, 783.

8. Madgwick, W.J., MacLean, N., Baynes, Y.A., 1972, Nature New Biol. 238, 137.

9. MacLean, N., and Madgwick, W., 1973, Cell Diff. 2, 271.

10. MacLean, N., Hilder, V.A., Baynes, Y.A., 1973, Cell Diff. 2, 261.

11. Zentgraf, H., Scheer, U., Franke, W.W., 1976, Exptl. Cell Res., in press.

12. Zentgraf, H., Deumling, B., Jarasch, E.D., Franke, W.W., 1971, J. Biol. Chem. 246, 2986.

13. Zentgraf, H., Laube-Boichut, E., Franke, W.W., 1972, Cytobiologie 6, 51.

14. Zentgraf, H., Falk, H., Franke, W.W., 1975, Cytobiologie, 11,1o.

15. Zentgraf, H., and Franke, W.W., 1974, Beitr. Pathol. 151, 169.

16. Bernhard, W, 1969, J. Ultrastruct. Res. 27, 250.

17. Schweiger, H.G., Bremer, H.J., and Schweiger, E., 1963, Hoppe Seyler's Z. Physiol. Chemie 332, 17.

18. Pemberton, R.E., Houseman, D., Lodish, H.F., and Baglioni, C. 1972, Nature New Biol. 235, 99.

19. Robberson, D.L., Aloni, Y., Attardi, G., and Davidson, N., 1971, J. Mol. Biol. 60, 473-484.

20. Perry, R.P., and Kelley, D.E., 1970, J. Cell Physiol. 76, 127.

21. Schechter, N.M., 1973, Biochim. Biophys. Acta 308, 129.

22. Scheintaub, H.M., and Fiel, R.J., 1973, Exptl. Cell Res. 80, 442.

23. Mandal, R.K., Mazumder, L., and Biswas, B.B., 1974, in: Control in Transcription, eds. B.B. Biswas, R.K. Mandal, A. Stevens, and W.E. Cohn (Plenum Press, New York and London), p. 295.

*Progress in Differentiation Research, ed. N. Müller-Bérat et al.*



# GENE EXPRESSION IN THE SALIVARY GLANDS OF CHIRONOMUS TENTANS

B. Daneholt, S.T. Case and L. Wieslander

Department of Histology,
Karolinska Institutet, Stockholm

## 1. Introduction

The polytene chromosomes in insect cells have proven useful in studies of gene expression in eukaryotic cells[1]. Due to their characteristic structure and large size, morphological changes as well as synthetic events taking place at various locations along the chromosomes have been possible to analyze. In the early 50's it was noted that some chromosome regions were swollen in certain tissues and developmental stages while not in others[2]. These were designated as puffs and were looked upon as morphological manifestations of gene activity. This interpretation of the puffing phenomenon later on got support from the observation that RNA synthesis was largely confined to the very same chromosome segments[3]. In our own studies of the salivary glands of Chironomus tentans we have tried to further test this puffing concept and to characterize and follow puff RNA from its site of synthesis to its site of function within the cell. It has then been important that one defined chromosome region, Balbiani ring 2 (BR 2) on chromosome IV, has been particularly suitable for individual analysis. It was shown earlier that a defined RNA species, 75 S RNA, is generated in this region[4]. 75 S RNA appears first in the nuclear sap and is subsequently found in the cytoplasm, where it accumulates and constitutes a major, non-ribosomal, non-transfer RNA species in the salivary gland cell[5,6]. It is also likely that another giant puff, BR 1, behaves in a similar way, delivering 75 S RNA to cytoplasm[6]. These observations on the behaviour of BR 75 S RNAs are in good agreement with the idea that a puff represents gene(s) being expressed, but they do not demonstrate that cytoplasmic 75 S RNA does participate in protein synthesis. We have therefore extended our studies in order to search for a function for the BR products in the cytoplasm. This has been done in studies of 75 S-RNA-containing ribonucleoprotein particles applying electron microscopic as well as biochemical techniques.

## 2. Production of BR granules and their fate

In electron microscopic studies of BR 1 and BR 2, it has been shown that the predominant elements are brushlike configurations consisting of several ribonucleoprotein (RNP) granules, each attached by an RNP stalk to a central

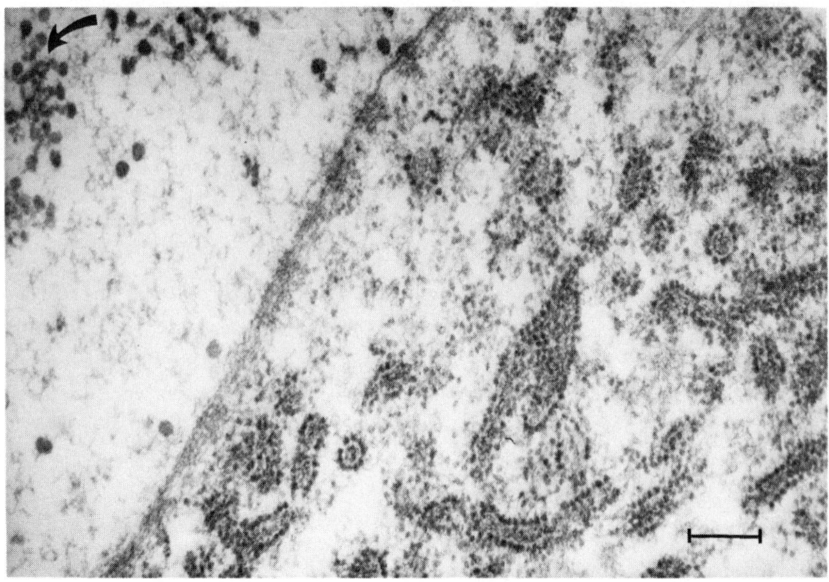

Fig. 1. Electron micrograph of a salivary gland cell in Chironomus tentans. The cell nucleus appearing to the left, contains segments of putative BR transcription complexes (arrow) and BR granules released from the complexes. The cytoplasm is rich in tubular endoplasmatic reticulum but is deficient in granules of the conspicuous size recorded in the large BRs and in the nuclear sap. The bar represents 0.2 µ.

deoxyribonucleoprotein (DNP) axis[7-9]. This structure is likely to represent part of a transcription complex, the DNP axis being the template for RNA synthesis and each RNP granule with its stalk representing a growing 75 S RNA chain with associated proteins[10]. Such a putative transcription complex is partly revealed in Fig. 1. Granules lacking stalks are observed in the surrounding nuclear sap and are likely to correspond to completed and released BR products. The granules occur frequently in the nuclear sap, also appear in the nuclear pores, but are scarce or absent in the cytoplasm[8] (fig. 1). While the granules are almost exclusively present in the nucleus, by far, most of the 75 S RNA molecules are present in cytoplasm[5]. If it is assumed that each 75 S RNA molecule is generated as part of a BR granule, then it can be concluded that the 75 S-RNA-containing particle has to change conformation and/or structure in connection with the transfer of 75 S RNA from the nucleus to the cytoplasm. This change of the ribonucleoprotein particle can take place when 75 S RNA passes the nuclear pore, or alternatively just after it has been transferred to cytoplasm. One plausible drastic change would be that 75 S RNA, with or without the associated proteins, enters polysomal complexes and acts as messenger RNA. This possibility has been directly tested in biochemical experiments.

3. Polysomal location of 75 S RNA delivered from BR 1 and 2

For the biochemical studies we chose larvae kept for about three days in ordinary culture medium supplemented with tritiated cytidine and uridine. During these conditions more than 90 % of the labelled RNA was found in cytoplasm and displayed the electrophoretic profile depicted in fig. 2. The dominating RNA species are the ribosomal 28 S and 18 S RNA and the transfer, 4 S RNA. Furthermore, there is a prominent 75 S RNA peak as well as a heterogeneous distribution of RNA molecules ranging from low molecular weight RNA up to the 75 S RNA. In order to analyze whether 75 S RNA is likely to have a messenger function, we have first tried to decide whether or not high molecular weight, non-ribosomal RNA (30-75 S) is present in polysomes.

Conventional homogenization techniques can be applied to *Chironomus* salivary glands in order to obtain polysomes but during such an extraction some degradation of the most high molecular weight RNA cannot be avoided[11]. In order

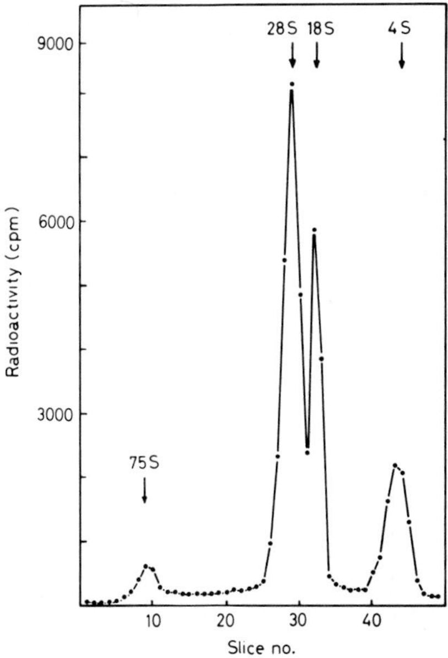


Fig. 2. Electrophoretic analysis in 1 % agarose of salivary gland RNA labelled *in vivo* for three days.

to reduce this degradation we developed a new method, which enabled us to keep a considerable amount of 75 S RNA intact. The salivary glands were placed in a detergent mixture and torn open with two siliconized, steel needles on a cold stage (2-4$^{o}$C) under a dissection microscope in order to achieve an efficient release of the polysomes. After completion of the extraction the remaining glandular material (containing the nuclei), was easily removed with the needles, and the polysome extract was directly layered on top of a 15-60 % sucrose gradient. The results show that there is a distinct monosome peak and a broad polysome region with the peak between 500 and 1,000 S (fig. 3). If EDTA was added to the extract, in order to selectively disassemble the polysomes, essentially all the activity in the polysome region was shifted to the polysomal supernatant, confirming that the rapidly sedimenting material represents true polysomes.

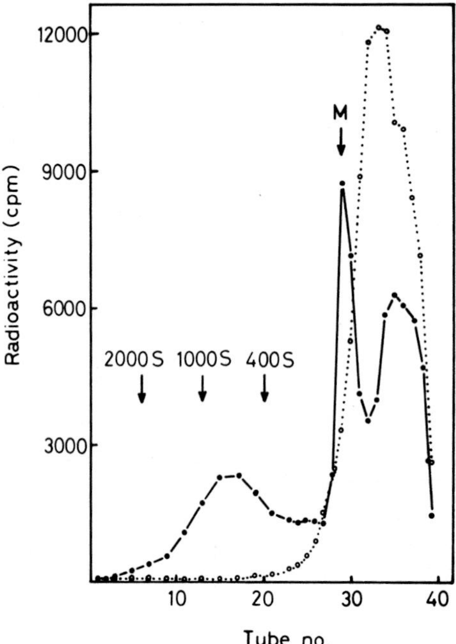


Fig. 3. Sucrose gradient sedimentation analysis of ribonucleoproteins extracted from salivary glands labelled for three days in vivo. The glands were extracted at 2-4$^{o}$C in a detergent solution (0.5 % DOC, 0.5 % Tween 80 in 0.1 M KCl, 0.003 M $MgCl_2$ and 0.02 M triethanolamine-HCl, pH 7.6). One half of the extract was treated with EDTA (0.02 M final concentration). Each sample was put on a 15-60 % gradient and spun at 40,000 rpm for 30 min at 4$^{o}$C. Untreated sample: closed circles; EDTA-treated: open circles.

To analyze whether high molecular weight, non-ribosomal RNA was present in the polysomes, and if so, to what extent, the following experiment was carried out. A polysomal extract was prepared and split into two portions, one of which was treated by EDTA. The two samples were put on sucrose gradients and fractionated as shown in Fig. 3. From both gradients certain fractions were pooled to form a polysome sample (fraction 4,6....22) and a polysomal supernatant sample (fraction 24,26....38), while the remaining odd-numbered fractions were used to establish the radioactivity profile. The labelled RNA in the four samples was released by Sarkosyl-pronase treatment, precipitated by ethanol, redissolved and analysed by electrophoresis in agarose gels (Fig. 4). It could then be observed that as much as 84 % of the high molecular weight, non-ribosomal RNA is present in the polysome region. After EDTA treatment, however, most of the high molecular weight RNA (77 %) is recorded in the supernatant. It can therefore be stated that cytoplasmic, high molecular weight, non-ribosomal RNA (30-75 S) is at least predominantly (more than 60 %) located in polysomes.

Although it is evident that 75 S RNA molecules are present in polysomes, it is necessary to directly demonstrate that nucleotide sequences transcribed from BR 1 and 2 are incorporated into polysomes. This could be done by _in situ_ hybridization[12] of labelled polysomal RNA. Grains were readily detected over BR 2, and to a lesser extent, over BR 1 (fig. 5), while in EDTA treated controls only very few grains were revealed. Since it is known that cytoplasmic RNA sequences that hybridize to BR 1 and 2 are exclusively located in the 75 S region[6], it can be concluded that at least some 75 S RNA molecules generated in BR 1 and 2, are incorporated into polysomal structures in the cytoplasm and, therefore, are likely to act as messenger RNA molecules.

## 4. Translation of BR RNA

If BR 75 S RNA is accepted as mRNA, the most immediate question is to what extent it is translated. It is conceivable that the polysome experiments can provide some information on this issue. The sedimentation profile of the salivary gland polysomes displays an ordinary shape but the sedimentation values are unusually large, indicating polysomes of considerable sizes (Fig. 2). The peak value of about 700 S should correspond to 30-40 ribosomes and the more extreme value of 2,000 S corresponds to as many as 200 ribosomes per polysome[13]. Considering the unusually large sedimentation values of these polysomes, it is particularly important to determine whether these are giant polysomes or some kind of artefactual aggregates. Therefore, various polysome containing fractions from the sucrose gradients have been analyzed by electron microscopy. It can be stated that the polysomes appear free from membranes and clumps of salivary

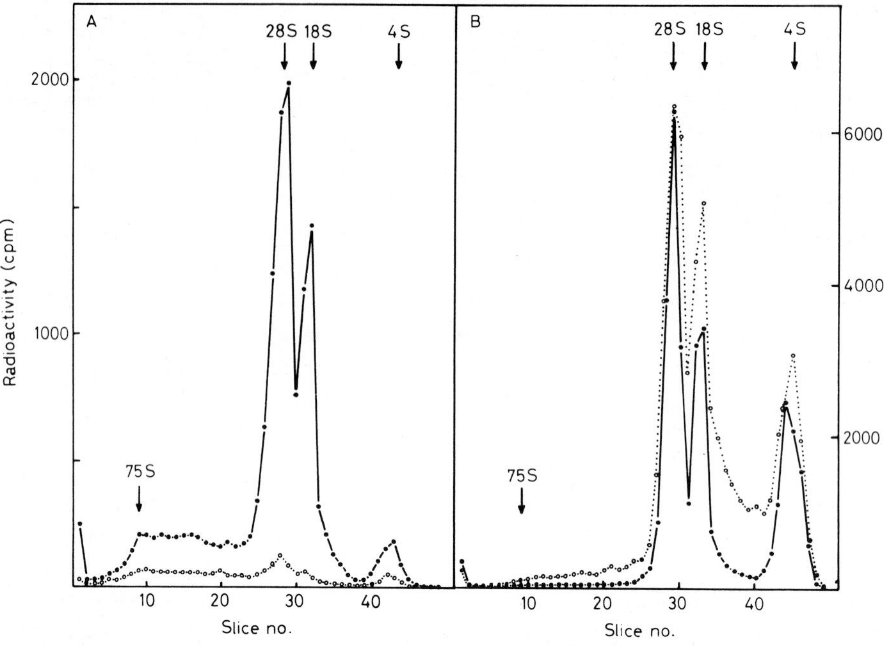


Fig. 4. Electrophoretic analysis of long term labelled RNA pooled from the polysome region (A) and polysomal supernatant (B) in an experiment analogous to that shown in fig. 3. Prior to sedimentation the polysomal extract was split into two portions, one of which was treated with EDTA (0.02 M). Polysome fractions (4,6....22) as well as polysomal supernatant fractions (24,26....38) were pooled from both gradients and RNA released by Sarkosyl-pronase treatment (0.5 % Sarkosyl, 1 mg/ml pronase) before electrophoresis in 1 % agarose. The untreated samples: closed circles; EDTA treated: open circles.

secretion. Moreover, the sizes of the polysomes observed in the electron microscope seem to fit very well with the values predicted from the sedimentation data. The 700 S polysomes are frequently displayed in an outstretched linear form with 30-40 ribosomes. However, polysomes of larger size (100 ribosomes or more) cannot be completely untangled. Although, it is difficult to unequivocally state that the most rapidly sedimenting polysomes do represent true polysomes of giant size, two observations argue in favour of such an interpretation. First, if the most high molecular weight RNA is partially degraded during the extraction, the most rapidly sedimenting polysomes are not obtainable, implying that the integrity of these polysomes is dependent upon intact RNA of very high molecular weight. Second, if labelled RNA from various parts of the polysome region is analysed by electrophoresis, it has been found that relatively more RNA of very high molecular weights (50-75 S) is present in the faster sedimenting polysomes (fraction 4-12) than in the slower sedimenting ones (fraction 14-22). If the most rapidly sedimenting polysomes simply consisted of aggregates of smaller polysomes, such a difference is not to be expected without any further assumptions. However, a size difference in the putative mRNA is consistent with the interpretation that the most rapidly sedimenting polysomes represent polysomes of giant size, while the slower sedimenting ones those of smaller size. To sum up, from the analysis of the salivary gland polysomes it seems reasonable to assume that the polysomes containing BR 75 S RNA harbour at least 30-40 ribosomes and probably as many as 100-200 per polysome. If the ribosome density on these polysomes are comparable to that found in other systems, e.g. globin mRNA[14], it can be inferred that the translated portion of the 75 S RNA must correspond to a segment far exceeding that required for the synthesis of an average sized polypeptide (such a segment should harbour about 10 ribosomes).

Cytological, cytogenetic and biochemical evidence suggest that the polypeptides translated from BR 75 S RNA are likely to be salivary polypeptides, the major polypeptide products synthesized in these cells[1]. It has proven to be a difficult task to characterize these salivary polypeptides. Grossbach[15] reported that they are of giant size (the most abundant one is as large as 500,000 daltons). The presence of long translated segments in 75 S RNA could then simply reflect the synthesis of single polypeptides of unusual size, the

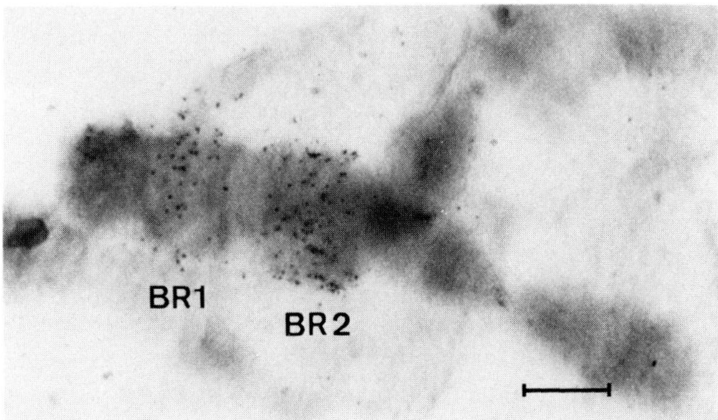


Fig. 5. _In situ_ hybridization of polysomal RNA obtained from the polysome region as described in fig. 4. The picture displays a chromosome IV with grains above BR 1 and BR 2. The bar represents 10 $\mu$.

largest requiring a segment carrying about 130 ribosomes. However, Rydlander and Edström[16] have recently pointed out that during the isolation procedure used by Grossbach, isopeptide bonds can be formed and giving rise to polypeptide aggregates. If they prevented the formation of isopeptide bonds by glycine ethyl ester, they obtained polypeptides of much smaller size (30,000--60,000 daltons), which they then regard as the primary translational products. If this latter view is correct, the recorded polysomes of giant size could then favour the possibility that a 75 S RNA molecule contains genetic information for more than one polypeptide, i.e. it should then be a polycistronic mRNA. Evidently more precise information is needed on the size of the amino acid coding segments of the 75 S RNA molecules from BR 1 and BR 2, as well as on the nature of the primary translational products, before it can be firmly stated whether each 75 S RNA molecule is a mono- or poly-cistronic mRNA.

## 5. Concluding remarks

All the available information on the synthesis, transport and functional role of the 75 S RNAs transcribed in BR 1 and BR 2 support the puffing concept, i.e. that a puff represents gene(s) being expressed. As a result of these investigations we have obtained a useful basis for further studies on gene expression in this differentiated cell. First, the generation of the primary transcription product can now be measured and reasonably well quantitated[10]. It is also of interest in this context that the activity of the BRs can be regulated by galactose[17]. Furthermore, the transcripts can be followed from the chromosomes into the nuclear sap and cytoplasm and finally into polysomes. Posttranscriptional regulation can be looked for in kinetic experiments designed to show the flow of BR 75 S RNA in the cell and to detect possible sites for degradation or inactivation of the molecules.

## Acknowledgments

The technical assistence of Miss Eva Mårtenzon and Mrs Sigrid Sahlén is gratefully acknowledged. The present work was supported from the Swedish Cancer Society, Magnus Bergvalls Stiftelse and Karolinska Institutet (Reservationsanslaget). S.T.C. is a recipient of a National Research Service Award from the National Institutes of Health (U.S.A.).

References


1. Daneholt, B., 1974, Int. Rev. Cytol. Suppl. 4, 417.
2. Beermann, W., 1952, Chromosoma 5, 139.
3. Pelling, C., 1964, Chromosoma 15, 71.
4. Daneholt, B., 1972, Nature New Biol. 240, 229.
5. Daneholt, B. and Hosick, H., 1973, Proc. Natl. Acad. Sci. 70, 442.
6. Lambert, B. and Edström, J.-E., 1974, Mol. Biol. Reports 1, 457.
7. Beermann, W. and Bahr, G.F., 1954, Exptl. Cell Res. 6, 195.
8. Stevens, B.J. and Swift, H., 1966, J. Cell Biol. 31, 55.
9. Vazquez-Nin, G. and Bernhard, W., 1971, J. Ultrastruct. Res. 36, 842.
10. Daneholt, B., 1975, Cell 4, 1.
11. Hosick, H. and Daneholt, B., 1974, Cell Differentiation 3, 273.
12. Lambert, B., Wieslander, L., Daneholt, B., Egyházi, E. and Ringborg, U., 1971, J. Cell Biol. 53, 407.
13. Reisner, A.H., Askey, C. and Aylmer, C., 1972, Analyt. Biochem. 46, 365.
14. Slayter, H.S., Warner, J.R., Rich, A. and Hall, C.E., 1963, J. Mol. Biol. 7, 652.
15. Grossbach, U., 1973, Cold Spring Harb. Symp. Quant. Biol. 38, 619.
16. Rydlander, L. and Edström, J.-E., 1975, Proc. 10th FEBS Meeting (Paris), in press.
17. Beermann, W., 1973, Chromosoma 41, 297.

*Progress in Differentiation Research, ed. N. Müller-Bérat et al.*



# REGULATION OF ERYTHROID DIFFERENTIATION IN FRIEND ERYTHROLEUKAEMIC CELLS

P. R. Harrison, N. Affara, D. Conkie, T. Rutherford, J. Sommerville and J. Paul

The Beatson Institute for Cancer Research,
Glasgow, Scotland

## 1. Introduction

This paper describes some recent molecular - genetic studies on the induction of erythroid markers in mouse erythroleukaemic (Friend) cells. These cells were obtained by culturing fragments of spleens from DBA/2 mice infected with the Friend virus complex[1], and seem to be derived from transformed stem cells committed to the erythroid line (i.e. CFU-E's)[2], which have become arrested at an early (probably proerythroblast) stage of erythroid differentiation. Unlike normal CFU-E's, they do not respond to the hormone erythropoietin[3]. However, after treatment with dimethyl sulphoxide (DMSO)[4], they do exhibit a whole range of erythroid characteristics, such as synthesis of haem[4], mouse α and β globins[5,6] and globin messenger RNAs[7,8], erythrocyte surface antigens[9], carbonic anhydrase[10] and morphological differentiation to the orthochromatic stage[4]. In this respect they represent a useful model system for elucidating the regulation of a co-ordinated series of functions within a single differentiated cell type.

This report concerns three integrated aspects of our work: (1) the isolation of both inducible and non-inducible variants of these cells; (2) the characterisation of these variants at the molecular level; and (3) the study of interactions of genes by dominance analysis in cell hybrids formed between variants.

## 2. Methods

The relevant methods used in these studies have been described elsewhere[7,8,11-13]. Nuclear RNA was isolated from nuclei prepared by the citric acid method[14] according to an adaptation of the method of Glisin et al.[15] in which purified nuclei suspended in sodium dodecyl sarcosinate are treated with proteinase K before centrifugation in caesium chloride. The content of globin-specifying sequences in these RNA preparations was measured by titration with globin cDNA[16,17] prepared by transcription of purified and characterised reticulocyte globin messenger RNA[18].

To facilitate the isolation of hybrid cells, 5'bromodeoxyuridine (BUdR)-resistant and thioguanine/azaguanine-resistant markers were introduced into parental lines and the HAT selection system was employed[19] to select hybrid cells from mixtures of parental cells fused with inactivated Sendai virus[20].

## 3. Results

### (a) Origin of cell lines

The relationships among the clones used in these studies are indicated in Figure 1.

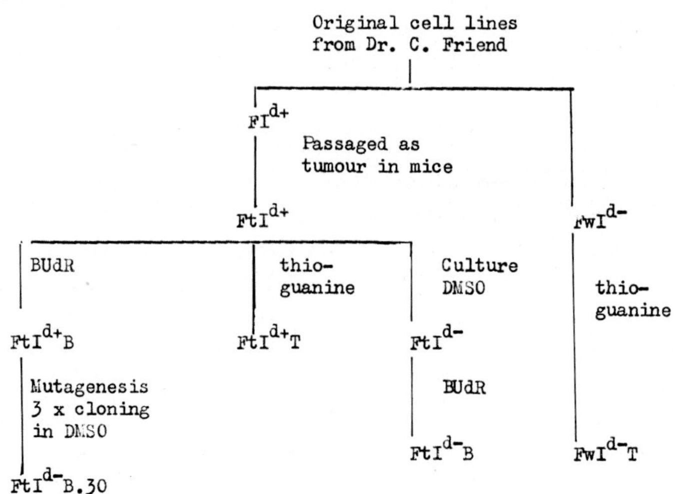


Fig. 1. Origin of Friend cell variants used in these studies. $I^{d+}$ or $I^{d-}$ indicate whether or not the variant can be induced to form haemoglobin by treatment with DMSO. B or T indicate that cell is resistant to BUdR or thioguanine/azaguanine.

The first inoculum received from Dr. Friend six years ago was designated Fw by ourselves. When the inducibility of haemoglobin synthesis with DMSO was reported, we found that this cell line was not inducible ($FwI^{d-}$); subsequently, a thioguanine-resistant clone was prepared ($FwI^{d-}T$)[11]. Another inoculum received later from Dr. Friend (clone 707 in her terminology) was found to be inducible ($FI^{d+}$). This was passaged as a tumour in mice and the re-isolated line was designated $FtI^{d+}$. BUdR-resistant ($FtI^{d+}B$) and thioguanine-resistant ($FtI^{d+}T$) clones were isolated from this[11].

Another line was obtained by growing cells continuously in DMSO and proved to be non-inducible ($FtI^{d-}$)[13]; a BUdR-resistant clone was then isolated ($FtI^{d-}B$). More detailed studies have shown that DMSO-resistant variants can be obtained by cloning inducible Friend cells in 1.5-1.7% DMSO, at a frequency of about $10^{-4}$-$10^{-5}$. This frequency can be increased by mutagenesis of Friend cells before cloning in DMSO (for example, 5-10 fold with ICR 191). However, only a small proportion of these variants reclone successively in DMSO at high frequency (10-40%) and remain stable on culture without DMSO.

As a control, a non-erythroid line (Ly) derived from a lymphoma (L5178Y) of DBA/2 mice was studied; to facilitate the preparation of cells, a thioguanine-resistant mutant was isolated (LyT) after mutagenesis with EMS.

(b) Haemoglobin and Haem synthesis in variants and their hybrids

All our inducible lines behave in the same way with respect to timing of synthesis of haemoglobin, as judged by benzidine staining. The percentage of cells staining with benzidine increases rapidly after about three days of treatment with DMSO reaching a value of 80–85% after the fifth day. In contrast, the maximum rate of incorporation of $Fe^{59}$ into haem seems to occur after three days treatment with DMSO. This may mean that the globin chains and haemoglobin are synthesised earlier than is detected with benzidine staining. Since globin messenger RNAsseem to be fully induced after treatment with DMSO for 3 days[8,9], this suggests there is no significant delay between synthesis of globin messenger RNAs and their translation.

Table 1

Synthesis of haem and haemoglobin in Friend cell variants and their hybrids.

| Inducible Status | Other Markers | % cells haemoglobin⁺ -DMSO | +DMSO | Rate haem synthesis -DMSO | +DMSO |
|---|---|---|---|---|---|
| $I^{d+}$ | Ft | 1 | 80 | - | - |
| | FtB | 1 | 80 | 0.8 | 13 |
| | FtT | 1 | 80 | 1.2 | 8 |
| $I^{d-}$ | FtB | 0.01 | 0.05 | 0.8 | 0.4 |
| | FtB.30 | 0.01 | 0.05 | - | - |
| | FwT | 0.1 | 0.5 | 0.9 | 1.5 |
| | LyT | Nil | Nil | 0.2 | 0.2 |
| $I^{d+} \times I^{d+}$ | FtB x FtT | 1 | 65-80(4) | 0.7 | 17 |
| $I^{d+} \times I^{d-}$ | FtT x FtB | 0.03 | 0.1 | 1.4 | 2.8 |
| | FtB x FwT | 1 | 15-30(3) | 1.0 | 1.5-6 |
| | FtB x LyT | Nil | Nil | 1.0 | 1.3 |
| $I^{d-} \times I^{d-}$ | FtB x FwT | 0.1 | 5 | 1.4 | 3.6 |

Haemoglobin synthesis was assayed by benzidine staining after 0 and 5d DMSO. The rate of haem synthesis (p mol $FE^{59}$ incorporation into total haem/$10^6$ cells/h) was determined at 0 and 3-4d. The figures in brackets denote the number of independent fusion events studied.

The non-inducible lines synthesise little haemoglobin or haem (Table 1). The $FwI^{d-}$ line may show slightly increased haemoglobin and haem synthesis after DMSO treatment, which is reflected in a detectable rate of synthesis of globin chains (Table 5).

In order to investigate the genetic control of haemoglobin synthesis in Friend cells, cell hybrids were formed to analyse dominance and complementation relationships at the phenotypic level. The karyological evidence concerning the hybrid status of these cell lines is summarised in Table 2. Chromosome losses in the hybrid cells from the values predicted from the parental chromosome numbers are small (3-14%).

For all the hybrids except one, this karyological evidence is unequivocal. However, since the marker biarmed chromosomes in the $FtI^{d+}T$ and $FtI^{d-}B$ lines cannot be distinguished by our present methods the hybrid status of the cells obtained by fusion of these lines is not rigorously proven by this evidence alone. However, these cells incorporate both hypoxanthine and thymidine at the expected rates, unlike the parental lines or parental revertents selected at low frequency in HAT (see reference 21 for details). This proves the hybrid status of the $FtI^{d+}T$ X $CtI^{d-}B$ line.

Table 2

Dominance and complementation analysis in Friend cell hybrids

| Status | Parental Markers | Chromosome number Expected | Actual |
|---|---|---|---|
| $I^{d+}$ | FtB | | 36-37 (4) |
| | FtT | | 49 (4) |
| $I^{d-}$ | FtB | | 38 (4) |
| | FwT | | 37 (5) |
| | LyT | | 41 (1) |
| $I^{d+}$ x $I^{d+}$ | FtB x FtT | 76 (8) | 70-75 (8*) |
| $I^{d+}$ x $I^{d-}$ | FtT x FtB | 77 (8) | 69-71 (8) |
| | FtB x FwT | 74 (9) | 67-73 (9**) |
| | FtB x LyT | 78 (5) | 65-68 (5*) |
| $I^{d-}$ x $I^{d-}$ | FtB x FwT | 75 (9) | 69-71 (9*) |

Figures in brackets give the numbers of biarmed chromosomes. One or two asterisks denote that biarmed chromosomes can be identified positively to one or both parents.

To determine whether the act of fusion might, by itself, affect the response to DMSO, hybrids were prepared from four independent fusion experiments between two inducible cells. As shown in Table 1, these hybrid lines behaved very like parental inducible cells. Hybrids prepared between inducible cells and the three different non-inducible cells revealed interesting differences. In the cross involving the non-inducible $FtI^{d-}$, non-inducibility appeared to be dominant. In contrast, the cross between the $FtI^{d+}$ and $FwI^{d-}$ showed inducibility to an intermediate degree. The figures given are the averages of observations of hybrids derived by three independent fusion experiments and confirm the observations reported previously[11]. This result implies that inducibility is dominant over non-inducibility but to a partial extent only. Finally, fusion of the inducible Friend cell with a non-erythroid lymphoma cell gave a hybrid in which no haemoglobinised cells could be recognised with or without DMSO treatment. In this instance, therefore, there was also evidence for trans-dominant repression of the erythroid phenotype by the genome of the non-inducible cell. This table also shows the behaviour of a hybrid between two non-inducible cells (Ft and Fw), which exhibited a rather low level of haemoglobinised cells. This lack of complementation is perhaps not surprising, since non-inducibility in $FtI^{d-}$ is dominant.

Finally, it is significant that each variant or its hybrid behaved identically with respect to both haemoglobin and total haem synthesis. This implies that these two functions are tightly coupled.

In order to analyse more precisely the basis of these phenomena at the phenotypic level, assays were performed of the levels of globin-specifying sequences in the cytoplasms and nuclei of these variants and their hybrid cells.

(c) The $FtI^{d-}$ variant

As shown in Table 3, treatment with DMSO induces a very considerable increase in the numbers of globin RNA molecules in both cytoplasm and nucleus of a typical inducible cell. In contrast, in the non-inducible $FtI^{d-}B$ variant, the level of globin messenger RNA is very low, and it is not increased significantly (if at all) by DMSO treatment.

Table 3

Globin RNA levels in the inducible $FtI^{d+}T$ parent, the non-inducible $FtI^{d-}B$ variant and the corresponding hybrid.

| Cell | % cells Hb+ +DMSO | Cytoplasm | | Nucleus | |
|---|---|---|---|---|---|
| | | -DMSO | +DMSO | -DMSO | +DMSO |
| $I^{d+}$ FtT | 80 | 300±150 | 9000±3000 | 100,110 | 1400 |
| $I^{d-}$ FtB | ~ 0.5 | 100±40 | 150±80 | 15,15 | 35±10 |
| $I^{d+} \times I^{d-}$ | ~ 0.1 | 900±300 | 2000±1000 | 80±40 | 90±70 |

Values quoted are those of single experiments or the averages of three or more (usually 4-6) independent experiments ± the standard deviation. DMSO (1.5%) treatment was for 5d.

In the hybrid formed between the inducible parent and the $FtI^{d-}$ variant, globin RNA sequences are present at a significantly high level than either of the parental cell lines. However, DMSO-induction of globin messenger RNA sequences in the hybrid above the basal level is small (two-fold) in comparison with the inducible cell (30-fold). A similar conclusion follows from consideration of the nuclear globin RNA level. Thus the non-inducibility of haemoglobin synthesis in this variant seems to be due to a defect in globin gene expression at the nuclear level and this defect in inducibility acts in a trans-dominant manner, although perhaps not completely so. The basal level of globin gene expression may be regulated independently.

(d) The $FwI^{d-}T$ variant

The $FwI^{d-}$ line has a somewhat higher cellular content of globin messenger RNA molecules than the $FtI^{d-}$ variant discussed above (Table 4 cf. Table 3). DMSO treatment does not increase this level significantly. When this variant was tested for dominance against the $FtI^{d+}B$ parent, it was found that after DMSO treatment the hybrid cells contained about the same number of globin messenger RNA molecules as the inducible parental line; however only 20-30% of the cells contained haemoglobin.

Table 4

Globin mRNA levels in the $FtI^{d+}B$ parent and the $FwI^{d-}T$ variant and the corresponding hybrid.

| Cell | % cells containing haemoglobin +DMSO | molecules globin mRNA -DMSO | molecules globin mRNA +DMSO |
|---|---|---|---|
| $I^{d+}FtB$ | 80 | 170±70 | 6000±2000 |
| $I^{d-}FwT$ | ~ 1 | 500±250 | 500±200 |
| $I^{d+}$ x $I^{d-}$ uncloned | 15-30 | - | 6000,8000 |
| clone 1/4 | 55 | 600,800 | 20,000±8000 |
| clone 1/5 | 6 | 600 | 1200,1500 |

See Table 3 for details.

In situ hybridization of globin cDNA showed subsequently that the uncloned hybrid line contained a subpopulation of cells after DMSO treatment which possessed high levels of globin messenger RNA, whereas the majority of the cells had much lower levels (see reference 21 for details). This conclusion has been confirmed by cloning experiments; clones of the $FtI^{d+}B$ x $FwI^{D-}T$ hybrid have been isolated which show good or poor responses to DMSO (Table 4). Significantly, the response in terms of accumulation of globin messenger RNA correlates with that of haemoglobin itself, as measured by the proportion of the cells staining with benzidine (Table 4, see also ref. 21) or by the rate of $Fe^{59}$ incorporation into haem. This is consistent with a genome-dosage effect. The high cellular content of globin messenger RNA molecules in those clones (e.g. 1/4) giving a good response to DMSO suggests that in these cells both parental sets of globin genes are active. This evidence therefore supports the interpretation that the defect responsible for the non-inducibility of the $FwI^{d-}$ variant can be recessive or dominant depending on the overall chromosome balance in a particular hybrid clone.

Table 5
Karyotype analysis of $FwI^{d-}T \times FtI^{d-}B$ hybrid clones

| Clone | % cells $Hb^+$ +DMSO | modal chromosome number | number biarmed chromosomes | % mitotic figures with small Fw biarmed chromosome |
|---|---|---|---|---|
| 1/5 | 5% | 68 | 9 | 100% |
| 1/4 | 55% | 62 | 6-8 | 0% |
| 2/1 | 4% | 75 | 9 | 100% |
| 2/3 | 50% | 65 | 7-9 | 53% |

The expected complement of chromosomes for this hybrid is 74-75 chromosomes, including 9 biarmed chromosomes (Table 2).

In fact, more detailed karyotype analysis (Table 5) has revealed that two clones of the $FwI^{d-} \times FtI^{d+}$ hybrid showing good response to DMSO have fewer chromosomes (including biarmed chromosomes) than two clones which are not inducible by DMSO. In particular, a small biarmed chromosome derived from the $FwI^{d-}$ parent seems to be lost frequently (but not always) in those clones which respond well to DMSO.

Since the $FwI^{d-}$ variant exhibited poor inducibility of haem as well as haemoglobin synthesis (Table 1), it was supposed that a defect in haem metabolism might be responsible for lack of inducibility by DMSO. When $FwI^{d-}$ cells were

Table 6
Effect of haemin on the rate of globin synthesis in Friend cell variants.

| Inducible Status | Other Markers | Treatment of cells: Nil | 1.5% DMSO | 0.1mM Haemin | Haemin + DMSO |
|---|---|---|---|---|---|
| $I^{d+}$ | FtB | 1% | 25% | 5% | - |
| $I^{d-}$ | FwT | not detectable | 2% | 3% | 7% |

Cells were treated with DMSO and/or haemin for 4d and then labelled for 18h with [$^3$H] leucine in leucine-free medium containing 15% serum. Cytoplasmic proteins were co-precipitated with erythrocyte lysate by acid/acetone and then chromatographed on CM/urea. The values shown give the percentage of the total radioactivity co-chromatographing with the β and α globin peaks.

treated with haemin plus DMSO (or to a lesser extent haemin alone) the rate of globin synthesis was increased very significantly (Table 6), to about one third of the level in DMSO-treated inducible cells. However, treatment of inducible lines with haemin did not increase the rate of globin synthesis significantly (Table 6). These results suggest that DMSO and haem act by different mechanisms. Further work (Table 7) showed that treatment of $FwI^{d-}$ cells with haemin plus DMSO also increased the cellular level of globin messenger RNA molecules very significantly. However, such treatment had only small effects on globin messenger RNA levels in inducible lines or other Friend cell variants (Table 7); haem is not able to induce cells as effectively as DMSO itself.

Table 7

Effect of DMSO and/or haemin treatment on accumulation of globin mRNA in Friend cell variants.

| Inducible Status | Other Markers | Treatment of cells: Nil | 1.5% DMSO | 0.1mM Haemin | Haemin + DMSO |
|---|---|---|---|---|---|
| $I^{d+}$ | Ft | 70±30 | 800±300 | 250±100 | 800±300 |
| | FtB | 7±3 | 450±150 | 55±30 | 700,1000 |
| $I^{d-}$ | Lyt | < 2 | < 2 | < 2 | < 4 |
| | FtB | 4±2 | 5±3 | 4,8 | 13,20 |
| | FwT | 20±10 | 35±10 | 50±15 | 300±80 |

Values (parts per million (ppm) of cytoplasmic RNA) are quoted for independent experiments, or the average± the standard deviation of the results.

Yet the FwI$^{d-}$ line seems to be defective in some function which is reversible by haem and which prevents globin messenger RNA accumulation and translation. Moreover, in the FwI$^{d-}$ x FtI$^{d+}$ hybrid clone showing a poor response to DMSO (clone 1/5), treatment with haem in addition to DMSO produces a greater production of cells containing haemoglobin (results not shown).

(e) The lymphoma variant

The lymphoma variant (LyT) accumulates little, if any, globin-specifying sequences in the cytoplasm or nucleus after DMSO treatment (Table 8).

Table 8

Globin RNA levels in the FtI$^{d+}$B and LyI$^{d-}$T cells and the corresponding hybrid

| Cell | % cells Hb+ | Molecules of globin RNA: Cytoplasm | | Nucleus | |
|---|---|---|---|---|---|
| | +DMSO | -DMSO | +DMSO | -DMSO | +DMSO |
| $I^{d+}$ FtB | 80 | 170±70 | 6000±2000 | 25,35 | 1600 |
| $I^{d-}$ LyT | Nil | < 20 | < 20 | < 5 | < 10 |
| $I^{d+}$x $I^{d-}$ | Nil | 900±200 | 8000±2500 | 160±80 | 700±200 |

See Table 3 for further details.

However, the hybrid formed by fusing the LyT line to the appropriate inducible Friend cell parent (FtI$^{d+}$B) shows a large increase in the cellular content of globin RNAs in both cytoplasm and nucleus after DMSO treatment, although very little, if any, of this messenger RNA appears to be translated into globins (Table 9). The numbers of globin RNA sequences in the cytoplasm and nucleus of this hybrid cell after DMSO treatment are consistent with only one set of globin genes being active (most probably that derived from the Friend cell parent). Alternatively, both sets of parental globin genes may be active but subject to dosage compensation. It is impossible to distinguish between these possibilities at the present time. However, there is little doubt that the lymphoma cell is trans-dominant with respect to translation of these globin RNA sequences. Significantly, as was found for the $FtI^{d+}T$ x $FtI^{d-}B$ hybrid (Table 3), the numbers of globin RNA molecules in the nucleus or cytoplasm of the untreated lymphoma x Friend cell hybrid are much higher than in either of the parental lines (Table 8).

In view of our previous findings with the $FwI^{d-}$ variant, the effect of haem on the Friend cell x lymphoma hybrid was investigated (Table 9).

Table 9

Effect of haemin and/or DMSO treatment on the lymphoma and Friend cell hybrid ($LyI^{d-}$ x $FtI^{d+}$)

| Assay | Nil | 1.5% DMSO | 0.1mM Haemin | DMSO+ Haemin |
|---|---|---|---|---|
| % globins | Not detectable | < 1%(3) | 4%, 5% | 9%, 10% |
| globin mRNA | 20±5 | 210±70 | 110±25 | 600±140 |
| nuclear globin RNA | 30 | 100 | 130 | 300 |

Globin synthesis (% total protein synthesis) and globin-specifying sequences (ppm) were determined as described in previous Tables.

As was found in the $FwI^{d-}$ variant, treatment of the Friend cell x Lymphoma hybrid with haem increased the rate of synthesis of globins and the accumulation of globin RNA sequences in both nucleus and the cytoplasm. Moreover, haem potentiated the effect of DMSO on globin RNA levels. Furthermore, treatment of this hybrid with haemin plus DMSO increased the proportion of the cells which differentiated as far as the orthochromatic stage.

## 4. Discussion

A basic element in our strategy has been the isolation of non-inducible variants by selecting directly against specific erythroid markers. Selection against the ability of inducible Friend cells to differentiate terminally by growth or cloning in DMSO is an example of this approach. In so far as other erythroid functions (for example haem and globin synthesis) remain coupled to terminal differentiation, DMSO-resistant variants may also be non-inducible for these functions. One such stable variant, $FtI^{d-}B$, has been characterised in some detail in the present studies. Induction of globin RNA accumulation in this variant seems to be suppressed, either due to a transcriptional defect in the chromatin or to a fault in processing the initial RNA transcript which renders the globin RNA sequences unstable. Much further work will be required to distinguish these alternatives, for example by a direct investigation of chromatin transcription in these cells. The genetic studies strongly suggest that the mechanism of this loss of inducibility in the $FtI^{d-}B$ variant involves anaction of a diffusible, trans-dominant suppressor. However, it is impossible at present to exclude the alternative but unlikely interpretation that the hybrid clone studied in these genetic experiments represents a hybrid between an $FtI^{d-}$ cell and a rare non-inducible cell in the $FtI^{d+}T$ parental population.

A comparison of this $FtI^{d-}$ non-inducible variant with the haemopoetic, but non-erythroid, lymphoma line (LyT) poses important questions concerning the manner in which globin genes can be regulated in different cell types. The lymphoma line contains few, if any, globin RNA sequences in the nucleus or cytoplasm, although recent work in this Institute suggests that the globin genes are not entirely inactive in this cell[22]. The genetic studies do not distinguish whether control of transcription in the lymphoma cell is cis-dominant or alternatively recessive with dosage compensation. Nevertheless, the lymphoma genome seems able to prevent DMSO-induced accumulation of globins in the Friend cell x lymphoma hybrid, unless the hybrid is also treated with haem. Under the latter conditions, globin messenger RNAs accumulate in both nucleus and cytoplasm, and are translated into globins. Erythroid maturation to the orthochromatic stage is also enhanced by haem plus DMSO, but not by DMSO alone. This is in contrast to the situation in inducible Friend cells or the $FtI^{d-}$ variant, where haem has only a small or no effect (see also reference 23).

There could be many explanations for this phenomenon. Assuming that globin chains are not rapidly degraded in the lymphoma x Friend cell hybrid in the absence of haem, the results are most readily interpreted in terms of a trans-dominant block of translation of DMSO-induced globin messenger RNAs in these hybrid cells, which is reversed by treatment with haem. By increasing globin messenger RNA translation, haem could render it more stable by incorporation into polysomes. However this situation would not entirely be analagous to that in the reticulocyte lysate where treatment with haemin enhances the synthesis of all proteins rather than of globins preferentially[24]. Alternatively haemin may alter the rate of transcription of the globin gene directly or indirectly, or stabilise globin RNA sequences non-specifically, for example by inhibiting nucleases. Further work is in progress to clarify the situation.

The third non-inducible variant, $FwI^{d-}T$, seems to be non-inducible due to some defect which can be partly relieved by haem in a similar manner to that described above in connection with the lymphoma x Friend cells hybrid. The genetic studies seem to show that non-inducibility is dominant unless certain chromosomes are lost, in which case both parental sets of globin genes seem to be active as judged from the numbers of globin messenger RNAs produced. This suggests that a negative regulator gene is involved in the process, although its chromosomal location has not yet been ascertained.

These results are not yet sufficiently complete to permit making all but the most general conclusions. However, they do show (1) that mutants can be obtained with which to analyse the regulation of differentiation in Friend cells; (2) that these mutants reveal different molecular levels of control and have different dominance relationships implying different genetic mechanisms of action; (3) that the induced and basal states of globin gene expression may be regulated by independent mechanisms; (4) that complex feedback interactions exist between independent differentiated pathways (i.e. the globin and haem pathways and terminal differentiation) and (5) nevertheless, these co-ordinated markers can be uncoupled under certain circumstances. These general conclusions are summarised in schematic form in Figure 2.

Fig. 2. Genetic dissection of Friend cell differentiation.

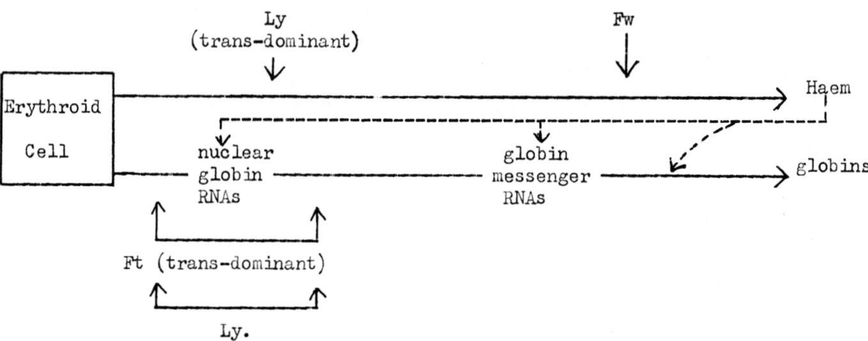


This figure summarises the genetic and molecular evidence discussed in the text. The solid arrows give the probable sites of defects in induction of differentiation in the three non-inducible variants, Fw, Ft and Ly. The dotted arrows show a scheme to explain how the haem pathway interacts with the globin pathway.

## 5. Acknowledgment

Mr. Alastair MacNab contributed excellent technical assistance in these studies, which were funded by grants from The Medical Research Council and the Cancer Research Campaign.

## 6. References

1. Friend, C., 1957, J. exp. Med. 105, 307.
2. Tambourin, P.E. and Wendling, F., 1975, Nature 256, 320.
3. Preisler, H.D. and Giladi, M., 1974, Nature 251, 645.
4. Friend, C., Scher, W., Holland, J. G. and Sato, T., 1971, Proc. Nat. Acad. Sci. U.S.A., 68, 378.
5. Ostertag, W., Melderis, H., Steinheider, G., Kluge, W. and Dube, S., 1972, Nature New Biol., 239, 231.
6. Boyer, S.H., Wuu, K.D., Noyes, A.W., Young, R., Scher, W., Friend, C., Preisler, H.D. and Bank, A., 1972, Blood, 40, 823.

7. Gilmour, R.S., Harrison, P.R., Windass, J., Affara, N. and Paul, J., 1974, Cell Differentiation, 3, 9.

8. Harrison, P.R., Gilmour, R.S., Affara, N., Conkie, D. and Paul, J., 1974, Cell Differentiation, 3, 23.

9. Fijinami, W., Sugimoto, Y. and Hagiwara, A., 1973, Development Growth and Differentiation, 15, 141.

10. Kabat, D,, Sherton, C.C., Evans, L.H., Bigley, R. and Koler, R.D., 1975, Cell, 5, 331.

11. Paul, J. and Hickey, I., 1974, Exp. Cell Res., 87, 20.

12. Harrison, P.R., Conkie, D., Affara, N. and Paul, J., 1974, J. Cell Biol., 63, 402.

13. Conkie, D., Affara, N., Harrison, P.R., Paul, J. and Jones, K., 1974, J. Cell Biol., 63, 414.

14. Getz, M.J., Birnie, G.D., Young, B.D., MacPhail, E. and Paul, J., 1975, Cell, 4, 121.

15. Glisin, V., Crkvenjakov, and Byus, C., 1974, Biochemistry, 13, 2633.

16. Harrison, P.R., Birnie, G.D., Hell, A., Humphries, S., Young, B.D. and Paul, J. 1974, J. Mol. Biol., 84, 539.

17. Young, B.D., Harrison, P.R., Gilmour, R.S., Birnie, G.D., Hell, A., Humphries, S., and Paul, J., 1974, J. Mol. Biol., 84, 555.

18. Williamson, R., Marrison, M., Lanyon, W.G., Eason, R. and Paul, J., 1971, Biochemistry, 10, 3014.

19. Littlefield, J.W., 1964, Science, 145, 709.

20. Harris, H. and Watkins, J.F., 1965, Nature, 205, 640.

21. Paul, J., Harrison, P.R., Conkie, D. and Affara, N., 1975, Proc. ICN - UCLA Symp., Squaw Valley, U.S.A., in press.

22. Humphries, S., Windass, J. and Williamson, R., 1975, Cell, in press.

23. Cimadevilla, J.M. and Hardesty, B., 1975, Biochem. Biophys. Res. Commun., 63, 931.

24. Mathews, M.B., Hunt, T., and Brayley, A., 1973, Nature New Biol., 243, 230.

*Progress in Differentiation Research, ed. N. Müller-Bérat et al.*
*© 1976, North-Holland Publishing Company - Amsterdam, The Netherlands.*



# GENETIC DIFFERENCE BETWEEN GERM LINE AND SOMATIC DNA IN ASCARIS LUMBRICOIDES

Heinz Tobler
Institute of Zoology, University of Freiburg,
1700 Freiburg, Switzerland

## 1. Introduction

Cell biologists generally agree on the notion that development and differentiation of a multicellular organism is based on a constant genome of the different cell types rather than on changes in the numbers or kinds of genes. This so-called DNA constancy rule and the concept of variable gene activity are supported by a variety of experimental results and observations[1-3]. However, there are important exceptions to the DNA constancy rule. Gene amplification[4], gene magnification[5], and gene underreplication[6] lead to quantitative differences between the DNA content of various cell types. Other violations of this doctrine are represented by the complete loss of the nuclear genetic material as observed in certain terminally differentiated cells such as mammalian erythrocytes, eye lens fiber cells or sieve elements in plants, as well as the partial expulsion of chromatin from the presumptive somatic cells that occurs during early cleavage stages of some nematodes, insects, and crustaceans[7]. The present communication is directed at elucidating the informational content encoded in the nuclear DNA before and after chromatin elimination in Ascaris lumbricoides.

## 2. The process of chromatin elimination in Ascarids

In 1887, Boveri[8] discovered that chromatin diminution occurs in early cleavages of Ascaris megalocephala. During the second cell division, the chromosomes in the ventral $P_1$ cell are normally distributed to the two daughter cells (see Fig. 1). This is in contrast to the situation in the dorsal $S_1$ cell, where the central region of the chromosomes break up into a large number of small chromosomes. Only these diminuted chromosomes are distributed to the two daughter nuclei, the distal ends remain in the cytoplasm where they eventually degenerate. Thus the two nuclei of the $S_{1a}$ and $S_{1b}$ cell contain less DNA than the nuclei of the cells $S_2$ and $P_2$. During the following cleavage from the 4-to the 8-cell stage, the same process takes place in the $S_2$ cell. Boveri showed by cell lineage analysis that all cells with the reduced amount of chromatin become somatic cells, whereas nuclei with the original integrity and full quantity of DNA give rise to germ line cells (see Fig.2). Basically the same process occurs also in Ascaris lumbricoides[9], although it differs somewhat in detail from that described for Ascaris megalocephala. Since it has become increasingly difficult to get enough Ascaris megalocephala worms for experimental investigations, all experiments that are described have been performed with the pig intestinal parasite Ascaris lumbricoides (A. suum).

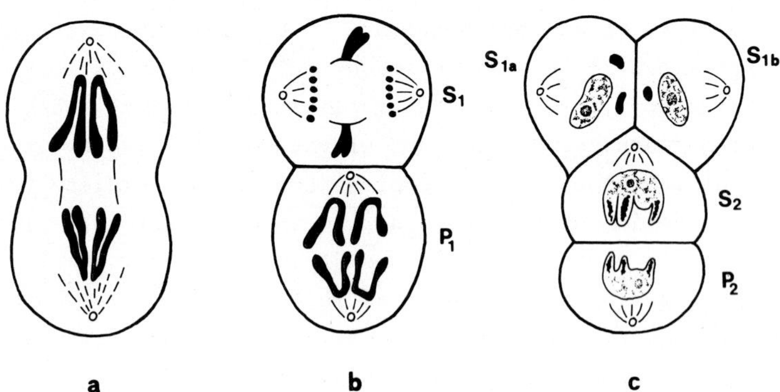


Fig.1. Chromatin elimination in Ascaris megalocephala. a) First cleavage division, b) second cleavage division. Chromatin diminution occurs in the top $S_1$ cell but not in the lower $P_1$ cell. c) 4-cell stage after completion of the 2nd cell division. The cells $S_{1a}$, $S_{1b}$ and $S_2$ give rise to all somatic cells whereas the $P_2$ cell yields the germ line cells. (After Boveri[10]).

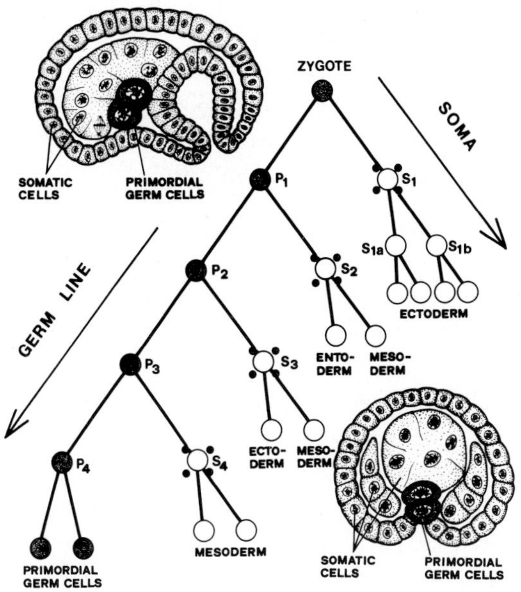


Fig.2. Cell lineage in Ascaris megalocephala. Chromatin elimination (:O:) takes place in the presumptive somatic cells from the 2nd to the 5th cleavage division. These cells eventually form in a precisely determined manner the various parts of the adult body, whereas the primordial germ line cells with the full complement of DNA give rise to the germ cells. (After Boveri[14]).

## 3. Genome size and amount of eliminated DNA in Ascaris lumbricoides

The genome size before and after elimination has been determined by isotope dilution and gave values of $0.63 \times 10^{-12}$g DNA per haploid germ line genome and $0.46 \times 10^{-12}$g DNA per haploid somatic genome[11]. Thus, chromatin elimination leads to a reduction of 27% of the organism's DNA. Although our estimations of the genome sizes are slightly larger than the values reported by two other laboratories[12,13], the amount of 22% eliminated DNA determined by Moritz[13] using different methods is in good agreement with our own measurements.

## 4. Characterization of the genome before and after elimination

DNA was prepared from spermatids and 4-cell stages and used as germ line DNA. In contrast to Ascaris megalocephala, chromatin elimination in Ascaris lumbricoides does not take place before the third cleavage division[9]. DNA extracted from 12-day-old larvae will be referred to as retained or somatic DNA.

In the first series of experiments, germ line and somatic DNA were analyzed by analytical CsCl density gradient centrifugations. Both DNA samples showed a major peak with a buoyant density of 1.697 g/cm$^3$. Since germ line DNA prepared from spermatids and somatic larval DNA were indistinguishable in their CsCl density profiles, we concluded that the eliminated DNA does not differ in its buoyant density from the somatic DNA. In contrast, egg DNA[15] as well as 4-cell stage DNA[11] are characterized by an additional light satellite banding at a buoyant density of 1.686 g/cm$^3$. Electron microscopy revealed that this DNA comprises supercoiled, covalently closed as well as some double-forked circular molecules with an average contour length of 4.64 µm[16]. Such molecules have been described in a large number of metazoan animals and shown to be of mitochondrial origin (see Borst[17], for review). On the average, mitochondrial DNA amounts to about 40% of the total cellular DNA of Ascaris lumbricoides 4-cell stages.

Figure 3 summarizes the data of renaturation experiments with larval, spermatid and 4-cell stage DNA. About 10% of the retained somatic DNA is repetitious with an average family size of 5500-7000 copies, the rest appearing to consist of unique sequences. On the other hand, germ line DNA from spermatids contains 23% fast-renaturing DNA with a family size in the order of 7000-10,000 copies. Since 27% of germ line DNA is expelled from prospective somatic nuclei during chromatin elimination, the eliminated DNA must consist of repetitious and unique sequences in a ratio of about 1:1. In 4-cell stages, the fraction of repetitious DNA sequences is increased by about 25% as compared to spermatid DNA. As has been discussed before, DNA extracted from whole 4-cell stages contains a sizable fraction of mitochondrial DNA. Due to the relatively minor complexity of its sequences, metazoan mitochondrial DNA renatures very rapidly. Therefore, it seems reasonable to conclude that the

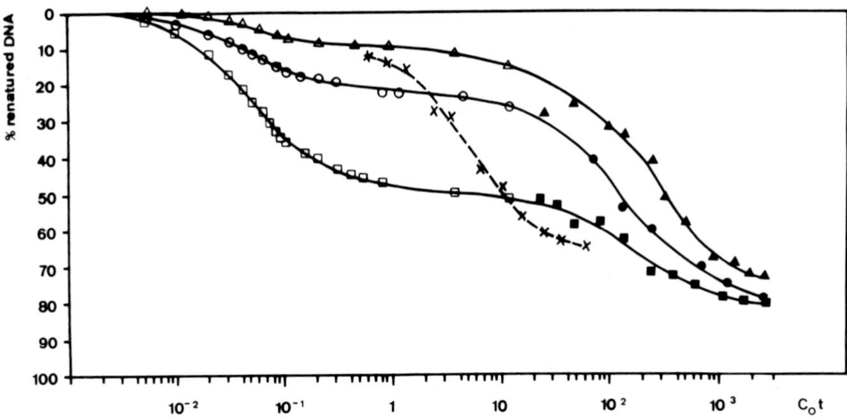


Fig.3. Kinetics of reassociation of DNA from Ascaris lumbricoides. All DNAs were sheared, heat denatured, incubated at 60°C in 0.12 M PB, 0.001 M EDTA, and their rate of renaturation determined. $C_ot$ points up to a $C_ot$ of 10 were determined in a spectrophotometer (open symbols). The DNAs were reassociated at concentrations of approximately 50 µg/ml. $C_ot$ points above 10 were determined by hydroxyapatite binding (filled symbols). The values were corrected for partially reannealed sequences according to Britten and Kohne[18]. Larval DNA (▲—▲) was reassociated at 400 µg/ml, spermatid DNA (●—●) was reassociated at 200 µg/ml, and 4-cell stage DNA (■—■) was reassociated at 165 µg/ml. The dashed line represents the renaturation of $^{32}P$-labeled Escherichia coli DNA at a concentration of approximately 10 µg/ml. (From Tobler et al.[11]).

difference in the amount of fast renaturing DNA sequences between 4-cell stage and spermatid DNA resides in the high content of mitochondrial DNA of 4-cell stages.

It has been proposed that chromatin elimination serves the purpose of discarding large amounts of rRNA genes that might have been selectively amplified during oogenesis[19]. In order to test this hypothesis, saturation hybridization experiments of 18S and 28S rRNA with spermatid and larval DNA were carried out[20]. The results clearly show that the eliminated DNA is not enriched for 18S and 28S rRNA genes, but contains rDNA in the same proportion as germ line and somatic DNA.

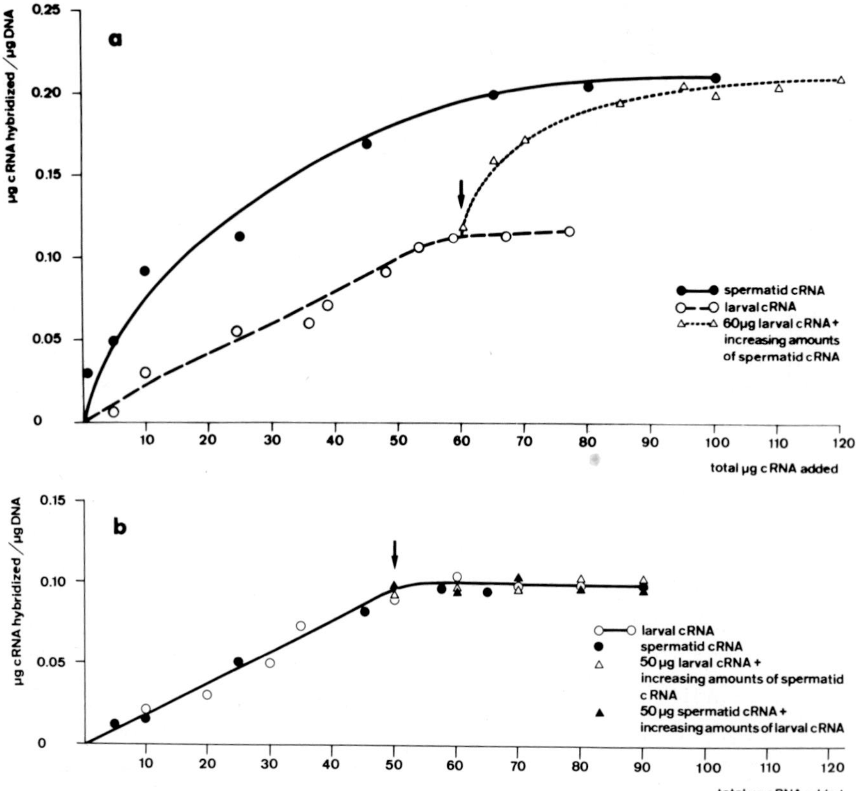


Fig.4. (a) Saturation of Ascaris spermatid DNA with increasing amounts of larval (o---o) and spermatid (●——●) cRNA. After the saturation value for larval cRNA was determined (60 µg/larval cRNA), increasing amounts of spermatid cRNA (△·····△) were hybridized in the presence of saturating amounts of larval cRNA. (b) Saturation of Ascaris larval DNA with increasing amounts of larval (o——o) and spermatid (● ) cRNA. In additional experiments, saturating amounts of larval cRNA in the presence of increasing amounts of spermatid cRNA ( △ ) or saturating amounts of spermatid cRNA with increasing amounts of larval cRNA ( ▲ ) were hybridized to larval DNA. (From Tobler et al.[11]).

The most important question, however, is whether the eliminated DNA differs in a qualitative or merely in a quantitative way from the retained DNA. In order to tackle this problem, sequential hybridization experiments were performed (Fig.4). Since radioactively labeled DNA is not easily available from Ascaris, we utilized complementary RNA (cRNA) derived from spermatid and larval DNA in our hybridization experiments. Both cRNAs were transcribed in vitro from the appropriate DNA templates using RNA polymerase from E. coli. The two cRNAs were hybridized in increasing amounts to either spermatid or larval DNA. Fig. 4a shows that at saturation, larval cRNA anneals with about 11% of the spermatid DNA, and spermatid cRNA with about 21%. If saturating amounts of larval cRNA are hybridized together with increasing amounts of spermatid cRNA, the extent of hybridization rises again until the same saturation level as with spermatid cRNA alone is reached. This result clearly indicates that germ line DNA does contain sequences which are no longer present in the larval DNA. However, for technical reasons, only the repetitious sequences of the genome have been analyzed in these experiments. Therefore, at the present time we can merely state that the eliminated repetitious sequences are distinct from the repetitious sequences retained in somatic cells.

In control experiments, larval or spermatid cRNA was annealed to larval DNA (Fig. 4b). As was anticipated, the saturation plateau is the same for both cRNAs. Furthermore, the saturation level is not changed by the addition of increasing amounts of spermatid cRNA to saturating amounts of larval cRNA or vice versa. This result is to be expected since all somatic DNA sequences have to be contained in the genome of germ line cells.

## 5. Summary

The process of chromatin elimination which takes place during early cleavage stages of some nematodes, insects, and crustaceans leads to presumptive germ line cells containing more DNA than the presumptive somatic cells. In the nematode worm Ascaris lumbricoides, about 73% of the total germ line DNA is retained in somatic cells. The eliminated DNA does not differ from the retained DNA in base composition, but is enriched for repetitious DNA sequences. Germ line and somatic DNA contain the same percentage of genes coding for 18S and 28S rRNA. This excludes the possibility that chromatin elimination serves the purpose of discarding large amounts of ribosomal cistrons. The repetitious DNA sequences of germ line cells are qualitatively different from the repetitious DNA sequences retained in somatic cells.

Acknowledgements
This research was supported by the Swiss National Science Foundation, grants no. 3.701.72 and 3.119.73 and by the SANDOZ-Stiftung zur Förderung der medizinisch-biologischen Wissenschaften. I am very thankful to my co-workers in our laboratory for the critical comments on the manuscript and to Mr. P. Geinoz for drawing the illustrations.

References

1. Davidson, E.H., 1968, Gene activity in early development (Academic Press, New York).
2. Tobler, H., 1972, in: Nucleic acid hybridization in the study of cell differentiation, ed. H. Ursprung (Springer-Verlag, Berlin and New York), p.1.
3. Gurdon, J.B., 1974, The control of gene expression in animal development (Clarendon Press, Oxford).
4. Tobler, H., 1975, in: The Biochemistry of animal development, vol. 3, ed. R. Weber (Academic Press, New York), p.91.
5. Ritossa, F.M., 1968, Proc.nat.acad.sci. U.S. 60, 509.
6. Gall, J.G., Cohen, E.H., and Polan, M.L., 1971, Chromosoma 33, 319.
7. Beams, H.W. and Kessel, R.G., 1974, in: International review of cytology, vol. 39, eds. G.H. Bourne, J.F. Danielli and K.W. Jeon (Academic Press, New York), p.413.
8. Boveri, T., 1887, Anat. Anz. 2. 688.
9. Meyer, O., 1895, Z. Naturwiss. (Jena) 29, 391.
10. Boveri, T., 1899, Festschr. Kupffer (Fischer, Jena).
11. Tobler, H., Smith, K.D. and Ursprung, H., 1972, Develop. Biol. 27, 190.
12. Searcy, D.G. and MacInnis, A.J., 1970, Evolution 24, 796.
13. Moritz, K.B., 1975, pers. communication.
14. Boveri, T., 1910, Festschr. R. Hertwig, Bd. 3 (Fischer, Jena).
15. Bielka, H., Schultz, I. and Böttger, M., 1968, Biochim. Biophys. Acta 157, 209.
16. Tobler, H. and Gut, C., 1974, J. Cell Sci. 16, 593.
17. Borst, P., 1972, Ann. Rev. Biochem. 41, 333.
18. Britten, R.J. and Kohne, D.E., 1966, in: Carnegie Inst. Wash. Yearbook 65, 78.
19. Wallace, H., Morray, J. and Langridge, W.H., 1971, Nature New Biology 230, 201.
20. Tobler, H., Zulauf, E. and Kuhn, O., 1974, Develop. Biol. 41, 218.

Progress in Differentiation Research, ed. N. Müller-Bérat et al.



# MICRONUCLEATION, ENUCLEATION AND CELL RECONSTRUCTION EXPERIMENTS

T. Ege, P. Elias, N.R. Ringertz, and E. Sidebottom
Institute for Medical Cell Research and Genetics,
Medical Nobel Institute, Karolinska Institutet
S-104 01 Stockholm 60, Sweden

## Introduction

Recent technical developments in Cell Biology have made it possible to obtain large numbers of anucleate cells (also known as cytoplasts) and minicells, nuclei surrounded by a thin rim of cytoplasm and an intact plasma membrane (karyoplasts). Both types of sub-cellular fragment are viable for periods varying between a few hours and a few days and both can be successfully employed in cell fusion experiments to produce reconstituted cells. Similar methods of enucleation and fusion have also been applied to "micronucleated" cells, cells in which the genome has been fragmented into many small micronuclei by prolonged exposure to anti-mitotic agents such as colchicine.

These new methods of cell reconstruction promise to complement the existing techniques of cell hybridization and nuclear transplantation and provide further ways of investigating problems of current interest and importance such as the role of the cytoplasm in the control of nuclear activity, mechanisms of regulation of gene expression and the control of determination and differentiation. In addition these methods should assist studies of the interactions between viruses and host cells and between mitochondria and nuclear genes.

## Anucleate cells and Minicells

Several methods of enucleating large numbers of cells with a high efficiency have now been described (2, 3, 4, 5). These methods all employ centrifugation in the presence of cytochalasin B. (1). Enucleation is usually achieved with cells adhering to glass or plastic surfaces (2,3,4) but enucleation of cells in suspension has now been described (5). We most commonly use discs of 25 mm diameter and centrifuge, with the cells inverted, in PBS containing 10 μg/ml cytochalasin B and 10% calf serum. Different types of cells vary in the conditions required for optimal enucleation. Centrifugal forces of between 3 000 - 48 000 g for periods of 20′- 60′ have been used. In favourable circumstances > 98% enucleation is achieved (6, 7).

The anucleate cells remain attached to the culture surface and although they appear grossly abnormal immediately after centrifugation they quickly recover so that they soon resemble the parent cell morphologically (Fig. 1). They are motile and initially can synthesize protein at rates similar to the intact parent cells, but these activities steadily diminish and most cells round up and die within 72 h. The dry mass of anucleate cells varies from about 35% of that of the whole cell in some cell lines to nearly 60% in others.

The nucleated fragments of the cells (minicells) are recovered from the bottoms of the centrifuge tubes. Minicells from different cell lines contain between 10% and 25% of the original cytoplasm of the parent cell. Most of them exclude trypan blue and synthesize RNA; some even replicate their DNA. They do not, however, spread on surfaces or regenerate their lost cytoplasm and almost all lyse within 48 h (7).

## Micronucleation and Microcells

It has been known for some time that when cells are exposed for long periods to antimitotic agents such as colchicine, metaphase is grossly deranged and many cells undergo micronucleation and polyploidization (14), (Fig. 2). The mechanism of micronucleation is not yet clear but cytogenetic, cytochemical and time lapse cinemicrographic studies suggest that in the absence of a normal mitotic spindle metaphase chromosomes are distributed at random in the cell, and after a prolonged mitosis, nuclear membranes reassemble around groups or even individual chromosomes to give rise to micronuclei. It appears that the number of micronuclei formed in a cell can approach, though not exceed, the number of chromosomes in the cell. It is not clear whether micronuclei can be formed from less than a complete chromosome, although evidence from other experimental systems, such as those using X irradiation, suggests that they can.

If micronucleated cells are centrifuged in the presence of cytochalasin B the micronuclei, surrounded by a rim of cytoplasm within a plasma membrane, are drawn out from the cell to give rise to microcells (12). As with minicells most of the microcells are viable in the sense that they persist in culture for a short time, and exclude trypan blue (8).

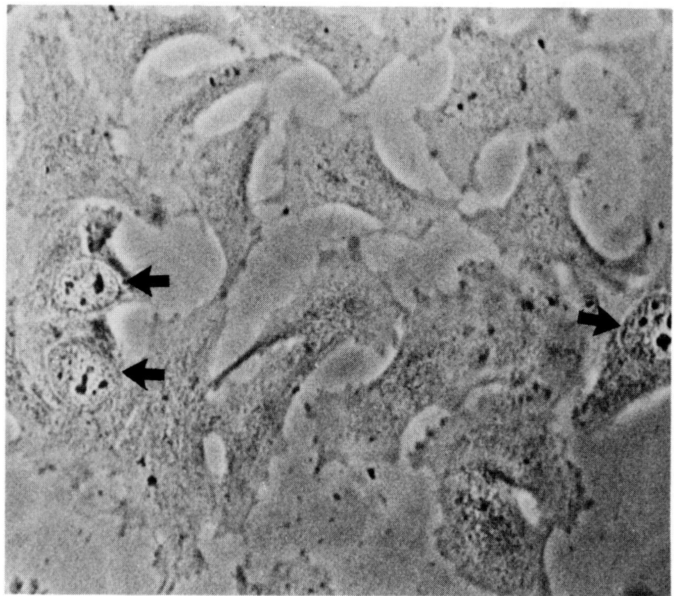

Fig. 1. Phase contrast microphotograph of enucleated HeLa cells together with a few nucleated cells (arrows).

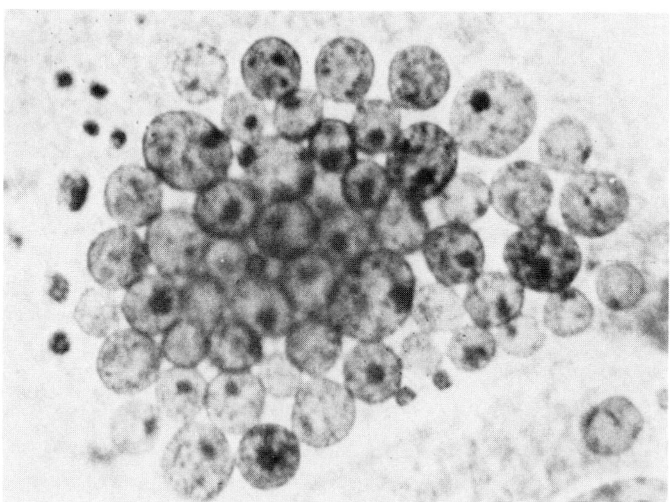

Fig. 2. Giemsa stained preparation showing a micronucleated hamster cell (C15S) containing many micronuclei. The majority of the micronuclei contain nucleoli. The cells were treated with 1 μg/ml colcemid for 48 h.

Cell Reconstruction

Three different types of reconstructed cells have been prepared, using Sendai virus to induce fusion.

1. Reconstituted cells
   a) anucleate cells x minicells
   b) anucleate cells x nucleated erythrocyte ghosts
2. Cybrids. Anucleate cells x intact cells
3. Microcell heterokaryons. Microcells x intact cells

One of the chief problems of these experiments lies in the recognition of, and distinction between, the intact parental cells, the reconstituted cells and cybrids. Two schemes which assist identification are illustrated schematically in figs. 3 & 4. They employ autoradiography after isotope incorporation, alone (8) or in combination with the use of engulfed plastic beads as markers (9, 10).

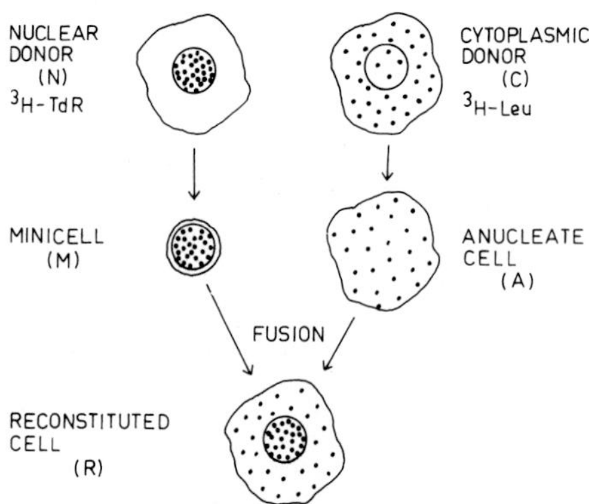


Fig. 3. Identification of reconstituted cells by autoradiography. The nuclear donor cell is heavily labelled with $^3$H-thymidine. In autoradiograms the nucleus will be covered by many silver grains whereas the cytoplasm remains unlabelled. The cytoplasmic donor is exposed to $^3$H-leucine at a dose which causes a light labelling of both nucleus and cytoplasm. Reconstituted cells formed by the fusion of minicells with anucleate cells can then be identified as cells having a strong nuclear labelling ($^3$H-TdR) and a weak cytoplasmic label (see reference 8).

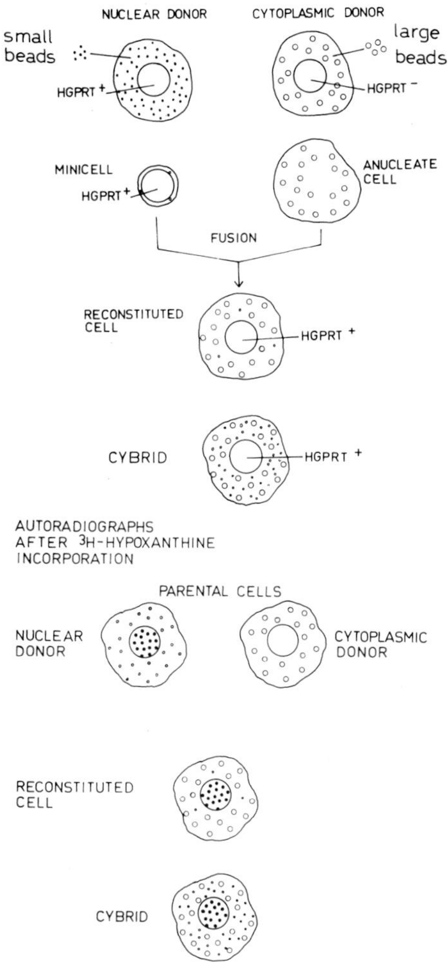


Fig. 4. Identification of reconstituted cells and cybrids with the aid of engulfed plastic beads and autoradiography. The cytoplasmic donor is labelled in the cytoplasm with 30-70 large plastic beads (1,0 μ). This cell is also deficient in hypoxanthine guanine phosphoribosyl transferase because of a nuclear gene mutation ($HGPRT^-$) and therefore unable to incorporate exogenous $^3$H-hypoxanthine. The nuclear donor cell is labelled in the cytoplasm with 150 - 300 small plastic beads (0.4 μ). This cell is $HGPRT^+$ and can therefore use exogenous $^3$H-hypoxanthine. After fusion of minicells and anucleate cells the preparations are exposed to $^3$H-hypoxanthine and subjected to autoradiography. Reconstituted cells, cybrids and parental cells can then be distinguished from each other.

1. Reconstituted cells. Anucleate cells x minicells.

Reconstructions of this type have been achieved with the two cell components taken from the same, or from different, cell lines. Experiments in which chick erythrocyte ghosts are fused with anucleate cells may be considered as a special case of this type of reconstruction (Fig. 5).

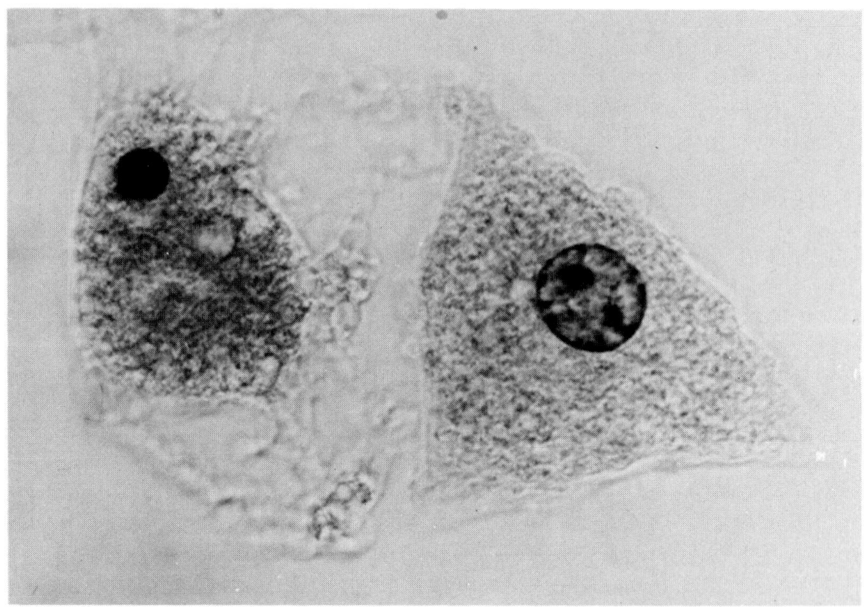

Fig. 5. UV microphotograph (260 nm)showing two reconstituted cells 73 h after fusion. The left cell is beginning to round up and has a small condensed nucleus which appears to have undergone pycnosis. The cell to the right remains well attached and has an enlarged nucleus containing nucleolus-like bodies and a dispersed chromatin. The nucleus is derived from a chick erythrocyte and the cytoplasm from a mouse L cell.

The different cell combinations that have been successfully used in reconstitution experiments to date are given below:

| Minicell | | Anucleate cell |
|---|---|---|
| rat L6 myoblast | x | rat L6 myoblast |
| rat L6 myoblast | x | rat L6TG$^R$ myoblast*) |
| mouse B82 cell **) | x | rat L6TG$^R$ myoblast*) |
| chick erythrocyte | x | mouse L cell |
| chick erythrocyte | x | chick fibroblast |
| human lymphocyte | x | mouse B82**) |

*) rat L6TG$^R$ myoblast is deficient in hypoxanthine guanine phosphoribosyl transferase (HGPRT$^-$) and resistant to thioguanine (TG$^R$)

**) mouse B82 cell is deficient in thymidine kinase (TK$^-$) and resistant to bromodeoxyuridine (BUdR$^R$)

The properties of the reconstituted cells can be summarized in the following statements:

a) Normal subcellular architecture is established around the incoming nucleus. Electron microscopic studies of cells reconstituted from enucleated rat myoblasts and chick erythrocytes show that the incoming nucleus is well integrated into the rat cytoplasms. The nuclear membrane and pores have a normal appearance and the Golgi complexes of the anucleate cell reorient into a normal spatial relationship with the nucleus (13).
b) Antigens migrate from the anucleate cell into the incoming nucleus (6)
c) RNA is synthesized and transported out of the nucleus (10)
d) The incoming nucleus responds to regulatory signals from the cytoplasm. Chick erythrocyte nuclei show many of the morphological characteristics of reactivation including the appearance of nucleoli (6, 11)
e) Reconstituted cells survive longer than do the unfused cell fragments. Cells reconstituted from chick erythrocytes and enucleated chick fibroblasts occasionally survive to 120 h, whereas all free anucleate cells in the same preparations have disappeared by 72 h
f) Observations of reconstituted cells in mitosis and the occurrence of pairs of cells in sparse cultures suggest that at least a small proportion of these cells divide (10).

2. Cybrids.

These cells are formed either deliberately by fusion of anucleate with intact cells or as "contaminants" in minicell x anucleate cell reconstitutions where minicell preparations contain a small number of intact nuclear donor cells. Our interest has not to date focussed on these cells.

### 3. Microcell Heterokaryons

Microcell heterokaryons can be clearly identified when microcells prepared from $^{3}$H-thymidine labelled cells are fused with unlabelled cells (8). It has been shown that the micronuclei in such heterokaryons synthesize RNA. We are at present attempting to isolate mononucleate hybrid cells in selective systems in which the microcells are necessary to provide genes to complement genetic defects in the host cells.

The various methods of cell reconstruction (including cell fusion with intact cells) are summarized schematically in fig. 6, and the main applications of the new technique are listed in the table below:

MAIN APPLICATIONS OF CELL RECONSTRUCTION TECHNIQUES

A. Minicells x anucleate cells
   1. Nucleo-cytoplasmic interactions
   2. Regulation of gene expression.(Stability of epigenotype)
   3. Dependence of mitochondria on nuclear genes

B. Microcells x intact cells
   1. Chromosome mapping (including virus integration sites)
   2. Gene complementation analysis

It should be noted that the cell reconstitution experiments have some advantages over conventional cell fusion experiments because the interactions of a single nucleus in a foreign cytoplasm can be studied and the analysis is therefore less complex than the usual situation in heterokaryons where two nuclei and two cytoplasms are mixed.

Hybrid cells made with microcells also have an advantage over conventionally produced hybrids in that only a small part of the genome of one cell type (down to one chromosome) is introduced into the second cell type so that it is not necessary to wait for chromosomal segregation to occur before the cells can be used for gene mapping.

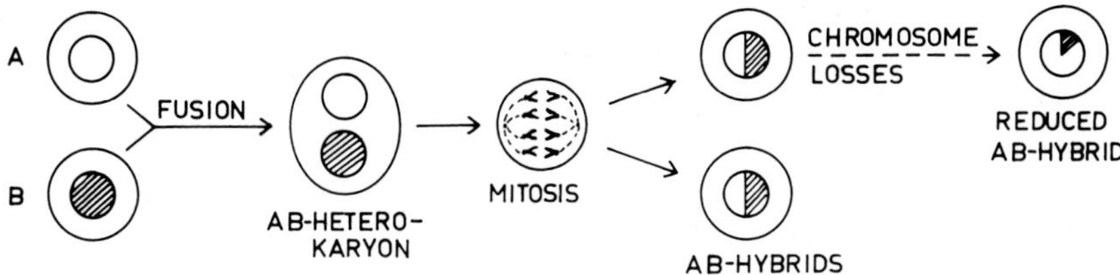


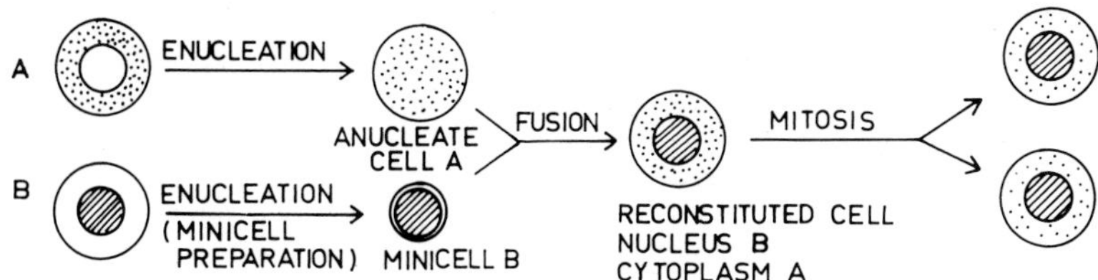


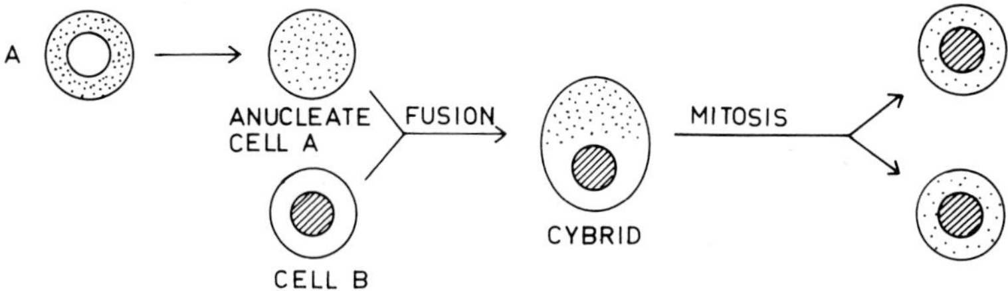


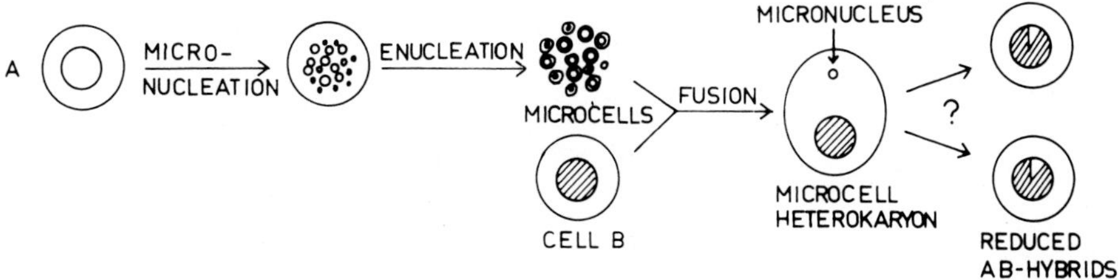


Fig. 6. Schematic summary of cell genetic experiments performed by virus induced fusion of cells and cell fragments.

References:

1. Carter, S.B., 1967, Nature 213, 261.
2. Prescott, D.M., Myerson D., Wallace, J., 1972, Exptl. Cell Res. 71, 480.
3. Wright, W.E., Hayflick, L., 1972, Exptl. Ce.. Res. 74, 187.
4. Poste, G., 1973, Methods in Cell Biology VII, ed. D. Prescott, Academic Press, p. 273.
5. Wigler, M.H., Weinstein, I.B., 1975, Biochem. Biophys. Res. Comm. 63, 669.
6. Ege, T., Zeuthen, J., Ringertz, N.R., 1973, Nobel Symp. 23, on "Chromosome identification", eds. T. Caspersson, L. Zech, Academic Press, p. 189.
7. Ege, T., Hamberg, H., Krondahl, U., Ericsson, J., Ringertz, N.R., 1974, Exptl. Cell Res. 87, 365.
8. Ege, T., Krondahl, U., Ringertz, N.R., 1974, Exptl. Cell Res. 88, 428.
9. Veomett, G., Prescott, D.M., Shay, J., Porter, K.R., 1974, Proc. Natl. Acad. Sci., U.S., 71, 1999.
10. Ege, T., Ringertz, N.R., 1975, Exptl.Cell Res. 94, in press.
11. Ege, T., Zeuthen, J., Ringertz, N.R., 1975, Somatic Cell Genetics, 1, 65.
12. Ege, T., Ringertz, N.R., 1974, Exptl. Cell Res. 87, 378.
13. Dupuy-Coin, A.-M., Ege, T., Bouteille, M., Ringertz, N.R., 1975, Exptl. Cell Res., in press.
14. Stubblefield E., 1964, Cytogenetics of cells in culture, p. 223, ed. R.J.C. Harris, Academic Press.

*Progress in Differentiation Research, ed. N. Müller-Bérat et al.*



# PROPERTIES OF HUMAN MYELOMA x NON-LYMPHOID CELL HETEROKARYONS AND PROLIFERATING HYBRIDS

Jesper Zeuthen
Institute of Human Genetics
The Bartholin Building
University of Aarhus
DK-8000 Aarhus C, Denmark

## 1. Introduction

Cells can be induced to undergo cytoplasmic fusion by which cells containing several nuclei are formed. The first product of cell fusion is a bi- or multinucleate cell called a _homokaryon_ if the parental cells are identical, or a _heterokaryon_ if the parental cells are different. Nuclear fusion in a heterokaryon results in a cell with a single nucleus containing both the parental genomes in a single nucleus, a _synkaryon_ or _proliferating hybrid cell_. The conversion of a heterokaryon into a hybrid cell requires synchronization of the cell cycles and fusion of the nuclei by processes which are still not completely understood but may involve simultaneous mitosis and the formation of a single spindle rather than the fusion of interphase nuclei. For every heterokaryon formed only a small fraction (rarely more than one in 10.000) succeeds in forming a common nucleus and a growing synkaryotic hybrid cell, so it is evident that a strong selective pressure is acting at this stage. The isolation of hybrid cells is only feasible if they grow more vigorously than the parental cells or if selective conditions can be applied.

Studies on heterokaryons have the advantage that (1) all the parental chromosomes may be assumed to be present, (2) the exact ratio of parental genomes in the heterokaryon can easily be determined, and (3) the relatively high frequency at which heterokaryons are formed makes it possible to assay various characteristics immediately after fusion. The regulatory mechanisms which were present in the parental cells at the time of fusion must be expected to be present in heterokaryons. Each of the two different nuclei in the fused cell must react with these stimuli and this interaction must result in the _activation_, _retention_ or _repression_ of the properties of interest. Disadvantages in studies of heterokaryons are that (1) the multinucleated state persists only for a limited span of time which might be too short for interactions to be expressed, and (2) it is possible that some macromolecular "activators" or "repressors" are not able to pass nuclear membrane boundaries. For these reasons studies on synkaryotic hybrid cells are a valuable complement to studies on heterokaryons.

Synkaryotic hybrid cells have the advantages that (1) chromosome loss can permit the assignment of genes to chromosomes, (2) synkaryons form colonies of cells which can be grown to high densities to allow biochemical analyses, (3) these rare synkaryotic clones can be detected by their outgrowth in selective media, and (4) the two parental genomes can interact freely within the same nucleus.

We have investigated the fate of immunoglobulin (IgE) synthesis after fusion of a line of human myeloma cells, 266Bl that synthesizes complete IgE[1] with different types of mouse and human nonlymphoid cells. Using a cytochemical assay, immunofluorescence staining for the cytoplasmic content of the lambda light chain component, we were able to investigate if single heterokaryons formed from the fusion of human myeloma cells and non-lymphoid cells contained detectable quantities of cytoplasmic lambda chains and found that initially these heterokaryons did, but that it was rapidly lost. The time kinetics of this process suggests a very rapid inhibition of new synthesis of lambda chains[2]. To investigate immunoglobulin production at later timepoints we produced a synkaryotic cell hybrid between the human myeloma cell line and a thymidine kinase less subline of mouse L cells and found that this hybrid, even very early after isolation where the majority of human chromosomes were conserved, was unable to produce complete IgE, lambda light chains or epsilon heavy chains[3]. Interestingly, the degree of contact inhibition of the hybrid cells was much greater than the parent mouse cells suggesting that regulated growth had been restored by a form of complementation[3].

## 2. Materials and Methods

Myeloma cells. The continuous human myeloma line 266Bl secretes complete IgE[1], and can be cultivated in suspension in the presence of feeder layers of allogeneic skin fibroblasts or glia-like cells[4]. The karyotype is near-diploid with a modal number of 44 and 4 aberrant marker chromosomes (Fig. 1). The synthesis of complete immunoglobulin molecules is in contrast to another human myeloma cell line RPMI 8226[5] that only produces free lambda chains. Analysis of the secreted immunoglobulin by gel filtration demonstrates, however, that the light chains are produced in excess (Fig. 2). The tendency of myeloma cells in culture to convert into production of only the light chain is well known and occurs at a high rate[6], but our cultures of human myeloma cells still secreted the complete immunoglobulin molecule. The rate of production of complete IgE molecules by the 266Bl human myeloma cell line was 4 μg/$10^6$ cells /day which is somewhat less than for the RPMI 8226 line which produces as much as 18 μg of free lambda chains/$10^6$ cells/day[5].

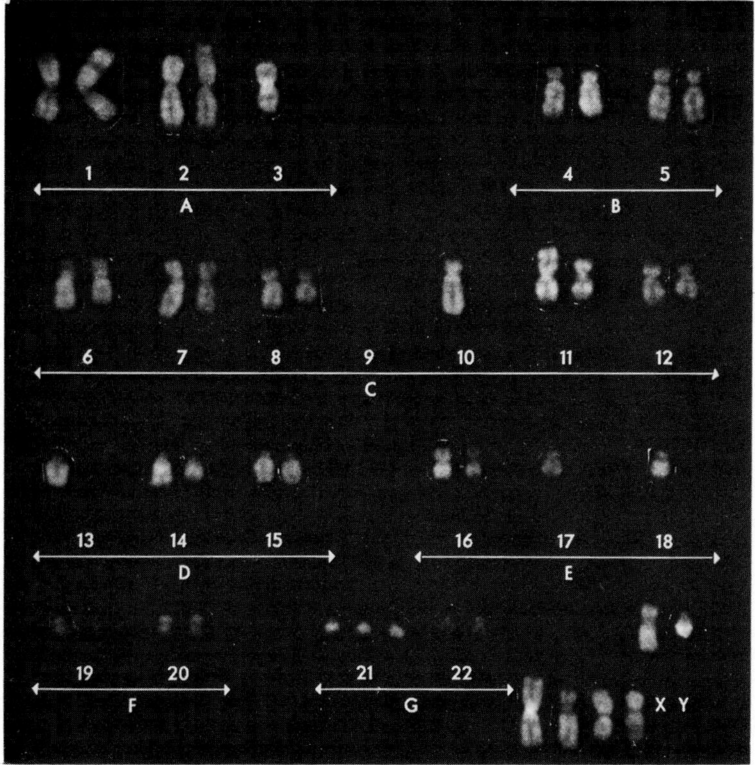


Fig. 1. Karyotype of 266Bl myeloma cells (Q-banding). The karyotype is near-diploid with a modal chromosome number of 44 and 4 marker chromosomes (bottom left).

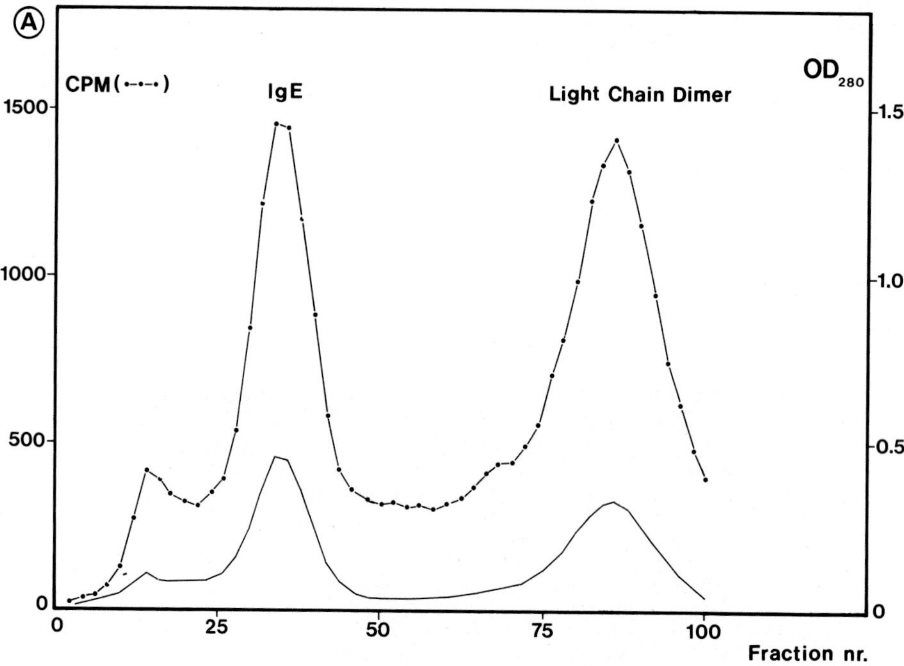


Fig. 2. (a) Synthesis of immunoglobulin by 266B1 myeloma cells. The culture was labelled with ($^{35}$S)-cysteine and the supernatants analyzed by gel filtration on Sephadex G200 after addition of an excess of IgE(ND) and Bence Jones Protein to serve as markers at $OD_{280}$. The distribution of radioactivity shows that in addition to complete IgE a second peak at the same position as the Bence Jones Protein ( light chain dimer) is produced by the 266B1 myeloma cells. Since complete IgE contains 40 half-cystine residues and the lambda light chains 5 half-cystine residues the amount of light chain molecules produced by the myeloma cells corresponds to about a 10 - fold molar excess as compared to the complete IgE molecules ( H. Bennich, unpublished results).

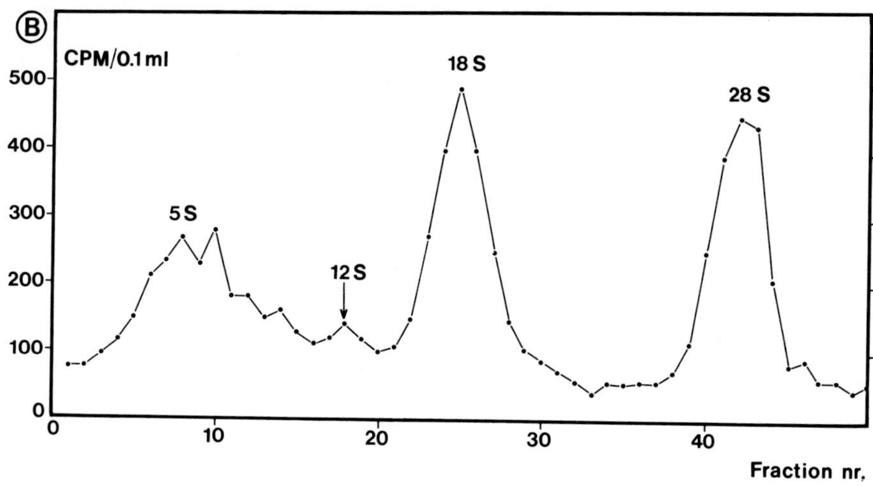


(b) Sucrose gradient centrifugation profile of total polysomal RNA isolated from a culture of 266B1 myeloma cells after 6 hrs labelling with ($^3$H)-uridine. The 12S peak corresponding to light chain mRNA is indicated by the arrow while the mRNA corresponding to the epsilon heavy chain is sedimenting at the same position as 18S rRNA ( J. Zeuthen and A. Leth Bak, unpublished results).

Production and identification of myeloma x non-lymphoid cell heterokaryons. In comparison with several other human lymphoid lines tested for their yield of heterokaryons with several primary and established lines of non-lymphoid cell induced by inactivated Sendai virus, the human myeloma cell line 266Bl was found to give the highest yield of heterokaryons (about 5% of the total cell number). To achieve this yield the tricine-buffered fusion medium (pH 8.0) described by Croce et al.[7] was used. To identify heterokaryons the non-lymphoid partner cell was prelabelled with ($^3$H)-thymidine for 3 days prior to fusion[2]. After scoring individual bi- and multinucleate cells for their immunoglobulin content by immunofluorescence these could later be identified as homo- or heterokaryons due to the labelling of their nuclei. In order to localize cells, maps of preparations were obtained by photography of overlapping fields of the preparation at low magnification on Polaroid film and on these maps individual bi- or multinucleate cells could be localized (Fig. 3).

Production and identification of myeloma x mouse L cell hybrids. Human myeloma cells were fused with thymidine kinase deficient Cl-1D cells derived from the LM subclone of $C_3H$ mouse L cells[8] by means of UV-inactivated Sendai virus. Hybrid cells were isolated in selective medium containing hypoxanthine, aminopterin and thymidine (HAT medium) as cells growing in monolayer[3].

The isolated hybrid cells could be identified by chromosome preparations stained with quinacrine mustard, or stained for centromeric heterochromatin using the C-banding procedure or the fluorescent benzimidazol derivative "Hoechst 33258".

For further confirmation of the hybrid nature of these cells they were also studied for their enzyme content using electrophoresis on cellulose acetate gels.

## 3. Extinction of immunoglobulin synthesis in heterokaryons

Human myeloma cells were fused with ($^3$H)-thymidine prelabelled L929 mouse cells. Samples were taken from the cultures at various intervals after fusion and individual bi- or multinucleate cells were localized, mapped and scored for immunofluorescence after staining with fluorescein isothiocyanate conjugated rabbit anti-human lambda light chain serum. After autoradiography homo- and heterokaryons could be identified with certainty due to the nuclear prelabelling of the L929 cells. Initially, heterokaryons were scored as fluorescent, but as soon as 4 hrs after fusion of the cells the first lambda chain negative heterokaryons start to appear and this absence of fluorescence persisted at later timepoints (Fig. 4). The myeloma x myeloma homokaryons present in the same preparations were scored as

Fig. 3. Mapping of bi- and multinucleate cells in a preparation of 266B1 myeloma cells fused with ($^3$H)-thymidine prelabelled mouse L 929 cells. Bi- or multinucleate cells were localized and numbered (see figure), scored for immunofluorescence after staining for lambda light chains, and subsequently identified by autoradiography as homo- or heterokaryons (dark-field illumination, approx. 50 x magnification).

fluorescent throughout the experiment though a small fraction had lost their fluorescence at early timepoints after the termination of fusion (Fig. 4).

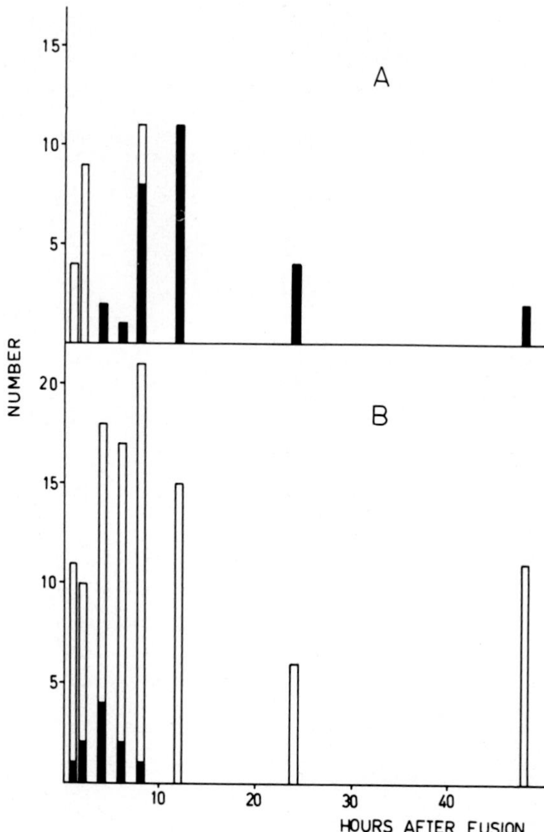


Fig. 4. Immunoglobulin content (lambda light chain) of 266B1 myeloma x mouse L929 heterokaryons (A) and myeloma x myeloma homokaryons (B) as determined by immunofluorescence staining of individual binucleate cells. Abscissa: hours after fusion, ordinate: absolute number of binucleate cells studied for each timepoint. Open bars represent fluorescent cells, filled bars non fluorescent cells. (Data from Zeuthen et al.[2])

In addition to the binucleate homo- and heterokaryons scored in Fig. 4 a few multinucleate heterokaryons were found in the preparations. The tendency of retention of fluorescence in multinucleate cells with one L929 nucleus and more than one myeloma nucleus might indicate a gene dosage effect (Table I).

TABLE I

FUSION OF 266Bl HUMAN MYELOMA CELLS WITH MOUSE L929 CELLS

Immunofluorescence of multinucleate heterokaryons[1)]

| Time after fusion | Nuclear constitution[2)] | Immunofluorescence[3)] |
|---|---|---|
| 4 hrs. | 1 (2,1) | n.fl. |
| 8 hrs. | 1 (1,2) | fl. |
| | 1 (2,1) | n.fl. |
| 12 hrs. | 2 (2,1) | n.fl. |
| 24 hrs. | 1 (2,1) | n.fl. |
| | 1 (1,2) | n.fl. |
| 48 hrs. | 3 (2,1) | 3 n.fl. |
| | 3 (1,3) | 2 n.fl. + 1 fl. |
| | 1 (1,2) | fl. |
| | 1 (2,2) | n.fl. |
| | 1 (4,1) | n.fl. |
| | 1 (5,2) | n.fl. |

Notes: 1) Multinucleate heterokaryons were scored in the same experiment as the binucleate cells in Fig. 4.
2) No.L929 nuclei, no. of myeloma nuclei.
3) Fluorescent = fl., Non fluorescent = n.fl.

Similar results were obtained in fusion experiments of 266Bl with human Hela cells (an established cell line) and with human fibroblasts or glia-like cells (secondary cultures) so it appears that the extinction of lambda chain production is neither sensitive to species differences, since the phenomenon is seen equally well with mouse or human non-lymphoid cells, nor to the transformed status of the non-lymphoid cell since extinction is seen for heterokaryons of myeloma cells with normal cells as well as tumor cells adapted to tissue culture.

The loss of lambda chains in the heterokaryons was in all cases very rapid being completed within 4 to 6 hrs after fusion. Since lambda chains were assumed to be secreted continuously this rapid loss could indicate an almost immediate inhibition of lambda chain synthesis. To test this, cultures of growing 266Bl myeloma cells were treated with inhibitors of protein synthesis to evaluate the rate of turnover of lambda chains if no new synthesis of lambda chains could take place. Two inhibitors were used: puromycin, which acts at the initiation of protein synthesis, and cycloheximide, which acts on chain elongation. Using these inhibitors essentially the same result was obtained: the cytoplasmic content of lambda chains as scored by immunofluorescence as in the heterokaryon experiments was lost after 3 to 4 hrs which is similar to the loss of lambda chains 4 to 6 hrs after fusion (Fig. 5).

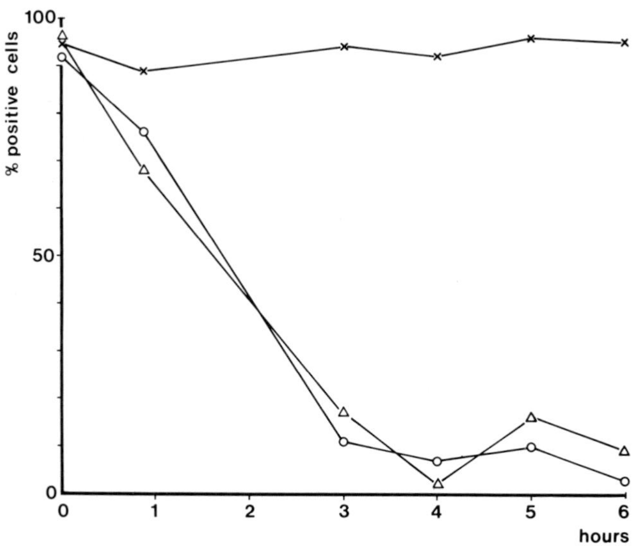


Fig. 5. Turnover of lambda light chains in proliferating 266Bl myeloma cells. Cultures of myeloma cells were treated with 50 $\mu$g/ml puromycin (-Δ-Δ-Δ-), 50 $\mu$g/ml cycloheximide (-O-O-O-), while the third culture (-X-X-X-) served as a control. The percentage of fluorescence positive cells was scored after immunofluorescence staining for lambda light chains[2].

Though it cannot be proved from these experiments, the extinction of the differentiated function (lambda chain synthesis) in heterokaryons could therefore possibly be on the level of protein synthesis. In that case heterokaryons could contain untranslated, inactive mRNA for lambda chains.

4. Phenotype of human myeloma x mouse L cell hybrids

The synkaryotic hybrid cells were obtained by fusion using inactivated Sendai virus of 266Bl human myeloma cells with Cl-1D mouse cells and could be isolated by growth in monolayer in selective HAT medium. The human myeloma cell line grows in suspension and is of a fibroblast-like appearance. In contrast, the hybrid cells were of a more epitheloid morphology (Fig. 6). The growth behaviour of the hybrid cells was intermediate between that of the parent myeloma cells and the subline of mouse L cells, since the hybrid cells grew slowly like the parent myeloma cells, but in monolayer like the mouse cells. Karyotype analysis 5 to 6 generations after isolation confirmed the hybrid nature of these cells, since the metaphases contained a total of 105 to 120 chromosomes, of which 39 to 47 could be identified as telocentric mouse chromosomes, and more than 8 as biarmed mouse chromosomes, with the remainder predominantly human chromosomes, of which most could be positively identified in the hybrid by quinacrine banding, since the parent 266Bl myeloma cells had a near-diploid karyotype with a modal number of 44 and only 4 aberrant marker chromosomes. The identification of the majority of the human chromosomes in the hybrid cells could be confirmed by enzyme electrophoresis which separated human and mouse enzymes[3].

Immunofluorescence staining of the hybrid cell population immediately after its isolation (5 to 6 generations) for the lambda and epsilon chains of human IgE did not reveal any cells which contained detectable amounts of these immunoglobulin components. In subsequent analyses performed approximately 6 to 11 generations after the isolation of the hybrid cells their secretion of IgE into the medium was measured using a sensitive radioimmunosorbent test in which the supernatant isolated from the hybrid cells was concentrated and allowed to compete for binding to antibody with ($^{125}$I)-labelled purified IgE standard[9]. The hybrid cells were found to secrete IgE at a rate less than 0.1 ng/$10^6$ cells/48 hrs while the parent myeloma cells secrete IgE at a rate as high as 8.1 $\mu$g/$10^6$ cells/48 hrs, i.e. the secretion of IgE by the hybrid cells was more than 80.000-fold reduced compared to the parent myeloma cells[3].

Surprisingly, the growth behaviour with respect to density dependent regulation of growth appeared to have been changed in the hybrid cells as compared to the parent mouse Cl-1D line, which exhibit partially "transformed" characteristics in tissue culture. The hybrid cells grew to a significantly lower saturation density than the parent Cl-1D cells (Fig. 7), and this observation could indicate a form of complementation with respect to regulation of growth in the hybrid between the Cl-1D cells and myeloma cells which grow in suspension. A similar observation has been reported by Levisohn and Thompson[10] in their studies on a hybrid between HTC rat hepatoma cells and the mouse Cl-1D cells.

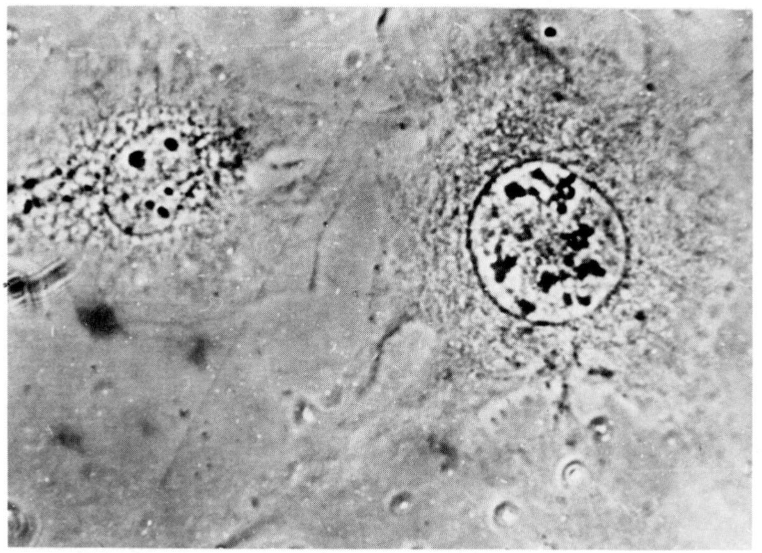

Fig. 6. Morphology of 266B1 myeloma x mouse Cl-1D hybrid cells after approximately 40 generations in culture. Note flat epitheloid appearance and several nucleoli (approx. 1000 x magnification, phase contrast)[3].

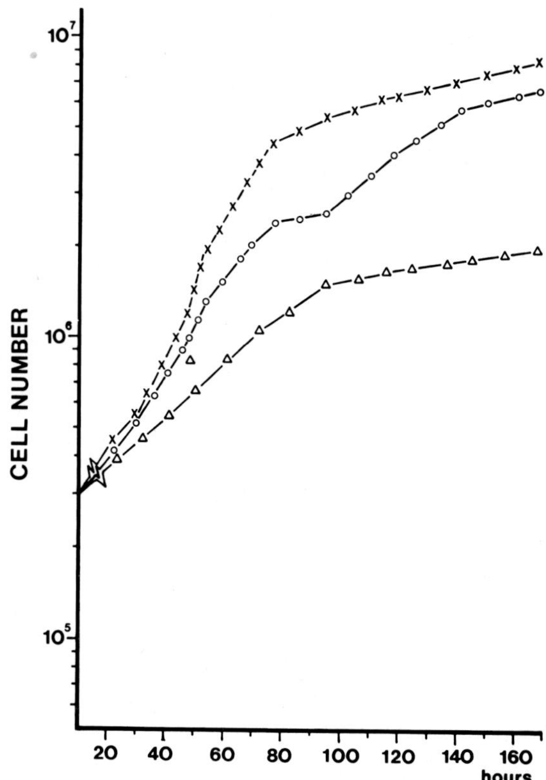


Fig. 7. Growth curves of Cl-1D mouse cells (- x- x- x-) and two passages of 266B1 myeloma x mouse Cl-1D hybrid cells, 20 generations in culture (-Δ- Δ- Δ-) and 40 generations in culture (- 0- 0- 0-). Note the slow growth rate and low saturation density of the early passage of hybrid cells in contrast to the higher growth rate and saturation density of the later passage of the hybrid cells more similar to the growth characteristics of the parental mouse Cl-1D cells[3].

5. Discussion

In the studies discussed we have found evidence that the synthesis of immunoglobulin (IgE or its lambda chain component) is suppressed after fusion of the human myeloma cell line 266Bl with several types of normal cells or established cell lines of both mouse and human origin. In the studies on heterokaryons we have observed this to be a very rapid phenomenon and the extinction of immunoglobulin synthesis in heterokaryons could therefore possibly be on the level of protein synthesis. In that case heterokaryons could contain untranslated, inactive mRNA for immunoglobulin. At present we are trying to obtain a cDNA probe directed against the immunoglobulin mRNA's to make it possible to test this hypothesis directly by _in situ_ hybridization and autoradiography of single myeloma x non-lymphoid heterokaryons. In the synkaryotic myeloma x Cl-1D hybrid cell line we have been able to quantitate the degree of extinction of immunoglobulin synthesis to be a reduction of 80.000-fold in the hybrid as compared to the parent myeloma cell line. This is a much greater reduction than in other studies reported on the extinction of immunoglobulin synthesis in hybrids between myeloma and non-lymphoid cells[11, 12].

In studies where different lymphoid cells have been fused together the production of immunoglobulin by the parental cells has been retained[13, 14], and in recent studies on hybrid clones formed from the fusion of human lymphocytes with mouse myeloma cells synthesis of human immunoglobulin has been activated due to the fusion event [15, 16, 17]. The distance in differentiation which separates the two parental cells thus appears to determine whether differentiated properties are expressed. Fusion of myeloma cells with other lymphoid cells permits the continued expression of immunoglobulin synthesis. The fusion of myeloma cells with fibroblasts or lymphoma cells[13] did not result in the initiation of immunoglobulin synthesis by the non-immunoglobulin synthesizing parent cell. Similarly, fusion of lymphoblastoid cells with fibroblasts resulted in the production of only the lymphoblastoid parental light chain [18]. The fusion of myeloma cells with lymphocytes resulted in the secretion of several molecular species of human immunoglobulin.

All these observations have been made on synkaryotic hybrid cells and the studies reported here on the extinction of immunoglobulin synthesis in heterokaryons[2] are the first to suggest that a similar mechanism might operate in the early stages after cell fusion. Preliminary experiments have indicated that extinction of human immunoglobulin production did not occur if human myeloma cells were fused with mouse myeloma cells which agrees with the observations on hybrid cells mentioned above. At present no studies have been carried out on the reexpression of immunoglobulin synthesis in heterokaryons between e.g. human lymphocytes and mouse myeloma cells. If this effect could be demonstrated in the heterokaryon state it would be a way to test the validity of the concept of the quantal cell cycle, suggested to be necessary for the expression of various differentiated properties in differentiating cells, since a heterokaryon per definition has not undergone cell division.

## Acknowledgements

I wish to thank Drs. Kenneth Nilsson, Svante Stenman and Hans-Åke Fabricius who participated in different phases of this work.

This study was supported in part by grant No. 521/18 from the Danish Natural Science Research Council.

## References

1. Nilsson,K.,Bennich,H.,Johansson,S.G.O. and Pontén,J.,1970, Clin. exp. Immunol. 7, 477-489.
2. Zeuthen,J.,Stenman,S.,Fabricius,H.-Å. and Nilsson,K.,1975, Cell Differentiation, accepted for publication.
3. Zeuthen,J. and Nilsson,K.,1975, Cell Differentiation, accepted for publication.
4. Nilsson,K.,1971, Clin. exp. Immunol. 9, 785-793.
5. Matsuoka,Y.,Takahashi,M.,Yagi,Y.,Moore,G.E. and Pressman,D., 1968, J. Immunol. 101, 1111-1120.
6 Coffino,P. and Scharff,M.D.,1971, Proc.Nat.Acad.Sci. U.S.A., 68, 219-223.
7. Croce,C.M.,Koprowski,H. and Eagle,H.,1972, Proc.Nat.Acad.Sci. U.S.A.,69, 1953-1956.
8. Kit,S.,Dubbs,D.R.,Piekarski,L.I. and Hsu,T.C.,1963, Exp. Cell Res.,31, 297-312.

9. Johansson,S.G.O.,Bennich,H. and Wide,L.,1968, Immunology, 14, 265-272.

10. Levisohn,S.R. and Thompson,E.B.,1973, J. Cell Physiol., 81, 225-230.

11. Perlman,P.,1970, Nature ,228, 1086-1087.

12. Coffino,P.,Knowles,B.,Nathenson,S. and Scharff,M.,1971, Nature ,231, 87-90.

13. Mohit,B.,1971, Proc.Nat.Acad.Sci. U.S.A.,68, 3045-3048.

14. Cotton,R.G.H. and Milstein,C.,1973, Nature ,244, 42-43.

15. Schwaber,J. and Cohen,E.P.,1973, Nature ,244, 444-447.

16. Schwaber,J. and Cohen,E.P.,1974, Proc.Nat.Acad.Sci. U.S.A., 71, 2203-2207.

17. Schwaber,J.,1975, Exp. Cell Res.,93, 343-354.

18. Orkin,S.H.,Buchanan,P.,Yount,W.J.,Reisner,H. and Littlefield, J.W.,1973, Proc.Nat.Acad.Sci. U.S.A.,70, 2401-2405.

*Progress in Differentiation Research, ed. N. Müller-Bérat et al.*



# THE ROLE OF GENE ACTIVITY IN CHROMOSOME CONDENSATION INDUCED BY CYTOPLASM OF PROGESTERONE STIMULATED AMPHIBIAN OOCYTES

David Ziegler and Yoshio Masui
Department of Zoology, University of Toronto
Toronto M5S 1A1 Ontario, Canada

## INTRODUCTION

The cytoplasm of a frog oocyte undergoing meiotic maturation possesses factors that are able to induce the chromatin of interphase brain nuclei to condense into chromosomes similar to those found at mitosis[1,2,3]. We refer to this activity of maturing oocytes as the chromosome condensation activity (CCA). In <u>Rana pipiens</u>, CCA is first detected shortly after germinal vesicle breakdown (GVBD), and persists until activation[3]. In the absence of the germinal vesicle (GV), CCA fails to appear[3]. Once the brain nuclei are introduced into the oocytes, two potentially active nuclear genomes are present, the genomes of the brain nuclei and of the oocyte. Also the GV contains a large store of extra-chromosomal material. Here we have examined the involvement of each of these components in the production and action of the chromosome condensation factors.

## MATERIALS AND METHODS

To induce meiotic maturation, defolliculated fully grown oocytes of <u>Rana pipiens</u> were treated with progesterone and incubated at 18°C in a frog Ringer solution containing antibiotics[4]. Synchronously maturing oocytes were selected[3]. These oocytes usually undergo GVBD between $T_{13}$ and $T_{17}$, reach first metaphase (MI) at $T_{24}$ and second metaphase (MII) at $T_{48}$, where the number subscripted to the letter T refers to the hours after progesterone treatment.

Nuclei were isolated from the brains of adult <u>Rana pipiens</u> and injected into the animal hemispheres of the recipient oocytes. When MII oocytes were injected, they were immersed in 0.05M $NaH_2PO_4$ solution to prevent activation[3]. Recipient oocytes were enucleated by squeezing the germinal vesicles (GVs) out of the ovarian oocytes with forceps or by eliminating MI oocyte chromosomes with a glass needle[5].

GV nucleoplasm was prepared as follows: GVs were isolated manually in paraffin oil with watchmaker's forceps from ovarian oocytes. Fifty to two hundred GVs were collected and homogenized by 5 passages through a silicone coated micropipette while submerged in the paraffin oil. The homogenate was layered on heavy silicone oil (General Electric SF-1150) in a capillary tube with closed end and centrifuged for 30 min at 17,000 x g to sediment the chromatin, membranes and other insoluble materials. Approximately 200 $\eta$l of the GV-nucleoplasm was injected into each recipient oocyte. The entire operation was carried out at 2 - 4$^{\circ}$C.

$\alpha$-Amanitin and cycloheximide were dissolved in frog Ringer solution. Oocytes were both injected with and incubated in these inhibitor solutions so that the concentrations were equal internally and externally. Treatment of the isolated brain nuclei with inhibitor was carried out by suspending them in the nuclear isolation medium containing an inhibitor for 2 hrs at 4$^{\circ}$C. The nuclei were injected into recipient oocytes after washing. All the oocytes injected with brain nuclei were fixed 3 hrs after nuclear injection and the behavior of the brain chromosomes was examined by histological section or by a squashing technique described recently[5] that has enabled us to count the number of nuclei or chromosome sets in an oocyte.

RESULTS

The role of nuclear genomes in MI oocytes (Table 1): When oocytes were treated with various doses of $\alpha$-amanitin from the beginning of the maturation period, an increasing proportion of the brain nuclei were prevented from undergoing chromosome condensation as the dose of inhibitor increased. On the other hand, complete inhibition was produced regardless of the dose if the brain nuclei were treated before injection.

When brain nuclei were introduced at $T_{24}$ into oocytes that had been injected with $\alpha$-amanitin at $T_{22}$, in 66% of the oocytes all the brain nuclei failed to undergo chromosome condensation. If $\alpha$-amanitin pretreated brain nuclei were used, no chromosome condensation occurred in any oocyte.

Since CCA has been shown to be present and active in the oocyte cytoplasm by $T_{20}$[3], the results above suggest that the action of the cytoplasmic factor(s) inducing chromosome condensation, but not their production, is dependent upon the synthesis of RNA during the period when the brain nuclei are exposed to the oocyte cytoplasm. To determine whether the oocyte genome is responsible for this RNA synthesis, the brain nuclei were introduced into oocytes 2 hrs after the oocytes' chromosomes had been removed at $T_{22}$. Brain chromosomes underwent condensation in the absence of the oocytes' chromosomes just as frequently as they did in the untreated controls. Therefore, it appears likely that the RNA in question is not synthesized on the oocyte chromosomes, but in the brain nuclei.

MI oocytes already possessing CCA were treated with cycloheximide at $T_{22}$. Brain nuclei were introduced at $T_{24}$. In this case, brain chromosome condensation was completely suppressed at all doses of the inhibitor used. It is suggested that the RNA synthesized during exposure of brain nuclei to MI oocyte cytoplasm needs to be translated if the brain chromosomes are to condense.

TABLE 1

THE ROLE OF BRAIN AND OOCYTE NUCLEAR GENOMES IN THE INDUCTION OF CHROMOSOME CONDENSATION BY MI OOCYTE CYTOPLASM

| Treatment* | Duration of Treatment | No. of Oocytes Examined UTr** | Tr*** | Oocytes with Condensed Brain Chromosomes (%) UTr | Tr |
|---|---|---|---|---|---|
| $\alpha$-Amanitin $\mu$g/ml | $T_0$-$T_{27}$ | | | | |
| 0 | | 20 | 20 | 100 | 100 |
| 5 | | 22 | 35 | 64 | 0 |
| 10 | | 34 | 30 | 29 | 0 |
| 15 | | 38 | 30 | 21 | 0 |
| 20 | | 35 | 31 | 0 | 0 |
| $\alpha$-Amanitin $\mu$g/ml | $T_{22}$-$T_{27}$ | | | | |
| 0 | | 40 | 40 | 100 | 100 |
| 10 | | 36 | 37 | 33 | 0 |
| Removal of MI Oocyte Chromosomes | $T_{22}$-$T_{27}$ | | | | |
| Present | | 30 | - | 100 | - |
| Absent | | 52 | - | 96 | - |
| Cycloheximide $\mu$g/ml | $T_{22}$-$T_{27}$ | | | | |
| 0 | | 30 | - | 100 | - |
| 1 | | 37 | - | 0 | - |
| 10 | | 40 | - | 0 | - |

* The treated oocytes were injected with brain nuclei at $T_{24}$ and fixed at $T_{27}$.
** Nuclei untreated before injection.
*** Nuclei treated with inhibitor before injection.

<u>The role of the nuclear genomes in (MII) oocytes (Table 2):</u> The MII oocytes were treated with $\alpha$-amanitin at $T_{48}$ in the same manner as MI oocytes while the oocytes were immersed in 0.05M $NaH_2PO_4$ solution to suppress activation. Two hrs later the brain nuclei were injected. In all cases, chromosome condensation was induced, even when the brain nuclei were pretreated with the inhibitors.

Furthermore, when MII oocytes were treated with cycloheximide in the same way as with the α-amanitin, brain chromosome condensation was still not suppressed. It is therefore apparent that activity of the brain nuclear genome is not needed for their chromosomes to condense in MII oocytes. The oocyte genome also appears to be unnecessary for brain chromosome condensation to occur in MII oocytes since brain nuclei injected into oocytes at $T_{48}$ always underwent chromosome condensation even when the oocyte chromosomes had been removed at $T_{24}$. This was corroborated in an experiment where oocytes were treated with α-amanitin (10μg/ml) throughout the maturation period and injected with brain nuclei at MII. It was found that these oocytes induced brain chromosome condensation as frequently as the untreated oocytes did.

TABLE 2

THE ROLE OF BRAIN AND OOCYTE NUCLEAR GENOMES IN THE INDUCTION OF CHROMOSOME CONDENSATION BY MII OOCYTE CYTOPLASM

| Treatment* | Duration of Treatment | No. of Oocytes Examined | | Oocytes with Condensed Brain Chromosomes (%) | |
|---|---|---|---|---|---|
| | | UTr** | Tr*** | UTr | Tr |
| α-Amanitin μg/ml | $T_{46}$-$T_{51}$ | | | | |
| 0 | | 20 | 20 | 100 | 100 |
| 5 | | 40 | 40 | 100 | 100 |
| 10 | | 40 | 40 | 100 | 100 |
| 15 | | 35 | 35 | 100 | 100 |
| α-Amanitin | $T_0$-$T_{51}$ | | | | |
| 0 | | 42 | - | 100 | - |
| 10 | | 35 | - | 100 | - |
| Cycloheximide μg/ml | $T_{46}$-$T_{51}$ | | | | |
| 0 | | 30 | - | 100 | - |
| 1 | | 25 | - | 100 | - |
| 10 | | 32 | - | 100 | - |
| 50 | | 30 | - | 100 | - |
| Removal of MI Oocyte Chromosomes | $T_{24}$-$T_{51}$ | | | | |
| Present | | 20 | - | 100 | - |
| Absent | | 25 | - | 100 | - |

* In all cases, the treated oocytes were injected with brain nuclei at $T_{48}$ and fixed at $T_{51}$.
** Nuclei untreated before injection.
*** Nuclei treated with inhibitor before injection.

<u>The role of the GV-nucleoplasm (Table 3):</u> Progesterone-treated oocytes were enucleated at $T_0$ and injected with brain nuclei at

$T_{24}$. The result clearly showed that all these oocytes failed to induce brain chromosome condensation. On the other hand, the contents of GVs obtained from ovarian oocytes were able to restore CCA in the enucleated oocytes. Here, the GV contents were injected at $T_{22}$ and the brain nuclei at $T_{24}$.

The following two experiments demonstrated that the capacity of the ovarian oocyte GVs to restore CCA resided in the nucleoplasm, not in the chromosomes of the GV. In the first, whole GV contents isolated from ovarian oocytes which had been treated with actinomycin D for 2 hrs were injected into enucleated, progesterone-treated oocytes at $T_{22}$. In the second experiment, progesterone-treated, enucleated oocytes were injected with GV-nucleoplasm, free of all structural components, at $T_{22}$. In both experiments, the oocytes were injected with brain nuclei at $T_{24}$. It was found that brain chromosomes were induced to condense in a majority of the oocytes in both experiments.

TABLE 3

THE ROLE OF THE GV NUCLEOPLASM IN THE PRODUCTION OF CCA OF MI OOCYTES

| Type of Oocyte* | No. of Oocytes Examined | Oocytes with Condensed Brain Chromosomes (%) |
|---|---|---|
| GV Present | 60 | 98 |
| GV Absent | 80 | 0 |
| GV Absent + $T_0$ GV Contents | 25 | 100 |
| GV Absent + ACT** Treated $T_0$ GV Contents | 35 | 100 |
| GV Absent + $T_0$ Nucleoplasm | 42 | 71 |

* In all specimens, enucleation was accomplished at $T_0$, GV transfers were made when the recipients reached $T_{22}$ and brain nuclei were introduced at $T_{24}$. The nuclei were exposed for a 3 hr period.

** ACT = Actinomycin D.

The brain nuclei behaved differently in enucleated $T_{48}$ oocytes than they did in enucleated $T_{24}$ oocytes. In nucleated, $T_{48}$ oocytes, when the recipients were examined at 1, 2 and 3 hrs after nuclear injection, the frequency of chromosome condensation continued to increase. However, in the enucleated specimens, while the frequency of chromosome condensation increased during the first 2 hrs of exposure, it decreased after 3 hrs of exposure (Fig. 1). Although the mitotic index dropped, the number of interphase brain nuclei remaining after 3 hrs of exposure to the cytoplasm was almost the same in both nucleated and enucleated recipients. These results suggest that although brain chromosome condensation is initiated by enucleated MII oocyte cytoplasm, when they are condensed and their exposure to the cytoplasm is prolonged, the chromosomes are destroyed. This implies that the presence of the GV-nucleoplasm is not obligatory for the induction of chromosome condensation in MII oocytes but is required for the stabilization of chromosomes once condensed.

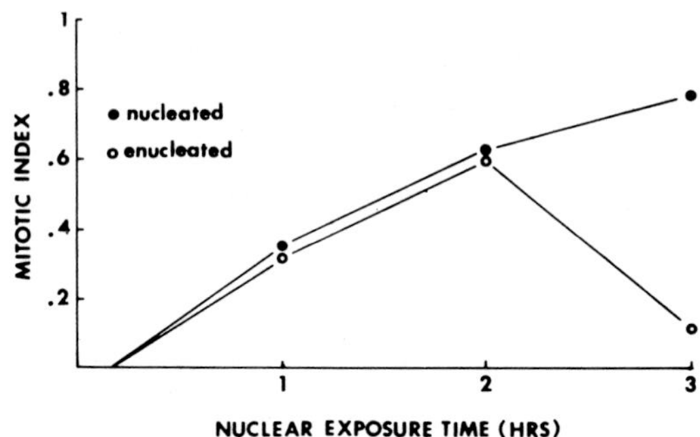


Fig. 1. The frequency of chromosome condensation (mitotic index) in nucleated and enucleated $T_{48}$ recipients is plotted as a function of the time the nuclei were exposed to the cytoplasm. Each data point represents determinations from a minimum of 20 oocytes.

## DISCUSSION

The cytoplasm of maturing or fully mature Rana pipiens oocytes is able to induce metaphase-like chromosome condensation in interphase brain nuclei exposed to the cytoplasm. The present study was undertaken to determine the necessity of genomic activity of the oocyte or brain nuclei for the production or action of the cytoplasmic factors responsible for this chromosome condensation activity (CCA).

It was found that in oocytes which reached MI, brain chromosome condensation was suppressed by RNA synthesis inhibitors, but not suppressed by removal of the oocyte chromosomes before introduction of the brain nuclei. Enucleated ovarian oocytes induced to mature and cultured for the same length of time as MI oocytes were also found to become capable of inducing chromosome condensation if ovarian oocyte GV-nucleoplasm was introduced at the time when the brain nuclei were injected. Since these oocytes were deprived of their chromosomes throughout their maturation history, it is concluded that neither the development nor function of CCA requires the oocyte genome. Therefore, it is highly likely that it is the inactivation of the brain nuclear genome that is responsible for the failure of brain chromosome condensation in MI oocytes under the influence of the RNA synthesis inhibitors. This contention may be supported by the fact that $\alpha$-amanitin suppression of chromosome condensation is made complete at all doses only when the brain nuclei are directly treated with the inhibitor.

In view of the fact that the protein synthesis inhibitors also suppressed chromosome condensation in MI oocytes, it appears likely that MI oocyte cytoplasm induces chromosome condensation in brain nuclei by eliciting their transcription and producing genetic messages that are then translated into the proteins necessary for chromosome condensation. In contrast to MI oocytes, chromosome condensation induced by MII oocyte cytoplasm is suppressed neither by RNA synthesis inhibitors nor by protein synthesis inhibitors. Apparently MII oocyte cytoplasm contains all the substances necessary for chromosome condensation in brain nuclei including some which are lacking in MI oocyte cytoplasm. Therefore, we conclude that the cytoplasmic state responsible for inducing chromosome condensation changes as an oocyte progresses through meiotic maturation from MI to MII.

Although development of CCA during meiotic maturation does not involve activity of the oocyte genome, it is dependent on the presence of GV nucleoplasm. Oocytes enucleated at the beginning of maturation fail to develop CCA when tested at $T_{24}$ but it is restored by reintroduction of the GV nucleoplasm. On the other hand, enucleated oocytes that were allowed to develop to $T_{48}$ were able to induce chromosome condensation, but the induced chromosomes were not maintained longer than 2 hrs. These results imply that the GV contributes substances necessary for the induction of chromosome condensation in MI oocytes. Alternatively, the substances required for the induction of chromosome condensation in MII oocytes are produced entirely in the cytoplasm. However, even in MII oocytes, some substance derived from the GV is involved in the maintenance of the condensed chromosomes.

The protein synthesis activity of maturing oocytes rises quickly after the onset of GVBD[6]. Our results suggest that some of these proteins synthesized during the latter half of the maturation process are responsible for chromosome condensation. We have recently found that proteins synthesized during this portion of the maturation period accumulate on the chromosomes of brain nuclei induced to condense in MII cytoplasm[7]. In view of our results indicating no involvement of the oocyte genome in the development of CCA, these proteins may arise from translation of maternal mRNA stored in the oocyte cytoplasm.

REFERENCES

1. Gurdon, J.B. (1967). On the origin and persistence of a cytoplasmic state inducing DNA synthesis in frogs' eggs. Proc. Nat. Acad. Sci. U.S. 58:545-552.
2. Gurdon, J.B. (1968). Changes in somatic cell nuclei inserted into growing and maturing amphibian oocytes. J. Embryol. Exp. Morphol. 20:401-414.
3. Ziegler, D. and Masui, Y. (1973). Control of chromosome behavior in amphibian oocytes. I. The activity of maturing oocytes inducing chromosome condensation in transplanted brain nuclei. Develop. Biol. 35:283-292.
4. Masui, Y. (1967). Relative roles of the pituitary, follicle cells and progesterone in the induction of oocyte maturation in Rana pipiens. J. Exp. Zool. 166:365-376.
5. Ziegler, D. and Masui, Y. (1976). Control of chromosome behavior in amphibian oocytes. II. The effect of RNA and protein synthesis inhibitors on the induction of chromosome condensation in transplanted brain nuclei by oocyte cytoplasm. J. Cell Biol. (in press).
6. Ecker, R.E. and Smith, L.D. (1968). Protein synthesis in amphibian oocytes and early embryos. Develop. Biol. 18:232-249.
7. Ziegler, D. and Masui, Y. (unpublished observation).

*Progress in Differentiation Research, ed. N. Müller-Bérat et al.*



# THYMIDINE KINASE DEFICIENCY IN CULTURED HAPLOID CELLS: GENE MUTATION OR GENE REGULATION?

Liselotte Mezger-Freed

The Institute for Cancer Research, Fox Chase Cancer Center,
Philadelphia, Pennsylvania, USA

## 1. Introduction

Cultured vertebrate cells deficient in thymidine kinase (E.C.2.7.1.75) have been investigated as possible examples of structural gene mutations since they were first reported by Morris and Fischer in 1960[1]. The enzyme deficient variants can be selected directly from bromodeoxyuridine (BUdR) because they do not incorporate this lethal analog of thymidine. The resistant cells multiply at a normal rate, presumably by utilizing a de novo pathway that includes thymidylate synthetase. The rapid multiplication is at variance with the notion that regulation of cell division is a function of thymidine kinase activity but, as a practical consequence, makes it possible to characterize the new phenotype. Many of the properties of the thymidine kinase deficient (TK-) phenotype are consistent with gene mutation: the enzyme deficiency itself, its stability through mitosis, the low frequency at which it occurs or reverts, the existence of intermediate stages reminiscent of a heterozygous state[2]. On the other hand, the same characteristics could result from epigenetic events; in fact, there is some evidence that mechanisms for inhibiting the expression of thymidine kinase exist and operate in tissues of the intact organism. For example, the specific activity of thymidine kinase in adult rodent liver is comparable to that of TK- cells in culture, whereas fetal liver resembles "wild-type" TK+ cultured cells in this respect[3]. Although the greater activity in fetal liver is partly due to the larger proportion of kinase-rich hematopoietic elements, it has been found that thymidine kinase activity is lost prior to changes in cellular composition following hydrocortisone treatment[4]. The analogy between the adult liver and TK- cells in culture extends to the cellular location of the residual enzyme activity, which is mitochondrial in both cases[5,6].

Because the nature of such changes in gene expression are not understood, it is difficult to determine whether similar epigenetic processes are responsible for thymidine kinase deficiency in cell cultures. In considering these problems, it may be helpful to describe some of the results of experiments with haploid amphibian cultures, the only vertebrate haploid lines now available[7]. They are haploid according to their karyotype and quantity of DNA per cell[7] and should, therefore, yield mutants for enzyme deficiency (recessives) with a much higher frequency than diploid cells. However, studies on thymidine kinase deficiency in amphibian cultures indicate that the frequency may not be as high as expected; two separate events are required to yield this phenotype even in haploid cells[8]. The first step is loss of a thermolabile kinase activity[9] followed by loss of a second form of the enzyme, probably coded for at a separate locus. The experiments discussed below concern themselves with the second step in the development of thymidine kinase deficient lines from haploid cells.

## 2. Materials and Methods

The frog cells are propagated as monolayer cultures at 25°C in diluted L-15 medium plus 10% fetal calf serum (referred to as MLM); the population doubling time is about two days. Procedures for initiating and propagating cultures and for determining growth curves have been published[10]. The ICR B20 cell line was derived from a colony isolated after exposing to 5 X $10^{-5}$ M BUdR the wild-type ICR 2A, a haploid cell line derived from haploid Rana pipiens embryos[11]. ICR B20 is resistant to $10^{-4}$ M BUdR although substantial thymidine kinase activity is present. Resistance is associated not only with loss of one of several forms of thymidine kinase but with a low uptake of thymidine (or BUdR) attributable to the absence of a thymidine specific saturable transport reaction[8].

For the selection of TK- cells, MLM containing $10^{-3}$ M BUdR was added to 3-day-old cultures that were at a density of 2 X $10^5$ cells per 25 $cm^2$ flask. Surviving colonies were harvested and characterized with regard to BUdR resistance and thymidine kinase specific activity. When it was observed that in some cases too many cells survived to allow formation of colonies, the contents of whole flasks were analyzed. The stability of the altered phenotype was tested by passaging the flask cultures once more in BUdR, followed by a passage in medium without the drug. Thymidine kinase activity was determined after each passage and in some cases after an additional five passages (20-25 generations) without BUdR (this procedure is abbreviated in tables 1 and 2 as: BUdR → BUdR → MLM → 5X MLM). In almost all cases, cultures remained TK- once they had become TK-, even though they were not derived from single cells but from many cells.

Thymidine kinase specific activity can be assayed with more reproducible results in ICR B20 cells than in the wild-type ICR 2A because activity is less sensitive to culture age and conditions[8]. The activity of cell extracts is assayed by incubation in a standard reaction mixture followed by measuring phosphorylation products of thymidine adsorbed to DEAE cellulose discs[8]. The thymidine kinase specific activity of ICR B20 cells measured in this way was between 2 and 3 picomoles thymidine phosphorylated per μg protein per 90 minutes. Thymidine kinase deficient cell lines have an activity that is consistently 0.3 picomoles per μg protein per 90 minutes, or less. This value is regarded as characterizing the TK- phenotype because cell lines with this level of activity are non-reverting and resistant to $10^{-3}$ M BUdR.

Some of the experiments were done with cultures derived from cells that had survived exposure to the acridine mustard compound ICR 191, a frameshift mutagen for bacteria[12]. The standard procedure consisted of applying MLM with 1 μg/ml ICR 191 to log-phase cultures for 48 hours (one generation time)[13], then decanting and allowing the cells to regain a normal doubling time in MLM, whereupon they were subcultured and BUdR was applied. During recovery from the lethal effects of ICR 191, the changes in cell population were traced in order to determine if such changes were in any way associated with the frequency of TK- cells after BUdR treatment.

## 3. Results

Cultures of line ICR B20 that had been exposed to the acridine mustard compound ICR 191 yielded colonies in $10^{-3}$ M BUdR at frequencies 30-50 times greater than control cultures; most of the isolates were thymidine kinase deficient (TK-)[8]. Such a result is consistent with the induction of frameshift mutations by ICR 191[12]. However, in subsequent experiments that had as their objective determining optimum conditions for mutagenesis, it was found that the behavior of cultures in BUdR, as well as the induction of thymidine kinase deficient variants, was unusually sensitive to culture conditions during and after application of ICR 191. An experiment illustrating the effect of two variables, cell density and subculture by trypsin upon removal of ICR 191, is shown in fig. 1. At low cell density ($2 \times 10^5$ cells per 25 $cm^2$ flask), the population halved during the 48 hours in 1 μg/ml ICR 191, whereas at a cell density ten-fold greater, cell number increased. After removal of ICR 191 from the densely-populated cultures, there was little change in cell number for several weeks, a situation that could reflect either failure of cells to cycle or a balanced cell attrition and cell multiplication. In contrast, cultures of low cell density continued to decrease in cell number for several weeks after ICR 191 was decanted unless they were subcultured by trypsin at this point; as a result of trypsin treatment, growth was resumed earlier, and cultures at 30 days had almost ten times as many cells as untrypsinized cultures.

Even more unexpected was the reaction to $10^{-3}$ M BUdR of the treated cultures (fig. 1); continuous exposure to this drug was begun following return to near normal growth rates after ICR 191 treatment and subculturing. It is evident that "resistance" to BUdR was affected by the ICR 191 treatment that had occurred more than three generations earlier; the magnitude of the difference between "mutagenized" and control cultures was greater than expected for events determined by gene mutation. Furthermore, resistance was not correlated with thymidine kinase deficiency (table 1); the activity in the most resistant culture (high density) was 4.4 prior to BUdR application, unusually high for ICR B20 cells. Although a large proportion of the cells survived in BUdR, the enzyme activity declined to

0.5 in BUdR and eventually became 0.2 after expansion in non-selective medium. The cultures in which the greatest attrition occurred in BUdR (selection?) remained enzyme positive. It should be noted that artificial mixtures of ICR B20 TK+ and TK- cells behave somewhat differently in BUdR; cultures with as high a proportion as 10% TK- cells show an earlier and steeper decline in cell number than the cultures in fig. 1. These observations taken together argue against selection of pre-existing TK- cells induced by ICR 191 mutagenesis.

Table 1

Effects of culture conditions during and after treatment with ICR 191 on thymidine kinase specific activity of cultures selected in $10^{-3}$ M BUdR

| Condition of ICR 191 treatment | Thymidine kinase specific activity* | | | | |
|---|---|---|---|---|---|
| | Before BUdR | After BUdR | → BUdR | → MLM | → 5 X MLM** |
| Control (no ICR 191) | | 1.8 | | | |
| Low cell density | | 1.8 | 3.8 | 1.2 | 2.2 |
| Low cell density plus trypsin | 3.8 | 0.4 | 0.4 | 0.2 | 0.2 |
| High cell density | 4.4 | 0.5 | 0.5 | 0.2 | 0.2 |

*picomoles thymidine phosphorylated/μg protein/90 minutes.
**See Materials and Methods section.

The application of 1 μg/ml ICR 191 to cultures of ICR B20 for periods of time as short as four hours may also change the subsequent reaction of cells to $10^{-3}$ M BUdR. In the experiments summarized in fig. 2 and table 2, ICR 191 was applied for four hours beginning at different times after subculture. After resuming a normal cell increase, the flasks were subcultured and exposed to BUdR following the usual procedure. Survival in BUdR was highest for cultures that had been exposed to ICR 191 for the period from 16 to 20 hours, and it was substantially less for cultures exposed for the same length of time immediately after subculture (fig. 2). Flask contents were assayed for thymidine kinase activity after resumption of growth in BUdR; also assayed were colonies harvested from sister flasks and propagated in MLM (without BUdR). The correspondence between thymidine kinase activity of colonies and of whole flasks after a passage in MLM was good, indicating that enzyme values for flasks do not conceal the presence of substantial proportions of TK- cells. There were no colonies for the 16 hour application because of the large proportion of the culture that survived in BUdR; in spite of this, the cultures declined from a thymidine kinase activity of 4.7 prior to BUdR treatment, to 0.4 and finally to a stable TK- state. In some cases, cultures with similar survival curves did not yield similar thymidine kinase activities; compare the control and the 0-4 hour treatment.

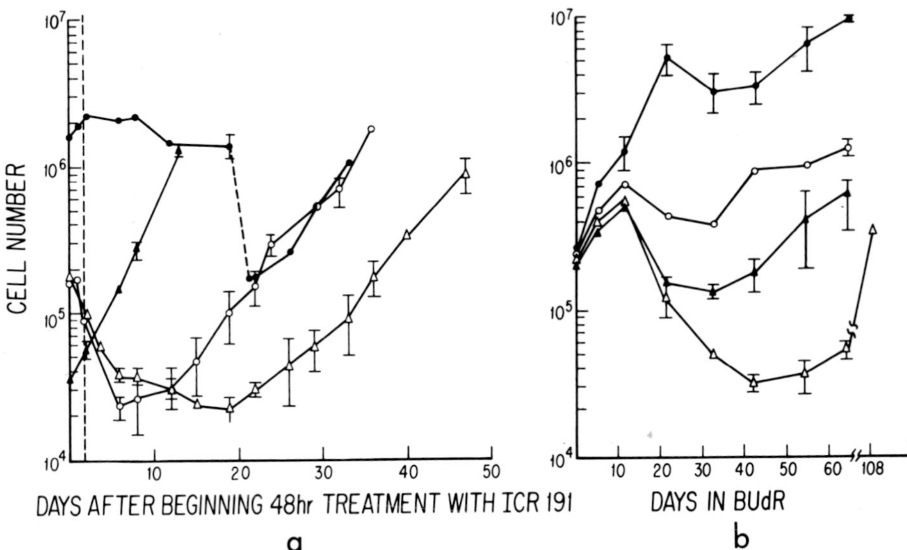


Fig. 1. Effects of culture conditions during and after treatment with ICR 191 on growth following ICR 191 treatment (a) and on growth in $10^{-3}$ M BUdR of the same cultures after subculturing (b). Conditions of ICR 191 treatment were: control, ▲ ; low cell density, △ ; low cell density, cultures subcultured by trypsin immediately following ICR 191 treatment, o ; high cell density, ●.

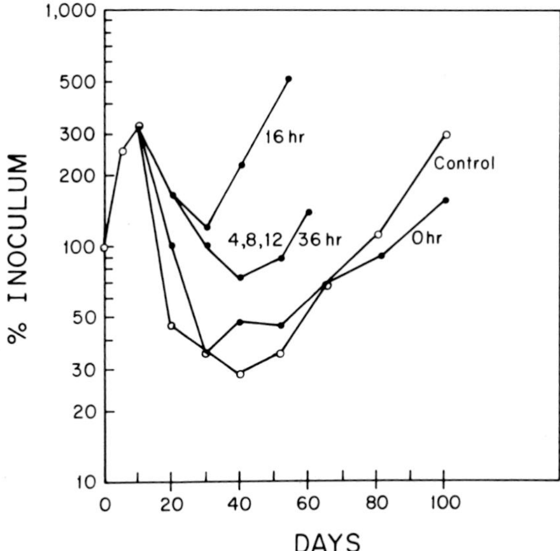


Fig. 2. Effects of length of period between subculturing and 4 hour treatment with ICR 191 on growth during subsequent exposure to $10^{-3}$ M BUdR.

## 4. Discussion

Structural gene mutations such as those leading to enzyme deficiency would be expected to be recessive and therefore to be detectable with a relatively high frequency in haploid cells. It was for this reason that cell lines were initiated from haploid embryos of the frog Rana pipiens; furthermore, the nuclear transplant technique developed for the same species by Briggs and King[14] held out the possibility of transferring a mutation from culture into the developing embryo. It has now been found that even in haploid cultures, two steps are required to obtain thymidine kinase deficient cells because of the existence of separate forms of the enzyme[9]. However, it is difficult to show whether or not either step is the result of gene mutations, as illustrated by observations on the loss of the second enzyme. The fully resistant, or TK-, cells are of a stable phenotype and their frequency can be increased with ICR 191, a compound mutagenic to bacteria.

Table 2

Effects of length of period between subculturing and 4 hour treatment with ICR 191 on thymidine kinase specific activity of cultures selected in $10^{-3}$ M BUdR

| ICR 191 treatment (hrs) | | Thymidine kinase specific activity | | | |
|---|---|---|---|---|---|
| | | After BUdR → | BUdR → | MLM → | 5 X MLM |
| Control | Flask | 0.4 | 0.3 | 0.2 | 0.2 |
| | Colonies | | | 0.1,0.1,0.1,0.2, 0.2,0.2,0.2 | |
| 0-4 | Flask | 2.2 | 3.9 | 2.4 | 1.2 |
| | Colonies | | | 1.7,1.9,2.3,2.4, 2.7,3.3,4.2 | |
| 4-8 | Flask | 3.0 | 0.3 | 0.3 | 0.3 |
| | Colonies | | | 0.2,0.3,0.3, 0.3,0.3,1.2 | |
| 8-12 | Flask | 3.8 | 0.8 | 0.4 | 0.4 |
| | Colonies | | | 0.3,0.3,0.3, 0.4,0.5,1.8 | |
| 12-16 | Flask | 1.7 | 0.8 | 0.5 | 1.6 |
| | Colonies | | | 0.2,0.2,0.3,0.4, 0.4,0.4,0.4 | |
| 16-20 | Flask | 0.4 | 0.3 | 0.3 | 0.1 |
| 36-40 | Flask | 1.1 | 0.4 | 0.3 | 1.8 |
| | Colonies | | | 0.2,0.3,0.4 | |

Nevertheless, the question of whether such a compound is acting as a mutagen in the induction of a phenotype is appropriate for somatic cells; experimental embryologists have demonstrated that the fate of embryonic tissues can be altered by a variety of compounds. Observations indicate that the culture conditions during and following application of ICR 191 significantly affect the number of TK- cells isolated after exposure to $10^{-3}$ M BUdR; in some cases at least 10% of the population has become TK- compared to a frequency of less than $10^{-3}$ for control cultures. Whether these TK- cells were present before selection and, if so, were induced by mutagenesis, has not been determined. However, observations suggesting that many of the variants were not present when BUdR was applied include: the unusually high thymidine kinase activity at this time; the growth curves, unlike those of artificial mixtures of TK- and TK+ cells; the occasional transformation of TK+ to TK- cultures under non-selective conditions.

If the possibility is considered that TK- cells originate from haploid frog cultures by means other than gene mutation, the function of ICR 191 in this process can now only be a matter of speculation. In tissues of intact organisms, thymidine kinase deficiency is associated with a decreased rate of cell division. For cultures in BUdR, following specific treatments with ICR 191, a relatively rapid rate of cell division is often followed by a period of minimal change in cell number, a sequence resembling the situation _in vivo_. The similarities suggest the hypothesis that specific changes in the patterns of cell division may somehow initiate loss of thymidine kinase activity. If thymidine kinase deficiency in haploid cell cultures is not the result of mutation and if reproducible conditions are found for its induction, this system would have many advantages for investigations of the molecular basis of gene expression.

Supported by U.S.P.H.S. grants CA-05959, CA-06927 and RR-05539 from the National Institutes of Health, Contract AT(11-1)3110 with the U.S. Energy Research and Development Administration (Report #COO-3110-19) and by an appropriation from the Commonwealth of Pennsylvania.

References

1. Morris, N.R. and Fischer, G.A., 1960, Biochim. Biophys. Acta 42, 183.
2. Littlefield, J.W., 1965, Biochim. Biophys. Acta 95, 14.
3. Machovich, R. and Greengard, O., 1972, Biochim. Biophys. Acta 286, 375.
4. Greengard, O. and Machovich, R., 1972, Biochim. Biophys. Acta 286, 382.
5. Adelstein, S.J., Baldwin, C. and Kohn, H.I., 1971, Develop. Biol. 26, 537.
6. Kit, S., Kaplan, L.A., Leung, W.-C. and Trkula, D., 1972, Biochem. Biophys. Res. Commun. 49, 1561.
7. Freed, J.J. and Mezger-Freed, L., 1970, Proc. Nat. Acad. Sci. U.S. 65, 337.

8. Freed, J.J. and Mezger-Freed, L., 1973, J. Cell Physiol. 82, 199.
9. Freed, J.J. and Hames, I.Z., 1975, manuscript in preparation.
10. Freed, J.J. and Mezger-Freed, L., 1970, in: Methods in cell physiology, vol. 4, ed. D.M. Prescott (Academic Press, New York) p. 19.
11. Mezger-Freed, L., 1972, Nature New Biol. 235, 245.
12. Creech, H.J., Preston, R.K., Peck, R.M., O'Connell, A.P. and Ames, B.N., 1972, J. Med. Chem. 15, 739.
13. Mezger-Freed, L., 1974, Proc. Nat. Acad. Sci. U.S. 71, 4416.
14. Briggs, R. and King, T.J., 1952, Proc. Nat. Acad. Sci. U.S. 38, 455.

*Progress in Differentiation Research, ed. N. Müller-Bérat et al.*



# PRESENCE OF DIVALENT CATIONS BOUND TO REITERATIVE DNA SEQUENCES IN HIGHER PLANT AND ANIMAL TISSUES

Igor Sissoëff, Jeanine Grisvard and Etienne Guillé
Laboratoire de Biologie Moléculaire Végétale
Bat. 430, Université Paris-Sud
91 405 Orsay, France

## INTRODUCTION

In the past years, numerous properties of reiterative DNA sequences present in eukaryotic genomes, have been described; particularly, they renature more rapidly than unique ones, they are dispersed all along the genome or clustered in specific areas, where they form the major part of constitutive heterochromatin[1]. In this last case, these DNA sequences give a satellite band in neutral CsCl gradient when their nucleotidic composition is sufficiently different from that of main band DNA. Sometimes, when reiterative DNA sequences are very short, their nucleotidic sequencing has been carried out[2].

The ability of some reiterative DNAs to form satellite bands in $Ag^+$ or $Hg^{2+}$-$Cs_2SO_4$ density gradients was frequently used for their preparative separation from the other sequences. At suitable rf , two reiterative DNA fractions are found on either side of main band DNA[3,4], one is heavier ($\rho$ = 1.55 to 1.57 g/ml) than bulk DNA, the other is lighter and has practically the density of total DNA when it is centrifuged in $Cs_2SO_4$ density gradients without added $Ag^+$ ions ($\rho$ = 1.42 to 1.44 g/ml); the main band DNA has a buoyant density ($\rho$ = 1.49 to 1.54 g/ml) depending on its mean G+C content, similar in this respect to bacterial DNAs[5].

To explain the presence of the light reiterative DNA fraction in $Ag^+$-$Cs_2SO_4$ density gradients, we propose that the "sites" able to bind $Ag^+$ ions in it, are occupied by other ions which do not increase the DNA density[6]. Different technical procedures including chelating treatments, emission phosphorescence and direct metal ion determination by anodic stripping voltammetry, were used to evidence the presence of metal ions in the DNA isolated from various biological materials.

## MATERIALS AND METHODS

At the beginning of this work, calf thymus DNA was used for three reasons; it is easily obtained in great amount; light DNA fraction can be obtained pure from other DNAs after one or two preparative ultracentrifugation in $Ag^+$-$Cs_2SO_4$ gradients at suitable rf; this DNA fraction exists in great relative amount (8%) and is well known at the biochemical level[7].

Dosage of metal ions by anodic stripping voltammetry was then applied to the DNA

This work was performed under the auspices of CNRS (L.A. 40 and RCP 845) and was supported in part by INSERM and Fondation de la Recherche Médicale Française
rf: ion/DNA-P molar ratio

from the bacteria *Agrobacterium tumefaciens* which is the oncogene agent of the crown-gall disease in higher plants, from mouse liver and rat hepatoma tissues and from plant materials including different tissues of the same plant (*Nicotiana tabacum*) and the same tissue of a given plant in various physiological conditions (tissue culture on agar medium, cell cultures in liquid medium, tissue and cell cultures from healthy and tumorous tissues).

*Extraction procedures*: Extraction procedure for calf thymus DNA has been described in details elsewhere[6]; *Agrobacterium tumefaciens*, mouse liver, rat hepatoma and plant material DNAs have been isolated by the method of Grisvard and Guillé[8].

*Preparative and analytical procedures*: CsCl, $Cs_2SO_4$ (with and without addition of $Ag^+$ ions) density gradients, dialysis procedures against chelating agents, phosphorescence measurements and anodic stripping voltammetry have been already described[6]. Sometimes, the metal ions were determined by anodic stripping voltammetry, directly after mineralization of the different fractions of the $Ag^+$-$Cs_2SO_4$ density gradients.

RESULTS

Studies on calf thymus DNA

*Chelating treatments*: Light fraction isolated from calf thymus DNA by $Ag^+$-$Cs_2SO_4$ density gradients was dialysed against EDTA (0.1 M, pH9), then against 0.1 M $Na_2SO_4$ to remove EDTA traces; after these dialysis, its buoyant density in $Ag^+$-$Cs_2SO_4$ density gradients according to various rf values was studied (fig.1).

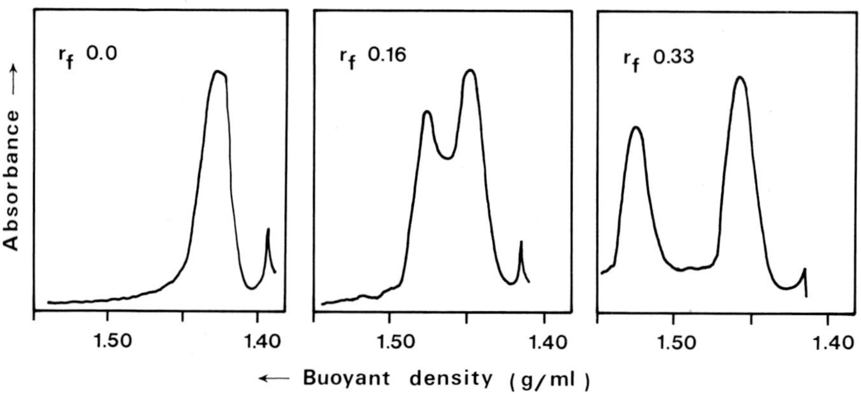


Fig.1. Microdensitometer profiles of light fraction from $Ag^+$-$Cs_2SO_4$ density gradient of calf thymus DNA, at different rf values

Even at low rf values (0.16 for instance), two fractions are clearly separated: the first one stays near the initial density of total light DNA component and the second acquires rapidly the density of the heaviest fraction of total thymus DNA. While total light fraction contains three reiterative components ($\rho$ = 1.700, 1.706 and 1.719 g/ml respectively, in neutral CsCl), the component which is made heavier after EDTA treatment is only constituted of satellite 1.719 g/ml (satellite II according to the designation of CsCl bands by Kurnit *et al*[9]).

*Phosphorescence measurements*: It is now well established that several metal ions like $Ag^+$ and $Hg^{2+}$ induce on polynucleotides and DNA an "heavy atom effect" by increasing the intercrossing system between singlet and triplet states[10]. This effect produces profound changes in emission properties and specially an increase in phosphorescence yield. Thus, by addition of $Ag^+$ ions on DNA sample, the 450 nm phosphorescence intensity is enhanced 20 fold approximatively and enable more precise measurements.

Titration of DNA with $Ag^+$ ions for main band and light fraction of an $Ag^+$-$Cs_2SO_4$ density gradient reveals a profound discrepancy between these two fractions (fig.2). For rf = 0, natural or intrinsic phosphorescence is always slightly greater for light DNA fraction; as rf is increased to 0.5, main band DNA binds more $Ag^+$ ions with respect to DNA-nucleotide than does light DNA fraction. This could be interpreted as the difficulty for $Ag^+$ ions to displace "endogeneous" metal ions bound to sites of the light DNA fraction.

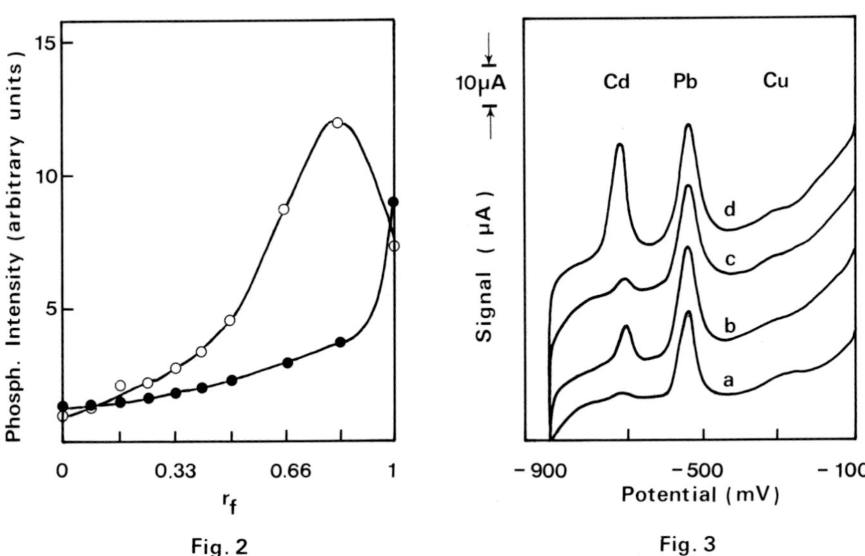


Fig. 2

Fig. 3

Fig.2 Phosphorescence intensity variations at 450 nm, versus rf for two calf thymus DNA components: o main band DNA; ● light DNA fraction; for all the samples, final DNA concentrations were 4μg/ml

Fig.3 Metal ions determination by anodic stripping voltammetry: 5 ml acetate + 1 ml digested fraction from a) main band DNA dialysate, b) main band DNA (350 μg), c) light DNA fraction dialysate, d) light DNA fraction (40 μg)

*Trace metal ions determination*: By anodic stripping voltammetry[11], only Cd, Pb, Cu were observed in appreciable amount in the various samples we have submitted to measurements.

Fig.3 shows records obtained for main band DNA, light DNA fraction and their corresponding dialysates. Quantitation of peak height realized by the method of standard additions was carried out immediatelyon the sample and gave the results expressed as rf values in table I. From these data, we can conclude that among the tested metals, at least two, Cd and Pb, exist in detectable amount in light DNA fraction roughly in an order of magnitude greater than in total DNA. By comparison, main band DNA reveals only Cd measurable peak.

TABLE I

Metal ions content (rf) of total DNA and DNA fractions from calf thymus

| metal | total DNA | main band DNA | light satellite DNA |
|---|---|---|---|
| $Cd^{2+}$ | $5\ 10^{-4}$ | $1\ 10^{-4}$ | $50\ 10^{-4}$ |
| $Pb^{2+}$ | $4\ 10^{-4}$ | - | $25\ 10^{-4}$ |
| $Cu^{2+}$ | $15\ 10^{-4}$ | - | - |
| $Bi^{2+}$ | - | - | - |

The rf values found for Cd and Pb in the light DNA fraction from calf thymus are not sufficiently high to account for the results obtained after chelating treatment of the same fraction. Other metal ions could be involved and explain these results; preliminary work using activation procedure, done in collaboration with the Centre de l'Energie Nucleaire de Grenoble (France), shows that Fe and Mg are also present in the DNA preparations.

Studies on other materials

Anodic stripping voltammetry method was applied to DNA from various biological materials (bacteria, calf thymus, mouse liver, rat hepatoma and numerous plant tissues), after a fractionation by $Ag^{+}$-$Cs_2SO_4$ density gradient. In all the tested materials, except bacteria, we found metallic cations in the light DNA fractions in greater amount than in the others (fig.4-7). We can observe that for a given species, for instance *Nicotiana tabacum*, differences exist in Cd content between different tissues; so, there is a progressive increase in Cd content from leaves, stems, healthy tissue cultures to tumorous tissue cultures (table II).

Differences exist also between the Cd content in healthy and tumorous tissues from various organisms; for example, in *Nicotiana tabacum*, the same DNA fraction, light in $Ag^{+}$-$Cs_2SO_4$ density gradients and frequently rich in G+C content, may contain 10 to 100 times more Cd in tumorous tissue than in healthy one (fig.6,7).

Sometimes, some sites in the dense DNA fractions after an $Ag^{+}$-$Cs_2SO_4$ density gradient contain more "endogeneous" ions than the main band DNA.

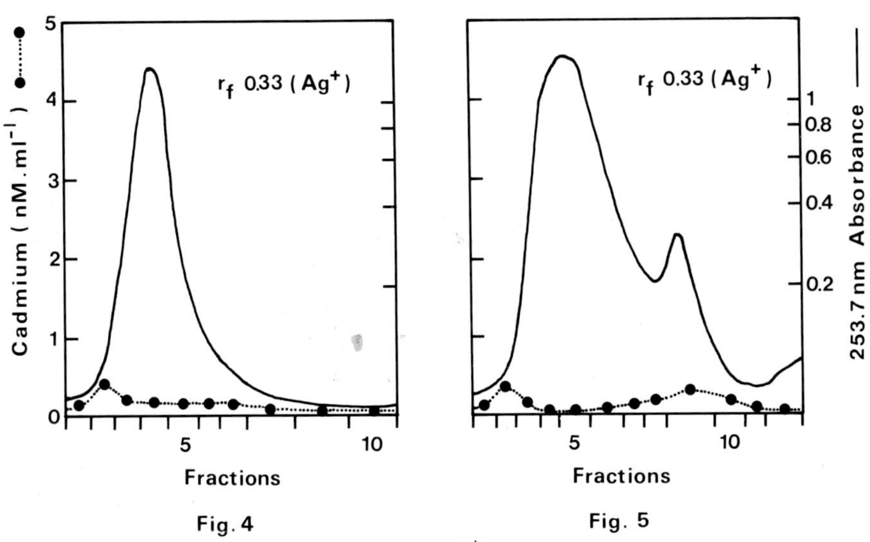


Fig.4,5. Cd content of DNA fractionated by $Ag^+$-$Cs_2SO_4$ density gradient from *Agrobacterium tumefaciens* (fig.4) and mouse liver (fig.5)

TABLE II

Cd content ($10^4$x rf) in light and main band DNAs isolated from various organisms and fractionated by $Ag^+$-$Cs_2SO_4$ density gradients

| Organisms | main band DNA | light DNA fraction |
|---|---|---|
| *Agrobacterium tumefaciens* | 5-10 | 5-10 |
| Calf thymus | 1 | 50 |
| Mouse liver (nuclei) | 1 | 14 |
| Rat hepatoma* | 3 | 240 |
| *Nicotiana tabacum* | | |
| stems | 14 | 1100 |
| leaves | 1 | 36 |
| healthy tissue cultures | 20 | 1500 |
| tumorous tissue cultures | 10 | 6600-11000 |
| *Parthenocissus tricuspidata**,** | | |
| healthy cell cultures | 2 | 7 |
| tumorous cell cultures | 7 | 190 |
| *Solanum tuberosum*** | | |
| starch parenchyma (nuclei) | 10 | 170 |

* Cd content determination in these materials have been realized directlyafter mineralization of the $Ag^+$-$Cs_2SO_4$ gradient fractions

** These materials were kindly provided by Drs Quetier and Vedel

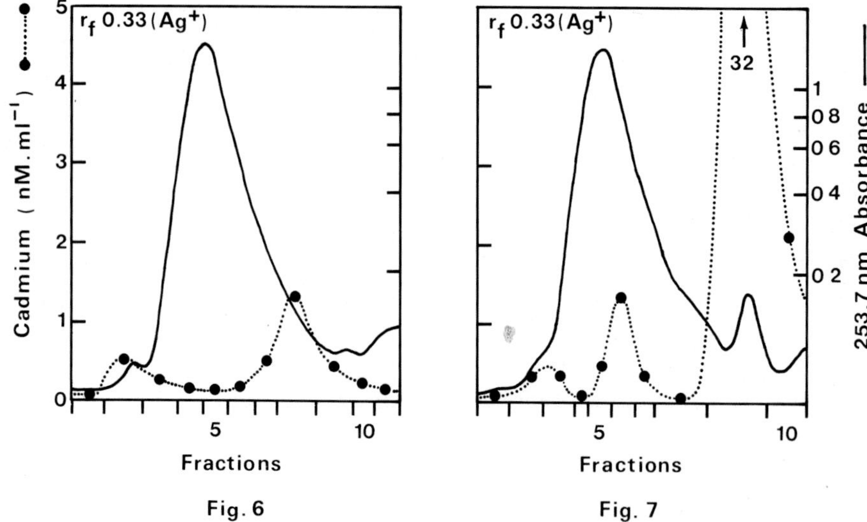


Fig.6,7. Cd content of DNA fractionated by $Ag^{+}$-$Cs_2SO_4$ density gradient from *Nicotiana tabacum* healthy (fig.6) and tumorous (fig.7) tissue cultures

DISCUSSION

From our results, two facts stand out:

- the preferential localization of metal ions in DNA fractions enriched in reiterative sequences
- the variations of metal ion content according to the tissue and/or the physiological state of this tissue.

Some metal ions are also present in the main band DNA, they may be bound to reiterative DNA sequences scattered in it; effectively, a fragmentation of the main band DNA, for instance by sonication, give rise to a new amount of heavy and light satellite fractions after $Ag^{+}$ fixation.

Our results may be confronted to properties already known for reiterative DNA sequences, for example, the fact that the helical structure of these sequences is generally destabilized[12] favours a metal ion intercalation.

Two aspects of this work will be more discussed, the origin of the metals detected in our preparations and the implications of our findings in the understanding of cancerous processes.

*Origin of metal ions*: These metal ions may be either exogeneous and/or endogeneous. In the first case, they may come from the mediums used for extraction, purification and dialysis of DNA, from various compartments of the cell, they may also come from the chromatin itself, since it is well known that the divalent cations are essential factors in maintaining the chromosomal integrity[13]. In the second alternative, they may be effectively bound *in vivo* to the DNA molecule. Exogeneous origin may be considered since, for instance, plant tissues contain Cd, Pb and Zn and even are able to accumulate them[14]. However, some facts may lead to reject at least partly this eventuality:

- check-tests done in the same condition as for $Ag^+$-$Cs_2SO_4$ density gradients never showed fixation of $Cd^{2+}$, $Pb^{2+}$ and $Cu^{2+}$ on the same sites as $Ag^+$ since competition was never detected after subsequent addition of $Ag^+$ ions;
- conditions used during extraction, purification and fractionation procedures favour the departure of metal ions especially those bound to the phosphate groups[15]. In the same conditions, divalent cations bound to the bases would be very well protected from the action of chelating agents as well as from high ionic strength[16].

Thus, we cannot definitively know if all metal ions we have detected are from exogeneous origin or if endogeneous ones are maintained on the sites they really occupied *in vivo*. However, it is well established that metal ions exist on DNA molecule[17] and some ones subsist after chelating treatment[18]. Whatever the origin of metal ions, great differences towards them are found in all the tested cases between reiterative DNA sequences and the other ones. Differences are also found according to the tissue and the physiological state of the tissue.

*The cancerous processes*

*The crown-gall disease*: the crown-gall disease is induced in susceptible plants by a virulent strain of the bacteria *Agrobacterium tumefaciens*. Two phases are clearly distinguishable during the instauration of the tumorous process: the "conditioning" step which arises from the injury of the susceptible plant tissue and the "induction" step during which the bacteria "transforms" the host cells prepared by the precedent step. The conditioning step was related to the transitory synthesis of a G+C rich satellite DNA[19,20]. The induction phase seems to result from the transfer of an exogeneous nucleic acid to the conditioned host cells; origin and real nature of foreign material is not clear (bacterial, phage and/or plasmid DNAs) even if recent results favour a plasmid role[21].

The great differences found in Cd content between tumorous and healthy tissue of *Nicotiana tabacum* and *Parthenocissus tricuspidata* may open new research area in the study of tumor induction processes; but these results may be irrelevant of the tumor mechanism itself, they may be interpreted as a general effect of division rate augmentation since a quantitative increase in Cd content has been found in reiterative DNA sequences according to the physiological conditions.

*Animal tumours*: in the case of animal tumours, we have actually tested only one material, rat hepatoma; similar phenomenons as those described for crown-gall disease may occur in animal tissues; so, during formation of nodules of mouse liver by chemical carcinogens - step which may be called conditioning phase - the appearance of an A+T rich satellite DNA has been described[22]. Cellular DNA synthesis may be required for establishment of the transformed state and during the process of transformation, one or more species of cellular DNA may be amplified[23].

Our results showing a large increase in Cd bound to reiterative DNA sequences of tumour DNA are to be confront with those of Andronikashvili *et al*[24], describing variations of numerous metal ions present in DNA preparations during the growth of transplanted sarcoma M.

We are presently looking for the metal ion binding to reiterative DNA sequences during the various steps of tumour induction, specially during the DNA amplification.

REFERENCES

1. Guillé,E. and Quetier,F. (1973) in Progress Biophys. Mol. Biol. (Butler,J.A.V. and Noble,J.,Eds.) vol.27, 121 Academic Press, London
2. Southern,E.H. (1970) Nature (London) 227, 794
3. Corneo,G. *et al* (1968) Biochem. 7, 4373
4. Yasmineh,W.G. and Yunis,J.J. (1971) Exp. Cell Res. 64, 41
5. Jensen,R.H. and Davidson,N. (1965) Biochem. 4, 17
6. Sissoëff,I. and Guillé,E; (1975) submitted to Arch. Biophys. Biochem.
7. Filipski,J. *et al* (1973) J. Mol. Biol. 80, 177
8. Grisvard,J. and Guillé,E. (1973) Prep. Biochem. 3, 83
9. Kurnit,D.M. *et al* (1973) J. Mol. Biol. 81, 273
10. McGlynn,S.P. *et al* (1969) in Molecular Spectroscopy of the triplet state, Prentice Hall, New Jersey
11. Barendrecht,E. (1967) in Electroanalytical Chemistry (Bard,A.J.,Ed.) vol.2, 53 Academic Press, New-York
12. Corneo,G. *et al* (1970) J. Mol. Biol. 48, 319
13. Steffensen,D.M. (1961) J. Amer. Chem. Soc. 90, 7323
14. Shanvrenkova,L.I. and Vorob'eva,L.I. (1974) Izv. Akad. Nauk. Turkm SSR 5, 22
15. Weser,U. and Bischoff,E. (1970) Europ. J. Biochem. 12, 571
16. Schlatterer,B. (1969) Master's thesis, Tubingen
17. Wacker,W.E.C. and Vallee,B.L. (1959) J. Biol. Chem. 234, 3257
18. Loring,H.S. and Waritz,R.S. (1957) Science 125, 646
19. Quetier,F. *et al* (1969) Biochem. Biophys. Res. Commun. 34, 128
20. Guillé,E. (1972) Doctoral thesis, Orsay, France
21. Zaenen,I. *et al* (1974) J. Mol. Biol. 86, 109
22. Epstein,S.M. *et al* (1969) Chem. Biol. Interactions 1, 113
23. Granboulan,N. and Tournier,P. (1965) Ann. Inst. Pasteur Paris 109, 837
24. Andronikashvili,E.L. *et al* (1974) Cancer Res. 34, 271

Progress in Differentiation Research, ed. N. Müller-Bérat et al.



# DIFFERONES: A SIMPLIFYING CONCEPT IN DIFFERENTIATION

Per O. Seglen

Department of Tissue Culture
Norsk Hydro's Institute for Cancer Research
The Norwegian Radium Hospital
Montebello, Oslo 3, Norway

The field of differentiation does not exactly suffer from a lack of terminology, but it does lack a fruitful working terminology related to the known molecular facts. The classical terms such as modulation, competence, determination, permission and instruction (to name a few) have still not been given much substance, and most of the more recent models for differentiation are hypothetical constructs carrying their own secluded set of terms.

In the following brief comment I do not want to present any new hypotheses, but merely try to emphasize some important features of differentiation in terms of well-known mechanisms and a few supplementary concepts. There are three main points I want to make:

## 1. A definition of differentiation

The first point regards the definition of differentiation. There exists no universally accepted definition; each worker tends to assemble his own collection of phenomena which he considers as being differentiations, and another collection which he thinks is not. My own suggestion is that differentiation be defined as a change in the pattern of gene expression in a cell or in a cell assembly (such as a tissue). This would include both the transcriptional and the post-transcriptional events which participate in determining the relative concentrations of individual protein or RNA molecules. (The most important aspects would be accessibility of the DNA template; RNA synthesis, processing, translocation and degradation; protein synthesis and degradation).

Notice that I regard any quantitative change in the pattern of gene expression as a differentiation, whereas many people insist that the concept should be reserved for qualitative changes. Consider a situation where the formation of a protein is dependent on an inducer in a dose-dependent fashion: if a small inducer dose raises the protein level from 0 to 0.1, and a second large dose raises it from 0.1 to 100, I cannot see any biological, practical or theoretical reason why the first induction should be called a differentiation and the second not. Establishment of a zero level would, at any rate, depend on the sensitivity of our assay. In other words, by an exclusively qualitative definition we can never be sure whether we are really dealing with a differentiation or not. Furthermore, in a system containing more than one cell, quantity may reflect quality in the sense that the number of responding cells may vary.

The simplest type of differentiation, then, is a selective change in the level of a single protein (e.g. enzyme) or RNA, generally known as an "enzyme induction". More complex differentiations can be regarded as assemblies of such individual inductions. The problem of differentiation which, by definition, lies at the cellular level, can therefore be reduced to the biochemical level as the problem of enzyme (protein) induction, about which much is known.

## 2. The dynamics of differentiation

The second point I want to make concerns the dynamics of differentiation. Most cells are traversing the cell cycle, although sometimes very slowly, or they are on the pathway to cell death, both of which involves continuous changes in gene expression. In addition, the whole organism is traversing its life cycle, and is subject to daily, monthly and seasonal environmental changes which affect the pattern of gene expression. Thus a cell is never in a constant state with regard to gene expression, it is always changing its pattern, - i.e. it is always differentiating.

This has two important implications: one is what we might call the relativistic aspect of differentiation - any differentiation is a change in a changing system, and it has to be defined relative to its background. A given differentiation is not as much a new process, as it is a change in the direction and content of a process already going on.

The fact that differentiation is a continuous process - this we might call the continuity aspect of differentiation - means that we can follow it backwards in time as long as we like. Each cell in the body is the result of events which took place in preceding generations of cells and organisms, back to the origin of life and even further, i.e. there may be many end points of differentiation, but there are no beginnings.

Thus, when observing a given differentiation, we must realize that it is rather arbitrarily defined by the observer: by defining the starting point of our differentiation, we perform a conceptual excision from a larger, continuous process of differentiation. Some people draw a distinction between differentiation and a preceding event of determination, the latter being supposed to create the conditions necessary for the differentiation. This distinction is completely unnecessary, since determination and differentiation are really continuous stages in a larger differentiation process, only separated by our arbitrary conceptual cut (the definition of the starting point).

## 3. Agents of differentiation: The differones

My third point concerns the effectors and biochemical mechanisms of differentiation. An enormous variety of molecules have been shown to induce differentiation (defined as a change in gene expression) in various biological systems. These include ions, gases, drugs, metabolites, hormones, matrix proteins, cyclic nucleotides, receptors, chromatin proteins and a host of chemically uncharacterized intra- and extracellular factors, detectable e.g. by cell hybridization or heterotypic tissue inductions. Several such effectors interact to induce a differentiation, such as in the well-known type of effector chain consisting of hormone, receptor, adenylate cyclase, cyclic AMP, protein kinase and nuclear phosphoproteins. Since we know that several effector chains may control the expression of one gene, and since one effector may affect the expression of several genes, it is appropriate to speak about effector networks rather than effector chains. To illustrate the concept, a tentative effector network controlling tyrosine aminotransferase synthesis is shown in fig. 1.

In order to simplify these very complex relations, at least conceptually, I suggest that all the effectors involved in differentiation be assigned equal importance, and classified together under the name of "differones". It follows from the continuity aspect of differentiation that any distinction between primary and secondary effectors is arbitrary, and the concept of effector networks furthermore abolishes any distinction between effectors and mechanisms. Thus, glucorticoid hormone and an acidic repressor protein in the liver chromatin both act as differones, and the appearance of one or the other during development may be equally important in determining the further course of differentiation.

A classification scheme for the differones may be convenient. Since differentiation is defined as a process at the cellular level, it seems logical to classify the differones according to their origin relative to the differentiating cell (table 1).

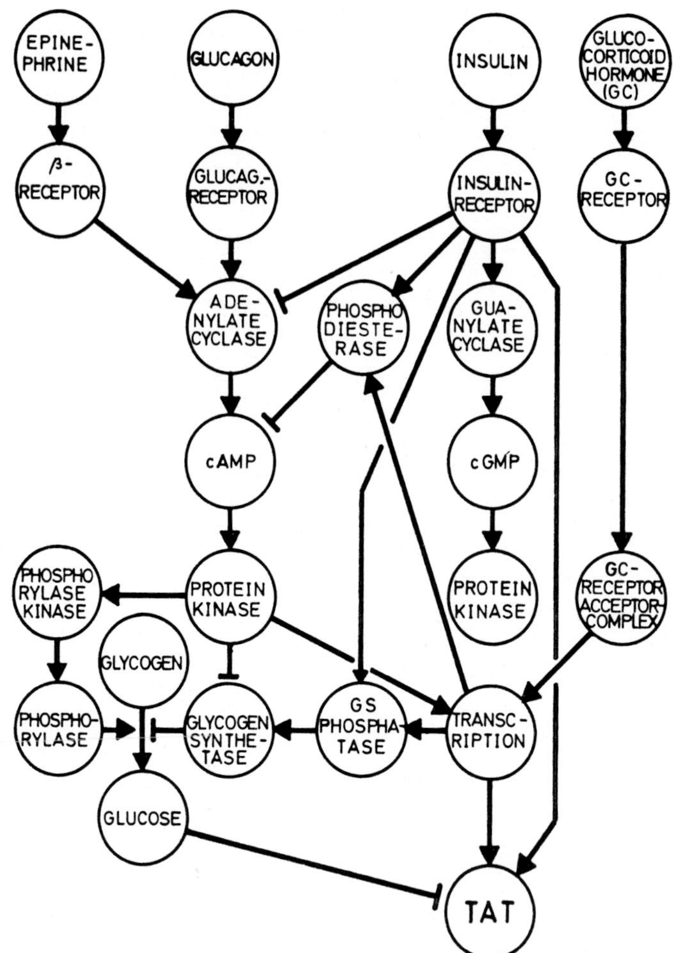


Fig. 1. A tentative effector network which may operate in the liver cell. The figure shows some of the differones which may be involved in the control of glycogenolysis and tyrosine aminotransferase (TAT) formation. TAT formation can be repressed by glucose and induced by epinephrine, glucagon, insulin and glucocorticoid hormone; the two latter hormones have also been implicated in glycogen synthetase phosphatase induction. Phosphodiesterase can be induced by cAMP and activated by insulin.

Pointed arrows indicate formation, activation or induction; flat-headed arrows indicate degradation, inactivation or repression.

Differones produced by the cell itself are called automones, while isomones are produced by other cells of the same type. Allomones are exogenous differones produced either by other types of cells (subclass hormones) or originating outside the body (subclass oecomones). The differone concept and the above classification has the advantage of being non-committed with respect to the molecular nature of the differone as well as its mechanism of action, both of which are often unknown.

Table 1. A classification of differones

1. Automones: Produced by the cell itself (or its predecessors, affecting only the cell itself.
2. Isomones: Produced by other cells of the same type.
3. Allomones: Produced by cells of other types, or of a non-cellular origin.
   - 3.1. Hormones: Produced by other cells within the individual.
   - 3.2. Oecomones: Originating outside the individual.
     - 3.2.1. Pheromones: Produced by other individuals of the same species.
     - 3.2.2. Xenomones: Produced by other species.
     - 3.2.3. Nutriomones: Biogenic substances in the food, or metabolites with a primarily nutritional function.
     - 3.2.4. Enviromones: Abiogenic environmental factors.

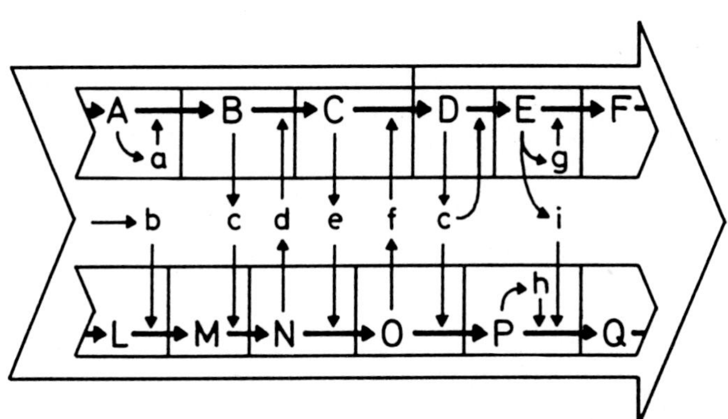


Fig. 2. A model of differentiation. The figure shows a complex differentiation system (outer frame) consisting of two differentiating cells (inner frames). The first cell differentiates from cell state A to cell state F through a series of intermediary cell states; the second cell differentiates from cell state L to cell state Q. Differones are indicated by small letters. Each cell state transition (change in gene expression pattern) is a separate differentiation, induced by one or several differones. A → B and E → F are automone-induced; P → Q requires the combined action of an automone and an allomone (hormone). D → E is isomone-induced; the remaining single-step differentiations are allomone (hormone)-induced. B → D and M → P show how differentiation can be driven forward by the dialectical interaction between two cells. Notice how the same differone (c) can have different effects in different cell states.

Differentiation, then, can be regarded as a continuous series of changes in gene expression, induced by alterations in the spectrum of differones acting upon the cell (fig. 2).

Fig. 2 shows a complex differentiation system consisting of two different types of cells. Each capital letter denotes a cell state, i.e. a specific pattern of gene expression. The transient cell state A includes the formation of the differone a, an automone which induces further differentiation of the cell into cell state B. In this state the upper cell produces the differone c (a hormone), which induces differentiation of the lower cell from state M to state N, in which state the hormone d is produced and induces further differentiation of the upper cell into state C. The point is that new cell states, i.e. changes in the pattern of gene expression, may include the formation of new differones, which provide the driving force for continued differentiation. A given cell state is determined by the spectrum of differones (past and present) acting upon the cell. If the expression of a gene at a given stage of differentiation is maintained by automones only, it will behave as a stable and heritable property in a controlled environment. If allomones are involved in the maintenance of the gene expression, the property may persist or get lost, depending on whether the environment (e.g. a tissue culture medium) contains the allomones in question. Several examples of each kind are known.

Reference: P.O. Seglen (1974). Differones. Control of gene expression and cellular differentiation by hormones and other agents, with particular emphasis on liver tissue. Norwegian J. Zool. 22, suppl. 1, pp. 1-131.

# Section 3.
# CELL MEMBRANES AND CELL SURFACES IN RELATION TO DIFFERENTIATION

*Progress in Differentiation Research, ed. N. Müller-Bérat et al.*



# SOME PRINCIPLES OF MEMBRANE DIFFERENTIATION

Werner W. Franke and Jürgen Kartenbeck

Division of Membrane Biology and Biochemistry,
Institute of Experimental Pathology,
German Cancer Research Center
D-69 Heidelberg, Federal Republic of Germany

Many cell differentiation processes are accompanied by, or even appear to consist of, characteristic changes in the cellular membrane system. In addition, a variety of transition forms exists within a specific membrane system as well as among different membranes that seem to indicate something like a "differentiation" of that membrane. The term "membrane differentiation", however, as it is frequently used in the literature, has diverse meanings and is applied to basically different phenomena. Therefore, it is sensible and helpful to organize and classify the different meanings and phenomena.

## I. CELL DIFFERENTIATIONS THAT ARE CHARACTERIZED BY CHANGES OF THE QUANTITIES OF MEMBRANES

A long list of examples of cell differentiation processes exists that are characterized by large increases of one type of membrane, either in terms of absolute mass or relative to other cell membranes. Such examples include (i) the dramatic increase in rough surfaced endoplasmic reticulum (rER) in the development and the stimulation of cells active in protein secretion, (ii) the formation of large amounts of specific forms of smooth ER (sER) in cell types that produce large amounts of steroid hormones or are induced to metabolize xenobiotic compounds, (iii) the development of a massive Golgi apparatus (GA) in cells that produce and secrete carbohydrate-rich materials, (iv) the formation of specific large vacuoles such as in differentiated plant cells, (v) the transitory accumulation or storage of specific vesicles containing secretory products, (vi) increases in the number of peroxisomes, lysosomes and similar vesicle structures as well as various alterations in the concentration of endo- and exocytotic vesicles, and (vii) a relative enlargement of cell surface membrane. Particularly striking changes in some cell systems are also (viii) increases in the numbers or sizes of mitochondria and plastids and changes in the ratio of inner versus outer membranes of these organelles and (ix) growth of nuclei, and therefore increase of nuclear envelope (NE) area. On the other hand, a variety of cell differentiations are characterized by absolute or relative decreases of membranes which may result in the complete disappearance of some specific membranes. This is, for example, observed during erythropoiesis and lymphopoiesis, during many cyst or spore formation processes, and in the development of seeds, oocytes and eggs, and sperm cells. Disappearance of membranes is also characteristic of a variety of cell pathological changes, such as during pyknosis. Such changes may be gradual or may take place during very brief time intervals.

---

Fig. 1. Schematic presentation of various forms of intramembrane heterogeneity which results in the formation of a mosaic of functional and structural units, each with a significant stability. The individual symbols represent different components in the specific membrane (double line), and the dotted regions denote the extracellular and the intracisternal (intravesicular) spaces, respectively. The arrangements sketched are as follows : (a) random distribution of (mobile) membrane components; (b)an area from which certain components are excluded (bracket; excluded components are symbolized by solid circles); (c) homogeneous cluster of a single type of membrane component (paracrystalline array); (d) heterogeneous clustering of several components; (e) region of randomly or non-randomly distributed components but with locally altered fluidity; (f) region in which clustered (as shown here) or randomly arranged components interact with a single type of cytoplasmic structure (small arrows symbolize interactions); (g) region in which randomly distributed (as shown here) or clustered membrane components interact with several extracellular (g) or cytoplasmic (g') components; (h) region with a stable membrane-to-membrane interaction and/or linkage, either between the surface and intracellular membrane (for example, subsurface cisternae, alveolar cisternae in ciliates) or between similar or different intracellular membranes (h'); (i) stable evaginations of a membrane, frequently characterized by the interaction with peripheral (membrane-associated) or with cytoplasmic structures (for example, microvilli); (j) membrane evagination of the bud type (for example, budding of viral envelopes); (k) invagination of a membrane, either as a transitory bleb or as a permanent infolding, again frequently in association with, and perhaps induced by, extracellular or cytoplasmic components; (l) region containing membrane gaps which are of transitory nature and/ or artificially induced; (m) association of membranes with nucleoprotein structures such as ribosomes (shown here, arrows indicate intimate interaction between these structures and the supporting membrane) and chromosomes; (n) regions of vesicle bud formation at intracellular membranes, frequently in association with, and/or induced by, cytoplasmic or intracisternal (shown here) components; (o) and (p) arrangement of membrane components in nonmembranous structures of, for example, filamentous, tubular, sheetlike, or hexagonal organization, either as heterogeneous (o) or homogeneous (p) aggregates.

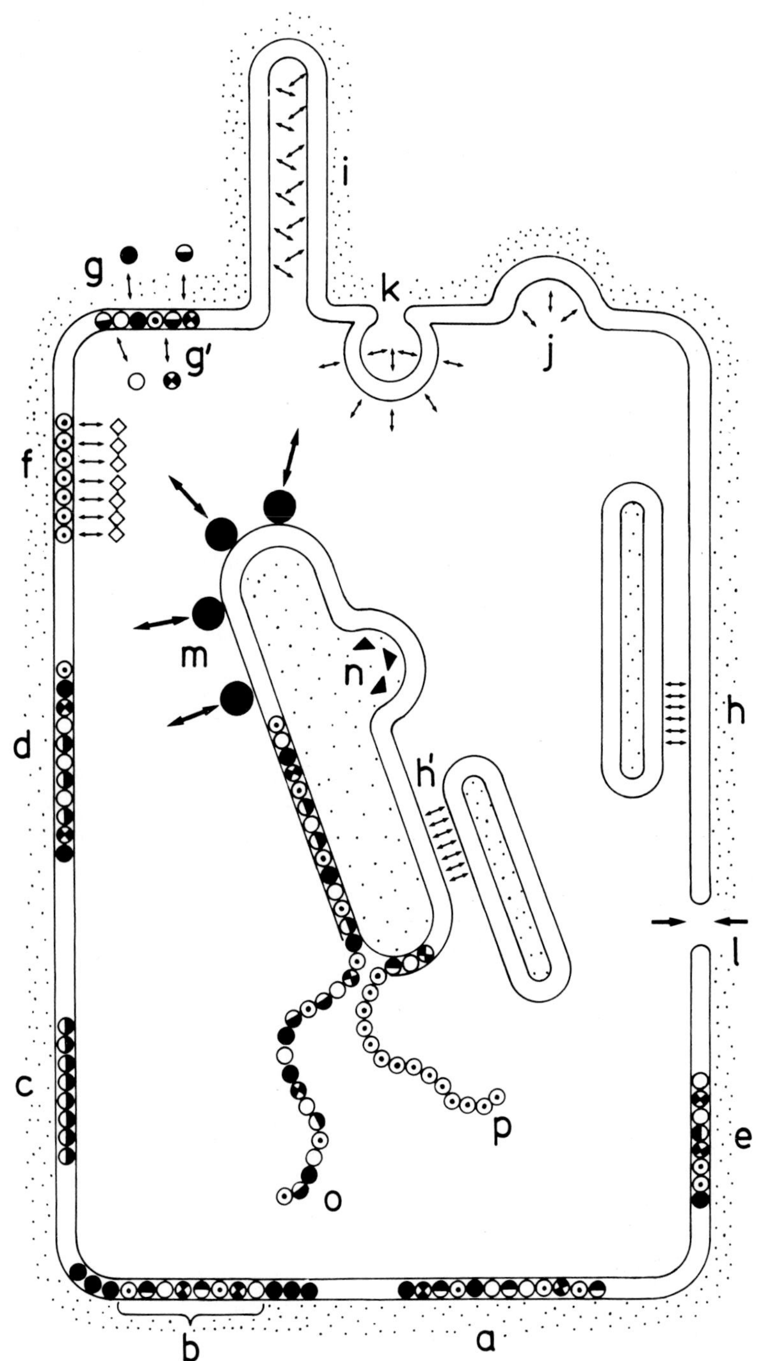


(1)

Although changes of specific membrane systems seem to be morphologically the dominant feature in a great many of such cell differentiation processes, it is somewhat inappropriate to use the term "membrane differentiation" to designate this situation. It is rather the change of the specific cell that is accompanied or caused by a change of the quantities of membranes which themselves may remain, at least structurally, unaltered (i.e. undifferentiated.

## II. DE NOVO APPEARANCE OR COMPLETE LOSS OF SPECIFIC MEMBRANE STRUCTURES IN THE COURSE OF CELL DIFFERENTIATION PROCESSES

In some cell differentiations one observes the appearance of morphologically novel membrane structures which then may become even characteristic for the specific differentiated state. Examples of this include vacuole formation in plants, formation of paired ER cisternae and annulate lamellae (AL)[1-3] in a variety of animal and plant cell systems, the transition from prolamellar bodies to thylakoids in greening plants, the de novo appearance of GA [4] and NE [3,5] as well as the appearance of storage or secretory vesicles that are specific and exclusive for the differentiated state. Correspondingly, complete, and in many examples even irreversible, losses of membrane structures such as plastids, ER, GA, and even NE, take place in other cell differentiation processes, including examples of cell degeneration such is in keratinization of epidermis, enucleation of mammalian erythrocytes, and in cell breakdown in the formation of ploem and xylem elements in plants. This group of phenomena would certainly merit the classification "membrane differentiations" in a broad sense, but it should not be confused with the meanings of this term described under IV-VI.

## III. QUANTITATIVE CHANGES OF SPECIFIC PARAMETERS AND PROPERTIES IN MEMBRANES

Such changes, which may have very profound influences on cellular functions, are morphologically described by quantitative changes of some substructures in a specific membrane system. Such changes include, for example, the number of polyribosomes per unit area of rER, the proportions of vesiculated and tubular sER or GA, the number of pore complexes per NE and AL, the density of ATPase containing particles on mitochondrial cristae, and the frequency of cell surface specializations (see below). Into this category of changes also fall the variations of the packing density of functional chemical moieties as well as the overall average figures for the frequency of interaction sites of membranes surfaces with peripheral proteins and cytoplasmic proteins including tubulin, spectrin, actin and myosin. Again, such quantitative changes do not necessarily "differentiate" the membrane character but it is obvious that, from a certain extent on, they may result in changes in the membrane nature itself and thus represent qualitative truly differentiating alterations (see VI).

## IV. INTRAMEMBRANE DIFFERENTIATIONS (LATERAL DIFFERENTIATIONS)

Changes that occur within a specific membrane continuum and result in the formation of local heterogeneity, i.e. in the formation of defined and stable mosaic patterns which are characteristic of the particular cell-type, differentiate one membrane locus from the bulk and/or from other loci. Some examples of such intramembrane differentiation phenomena, which do not involve chemical changes of the individual membrane components but may well involve changes of the intermolecular forces, are presented in the scheme of Fig. 1 and, in some selected morphological demonstrations, in the electron micrographs shown in Figs. 2-7. This concept emphasizes the occurrence of both fluid components, with a relatively high mobility and a tendency to random distributions (positive entropy), and fixed components which are present in a non-random distribution and constrained by specific forces (for reviews of the coincidence of both fluid and stable domains in membranes see 6-11). Such constraints might result from the mutual interaction of the constitutive membrane components or, on the other hand, from their interaction with peripheral membrane-associated components, cytoplasmic or nucleoplasmic structures, or other membranes (Fig. 1). As a result of such lateral intramembranous differentiations, most membrane systems cannot be regarded as uniform with respect to the distribution of membrane components but rather represent mosaics of structurally and functionally different units. This is particularly well demonstrated in plasma membranes and nuclear membranes (e.g. Figs. 1-7; see also 3, 5), but is also recognized in the differentiation of peroxisomes and transitional elements [12] from rER, in the enzymatic microheterogeneity of the ER system (for review, see 13), in dictyosomes (GA [4]), in the thylakoids of plastids, etc. Local structural heterogeneity is also recognized in various forms of disintegrations and rearrangements of membrane components;

Figs. 2-4. Examples of specific intramembrane domains and of associations of membranes with nonmembrane structures. Fig. 2a presents largely uniform aspects of nuclear envelope membranes (in the far left, two pore complexes are denoted by white arrows), of sER membranes, and the equidistantly spaced plasma membranes (denoted by the pairs of short arrows in the very top and the bottom) in spermatid cells of the newt, *Triturus alpestris*[3]. Note the occurrence of intermembrane cross-link threads between the two plasma membranes in some regions (e.g. in the bottom part). Note also the greater thickness and the clearer "dark-light-dark" pattern in the plasma membranes and in one of the vesicles (v), in contrast to the appearance of the nuclear membranes and the sER membranes and the other vesicles (for this comparison a nuclear membrane and an ER-profile adjacent to plasma membrane profiles are denoted by arrows).

Various forms of microtubule-membrane and microfilament-membrane interactions are shown in Figs. 2b - 4c. Microtubules can be closely associated with the inner surface of plasma membranes (third arrow from top in Fig. 2b which shows a cross-section through a pollen tube of the lily, *Lilium longiflorum*; (w, cell wall; c.f. 25) or are connected to the plasma membrane by lateral cross-bridge-structures (arrows No. 2, 5-7 from top in Fig. 2b), similar to the bridge-structures that connect the cortical membrane tubules and vesicles to this plasma membrane (arrows No. 1 and 4, Fig. 2b; c.f. 25).Such membrane-to-membrane connections are also observed (Fig. 3a) between the alveolar cisternae (AC) and the plasma membrane (PM) in the pellicula of the ciliate, *Tetrahymena pyriformis* (short arrows, 26, 27). Note that bundles of microtubules (denoted by the pair of arrows in the upper left of Fig. 3a) are linked to the filamentous layer ("epiplasmic layer , c.f. 27) that is associated with the inner aspect of the alveolar cisternae (arrow in the bottom of Fig. 3a) again by thread-like linkages. Similar to the microtubule-to-membrane cross-bridges (seen at higher magnification at the alveolar membrane in Fig. 3b, arrow, same preparation as in Fig. 3a) are cross-bridges that link the microtubules to the filamentous layer[27] denoted by the arrow in the bottom part of Fig. 3b). Fig. 3d shows the terminal attachment of microtubules of the spindle apparatus (arrows) of the lower fungus, *Phycomyces blakesleeanus*, at the polar plaques (P) of the nuclear envelope (NE, for refs., see 3,5,28-3o). Fig. 3c illustrates locally different aspects in the nuclear envelope of a rat spermatid cell (N, nucleus; CH, chromatin; PC, proximal centriole; SB, satellite body of this centriole containing the "matrix extensions" described by Fawcett and Philips[31], see there for further details). Note the altered structure of the nuclear envelope in the capitulum region which is characterized by the regularly spaced filamentous connections between the centriolar and pericentriolar dense matrix masses and the outer nuclear membrane (arrows).

Intermembrane fibrils that run between the dictyosomal cisternae of various organisms, especially of many plant cells, appear to connect the central regions of adjacent cisternae (arrow in Fig. 4a, lily pollen tube, FF, forming face of the dictyosome, c.f. 4) and sometimes reveal lateral cross-bridges to the membrane surfaces (for details see 32). Very regularly spaced membrane-to-membrane cross-bridges[26] are recognized (Fig. 4b) in the "pre-exocytotic attachment plaques" between casein containing vesicles (CV) and the apical plasma membrane of lactating rat mammary gland cells (double arrows in 4b; for details see 17; c.f. also 33. Note also the close association of microfilament bundles (long arrows in the left half of Fig. 4b) with the ridges of the apical surface.

(Legend Figs. 2-4 continued on p.222)

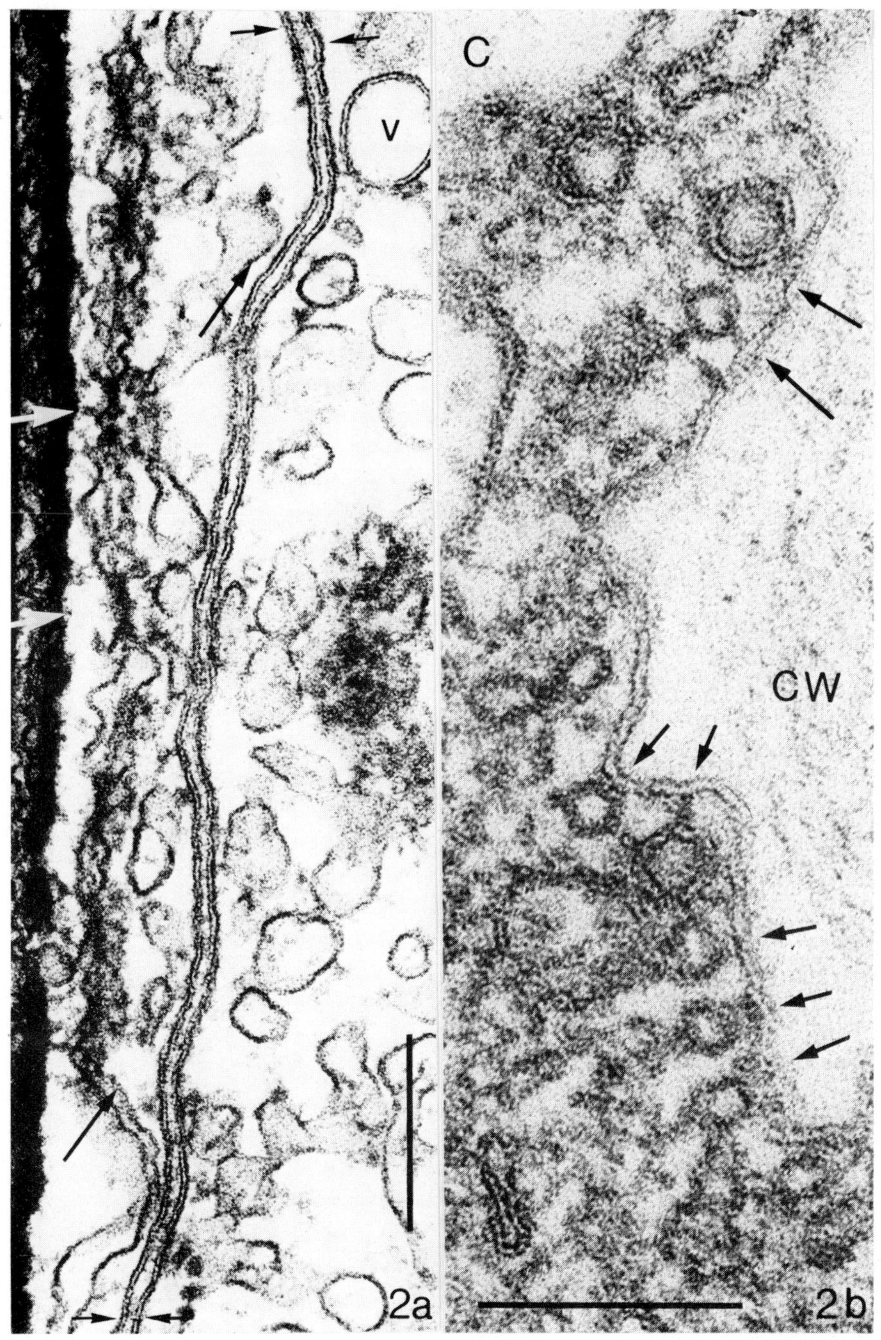
V
2a
C
CW
2b

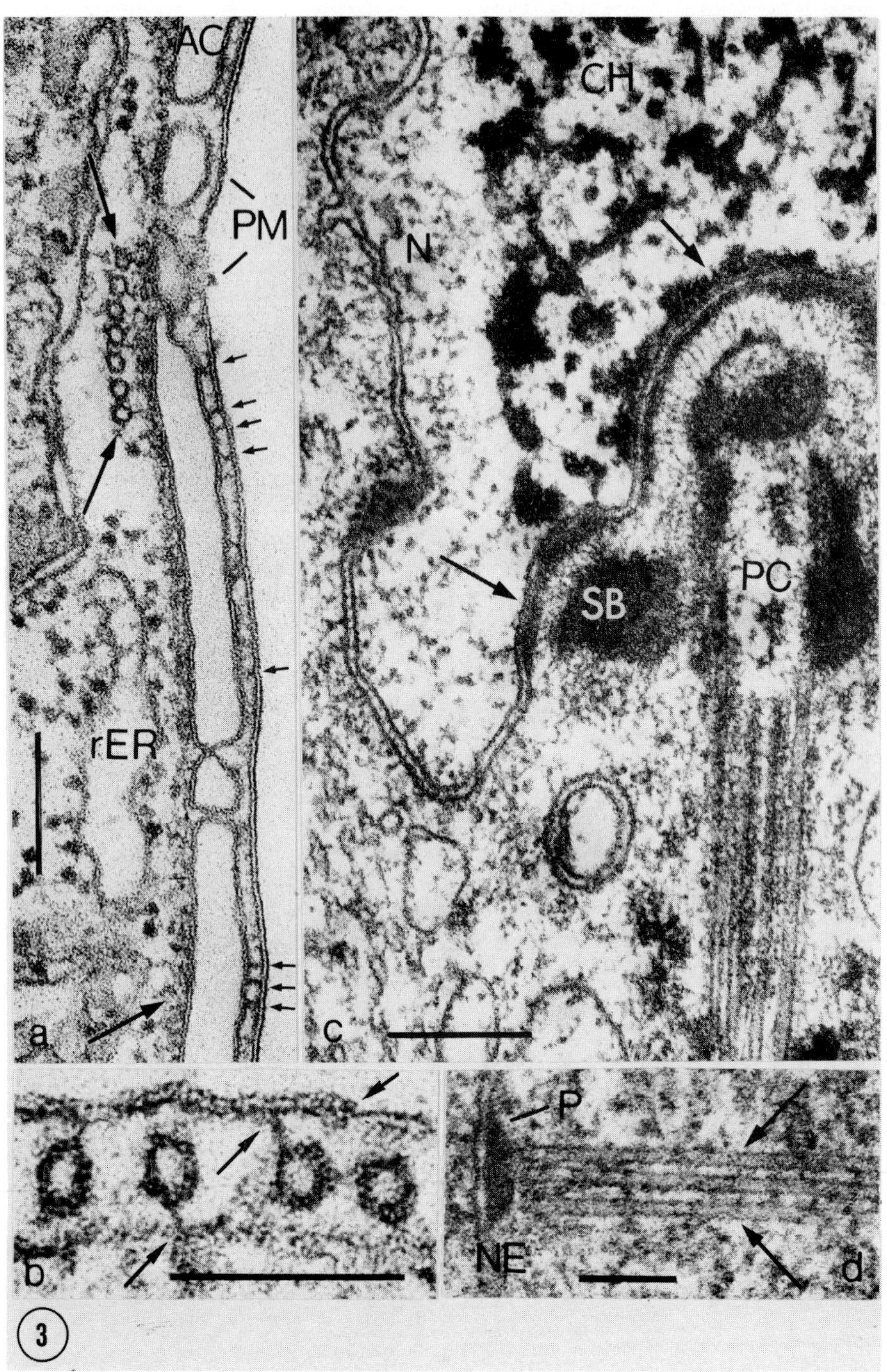
AC
PM
rER
a
N
CH
SB
PC
c
b
P
NE
d
3

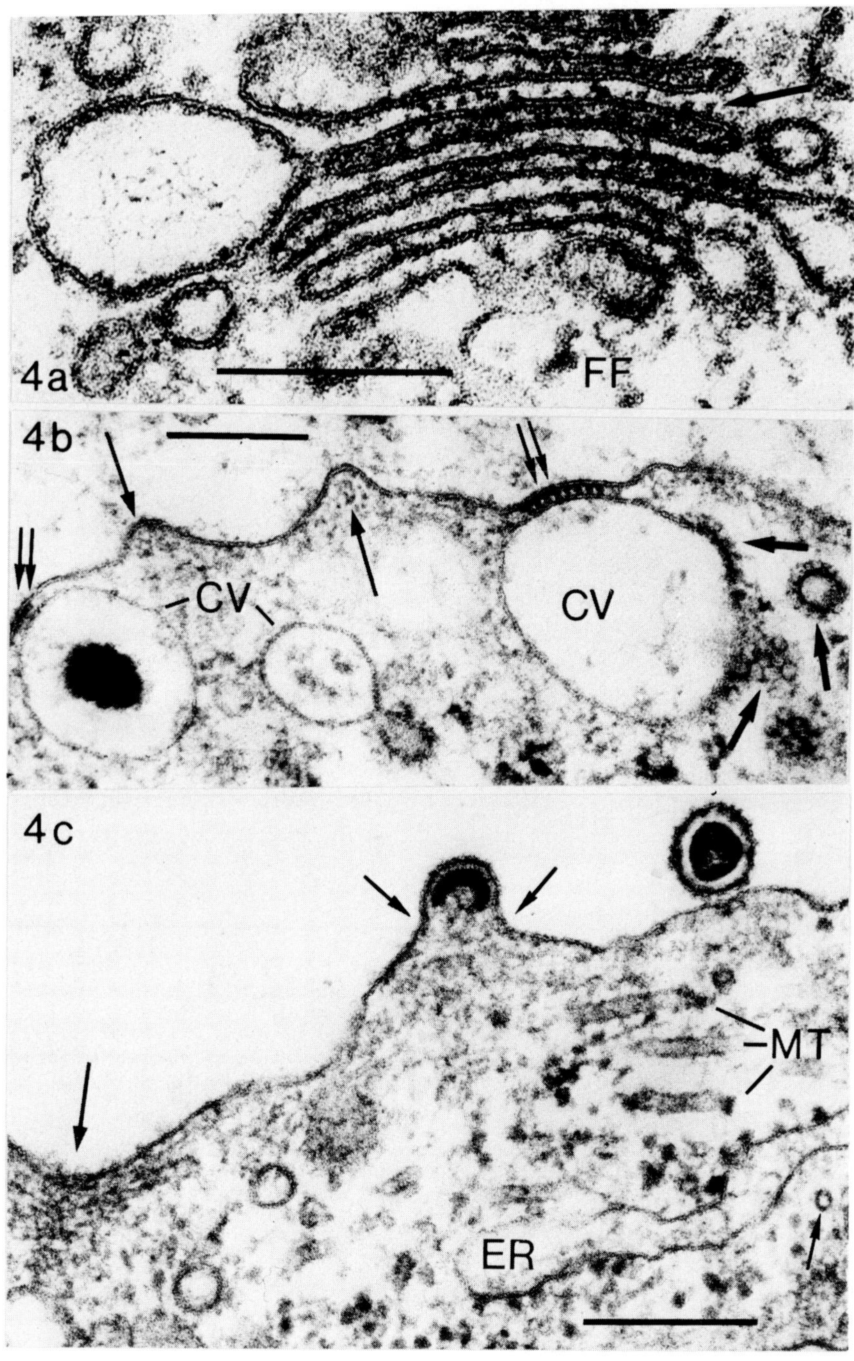
4a
FF
4b
CV
CV
4c
MT
ER

for example, a variety of cytopathological cases are characterized by the formation of "myelin-like configurations" and various filamentous or tubular structures which seem to be derived from "normal" membrane structures (for review see 14).

The special forms of membrane morphology in differentiated cells indicate that this lateral heterogeneity is functionally important. For example, in the plasma membrane of an epithelial gland cell, it is obvious that the basal plasma membrane is structurally and functionally different from apical surface areas, and that both are different from the lateral plasma membrane sides which contains differentiated regions specialized for intercellular association. Moreover, within one of these major differentiated surface areas one again notes, at a finer morphological level, a further differentiation into, for example, the different junctional complexes, microvilli, canalicular formations, etc. (Figs. 4-7). A particularly clear example of a structurally and chemically differentiated domain within one region of a cell membrane is the thickened plaques of hexagonally packed subunits that occur in the luminal plasma membrane of the mammalian urinary tract epithelium (for review see 15). It is also obvious that the steady state dynamics of different loci in the diverse membranes can greatly differ (see also VI). Pore complexes, for example, disappear in some regions of the nuclear surface of spermiogenic cells such as in the acrosome- or centriole-associated regions but are maintained in others (for refs. and further examples of heterogeneous distributions of pore complexes see 3,5). There exist, as another example, regions in the surface membrane in which endo- or exocyto-

---

Fig. 2.-4. cont. Typical "bristle coats" (thick arrows in Fig. 4b) over the entire surface of some vesicles and parts of the surfaces of others (A, alveolar space). Fig. 4c presents a region of the surface of an in vitro cultured murine sarcoma cell (strain 18o) in which microfilaments are closely associated with the plasma membrane (arrow in the left) adjacent to another region in which a viral envelope is formed by budding from the plasma membrane (center arrow pair). The latter situation is characterized by the intimate and equidistantly spaced attachment of the viral core to the membrane. A released mature C-type virus particle is seen in the upper right corner. Note also the frequency of microtubules (MT denote longitudinal sections, the arrows in the lower right indicates a cross-section) in this cortical zone. Scales indicate o.2 μm (in Figs. 2a; 3a, c, d; 4 a-c), and o.1 μm (in Figs. 2b, 3b).

tic events occur much more frequently, and membrane heterogeneity resulting form such events is frequently recognized in the form of pieces of membrane that are just being incorporated or are in the process of fusion, and *vice versa* , in surface invaginations that are in the process of endocytotic vesicle formation. In both processes this specific membrane differentiation is frequently correlated with the appearance of typical "bristle coat" structures (Figs. 4b, 5, 6, for refs. see 16,17). Corresponding surface protrusions and bud detachment processes such as during virus-budding and milk fat globule formation (see Fig. 4c and refs. 11,17-22) are also not distributed at random and, moreover, represent intramembranous islets of different turnover kinetics and fate. Therefore, all experiments which use total membranes, either native or isolated, or which use gross subfractions are not necessarily relevant to the functional situation present *in situ* since local heterogeneities are usually averaged out in such analyses.

As to the question of how such local and sometimes very selective and specific local heterogeneities within a membrane are produced, the recent analyses of "capping" and "clustering" phenomena provides some information for the formulation of hypotheses that encompass principles from two lines of observations which at first sight seem to be opposing: (*i*) the lateral mobility of individual membrane components, and (*ii*) the stabilization of certain intramembranous aggregates including nucleation sites as well as more permanent differentiations. Such differentiated domains within a membrane might be induced and stabilized by the association with extramembranous components (Figs. 1-7; for refs. see also 3,5-11). Examples for such interactions are probably the surface attachment of certain extracellular materials that induce endocytotic caveolae and subsequently their own uptake or local membrane fusion (c.f., e.g., 23), the association of viral cores of, e.g., togaviruses (arboviruses), myxoviruses, B- and C-type (oncorna)-viruses etc. with the inner aspect of the plasma membrane which may induce the local clustering of virus-specific membrane components and the subsequent formation and detachment of the viral envelope (for review see 11; for refs. on similar processes with the NE and various DNA-containing viruses see 3,5), and the associations of membranes with microtubules and/or microfilaments that have been shown to be cross-linked to specific membrane regions and, for example, influence the capping of surface

Figs. 5-7. Examples of transitory and more stable (Fig. 7) plasma membrane differentiations. Fig. 5 illustrates the transitory structural differentiations in the luminal plasma membrane of the epithelial cells of lactating rat mammary (for details see 17) as revealed in the exocytosis of casein micelles (the heavily stained micellar aggregates denoted by arrowheads in Figs. 5a-5c). Like normal small "coated vesicles" *sensu stricto* (arrow in Fig. 5c), secretory vesicles containing casein (CV) are also completely or partially surrounded by bristle coat structures (Fig. 5c). These structures are also recognized in plaque-like regions of the luminal plasma membrane (arrows in Figs. 5a and b), most probably as the result of the fusion of secretory vesicle membrane with plasma membrane. This interpretation (17) is strongly supported by the frequent attachments of casein micelles released into the lumen with such regions via fine thread-like connections (double arrow in Fig. 5a; A, alveolar space; MV, microvilli). Fig. 6 shows similar coated regions in the nascent plasma membrane during cell plate formation in a cultured plant cell, *Haplopappus gracilis* (Fig. 6a and b; some coat bearing regions and caps are denoted by the arrows; CW, cell wall material; for details see 16) and during plasmotomy in the course of cyst formation in the caps of the green alga, *Acetabularia mediterranea* (26, 34; Fig. 6c and d; PV, plasmotomic vacuoles, i.e. cytoplasmic vacuoles which grow and fuse which each other to form the walls around the cyst portions of the cap cytoplasm). The membranes of these vacuoles are the immediate precursor membranes to the plasma membranes of the forming cysts. Plasmotomic vacuoles seem to grow by incorporation of membrane material from coated vesicles (arrow and cv in Fig. 6d) and closely align with the outer membranes of the plastids ($P_1$ and $P_2$ in Fig. 6c; arrows indicate sites of alignment), again by the formation of somewhat regularly spaced membrane-to-membrane cross-bridges (denoted by the two arrows in the lower half of Fig. 6d; 26). Note the fibrillar, slime-like cell wall material contained in the plasmotomic vacuoles. Fig. 7 presents some of the "classic" examples of relatively stable structural differentiations in the lateral plasma membranes of epithelial cells such as in the lactating rat mammary as revealed in freeze-fracture preparations (Fig. 7a) and in the mouse intestine (Fig. 7b) as revealed in ultrathin sections. Note the different regions such as "tight junctions" (TJ in Fig. 7b, denoted by the two thick arrows in the upper part of Fig. 7a), the "zonula adherens" (ZA in Fig. 7b; most of the central, somewhat obliquely fractured membrane area in Fig. 7a), desmosomes (D in Fig. 7b; perhaps the area indicated at the second arrow from the bottom), and "gap junctions" (arrow in the bottom of Fig. 7a and the insert in Fig. 7a at higher magnification). For references and details, see 35-4o, for references on the formation of junctions, see 41,42,43. Note the frequency of associations of such plaques, as well as of the microvilli (MV, in the upper right corner of Fig. 7a and the upper left corner of Fig. 7b), with microfilaments (e.g. at the upper arrow in Fig. 7b) and thick filaments (e.g. at the lower arrow in Fig. 7b). Scales indicate o.2 $\mu$m (in Figs. 5 a-c; 6a, c,d; 7a,b) and o.1 $\mu$m (in Figs. 6b; inset Fig. 7a).

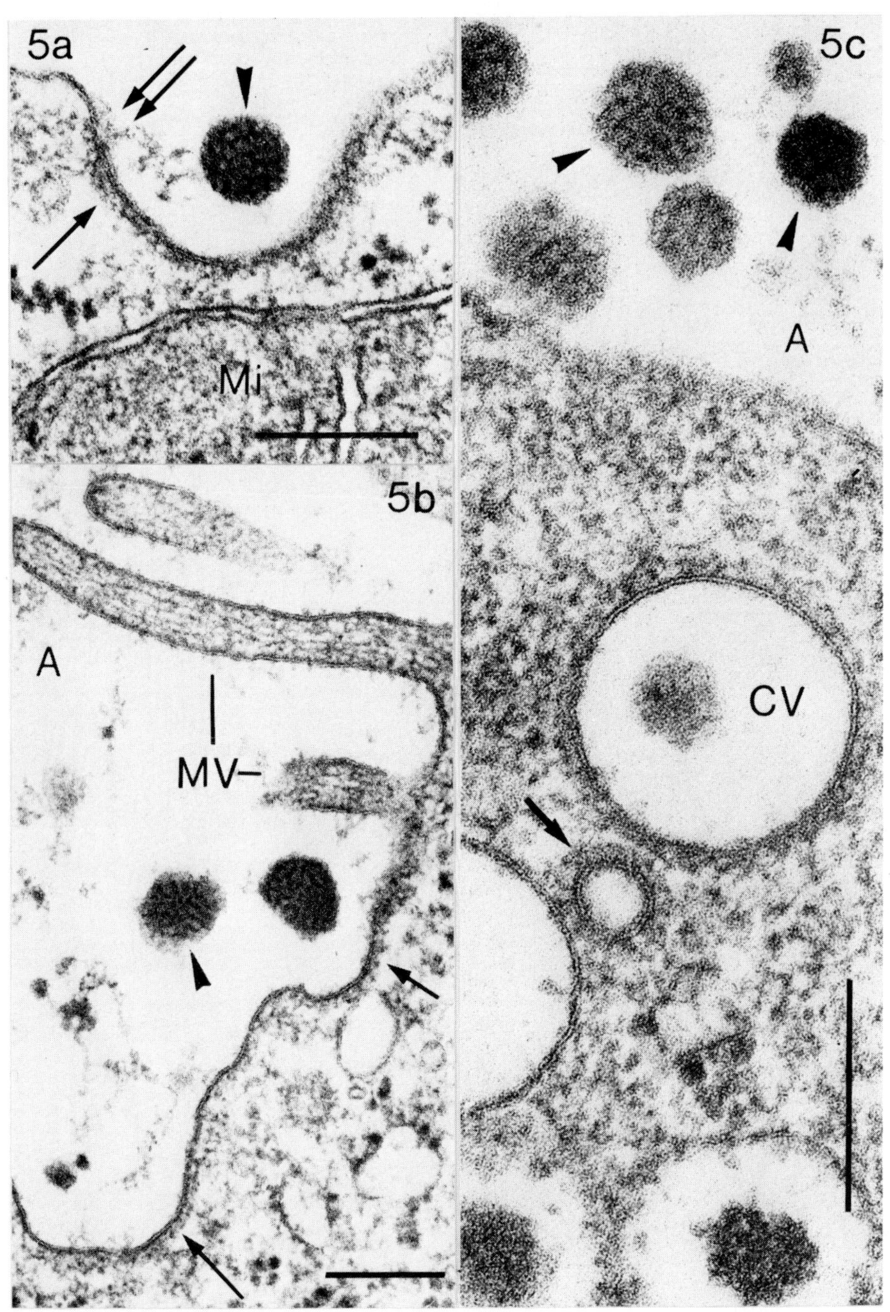
5a
Mi
5b
A
MV
5c
A
CV

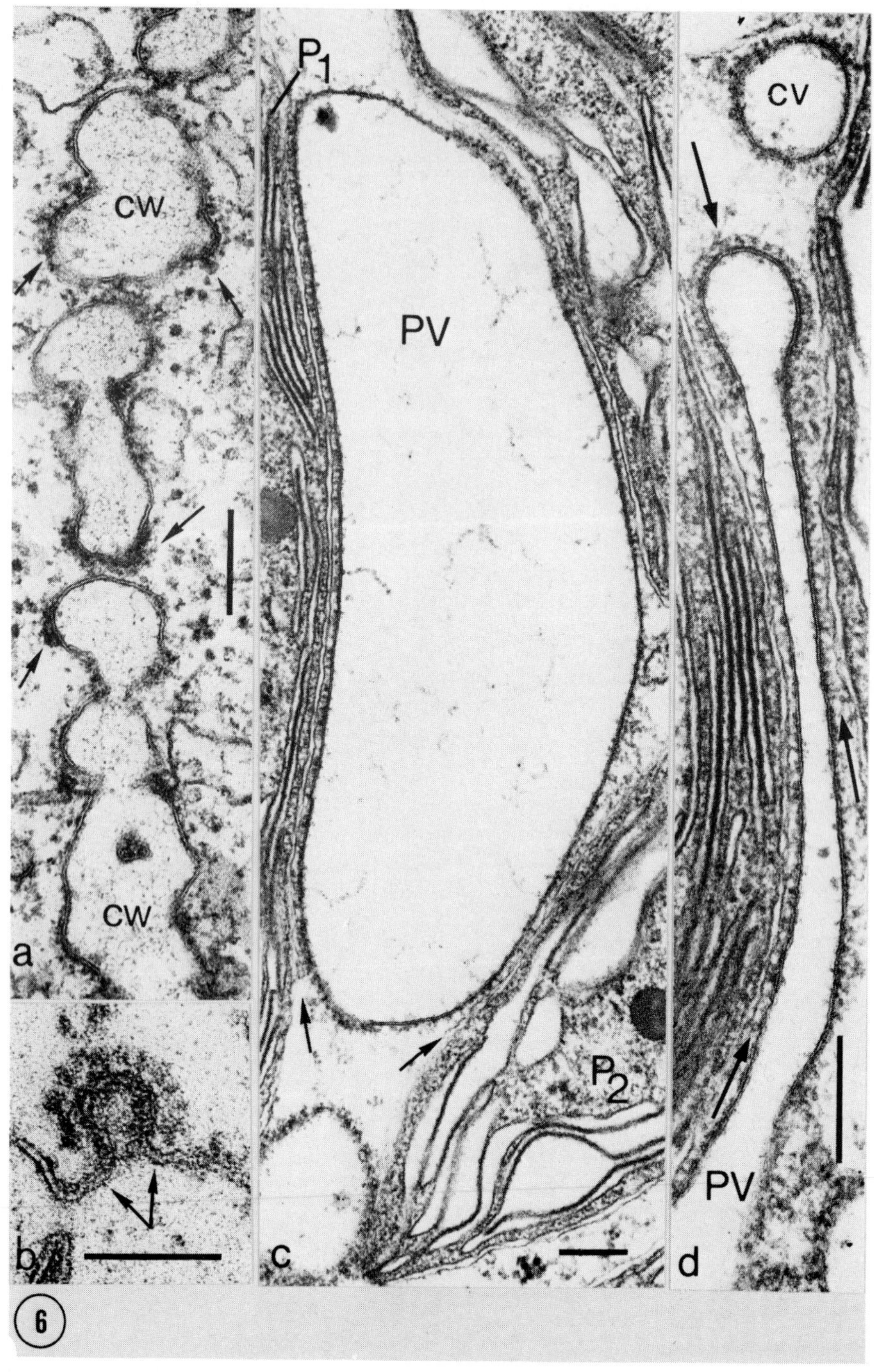
cw
cw
a
b
P1
PV
P2
c
cv
PV
d
6

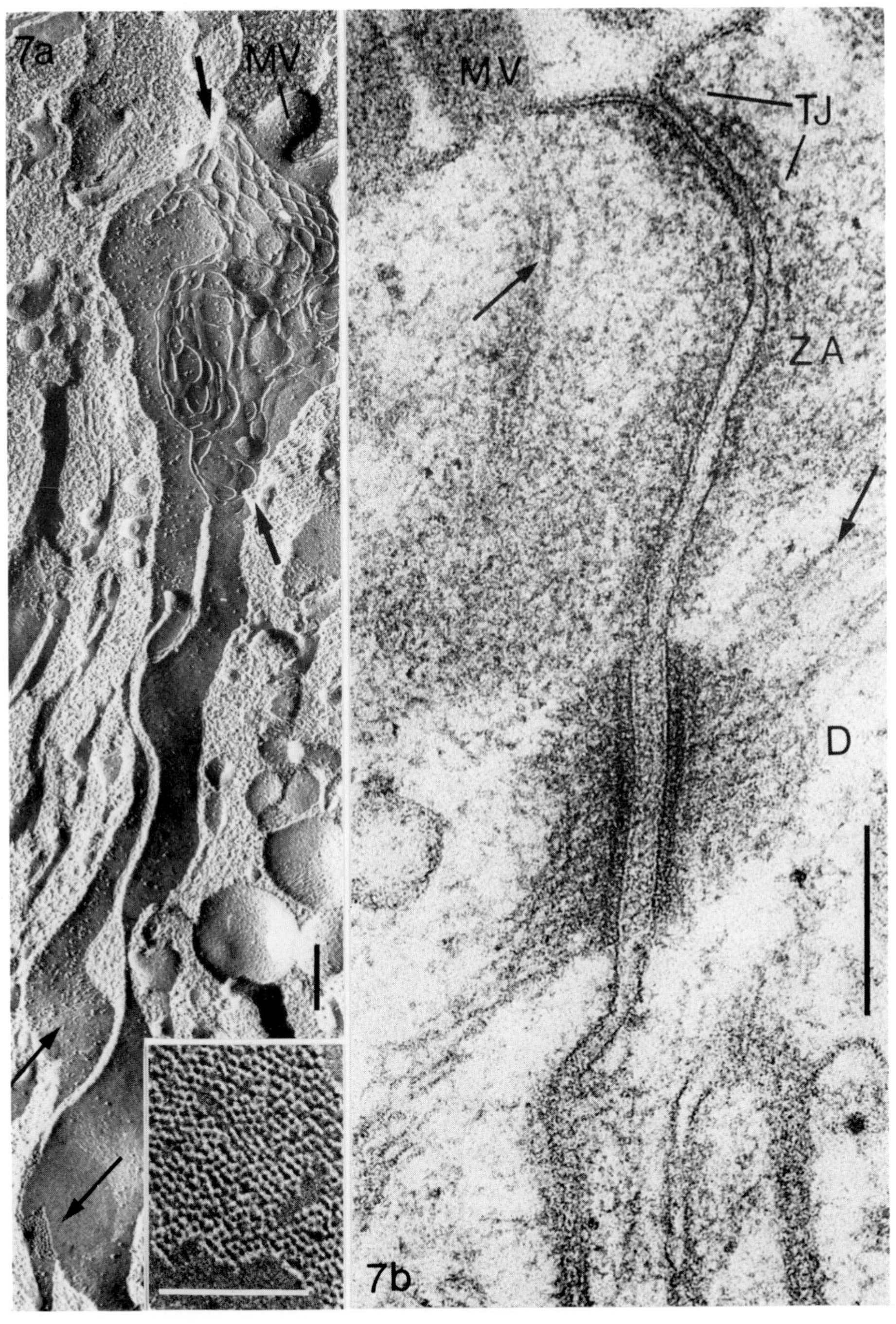
7a
MV
7b
MV
TJ
ZA
D

constituents or constitute functional membrane areas such as the plaques within the nuclear envelopes of various organisms [3,5-7,24]. The apparently widespread occurrence of such structural interactions might lead to the general concept that membranes are structurally and functionally not confined to what appears to be the membrane as such, that is the 6o-1oo Å thick layer demonstrable in the electron microscope, but are integrated into a cytoskeletal network and thus are linked and coupled to a variety of nonmembranous structures [25].

## V. MEMBRANE DIFFERENTIATION BY CHEMICAL ALTERATION OF COMPONENTS AND CHANGES IN THE CHEMICAL COMPOSITION

The changes described in this category of membrane differentiation involve neither alterations of the overall chemical composition of the specific membrane nor alterations of existing molecules, i.e. of the covalent chains, but rather chemical changes that occur in one and the same membrane system in different stages of cellular development. Such changes pertain to lipid moieties as well as to proteins and glycoproteins and have been particularly intensely studied with respect to surface antigens and the immune response (for refs. see 44-46), cell transformation (see, e.g.47-59) and developmental processes (e.g. 6o, 61). Since such changes are discussed in this meeting by Drs. L. Warren and A. Moscona in greater detail, (especially with respect to the phenomena of reduction of oligosaccharide chains in glycolipids (for review see 54) and the oversialylation of fucose-containing glycopeptides, and perhaps also fucoglycolipids, that occur a variety of transformed cells (e.g. 47,5o,53,56-58), they are only briefly mentioned here, in the context of membrane differentiation problems. It is especially important to know whether such qualitative changes can occur by a modification of membrane components *in situ*, for example by the surface-bound sugar transferases or phosphorylases, or whether they require the coordinate synthesis of whole novel complex membrane units (for the controversial discussion as to the significance of surface-bound glycosyltransferases, see 62,63). In view of the intimate interplay of membrane constituents, it is obvious that changes in the chemical composition of a specific membrane can result in considerable effects on other activities as well (see, for example, the influence of lipid compositions on membrane-bound enzymes as reviewed in 64).

## VI. MEMBRANE DIFFERENTIATION DURING MEMBRANE TRANSLOCATION (MEMBRANE FLOW)

As has been shown in various cell systems, in particular in mammalian liver and pancreas, the different membrane systems of the cell display differences in membrane structure (e.g. Fig. 2a, refs. 3-5, 67) and composition (see below). This is true not only for relatively unrelated membranes such as rER inner mitochondrial membrane, but also differences are noted among membranes which can exist in direct morphological continuity (e.g. Fig. 9a) and/or appear to be related with each other via membrane flow processes such as in the secretory pathway (Figs. 8 and 9). Two sorts of differences must be distinguished, quantitative versus qualitative. Gradual changes in the centrifugal direction of secretory membrane flow (as schematically arranged in Figs. 8 and 1o) have been noted with respect to structureal organization, lipid composition, patterns of polypeptide, carbohydrate composition and enzyme activities (Fig. 1o and 11; for details see 3-5, 12, 67, 77, 78). This has led Morré and collaborators to use the term "membrane differentiation" for the "change in composition or organization which accompanies the conversion from one type of membrane to another [67], in order to explain the differences in the various membrane systems that participate in the "membrane flow" of the morphologically defined membrane units surrounding the secretory products. As to the patterns of polypeptides identified by gel electrophoresis, various authors have emphasized the apparent differences between, for example, membrane fractions from ER, GA, and plasma membrane (for refs. see 12, 79-82) but detailed analyses show a variety of homologous bands as well as many others that are specific for the specific membrane (Fig. 11, for details see 78). However, it should be considered in this context that comparisons of bands of polypeptide-SDS complexes in polyacrylamide gels are not very meaningful since such patterns reflect merely size classes of polypeptides and might be disturbed by the degree and type of glycosylation of the various proteins. Large differences as well as some homologies are also observed among these membranes with respect to pigment contents and enzyme activities (see the reviews quoted above). When the kinetics of amino acids incorporated into membrane proteins are compared with those of the secretory products contained in the specific membranes (Fig. 12), one notes that the major part of the radioactivity contained in membrane protein does not parallel the

Fig. 8. Schematic presentation of diverse membrane types involved in the secretory pathway and their possible interrelationship via membrane continuities and membrane flow processes. (a) rER; (a') nuclear envelope; (b) sER or transitional elements in rER; (b') regions equivalent to transitional elements in the outer nuclear membrane; (c) ER-derived vesicles such as peroxisomes; (d) vesicles detached from ER; (d') vesicles detached from nuclear envelope; (e) dictyosomes of the Golgi apparatus; (f) smooth surfaced secretory vesicles derived from Golgi apparatus; (f') relatively large Golgi apparatus-derived vesicle partially covered by bristle coat ("condensing vacuoles" *sensu* Palade [12]); (g) and (g') free secretory vesicles of the smooth (g) or coated (g') type which are frequently smaller than (f) and (f'), respectively; (h) and (h') represent exocytotic stages of (g) and (g'), respectively; (i) lysosomal vesicles derived from Golgi apparatus or sER; (j) secondary lysosome or phagocytotic vacuole; (k) large smooth surfaced endocytotic vesicle during formation; (l) free endocytotic vesicle; (m) fusion of endocytotic vesicle with lysosome; (n) small smooth surfaced endocytotic vesicle; (n') coated endocytotic vesicle. Direct membrane continuities are observed between NE and ER (frequent in most cell systems; see however, 3,65,66), ER and Golgi apparatus (rarely observed, see Fig. 9a and refs. 67,68) ER and outer mitochondrial membrane (rare in most cell systems, for refs. see 69-72), ER and outer plastidal membrane (rare, see, e.g. 73). For detailed discussions of secretory pathways and for definitions,

Fig. 9. Some unusual aspects of rER-Golgi apparatus relationships as revealed in ultrathin sections through a rat mammary adenocarcinoma cell (yound tumor, induced by treatment with DMBA, for specific conditions see 74, for fixation conditions see 17, and in the ciliate, *Tetrahymena pyriformis* (for details see 75,76). Continuity of cisternae which are part of the rER (ER) as well as of a dictyosome (D) are denoted by arrows in Fig. 9a (N, nucleus). Small transitional element regions of the rER of *Tetrahymena* (pairs of arrows in Figs. 9b and d, see also at higher magnification in Fig. 9c) are identified. Note that the membrane intercept which participates in vesicle bud formation is distinguished from adjacent ER-regions by its greater thickness and/or stainability as well as by its associated, finely filamentous "fuzz" (for detailed discussion see 75). Note the aggregation of vesicles of variable sizes at such transitional elements which are interpreted as equivalents of dictyosomes [75]. Note also the differentiation of one ER-region into a smooth surfaced "rigid" cisterna (denoted by the arrow in the upper right of Fig. 9d; see 76). The arrow in the lower right of Fig. 9d denotes a smooth surfaced intermitochondrial "channel" connection which consists of a membrane which appears continous with the outer mitochondrial membranes of both mitochondria (Mi). Scales indicate o.2 μm.

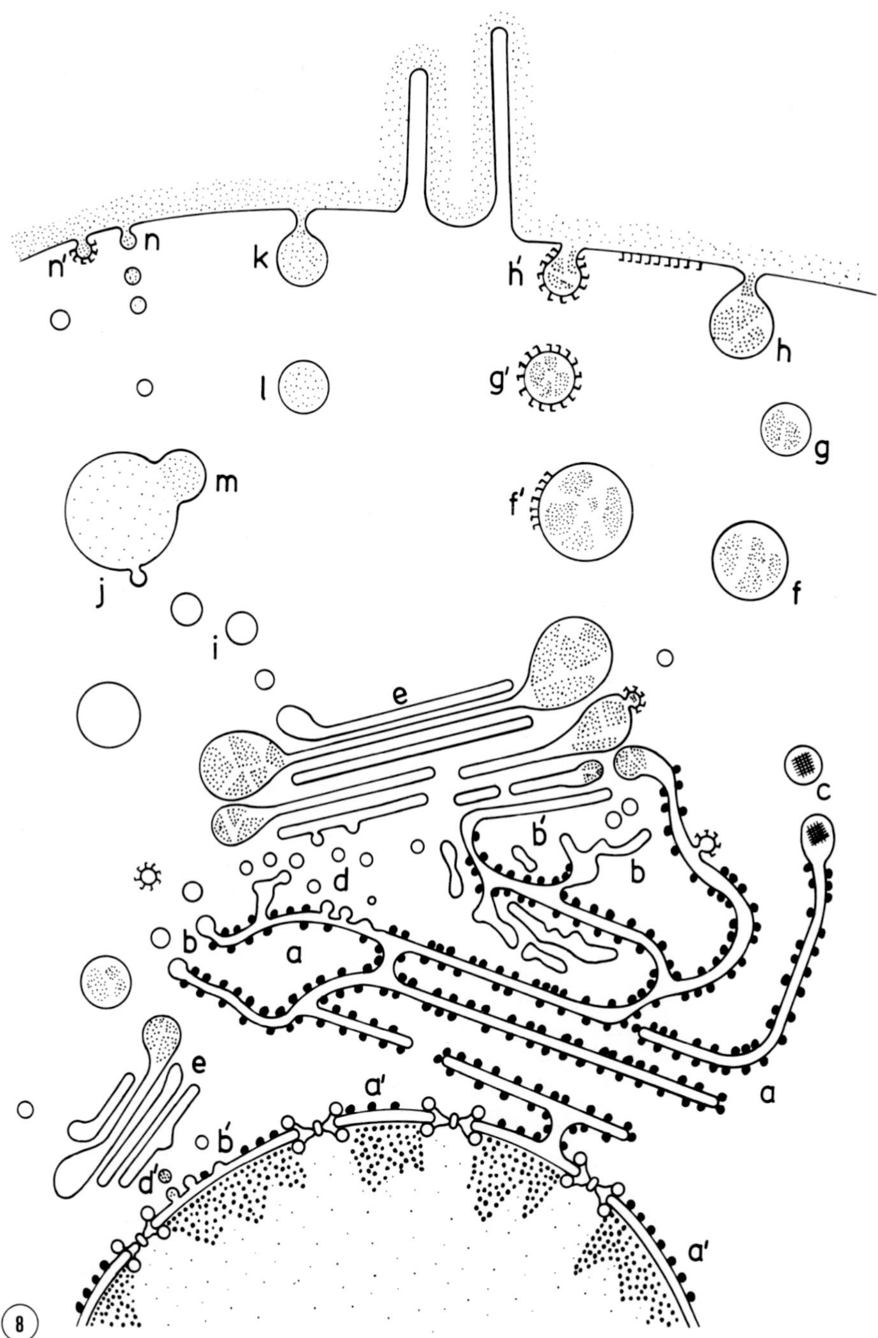
n'
n
k
h'
h
l
g'
g
m
f'
f
j
i
e
c
b'
b
d
a
e
a'
b'
d'
a'
a
8

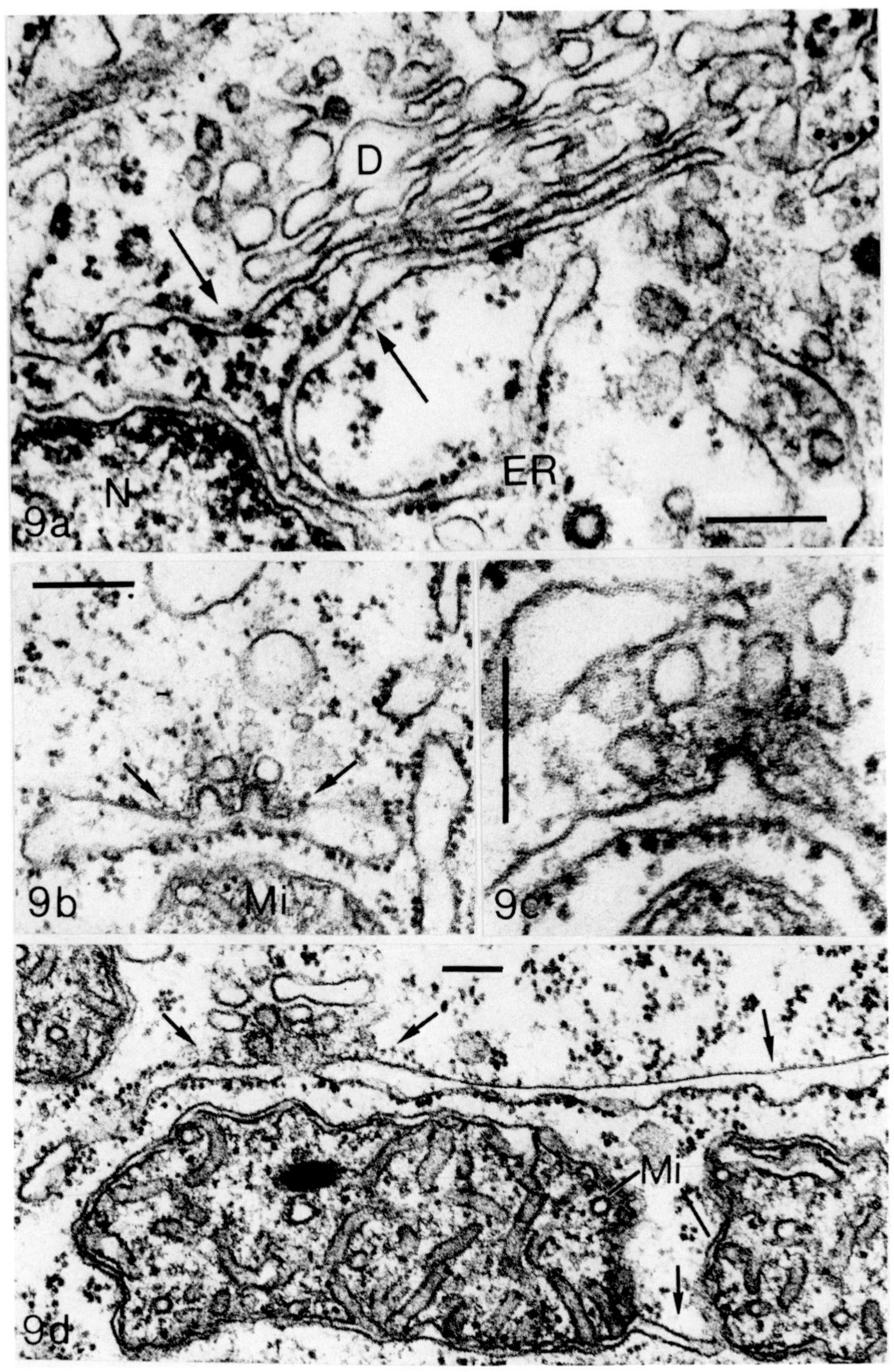
D
N
ER
9a
9b
Mi
9c
Mi
9d

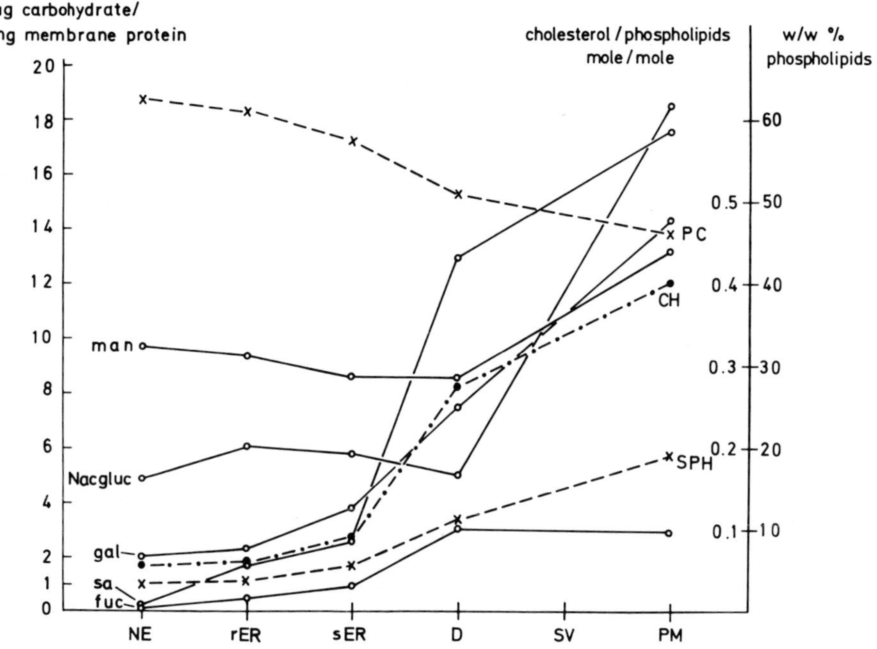


Fig. 1o. Diagram demonstrating changes in lipid and carbohydrate contents in relation to the direction of secretory membrane flow from nuclear envelope (NE), rER, sER, Golgi apparatus (D, dictyosome) to plasma membrane (PM) as determined in rat liver (data for fractions of secretory vesicles SV, are not included here. The diagram combines data obtained from membrane fractions treated with high salt concentrations, followed by washes in low salt buffers and sonication (for details see 5,77). Nacgluc, N-acetyl-glucosamine; man, mannose; gal, galactose; sa, sialic acids, determined as N-acetyl-neuraminic acid; fuc, fucose; PC, phosphatidyl-choline; CH, cholesterol, expressed in molar ratios relative to total phospholipids; SPH, sphingomyelin. The two phospholipids are represented by the dashed lines; the cholesterol by the dash-dot-line. Other phospholipids and carbohydrates that did not reveal such changes have not been incorporated into this diagram (c.f. 4,5).

kinetics of typical secretory proteins (see also 83) but rather reveals a much lower turnover (c.f. also the above quoted articles and refs. contained in 84-88). Only a relatively small portion of the membrane protein of rER, GA, and NE seems to parallel the kinetics of the vectorial flow of the secretory product (Fig. 12, c.f. 8o-83). When the kinetics of the individual bands of membrane polypeptides as revealed by gel electrophoresis (Figs. 11,13,14; c.f. 78) were compared, bands of a specifically high turnover were found

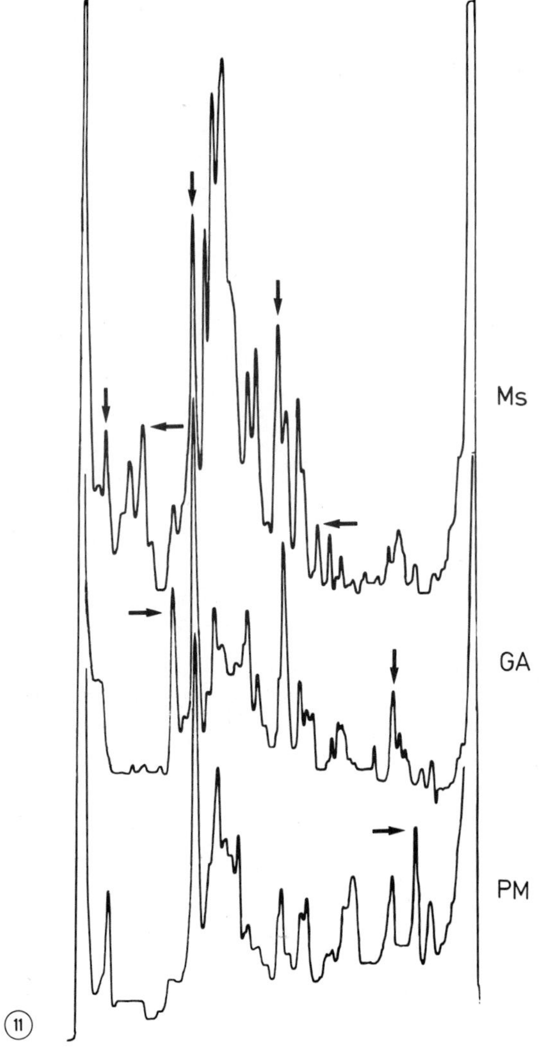


Fig. 11. Comparison of densitometric tracings of the membrane polypeptides of isolated plasma membrane (PM), rER microsomes (Ms), and Golgi apparatus (GA) from rat liver as revealed after treatment with high salt (1.5 M KCl, for details see 77,78) and low salt concentrations, followed by treatment with puromycin and low concentrations of deoxycholate (DOC, o.1%, for specific details see 77,78), solubilization by heating in 4% sodium dodecylsulfate (SDS) and electrophoresis in 7.5% polyacrylamide slab gels containing o.1% SDS, 78. The vertical arrows denote some polypeptide bands with identical mobility indicative of similar size. The purity of the fractions is demonstrated by the occurrence of specific polypeptide bands in some membranes but not in others (horizontal arrows).

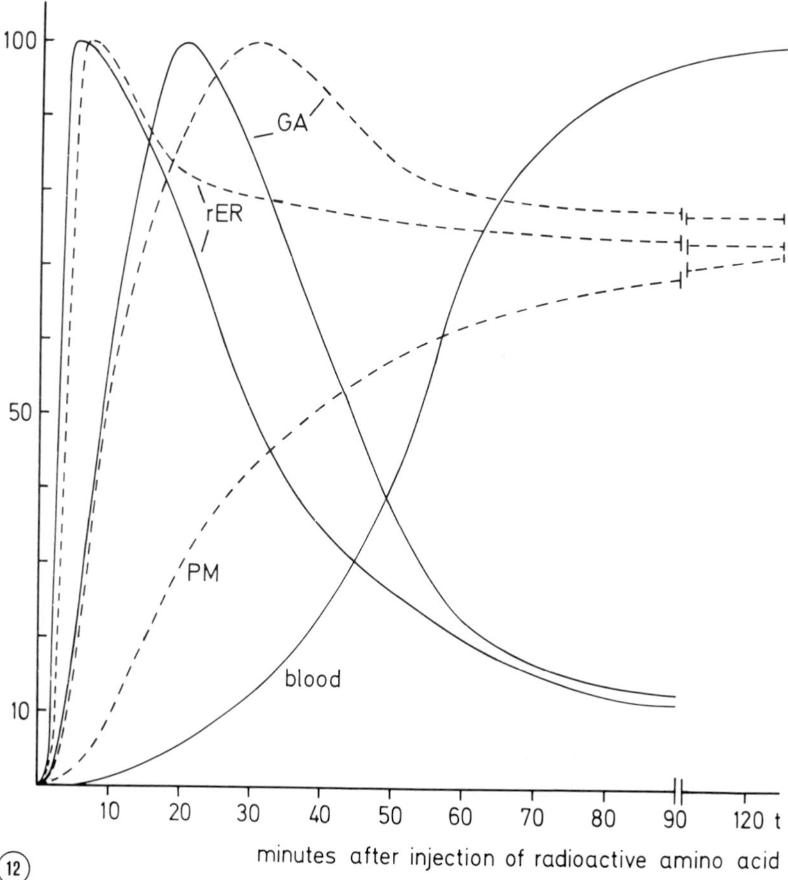


Fig. 12. Simplified diagrammatic combination of in vivo amino acid labelling kinetics of rat liver membrane proteins (symbols as used before; dashed lines) and one of the typical secretory proteins, serum albumin as prepared from the membrane fractions (rER and GA) and the blood system (all solid lines). The data presented are combined from 79,83,9o,91 and have been normalized to maximal (1oo%) specific radioactivity. Note that only a minor component of the radioactive membrane protein is kinetically in parallel with the secretory product, serum albumin. Most of the membrane protein-bound radioactivity shows relatively low turnover (half life times from ca. 6o hours in rER to ca. 4o hours in PM). The complex kinetics suggest heterogeneity of membrane proteins with respect to synthetic rates and, correspondingly, degradation and/or membrane flow rates.

besides others that showed moderate or even low turnover (Figs. 13 and 14). Although some of the high turnover bands which were identified might reflect residual contamination with membrane-associated protein, this is very unlikely for other bands which also show "flow type" turnover (for detailed discussions of contamination of membrane fractions by secretory proteins see 77,78,81,82,83,89). Morphological observations suggest the existence of specific sites of vesicle formation from endomembranes (e.g. Fig. 9) as well as of specific sites of fusion with other membranes. This is indicative of a functional heterogeneity; i.e., regions exist, or are formed by a clustering of components, within membranes which more frequently, if not exclusively, participate in membrane flow processes than other regions. In order to explain both the flow type kinetics of some membrane components and the low turnover of others, together with the observed gradual and abrupt changes in the composition of such morphologically related types of membranes, we propose that membrane flow is not random but is selective for specific membrane components and excludes others (Fig. 15). This selection of domains involved in flow processes might again be induced by extra membranous components (see IV), including the secretory products themselves. Selective secretory membrane flow would necessarily result in gradual or abrupt changes in membrane composition and thus in "membrane differentiation" sensu Morré et al.[67].

---

Figs. 13 and 14. Distribution of radioactivity in gelelectrophoretically separated polypeptides from rER (microsomes, rER, curves in Fig. 13 from rat liver) and plasma membrane (PM,curves in Fig. 14) isolated at different time points (1o min, 6o min, 96o min) after intraperitoneal injection of $^3$H-leucine (o.5 mCi per 15o g body weight rat; spec. radioact. 5o Ci/mmole). The two photographs in the figures represent central regions of both PM and rER membrane polypeptide patterns as revealed in the central parts of gels that were run in parallel on the same slab (note some band homologies). For details of isolation see 77, 78. Fractions were treated with 1 mM puromycin, high KCl concentrations, and o.1% DOC (for details of isolation and purity of fractions see 77). The stained gels were traced and then sliced into 1 mm fractions. Radioactivity was determined per slice fraction. The abscissa gives the slice number; the ordinate gives the cpm per %protein of total protein as identified by Coomassie Blue stain contained in the specific slice, which represents a close approximation to a determination of specific radioactivity per band. Note the heterogeneity of turnover of the different polypeptide size classes in both membrane fractions, indicative of heterogeneity in kinetics of synthesis and translocation (flow). Regarding the band in the central region of the rER membrane polypeptide pattern that contains the highest specific radioactivity and shows the highest turnover (Fig. 13), it cannot be excluded that this band represents contamination by serum albumin (probably in a membrane-attached state (the apparent molecular weight of this component is approximately 65 000 daltons).

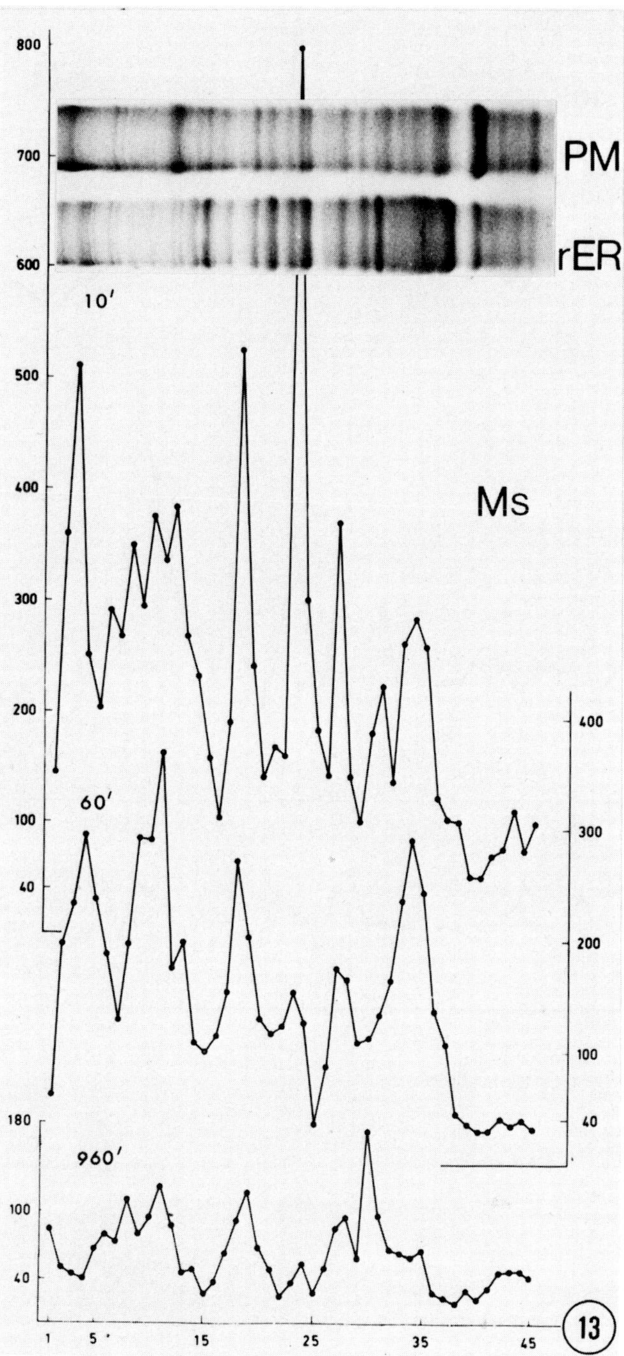
PM
rER
10'
Ms
60'
960'
13

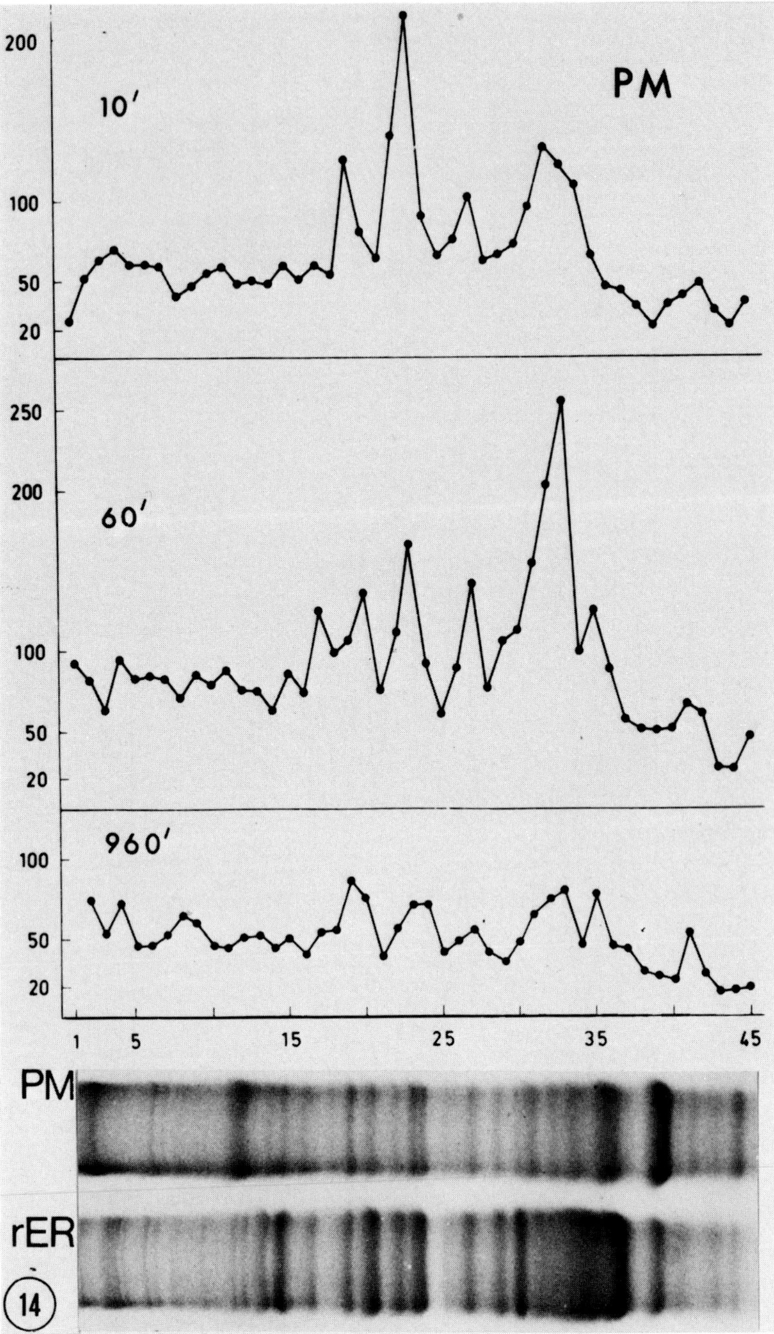

PM
10′
60′
960′
200
100
50
20
250
200
100
50
20
100
50
20
1
5
15
25
35
45
PM
rER
14

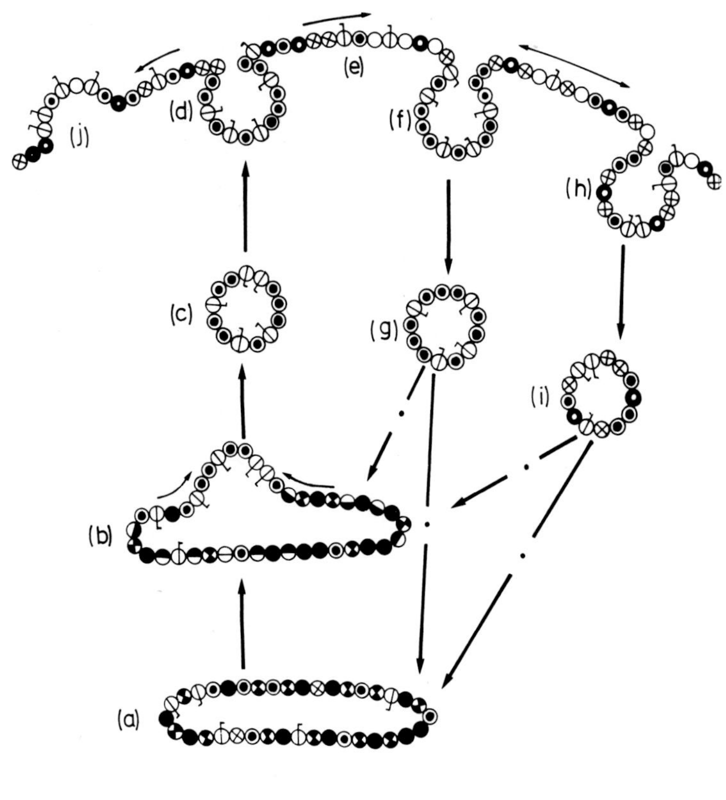


Fig. 15. Schematic drawing illustrating the concept of selective flow of membrane components. The concept assumes that the participation in vesicle formation and flow phenomena for each type of membrane component, particularly proteins and glycoproteins (components are represented by different symbols) is different. In this concept a specific membrane continuum is heterogeneous with respect to the turnvoer and the translocation frequency of its constituents. For example, specific components of an endomembrane cisterna (a) aggregate into a semistable (transitory) cluster that locally alters the character of the membrane, e.g. increases the plasticity or the fluidity (b) Components enriched in such a cluster then are contained in a vesicle bud which finally detaches and becomes a vesicle (c). After fusion of such a vesicle with another membrane (d) the components might become randomly distributed (e), again participate in aggregation and budding processes (e.g. as denoted by j), or remain together and recycle (f and g). Another possibility would be that the components mix to a certain degree with components characteristic for their new "host" membrane and recycle together in randomly or specifically aggregated moities (h and i). Thin arrows indicate lateral mobility; thick arrows indicate vesicle translocation.

References

1. Kessel, R.G., 1968, J. Ultrastruc. Res. Suppl. 1o, 1.

2. Wischnitzer, S., 197o, Int. Rev. Cytol. 27, 65.

3. Franke, W.W., and Scheer, U., 1974, in: The Cell Nucleus, ed. H. Busch (Academic Press, New York and London) p. 219.

4. Morré, D.J., Mollenhauer, H.H., and Bracker, C.A., 1971, in: Origin and Continuity of Cell Organelles, eds. J. Reinert and H. Ursprung (Springer Verlag, Heidelberg) p. 82.

5. Franke, W.W., 1974, Int. Rev. Cytol. 39, 1o2.

6. Yahara, I., and Edelman, G.M., 1975, Exptl Cell Res. 91, 125.

7. Yahara, I., and Edelman, G.M., 1975, Proc. Natl. Acad. Sci. USA. 72, 1579.

8. Edidin, M., 1974, in: Cellular Selection and Regulation in Immune Response, ed. G.M. Edelman (Raven Press, New York) p. 121.

9. Petit, J., and Edidin, M., 1974. Science (Wash.) 184, 1183.

1o. Nicolson, G.L., 1974, Int. Rev. Cytol., 39, 89.

11. Lenard, J., and Compans, R.W., 1974, Biochim. Biophys. Acta 344, 51.

12. Palade, G., 1975, Science (Wash.) 189, 347.

13. Schulze, H.U., and Staudinger, H. 1975, Naturwissenschaften 62, 331.

14. Franke, W.W., 1971, Protoplasma (Vienna) 73, 263.

15. Hicks, R.M., Ketterer, B., and Warren, R.C., 1974, Phil. Trans. Roy. Soc. Lond. B. 268, 23.

16. Franke, W.W., and Herth, W., 1974, Exptl. Cell Res. 89, 447.

17. Franke, W.W., Lüder, M.R., Kartenbeck, J., Zerban, H., and Keenan, T.W., 1976, J. Cell Biol., in press.

18. Schäfer, W., Bauer, H., Bolognesi, D.P., Fischinger, P., Frank, H., Gelderblom, H., Lange, J., and Nermut, M.V., 1974, in: Molecular Studies in Viral Neoplasia (25th Ann. Symp. Fundamental Cancer Research 1972 at Houston, Texas; The Williams & Wilkins Comp., Baltimore), p. 115.

19. Gelderblom, H., Bauer, H., Bolognesi, D.P., and Frank, H., 1972, Zbl. Bakt. Hyg. I. Abt. Orig. A 22o, 79.

2o. Garoff, H., Simons, K., and Renkonen, O., 1974, Virology 61, 493.

21. Hollmann, K.H., 1974, in: Lactation, eds. Larson, B.L., and Smith, V.R. (Academic Press, New York and London), p.3.

22. Patton, S., and Keenan, T.W., 1975, Biochim. Biophys. Acta 415, 273.

23. Bächi, T., and Howe, C., 1972, Proc. Soc. Exp. Biol. Med. 141, 141.

24. Kubai, D.F., 1973, J. Cell Sci. 13, 511.

25. Franke, W.W., Herth, W., VanderWoude, W.J., and Morré, D.J., 1972, Planta 1o5, 317.

26. Franke, W.W., Kartenbeck, J., Zentgraf, H.W., Scheer, U., and Falk, H., 1971, J. Cell Biol., 51, 881.

27. Franke, W.W., 1971, Cytobiologie 4, 3o7.

28. Franke, W.W., and Reau, P. 1973, Arch. Mikrobiol. 9o, 12o.

29. McCully, E.K., and Robinow, C.F., 1973, Arch. Mikrobiol. 94, 133.

3o. Bland, E.C., and Lunney, C.Z., 1975, Cytociologie 11,382.

31. Fawcett, D.W., and Phillips, D.M., 197o, in: Comparative Spermatology, ed. B. Baccetti (Academia Nationale dei Linzei, Rome, and Academic Press, New York & London) p. 13.

32. Franke, W.W., Kartenbeck, J. Krien, S., VanderWoude, W.J., Scheer, U., and Morré, D.J., 1972, Z. Zellforsch. 132, 365.

33. Satir, B., 1974, in : Transport at the Cellular Level, eds. M.A. Sleigh and D.H. Jennings (Cambridge: at the University Press, Cambridge, UK) p. 399.

34. Berger, S., Herth, W., Franke, W.W., Falk, H., Spring, H.,and Schweiger, H.G., 1975, Protoplasma (Vienna) 84, 223.

35. Elias, P.M., and Friend, D.S., 1975, J. Cell Biol. 65, 18o.

36. McNutt, N.S., Hershberg, R.A., and Weinstein, R.S., 1971, J. Cell Biol. 51, 8o5.

37. Friend, D.S., and Gilula, N.B., 1972. J. Cell Biol. 53, 758.

38. Humbert, F., Pricam, C., Perrelet, A., and Orci, L., 1975, J. Ultrastruct. Res. 52, 13.

39. Campbell, R.D., J.H., 1971, In: Origin and Continuity of Cell Organelles, eds. J. Reinert and H. Ursprung (Springer Verlag, Berlin), p. 261.

4o. Orci, L., and Perrelet, A., 1975, Freeze-Etch Histology (Springer Verlag, Berlin, p. 1-168).

41. Benedetti, E.L., Dunia, I., and Boemendal, H., 1974, Proc. Nat. Acad. Sci. USA, 71, 5o73.

42. Hudspeth, A.J., 1975, Proc. Nat. Acad. Sci. USA 72,2711.

43. Revel, J.P., 1974, In: The Cell Surface in Development, ed. A.A. Moscona (Wiley & Sons, New York) p. 51.

44. Goldschneider, J., 1974, In: The Cell Surface in Development, ed. A.A. Moscona (Wiley & Sons, New York) p. 165.

45. Vitetta, E.S., and Uhr, J.W., 1975, Biochim. Biophys. Acta, 415, 253.

46. Boyse, E.A., and Bennett, D., 1974, in: Cellular Selection and Regulation in the Immune Response, ed. G.M. Edelman (Raven Press, New York) p. 155.

47. Buck, C.A., Fuhrer, J.P., Soslau, G., and Warren, L., 1974, J. Biol. Chem. 249, 1541.

48. Shinitzky, M., and Inbar, M., 1974, J. Mol. Biol. 85, 6o3.

49. Vlodavsky, I., and Sachs, L., 1974, Nature 25o, 67.

5o. Steiner, S., Brennan, P.J., and Melnick, J.L., 1973, Nature New Biol. 245, 19.

51. Herschman, H.R., 1972, in: Membrane Molecular Biology, eds. C.F. Fox and A.D. Keith (Sinauer Ass. Inc. Publ., Stamford, Conn.) p. 471.

52. Grant, J.P., Bigner, D.D., Fischinger, P.J., and Bolognesi, D.P., 1974, Proc. Nat. Acad. Sci. 71, 5o37.

53. Emmelot, P., 1973, Eur. J. Cancer 9, 319.

54. Brady, R.O., and Fishman, P.H., 1974, Biochim. Biophys. Acta 355, 121.

55. Sakiyama, H., Gross, S.K., and Robbins, P.W., 1972, Proc. Nat. Acad. Sci. 69, 872.

56. Warren, L., Fuhrer, J.P., and Buck, C.A., 1972, Proc. Nat. Acad. Sci. 69, 1838.

57. van Beek, W.P., Smets, L.A., and Emmelot, P., 1973, Cancer Res. 33, 2913.

58. van Beek, W.P., Smets, L.A., and Emmelot, P., 1975, Nature 253, 457.

59. Wickus, G.G., and Robbins, P.W., 1973, Nature New Biol. 245, 65.

6o. Moscona, A.A., 1974, In: The Cell Surface in Development, ed. A.A. Moscona (Wiley & Sons, New York) p. 67.

61. Changeux, J.P., 1974, In: The Cell Surface in Development, ed. Moscona (Wiley & Sons, New York) p. 2o7.

62. Keenan, T.W., and Morré, D.J., 1975, FEBS Letters 55, 8.

63. Porter, C.W., and Bernacki, R.J., 1975, Nature 256, 648

64. Farias, R.N., Bloj, B., Morero, R.D., Sineriz, F., and Trucco, R.E., 1975, Biochim. Biophys. Acta 415, 231.

65. Franke, W.W., Berger, S., Falk, H., Spring, H., Scheer, U., Herth, W., Trendelenburg, M.F., and Schweiger, H.G., 1974, Protoplasma 82, 249.

66. Franke, W.W., Spring, H., Scheer, U., and Zerban, H., 1975, J. Cell Biol. 66, 681.

67. Morré, D.J., Keenan, T.W., and Huang, C.M., 1974, in: Advances in Cytopharmacology, eds. B. Ceccarelli, F. Clementi, and J. Meldolesi (Raven Press, New York) Vol. 2, p. 1o7.

68. Bracker, C.E., Grove, S.N., Heintz, C.E., and Morré, D.J., 1971, Cytobiol. 4, 1.

69. Franke, W.W., and Kartenbeck, J., 1971, Protoplasma 73, 35.

7o. Morré, D.J., Merrit, W.D., and Lembi, C.A., 1971, Protoplasma 73, 43

71. Bracker, C.E., and Grove, S.N., 1971, Protoplasma 73, 15.

72. Kellems, R.E., Allison, V.F., and Butwo, R.A., 1975, J. Cell Biol. 65, 1.

73. Crotty, W.J., and Ledbetter, M.C., 1973, Science 182, 839.

74. Rathke, P.C., Schmid, E., and Franke, W.W., 1974, Cytobiol. 1o, 366.

75. Franke, W.W., Eckert, W.A., and Krien, S., 1971, Z. Zellforsch. 119, 577.

76. Franke, W.W., and Eckert, W.A., 1971, Z. Zellforsch. 122, 244.

77. Franke, W.W., Keenan, T.J., Stadler, J., Genz, R., Jarasch, E.-D., and Kartenbeck, J., 1976, J. Cell Biol., in press.

78. Kartenbeck, J., 1974, Thesis, University of Freiburg i. Br. (Germany) p. 1-233.

79. Fleischer, B., and Fleischer, S., 1971, in: Biomembranes, ed. L.A. Manson (Plenum Press, New York) Vol. 2, p. 75.

8o. Wallach, D., Kirshner, N., and Schramm, M., 1975, Biochim. Biophys. Acta 375, 87.

81. Meldolesi, J., 1971, in: Advances in Cytopharmacology, eds. F. Clementi and B. Ceccarelli (Raven Press, New York), Vol. 1, 145.

82. Meldolesi, J. 1974, Phil Trans. R. Soc. Lond. B. 268, 39.

83. Franke, W.W., Morré, D.J., Deumling, B., Cheetham, R.D., Kartenbeck, J., Jarasch, E.-D., and Zentgraf, H., 1971, Z. Naturforsch. 26b, 1o31.

84. Riordan, J.R., Mitranic, M., Slavik, M., and Moscarello, M.A. FEBS Letters 47, 248.

85. Kawasaki, T., and Yamashina, I., 1971, Biochim. Biophys. Acta 225, 234.

86. Schimke, R.T., and Dehlinger, P.J., 1973, in: Biological Membranes, eds. D. Chapman and D.F.H. Wallach (Academic Press, New York) p. 115.

87. Schimke, R.T., and Dehlinger, P.J., 1971, in: Drugs and Cell Regulation, ed. E. Mihich (Academic Press, New York) p. 121.

88. Schimke, R.T., 1974, in: Advances in Cytopharmacology, eds. B. Ceccarelli, F. Clementi and J. Meldolesi (Raven Press, New York) Vol. 2, 63.

89. Castle, J.D., Jamieson, J.D., and Palade, G.E., 1975, J. Cell Biol. 64, 182.

9o. Glaumann, H., and Ericsson, J.L.E., 197o, J. Cell Biol. 47,555.

91. Peters, T., Fleischer, B., and Fleischer, S., 1971, J. Biol. Chem. 246, 24o.

*Progress in Differentiation Research, ed. N. Müller-Bérat et al.*



# MORPHOGENETIC CELL INTERACTIONS IN KIDNEY DEVELOPMENT

J. Wartiovaara, M. Karkinen-Jääskeläinen, E. Lehtonen, S. Nordling and L. Saxén

Third Department of Pathology, University of Helsinki

## 1. Introduction

Embryonic development is characterized by continuous changes in the organism; cells and cell populations within the developing embryo are motile and their position and relations with adjacent cells are constantly changing. To ensure synchronized, orderly development, various control systems must exist. The developing cells should have the following capabilities:

- to sense their position within the organism
- to recognize cells with similar developmental history, and
- to communicate with adjacent cells with different developmental background and destiny.

All these control systems have been shown to operate in a developing organism. Cells have been suggested to read positional information[1] apparently provided by organismal, diffusible factors[2]. They can also recognize like cells to which they adhere with specific ligands[3]. Finally, cells of different origin can communicate and exchange morphogenetic signals necessary for the development of both interactants[4,5].

The biological significance of the last-mentioned inductive tissue interactions has been known for over 50 years, yet our understanding of the molecular basis is poor. Very few specific signal substances, which are operative during normal interactive processes or can mimic their consequences, have been characterized. The vegetalizing factor acting on competent amphibian ectoderm[6], the perichordal proteoglycans stimulating somite chondrogenesis[7] and the mesenchymal factor exerting a morphogenetic action on embryonic pancreatic epithelium[8] are the few that can be mentioned. Because of this lack of chemical data it might be useful to go back one step and first ask some questions concerning the location and mode of transmission of such inductive signals.

From a conceptual standpoint three alternative ways of transmission of signal substances can be outlined (fig. 1):

1. By free diffusion of signal substances[9]
2. Through extracellular substances secreted by the interacting cells[10]
3. Through direct contacts between the interacting cells[11]

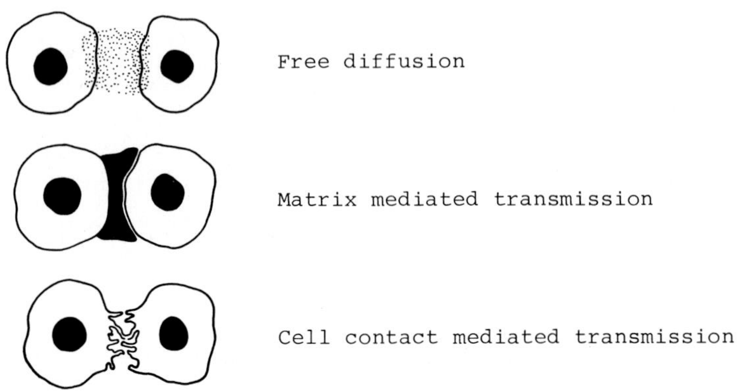


Fig. 1. Three different alternatives for transmission of inductive signals[14].

All these possibilities have been presented in connection with different model systems of tissue interaction. Our purpose has been to study the applicability of these concepts to kidney development.

## 2. Kidney development as a model system

The central morphogenetic feature in the development of the metanephric kidney is the formation of secretory tubules in the nephric mesenchyme. This involves aggregation and further differentiation of mesenchymal cells, a phenomenon which is triggered by an inductive influence from an ingrowing epithelial bud of the Wolffian duct. The interaction process has previously been characterized by various experimental means in mouse embryos[4,12,13,14]. A fruitful approach has been the use of the transfilter technique[12] by which kidney tubulogenesis can be achieved through thin porous filters especially when a potent heterotypic inductor, embryonic spinal cord is used.

## 3. Transmission characteristics of the signal

Grobstein[15] has shown that the transmission of the inductive signal is restricted in the kidney model system. The signal would only occasionally pass a Millipore filter with a nominal pore size of 0.1 µm. Furthermore, the signal never passed through three 0.4 to 0.8 µm filters with a total thickness of about 70 µm. Some years ago Crick[2] suggested that small molecular weight substances in a gradient could provide the information necessary for morphogenesis. Since this seemed contradictory to Grobstein's results, we investigated the effect of various filters on diffusion and induction and correlated these data to the ingrowth of cytoplasmic processes into the filter. It could be established that both with Millipore and Nuclepore filters a minimal pore size of appr. 0.15 µm was required for transmission to take place[16,17]. The pore size of the filter was found to have little effect on the diffusion of molecules investigated. Even the 0.1 µm pore filters allowed the passage of polio virus particles without much effect on the diffusion[17,18].

The velocity of the transmission of the inductive signal was measured by determining the minimal induction time needed in transfilter cultures before the responding mesenchyme can differentiate autonomously. When 0.8 µm pore size TA Millipore filter with a mean thickness of 28 µm was used the minimal induction time was about 18 hours. When the same experiment was done using double filters an increase of 12 hours in the minimal induction time was noticed. This was interpreted to mean that the transmission velocity of the signal was about 2 µm/h. This slow transmission was in sharp contrast to the diffusion measurements done with the same filters[18].

In a third set of experiments the dependence of the minimal induction time on the filter pore sizes was studied. The minimal induction time was prolonged by 12 hours when the filter pore was reduced from 0.5 µm to 0.2 µm. Also with these filters the measured free diffusion of molecules was affected very little by the decrease in the filter pore sizes[17].

These three sets of experiments demonstrating the small restriction on diffusion by different filters, the slow transmission of the inductive signal in the filter, and the prolongation of the minimal induction time when filter pore size was reduced, all seemed to speak against the hypothesis of free diffusion of the inducer (fig. 1).

## 4. Cell processes in filter

In order to investigate the validity of the two remaining hypotheses, electron microscopic studies of the transfilter induction situation were performed. When spinal cord tissue was used as inductor, cell processes were found deep in 0.5 and 0.2 µm Nuclepore filters[19] and penetrated filters with larger pore sizes (fig. 2). Inductor cell processes can even extend through the whole 14 um thick filter and form cytoplasmic networks visible by scanning electron microscopy on the side of the responding tissue (fig. 3). Penetration of cytoplasmic processes was only rarely found in 0.1 um pore size Millipore filters, which also prevented induction[16]. There seems to be a correlation between the minimal induction time and the time required for the inductor cells to send their processes through the filter. No accumulation of extracellular material, a prerequisite for a matrix-type of cell interaction, could be seen in the filter channels between the apposed membranes of the inducing and responding cells[17]. Furthermore, when another heterotypic inductor, embryonic salivary mesenchyme, was tested a positive correlation was observed between the passage of the inductive effect through Nuclepore filters of various pore sizes and penetration of cytoplasmic processes through the filter pores[20] (table 1).

## 5. Cell contacts in vivo

The results with the in vitro transfilter interaction seemed to speak in favour of a direct contact-mediated transmission of induction. Hence it was of interest to investigate whether the in vivo situation was the same. Using ruthenium red as an EM tracer substance for acid mucopolysaccharides[21], it was found that also in the developing kidney rudiment the interacting cells were at numerous sites in close contact with each other and were not separated

---

Fig. 2. Micrograph of transfilter culture of mouse metanephric mesenchyme (above) and spinal cord (below) in a case of positive tubule induction. Cytoplasmic processes are seen penetrating through the 3.0 µm pores of the interposed Nuclepore filter. 72 h cultivation.

Fig. 3. Scanning electron micrograph of cytoplasmic processes extending through a 0.5 um pore size Nuclepore filter. Processes from cells of spinal cord tissue, a potent heterotypic kidney inductor, located on the opposite side of the filter. x 1 500.

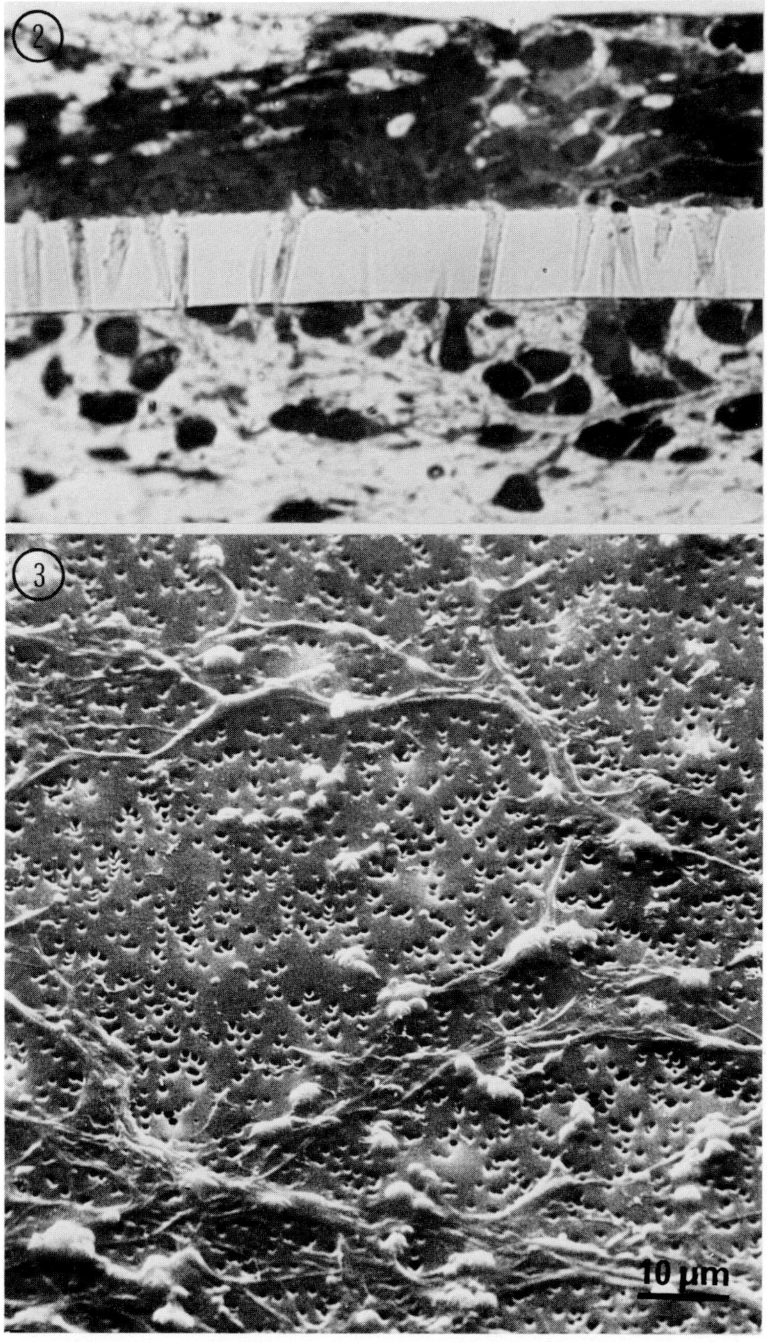
2
3
10 μm

Table 1

Formation of kidney tubules and penetration of inductor cells into filter pores in transfilter induction experiments as a function of inductor type and pore size of Nuclepore filter

| FILTER | INDUCTOR | | | |
|---|---|---|---|---|
| Nominal pore size | Salivary mesenchyme | | Spinal cord | |
| (μm) | Tubules | Penetration | Tubules | Penetration |
| 0.1 | . | . | ± | + |
| 0.2 | - | - | + | ++ |
| 0.6 | - | ± | + | ++ |
| 3.0 | ± | + | + | ++ |

. = not determined

by thicker layers of extracellular material[22]. The cells had surface protrusions that came to a distance of less than 10 nm from each other (figs. 4 and 5).

## 6. Concluding remarks

The transfilter experiments and the situation in vivo point to a direct contact-mediated type of cell interaction as being the most likely mode of transmission of inductive signals in kidney tubulogenesis. It still remains unclear what type of contacts are involved between the interacting cells. The question also arises whether the transfer of the induction can take place through interacting complementary surface molecules, a means of cell communication discussed lately by Roth[23], or whether the transfer of molecules occurs from cell to cell through specialized membrane junctions which have recently been demonstrated to exist in a number of developmental situations[24,25,26].

Fig. 4. Transmission electron micrograph of a 12-day mouse embryonic kidney rudiment. Inducing epithelial cells (E) are in close contact with responding mesenchymal cells (M). The area shown in fig. 5 is indicated with an arrow. Ruthenium red staining. x 12 000.

Fig. 5. Higher magnification of an area in fig. 4. Cell processes from epithelial (E) and mesenchymal (M) cells in close contact with each other (arrow). No separating larger amounts of extracellular material are visible. Some electron dense marker material (ruthenium red used as extracellular tracer material) is seen at places between the cell processes. x 150 000.

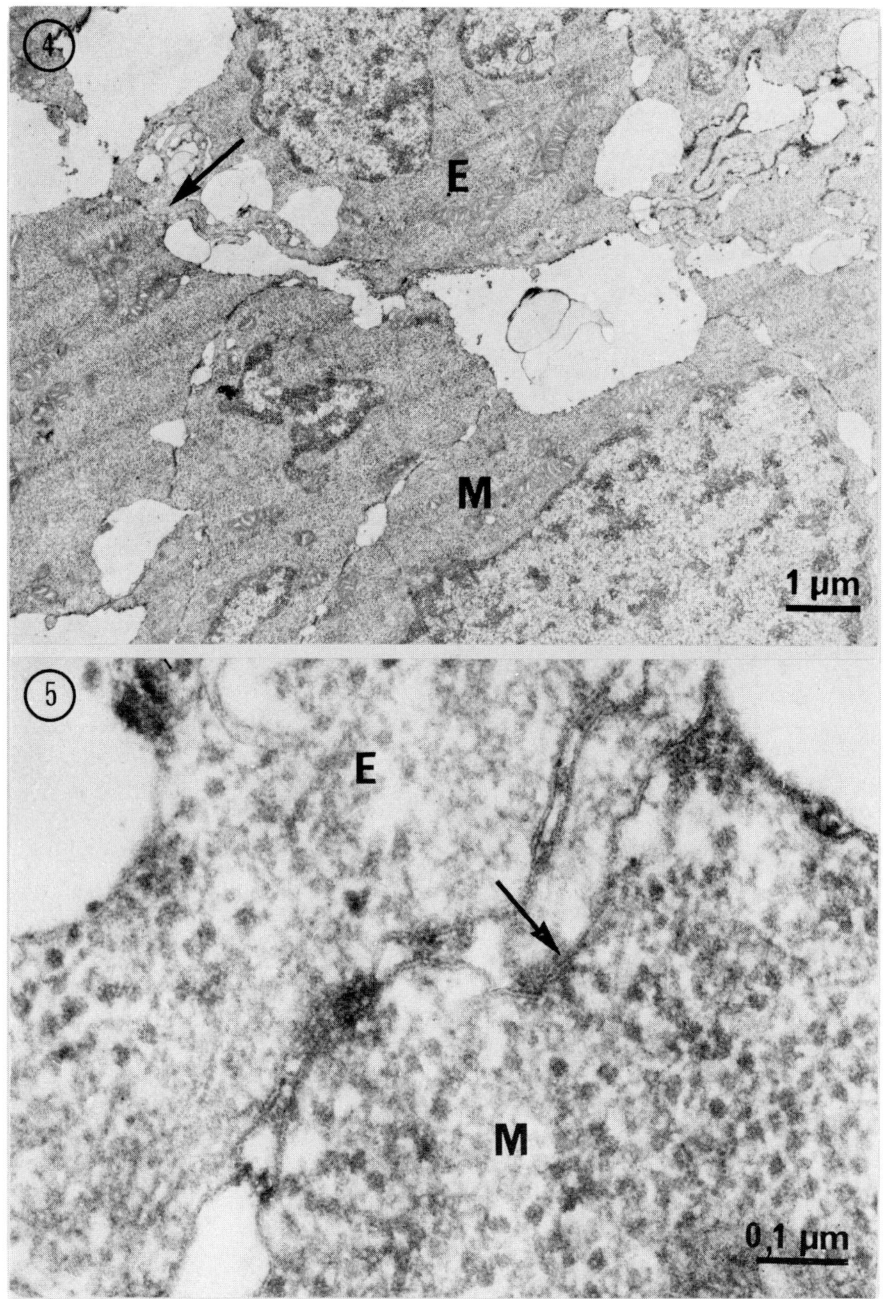
4
E
M
1 µm
5
E
M
0,1 µm

References


1. Wolpert, L., 1969, J. Theor. Biol. 25, 1.
2. Crick, F., 1970, Nature (Lond.) 225, 420.
3. Moscona, A.A., 1968, Develop. Biol. 18, 250.
4. Grobstein, C., 1967, Nat. Cancer Inst. Monogr. 26, 279.
5. Saxén, L., 1972, in: Tissue Interactions and Carcinogenesis, ed. D. Tarin (Academic Press, London) p. 49.
6. Tiedemann, H., 1968, J. Cell Physiol., Suppl. 1, 72, 129.
7. Kosher, R.A. and Lash, J.W., 1975, Develop. Biol. 42, 362.
8. Ronzio, R.A. and Rutter, W.A., 1973, Develop. Biol. 30, 307.
9. Holtfreter, J., 1955, Exp. Cell Res., Suppl. 3, 188.
10. Grobstein, C., 1955, J. Exp. Zool. 130, 319.
11. Weiss, P., 1947, Yale J. Biol. Med. 19, 235.
12. Grobstein, C., 1956, Exp. Cell Res. 10, 424.
13. Saxén, L., 1971, in: Control Mechanisms of Growth and Differentiation, eds. D.D. Davies and M. Balls (Cambridge University Press, Cambridge) p. 207.
14. Saxén, L., 1975, in: Extracellular Matrix Influences on Gene Expression, eds. H.C. Slavkin and R.C. Greulich (Academic Press, New York) p. 523.
15. Grobstein, C., 1957, Exp. Cell Res. 13, 575.
16. Lehtonen, E., Wartiovaara, J., Nordling, S. and Saxén, L., 1975, J. Embryol. Exp. Morphol. 33, 187.
17. Wartiovaara, J., Nordling, S., Lehtonen, E. and Saxén, L., 1974, J. Embryol. Exp. Morphol. 31, 667.
18. Nordling, S., Miettinen, H., Wartiovaara, J. and Saxén, L., 1971, J. Embryol. Exp. Morphol. 26, 231.
19. Wartiovaara, J., Lehtonen, E., Nordling, S. and Saxén, L., 1972, Nature (Lond.) 238, 407.
20. Lehtonen, E., Saxén, L., Wartiovaara, J., Nordling, S. and Karkinen-Jääskeläinen, M., (submitted)
21. Martínez-Palomo, A., 1970, Int. Rev. Cytol. 29, 29.
22. Lehtonen, E., 1975, J. Embryol. Exp. Morphol. (in press)
23. Roth, S., 1973, Quart. Rev. Biol. 48, 541.
24. Rifkind, R.A., Chui, D. and Epler, H., 1969, J. Cell Biol. 40, 343.
25. Revel, J.-P., 1974, in: The Cell Surface in Development, ed. A.A. Moscona (John Wiley & Sons, New York) p. 51.
26. Ducibella, T., Albertini, D.F., Anderson, E. and Biggers, J.D., 1975, Develop. Biol. 45, 231.

*Progress in Differentiation Research, ed. N. Müller-Bérat et al.*



# THE ROLE OF CELL SURFACE DYNAMICS IN CELLULAR INTERACTIONS

E.J. Ambrose

Chester Beatty Research Institute
Institute of Cancer Research: Royal Cancer Hospital
London SW3 6JB

## 1. Introduction

The biochemical constitution of the cell surface complex is now well known to play a major role in problems of cell contact formation, adhesiveness, electrical communication between cells etc. But it must be appreciated that the cell surface is a highly dynamic region of the cell and in many cells changes are continuously occurring in the microtopology of the cell surface. In this communication, an analysis of these movements in various types of tissue cells before and after malignant transformation is described. These studies have been carried out by time lapse filming both in monolayer tissue culture and in organ culture and by scanning electron microscopy. Although the malignant change is pathological these studies may well have a bearing on the nature of cell surface changes involved at certain stages of normal development as in gastrulation and in the migration of neural crest cells, episodes in the life of embryonic cells, which in certain respects closely resemble those of invading and metastasising cancer cells.

## 2. Cell surface dynamics in monolayer tissue culture

Fibroblasts from both embryonic and adult tissue are well known to spread uniformly on glass or plastic and to exhibit ruffled or undulating membrane activity on their leading edge. These wave-like movements also occur on the surface in contact with the substratum and cell locomotion is brought about by a combination of transverse waves and compressional waves generated along the surface[1]. This model for cell locomotion has been modified in recent

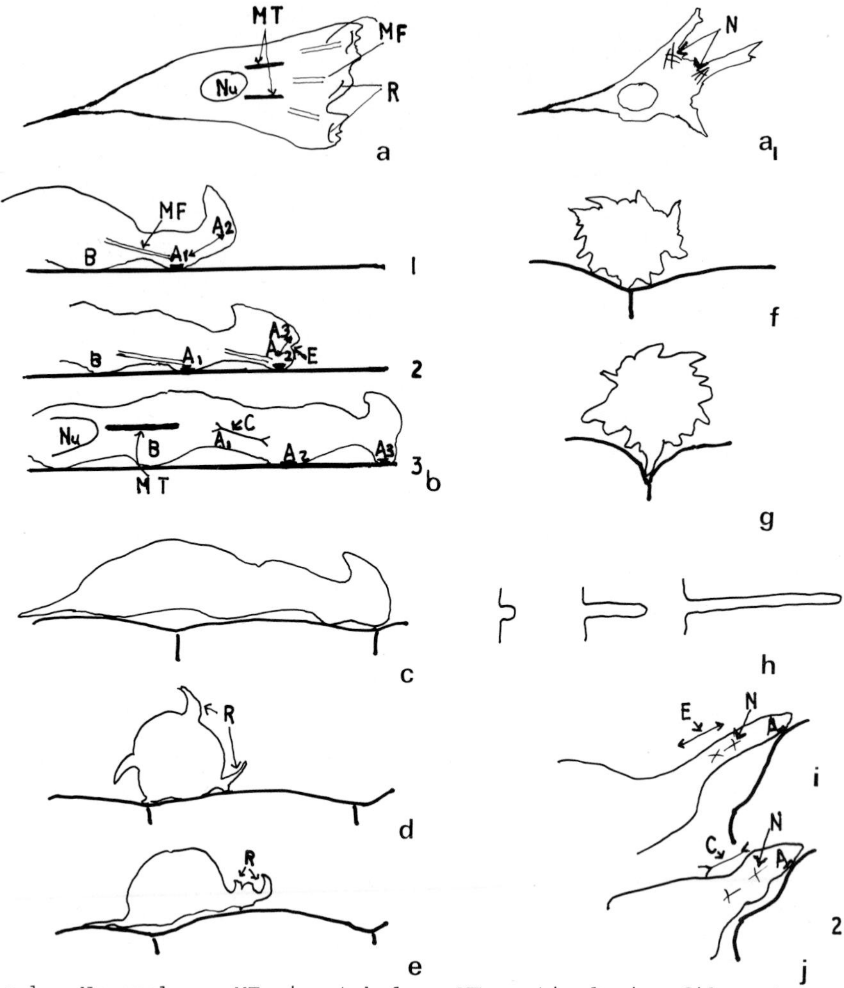


Fig.1. Nu-nucleus; MT-microtubules; MF-cortical microfilaments; R-ruffled membranes; N-network microfilaments; A-strongly adherent contact; B-weakly adherent contact; E-Expansion; C-Contraction.
a. Fibroblast on glass (plan); $a_1$. Sarcoma cell on glass.
b. 1,2 & 3.Migration of a fibroblast on glass (section)
c. Spreading of a fibroblast on epithelial tissue (section)
d. The same in 5 $\mu$gm/ml colcemid.
e. Macrophage on epithelial tissue, no colcemid(section).
f. Sarcoma cell on epithelial tissue (section).
g. The same after polypodial penetration
h. Evolution of a microvillus
i.1. Extension of a polypodial-like extension, followed i.2 by contraction in <u>Fundulus</u> cell (after Trinkaus).

years as a result of the work of Abercrombie, Heaysman and Pegrum[2], Harris and Dunn[3], and others. Two types of intermittent contact occur. We will call these A contacts (strong adhesions to the substratum formed relatively slowly and withdrawn slowly and B contacts (weak and fluctuating attachments). The behaviour of A contacts, which form at the leading edge of the advancing cell (region of lamellar cytoplasm), is illustrated in Fig. 1.b (1, 2 and 3). In Fig. 1.b.1 Contact $A_1$ has become well established on the substratum. Just within the cytoplasm, a dense deposit forms. Groups of microfilaments orientated towards the nucleus also develop in this region as shown[4]. The region between A.1 and A.2 becomes a region of cell surface expansion, A.2 not being attached to the substratum. A.2 now forms a contact (Fig.1.b.2). The region between A.1 and A.2 is now a region of active contraction so that attachment A.1 is broken. In the meantime, expansion between A.2 and A.3 has occurred and A.3 now forms a strong contact. The B contacts, on the other hand, are of low adhesiveness and are intermittent and irregular. They may play a key role in preventing the cell from becoming uniformly attached to the substratum over large areas when it would be impossible for the cell to pull itself along the surface. An examination of fibroblasts which have undergone transformation, either spontaneously due to long-term culture or as a result of treatment with polyoma or SV40 virus, does not reveal gross differences in cell surface dynamics. There is a progressive change as the cells become more anaplastic. The smooth fan-like lamellar cytoplasm on the leading edge of the cell becomes broken up into irregular pseudopodia, variable both in diameter and in length. These we now call polypodia. The smooth surface undulations characteristic of normal cells are gradually lost. This type of alteration in cell surface dynamics has been observed in all the cell types studied, i.e. in carcinomas, when compared with normal epithelium, in astrocytomas of various grades, and in melanomas.

In some of the cases of fibroblasts which have undergone spontaneous transformation and produce invasive tumours *in vivo*, the appearance of a ragged polypodial leading edge has also been observed, although the cells still line up and show contact inhibition of movement when studied in wounded edge cultures, i.e. production of polypodia is strikingly characteristic for all the malignant transformations investigated. The molecular basis for these changes has been investigated by scanning electron microscopy, by ion etching and by transmission electron microscopy[5,6,7,8]. In non-malignant fibroblasts bundles of sub-surface microfilaments, similar to the actin of smooth muscle, are seen to be orientated in the direction of locomotion in the lamellar cytoplasm. In addition, a network of filaments are observed in the regions of transient membrane disturbance. Myosin molecules are now known to be associated with the actin-like filaments in regions of contraction. The bundles of orientated sub-surface filaments (cortical microfilaments) are disturbed or completely absent in cells which exhibit polypodial activity. Network filaments which are thought to be responsible for surface contractions[9] are still present.

3. The dynamics of cell-surface movements within living tissues, organ culture studies.

By the use of an organ culture technique, using natural membranes such as the chick chorioallantoic membrane, the rat omentum, and chick amniotic membrane etc., as undamaged normal tissues[10], it has become possible, by developing a new time lapse filming technique[11], to observe the cell surface movements of cells in living tissues in considerable detail. The natural membrane is folded over and the folded edge is observed, using the new optical technique. 'Optical sections' of the living tissue can then be studied[12]. The various cell types already mentioned above have been placed on the surface of these membranes and their interactions with the normal epithelium and connective tissue has been investigated. The non-malignant and malignant cells used in the experiments have also been checked for their non-invasive and invasive character in vivo. The difference between the cell surface dynamics of non-malignant and malignant cells is more striking under these conditions. Non-malignant fibroblasts spread on normal epithelium by a process similar to that observed on glass or plastic; they eventually form a monolayer on the surface provided that the epithelium surface is smooth. Fibroblasts lying within the connective tissue itself have also been observed to produce ruffled membrane activity similar to that observed in monolayer cultures.

The sarcoma cells and carcinoma cells show a strikingly different behaviour; they become rounded on the surface of the epithelium and exhibit intense polypodial activity over most of the cell surface (Fig.1.f). After a short time one of the polypodial probes makes a point of attachment within the epithelium, apparently at a junction between epithelial cells. Further bombardment of the epithelial cells occurs and the tumour cell is eventually drawn between the epithelial cells (Fig. 1.g). That the polypodia play a key role in this process has been confirmed by a study of the effect of colcemid on the behaviour of cells placed on the normal epithelium. Fibroblasts in 5 μgm/ml of colcemid quickly lose their elongated form; this is due to a breakdown of the internal cytoskeleton formed by microtubules. This drug is well known to cause loss of the star-shaped form of astrocytes in monolayer tissue culture due to breakdown of microtubules[13]. But the sub-surface microfilaments are not affected and the cells continue to generate cell surface contractions and movements. The fibroblasts on epithelium when treated with colcemid assume a rounded form just like transformed cells as shown in Fig. 1.d. But the cell surface movements are identical with those of untransformed cells, wave-like movements and ruffled membranes gyrate round the cell circumference as shown in Fig. 1.d instead of moving along the flat surface of the cell. Polypodial activity is not observed. They do not penetrate epithelium. The colcemid-treated fibroblasts in fact look rather similar to mouse peritoneal macrophages (not treated with colcemid). Macrophages have a generally rounded shape when placed on epithelium but produce intense undulating membrane activity in the lamellar cytoplasm on their leading edge (Fig. 1.e). These results suggest that over-all cell shape is not the main factor affecting the ability of cells to penetrate normal epithelial. This has been confirmed by studies of the Harding-Passey melanoma cells. Melanocytes, being of neural crest origin, exhibit dendritic pseudopodia. These star-shaped extensions are still present in this melanoma. Treatment with colcemid causes retraction of the dendritic processes. A detailed study of Harding-Passey melanoma cells on epithelia by histological methods[10], and by time lapse filming[12], reveals that they penetrate equally well in the absence or in the presence of colcemid. In both cases, the intense surface polypodial activity plays a key role in the invasive process, the star-shaped dendritic processes tend to spread superficially.

## 4. Discussion

The studies outlined above support the view that cells which generate wave-like movements on the surface of the leading lamellar cytoplasm are basically spreading cells and will tend to form sheets on the surface of the other cell layers. On the other hand, cells which generate polypodia will tend to exhibit the capacity to penetrate normal tissues. The possibility cannot be excluded, however, that the penetrating capacity is also partly dependent on the local release of enzymes or other factors by the invading cells[14]. The polypodial activity is certainly essential. The polypodia have been carefully studied by stereoscan microscopy in the case of carcinomas; they probably arise by some degeneration of the numerous microvillae which occur uniformly on normal epithelia. But the microvillus is a narrow rod-like structure (~ 2000 Å in diameter); when extended it becomes rigid and is then withdrawn (Fig. 1.h). There is little evidence to suggest that it has marked motile function. It is here that the polypodium differs, being of variable and often of large diameter; plasma gel can enter the polypodium so that network microfilaments can generate contraction, i.e. a polypodium can enter a narrow space between other cells, can attach and then contract so drawing the cell forwards into the normal tissues somewhat as shown in Fig. 1.i. The basic molecular mechanism is similar to the intermittent contact type of movement of the normal fibroblast moving on a smooth surface as shown in Fig. 1.b.

Studies of cell movements in developing embryos <u>in vivo</u> by time lapse filming suggest that the cell surface activity observed is basically rather similar to that described above in the case of tumour cells which show polypodial activity in organ culture systems,[15,16,17] Studies of the deep cells of the Fundulus blastoderm by Trinkaus indicates that at the early blastula stage hemispherical surface protrustions develop on the surface and are then withdrawn; a few may adhere to other cells. They often elongate into lobopodia. At this stage cell locomotion begins and continues throughout gastrulation. When a lobopodium adheres to another cell or to the under-surface of the developing periblast it may retract, so moving the cell forwards as indicated in Fig. 1.i.

None of these types of cell movement are complete characteristic of a given cell type; there is a gradual transition in surface behaviour which is related to a new type of biological behaviour. It is comparatively simple to see how small changes in the number and degree of aggregation of sub-surface microfilaments may be all that is required to transform a stable spreading type of tissue cell into a migrating cell and back again to a stable tissue cell once it has achieved its correct location during morphogenetic movements in normal development. Such simple cell surface changes could be regulated either by nuclear or by cytoplasmic control during the critical developmental stages. This does not exclude the possibility that biochemical changes in the outer surface of the membrane leading to altered surface adhesiveness may not also be involved. But it is likely that the dynamic changes in surface topology as described in this paper play a dominant role.

References

1. Ambrose, E.J. 1961. Exp. Cell Res. Suppl. 8. 54

2. Abercrombie, M., Heaysman, J.E.M. and Pegrum, S.M. 1971. Exp. Cell Res. 67, 359.

3. Harris, A. and Dunn,G. 1972. Exp. Cell Res. 73, 519

4. Heaysman, J.E.M. and Pegrum, S.M. 1973. Exp. Cell Res. 78, 71.

5. Ambrose, E.J., Osborn, J.S. and Stuart, P.R. 1970. Liquid Crystals and Ordered Fluids.(Plenum Press) ed.Johnson,J.F. and Porter, R.S. p. 83.

6. Ambrose, E.J., Batzdorf, U. and Easty, Dorothy M. 1972. Jnl. Neuropath. & Exp. Neurology. 31, 596.

7. McNutt,N.S., Culp, L.A. and Black, P.H. 1971. J.Cell Biol.50,691

8. Vasiliev, Ju.M. and Gelfand, I.M. 1973. The Locomotion of Tissue Cells. (Elsevier, Amsterdam) ed. Abercrombie, M., p.311.

9. Wessells, N.K., Spooner, B.S. and Luduena, M.A. 1973. The Locomotion of Tissue Cells. (Elsevier,Amsterdam). ed. Abercrombie, M. p.53.

10. Easty, Dorothy M. and Easty, G.C. 1974. Br.J.Cancer, 29,36.

11. Ambrose, E.J. and Easty, Dorothy M. 1973. Differentiation,1,39
ibid. 1,277

12. Ambrose, E.J. and Easty, Dorothy, M. (In course of Publication)

13. Wessells, N.K., Spooner, B.S., Ash, J.F., Bradley, M.O. Luduena, M.A.,Taylor, E.L. , Wrenn, J.T. and Yamada, K.M., 1971. Science, 171, 135.

14. Easty, Dorothy M., Personal Communication.

15. Gustafson, T. and Wolpert,L., 1961. Exp. Cell Res. 24, 64.

16. Trinkaus, J.P. and Lentz, T.L. 1967. J.Cell Biol. 32, 139.

17. Trinkaus, J.P. 1973. Dev. Biol. 30, 68.

Acknowledgments

This investigation was supported by grants to the Chester Beatty Research Institute (Institute of Cancer Research: Royal Cancer Hospital) from the Medical Research Council and the Cancer Campaign for Research.

*Progress in Differentiation Research, ed. N. Müller-Bérat et al.* 


# INTERCELLULAR BRIDGES BETWEEN GERM CELLS DURING THE FIRST STAGES OF OOGENESIS IN ADULT LIZARD

S.Filosa and C.Taddei

Institute of Hystology and Embryology, Laboratory of Comparative Anatomy, University of Naples, Italy. (°)

## 1. Introduction

Intercellular bridges occurring between sister germ cells during embryonic development of the ovary have been found in different Vertebrates: Amphibia[1], Chicken[2], and Mammals[3]. These bridges, which connect oogonia or oocytes, lead to the formation of a syncytium, the function of which during oogenesis is not well defined. Some investigators[3] suggest that they are related to the synchronous development of all the germ cells constituting the cluster. Others[2,4] ascribe these bridges the function of transferring nutritive materials to the germ cell destined to become the egg.

This report concerns an investigation of the cellular interconnections in the germinal epithelium of the adult lizard, in which the oogenesis is known to be an uninterrupted process during the reproductive life[5].

## 2. Materials and Methods

Germinal epithelia were isolated under the dissecting microscope from 30 ovaries of adult females of Lacerta sicula Raf collected in the neighbourhood of Naples during the period from November to April.

The whole germinal epithelia (about 1 mm in diameter) were fixed in phosphate buffered formaldehyde-glutaraldehyde, pH 7.4, postfixed with phosphate buffered 2% osmium, dehydrated with alcohol and embedded in Epon. Ultrathin sections, obtained with an LKB Ultrotome, stained with uranyl acetate and lead citrate, were observed under a

(°) This work was supported by a C.N.R. grant no.CT 73/0591 and was carried out at the Electron Microscopy Center of the Faculty of Science of the University of Naples.

Siemens Elmiskop IA electron microscope.

## 3. Results

In the adult lizard the germinal epithelium is located in a small area close to the ilum of the ovary. It appears to be made of germinal elements at different stages of differentiation (oogonia and oocytes), generally organized in groups surrounded by stroma cells (Fig.1).

The germ cells are easily distinguishable under E.M.from the stroma cells by their larger size and round-shaped nucleus. The stroma cells are smaller, with an irregularly-shaped, generally elongated nucleus containing condensed chromatin (Fig.1).

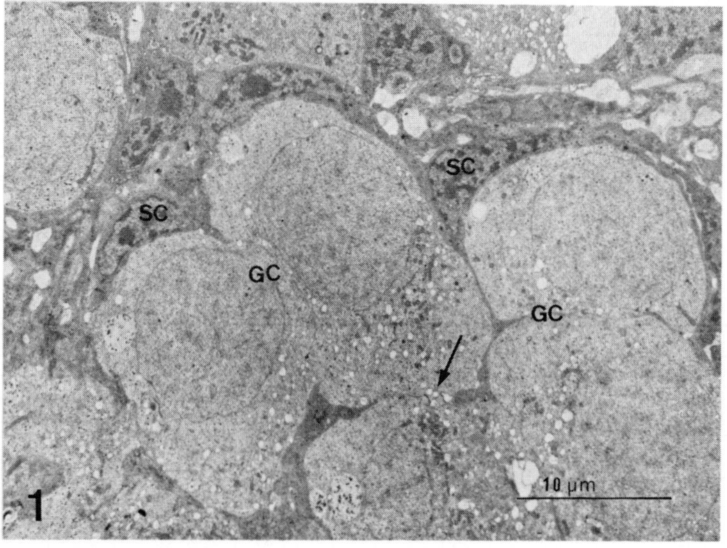


Fig.1.A portion of a germinal epithelium of adult lizard. A cluster of germ cells (GC), at the beginning of the meiotic prophase, surrounded by stroma cells (SC) is evident. Two germ cells are interconnected by an intercellular bridge (arrow).

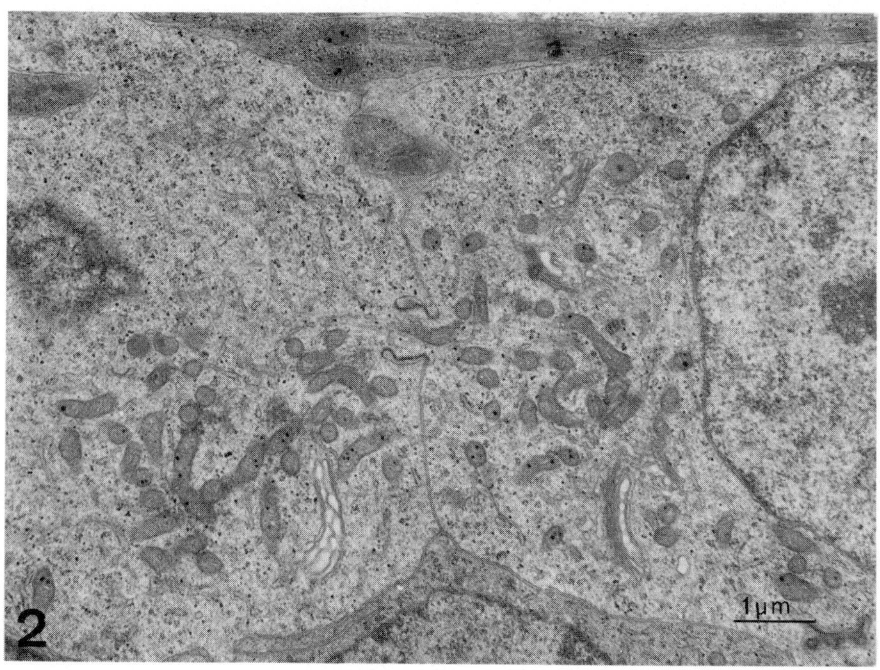


Fig.2. An intercellular bridge connecting two oogonia. A mitochondria is evident inside the bridge.

During the oocyte differentiation the first elements clustered in groups are oogonia, which are distinguishable by their round-shaped nucleus with a conspicuous portion of the chromatin condensed. In the cytoplasm, mitochondria, Golgi elements, ribosomes, and a few elements of endoplasmic reticulum are present (Fig.2).

The most remarkable feature of these cells is the presence of cytoplasmic interconnections between cells at the same stage of differentiation. In the region where the connection occurs, the plasma membrane of one cell is in continuity with that of the other cell, forminga sort of channel (about 600nm long and 500 nm large). In this region the opacity of the plasma membrane increases because of

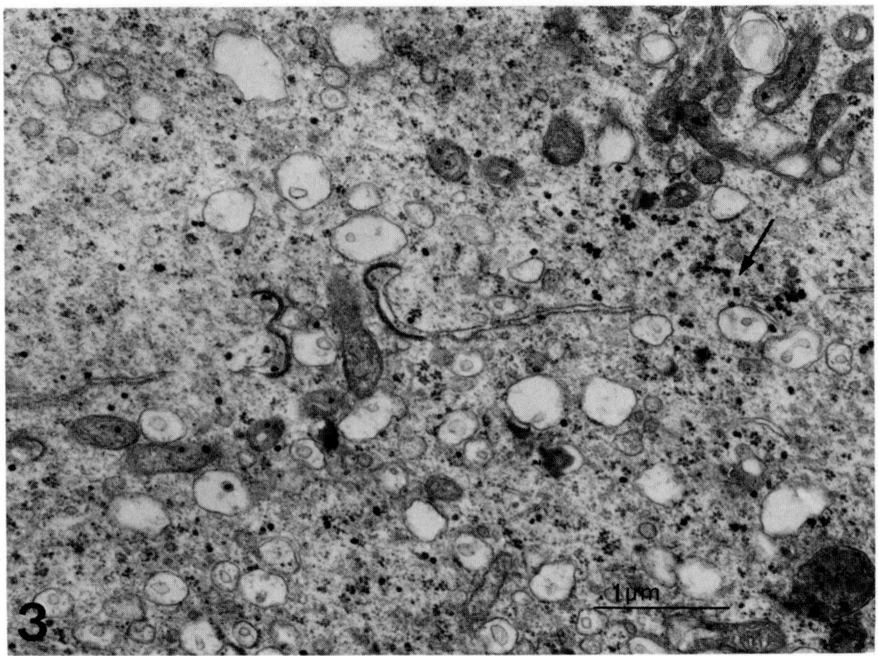


Fig.3. Cytoplasmic interconnections between two oocytes at zygo-pachytene. Beside the bridge, other interruptions of the plasma membrane (arrow) are evident.

the presence on the inner side of a layer (400 Å thick) of dense material. Cytoplasmic organelles such as mitochondria or vesicles can be found within the intercellular bridges.

At the beginning of the meiotic prophase (zygo-pachytene), characterized by enlargement of the nuclei and formation of synaptonemal complexes, the germ cells maintain the organization in clusters (Fig.1); intercellular bridges similar in structure to those observed between oogonia are present (Fig.3). It is noteworthy that, in addition to these channels, other interruptions of the plasma membranes are found, which connect the cytoplasms of the two cells (Fig.3).

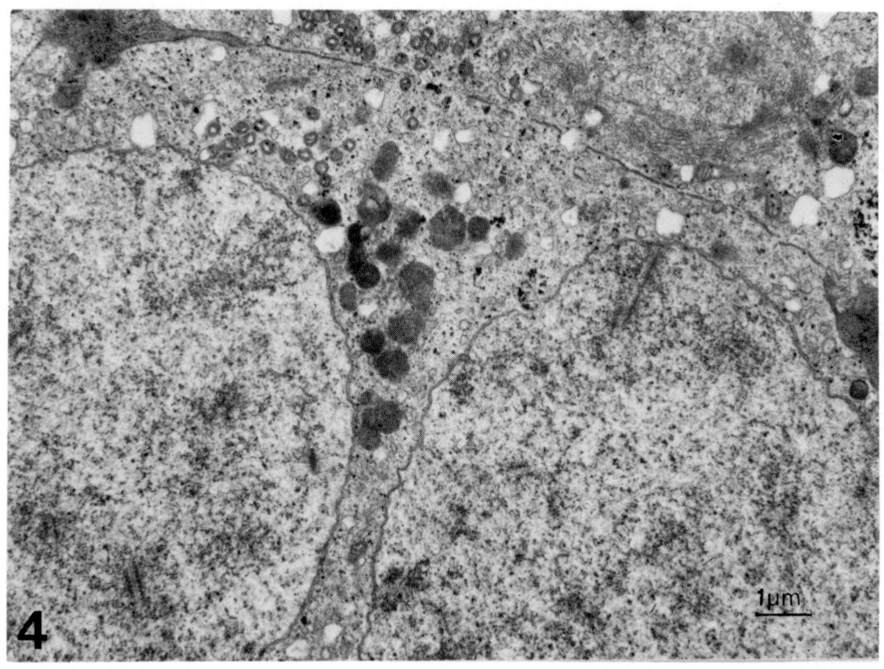


Fig.4. Two nuclei at zygo-pachytene. The synaptonemal complexes are evident. The plasma membrane is completely absent and the nuclei share a common cytoplasmic environment.

These interruptions are delimited by the plasma membranes of the two cells, but no structural organization, like that found at the level of the bridges, is evident. It is not infrequent to find germ cells at this stage (zygo-pachytene) in which the plasma membrane is totally absent and more nuclei share the same cytoplasm (Fig.4).

## 4. Discussion

This report shows that in the germinal epithelium of the adult lizard intercellular bridges occur between sister germ cells during the differentiation of oogonia and oocytes.

The formation of a syncytium constituted by all the germinal elements of a cluster[1,2,3,4] seems to be a general aspect of the first stages of oogenesis in the Vertebrates so far studied, and can be related to the differentiation of the germ cells.

The intercellular bridges found in the adult germinal epithelium of Lacerta are similar in structure to those described during embryonic development of the ovary in other Vertebrates. Nevertheless, in this species, while in the oogonia the cytoplasmic continuity is assured by the bridges, later on in the oocytes, at the beginning of the meiotic prophase, the disappearance of part or all of the plasma membranes delimiting two cells produces a more integrated syncytium.

Furthermore, in this species, in a later phase of oogenesis, at the beginning of auxocytosis, in follicles ranging from 100 μ to 2,000 μ in diameter, intercellular bridges establish themselves between particular follicle cells (pyriform cells) and the oocyte[6,7,8].

It seems that these two kinds of bridges found at different stages of oogenesis cannot be correlated, since the pyriform cells, differentiate inside the follicular epithelium around the oocyte, which have already left the original germinal epithelium.

## References

1) Ruby, J.R.; Dyer, R.F.; Skalko, R.G. and Volpe, E.P., 1970, Anat. Rec. 167, 1.

2) Skalko, R.G.; Kerrigan, J.M.; Ruby, J.R. and Dyer, R.F., 1972, Z.Zellforsch. 128, 31.

3) Zamboni, L. and Gondos, B., 1968, J.Cell Biol. 36, 276.

4) Ruby, J.R.; Dyer,R.F. and Skalko, R.G., 1969, J.Morph. 127, 307.

5) Franchi, L.L.; Mandl, A.M.; Zuckerman, S., 1962, "The Ovary" (S.Zuckerman, ed.) vol.1, p.1, New York, Academic Press.

6) Ghiara, G.; Limatola, E. and Filosa, S., 1968, Conference on Electron Microscopy Vol.2, p.331, Tip.Poliglotta, Roma.

7) Hubert, J., 1971, Z.Zellforsch. 116, 240.

8) Taddei,C., 1972, Exp.Cell Res. 72, 562.

# Section 4.
# ASPECTS OF CARCINOGENIC DISORDERS AND DIFFERENTIATION

*Progress in Differentiation Research, ed. N. Müller-Bérat et al.*



# ORIGIN OF NEOPLASTIC STEM CELLS

G. Barry Pierce
Department of Pathology,
University of Colorado Medical Center, Denver, Colorado

At the first conference of this society I discussed our observations, obtained in three systems, that malignant stem cells could differentiate[1]. In two of these situations, the differentiated progeny of the malignant stem cells was shown to be benign and indistinguishable from their differentiated normal counterparts[2,3]. As we will hear this afternoon, allophenic mice have been produced in three laboratories through incorporation of embryonal carcinoma cells in blastocysts[4]. It may be concluded from these experiments that the differentiated progeny of embryonal carcinoma cells are not only benign, but capable of integration with other cells to produce a functional animal. These observations all strengthen the concept that neoplasms are problems of developmental biology.

From studies on squamous cell carcinoma[3], adenocarcinomas of the breast[5], neuroblastoma[6], and leukemia[7], it is clear that tumors are a caricature of normal tissue. On the one extreme are malignant stem cells and on the other the benign differentiated progeny of those malignant cells. This cellular heterogeneity has not been taken into full account in considering the results of ultrastructural and biochemical studies of neoplasms.

Not only is a tumor a caricature of the normal tissue, more importantly, it is a caricature of the renewal process whereby normal tissues are maintained. Normal and malignant stem cells each have the same end point in that they can give rise to populations of differentiated functional cells. The neoplastic stem cells operate at different levels of control and evolve a mass, the biological characteristics of which may be incompatible with the life of the host. The lethality of malignant tumors is probably an incidental by-product of oncogenesis, and, as will be discussed, benign and malignant tumors probably have similar mechanisms.

Since a tumor is a tissue composed of cells of the body, and since all other tissues arise by the process of differentiation, we postulated years ago that a tumor was akin to a postembryonic differentiation[8]. The notoriously stable controls of differentiation would explain the heretability and stability of the neoplastic state which has usually been attributed to somatic mutation. In other words, it is postulated that carcinogenesis involves stable heritable changes in control of genome rather than structural changes in that genome.

In considering the problem of carcinogenesis as a problem of differentiation, the question was raised as to the origin of malignant stem cells. The theoretical implications of this problem have been considered before[9]. Briefly, if carcinogen is applied to the skin and a tumor develops in the skin it is invariably a tumor composed of skin cells. In other words, the responding cells are already determined for skin, and since they are mitotically active, we have theoretically pinpointed the normal stem cell as the target in carcinogenesis. This should not be surprising because the stem cells are the reactive cells of normal tissue and are responsible for hyperplasia and metaplasia as well as for maintaining the normal tissue. As for development, one can think of the process of differentiation as a process of sequentially evolving populations of stem cells. What has been confounding in studies of neoplasms has been the evolution of a morphologically less well differentiated tissue than the tissue of origin. These observations are not incompatible with the idea that a tumor is a differentiation in itself[9].

When polyoma virus is injected into a suitable animal, as many benign tumors develop as malignant ones[10]. When carcinogen is painted on the skin, papillomas appear long before carcinomas[11]. These observations could be explained by assuming that benign tumors are a stage in the development of malignant ones. As an alternative, it can be postulated that the phenotype of a tumor, whether benign or malignant, depends upon the state of differentiation of the cell responding to the carcinogenic stimulus, and the effect of the environment upon the responding cell. If the responding cell is almost completely differentiated, but still capable of one more cell division, then it could be postulated that upon transformation it would give rise to a benign tumor. Since these cells are very little different from the normal, the environmental conditions necessary for their maximal proliferation would be found in the tissue of origin. Thus, the benign tumor should appear relatively quickly. On the other hand, should the stem cell of the tissue respond to the oncogenic stimulus and become transformed, its biological needs which are now radically different from the normal would preclude rapid cell division and these initiated cells would lie dormant, possibly for years. This would account for the prolonged latent period in carcinogenesis. When a critical number of cells were produced and able to create an optimal microenvironment then they would be able to express the malignant phenotype[12].

If cytoplasm controls gene expression as could be deduced from nuclear transplant experiments[13], one could postulate that in chemical carcinogenesis cytoplasmic controls of gene expression are altered. The genome of the well differentiated cell is so locked-in for its particular differentiation that it cannot respond to these altered cytoplasmic controls except by an increase in cell division with lack of response to physiologic control. The stem cell, determined for a particular differentiation but not yet having acquired the stable cytoplasmic controls of the differentiated state could respond to the carcinogenic event by massive realignment of these cytoplasmic controls with expression of the phenotypic characteristics of malignancy.

If the above is true, then malignant stem cells should arise from normal stem cells.

Stevens has demonstrated the origin of the stem cells of teratocarcinoma to be from primordial germ cells[14]. Primordial germ cells and malignant stem cells were found to have similar ultrastructural manifestations of differentiation[15]. Franks and Wilson [16] have shown that normal and transformed fibroblasts have comparable degrees of differentiation. With this background, we postulated that normal stem cells and malignant stem cells should have comparable degrees of differentiation, and we studied and compared the state of differentiation of normal and malignant stem cells of breast[17], gut, yolk sac and cartilage. Stem cells were identified by autoradiography with electron microscopes after incorporation of tritium labeled thymidine into the cells.

In the case of the breast[17], the stem cells are contained in the rudimentary ducts found in the resting mammary gland. Ductal cells were found to be exceedingly undifferentiated, but so few of them were synthesizing DNA, the study was performed on the mammary glands of animals pregnant for 8 days. These cells differed from those in the resting gland only through the presence of vacuoles which have been described by many others, and their state of differentiation was comparable with that of the stem cells similarly identified in spontaneous adenocarcinomas of the breast. In either situation, the nuclei were often indented, the cytoplasm contained numerous polysomes not attached to membranes, and there was a paucity of rough endoplasmic reticulum with inactive appearing Golgi membranes with no evidence of secretion. It was concluded from this study that cells were present in the normal mammary gland that were as undifferentiated as the stem cells of mammary tumors.

Since we have used parietal yolk sac carcinomas as a source of basement membrane for studies in the biology of basement membrane [18], we compared the ultrastructure of the stem cells of this tumor with that of the distal endoderm, the normal counterpart of the tumor. Both the normal distal endoderm and its malignant counterpart proved to be well differentiated in the sense that the cytoplasm contained numerous profiles of rough endoplasmic reticulum distended with basement membrane antigens. Undifferentiated stem cells similar to those in the breast or teratocarcinomas were not observed in the yolk sac or its tumors. Cells of the normal and neoplastic tissues were directly comparable in their degrees of differentiation. The neoplastic cells contained many viral profiles budding from the rough endoplasmic reticulum. Again it was concluded that the stem cells of normal tissue were in all likelihood the cells of origin of the tumor cells.

Cells were found in normal rodent cartilage that were directly comparable in their degrees of differentiation to those found in chondrosarcomas. As in the case of distal endoderm and yolk sac carcinomas, the normal and neoplastic stem cells were well differentiated with numerous profiles of rough endoplasmic reticulum and well developed Golgi complexes distended with secretions. The cells were remarkably homogeneous in their degrees of differentiation, although those not synthesizing DNA were usually widely separated by chondromucoprotein presumably synthesized by the cells.

We are currently in the process of studying an adenocarcinoma of the bowel induced by instillation of chemical carcinogen in rats by Jerald Ward. From an ultrastructural standpoint the tumor contains mucous and vacuolated cells, each of which synthesize DNA. Chang and Leblond[19] have shown by autoradiography with the electron microscope that both mucous and vacuolated cells of the normal rodent colon are capable of synthesizing DNA. We have confirmed this observation and demonstrated that the state of differentiation of these normal stem cells is similar to that of their counterpart in the malignant tissue.

From these incomplete studies, it has become apparent that the stem cells of normal tissues which are responsible for maintenance of the adult phenotype, are as undifferentiated (or as differentiated) as their malignant counterparts. Since Stevens has shown the origin of one carcinoma from a normal stem cell and since we have now shown that normal tissues contain stem cells of comparable degrees of differentiation to their malignant counterparts, credence is lent to the postulate that the target in carcinogenesis is the normal stem cell. The parallels in neoplastic and normal development are now so strong that many of the basic tenets and approaches in neoplasia must be reconsidered in light of them.

References


1. Pierce, G.B., 1972, in: Cell differentiation, eds. R. Harris, P. Allin and D. Viza (Munksgaard, Copenhagen), p. 109.
2. Pierce, G.B., Dixon, Jr., F.J. and Verney, E.L., 1960, Lab. Invest. 9, 583.
3. Pierce, G.B. and Wallace, C., 1971, Cancer Res. 31, 127.
4. McBurney et al. These proceedings.
5. Wylie, C.V., Nakane, P.K. and Pierce, G.B., 1973, Differentiation 1, 11.
6. Goldstein, M.N. in: Cell differentiation, eds. R. Harris, P. Allin and D. Viza (Munksgaard, Copenhagen), p. 131.
7. Moore, M.A.S. These proceedings.
8. Pierce, G.B., 1967, in: Current topics in developmental biology, vol. 2, eds. A.A. Moscona and A. Monroy (Academic Press, New York), p. 223.
9. Pierce, G.B. and Johnson, L.D., 1971, In Vitro 7, 140.
10. Defendi, V. and Lehman, J.M., 1966, Int. J. Cancer 1, 525.
11. Berenblum, 1962, in: General Pathology, ed. H. Florey (Saunders and Co., Philadelphia and London), p. 528.
12. Pierce, G.B., Nakane, P.K. and Mazurkiewicz, J.E.. 1973, IVth International Symposium of the Princess Takamatsu Cancer Research Fund, eds., W. Nakahara, T. Ono, T. Sugimura and H. Sugano (University of Tokyo Press, Tokyo), p. 453.
13. King, T.J. and McKinnell, R.G., 1960, in: Cell physiology of neoplasia (University of Texas Press, Austin, Texas), p. 591.
14. Stevens, L.C., 1967, J. Nat. Cancer Inst. 38, 549.
15. Pierce, G.B., Stevens, L.C. and Nakane, P.K., 1967, J. Nat. Cancer Inst. 39, 755.
16. Franks, L.M. and Wilson, P.D., 1970, Europ. J. Cancer 6, 517.
17. Pierce, G.B. and Martinez-Hernandez, H., Submitted to Cancer.
18. Pierce, G.B., 1970, in: Chemistry and molecular biology of the intercellular matrix, vol. 1, ed., E.A. Balazs (Academic Press, New York), p. 471.
19. Chang, W.W.L. and Leblond, C.P., 1971, Am. J. Anat. 131, 73.

*Progress in Differentiation Research, ed. N. Müller-Bérat et al.*
*© 1976, North-Holland Publishing Company - Amsterdam, The Netherlands.*



# THE FATE OF TERATOCARCINOMA CELLS INJECTED INTO EARLY MOUSE EMBRYOS

V.E. Papaioannou, M.W. McBurney, R.L. Gardner, and M.J. Evans
Department of Zoology, Oxford University, England.

The stem cells of murine teratocarcinomas are called embryonal carcinoma. These cells can be propagated in vitro and can be cultured under conditions in which a variety of differentiated tissues are produced. The processes involved in this differentiation are probably similar to those processes which occur during normal embryonic development. In an attempt to provide evidence that pluripotential embryonal carcinoma cells respond to normal developmental signals, we have introduced lumps of cells from three different embryonal carcinoma lines into normal mouse embryos. One hundred and twenty-one animals were born from such embryos and 4 of them contained small amounts of apparently nonmalignant tissues derived from the injected cells. These animals remained apparently normal and healthy for at least 9 weeks. The embryonal carcinoma cells in these animals appear to have participated, like the normal embryonic cells, in the development of the fetus. Progeny of the embryonal carcinoma cells were detected in the skin and eyes as melanocytes, in the dermis of the skin as cells which produce the agouti hair phenotype, and in many internal organs. Seven other animals developed extra-gonadal teratocarcinomas which were apparent within 7 days after birth. In these animals, the embryonal carcinoma cells apparently did not respond normally to all developmental signals during embryogenesis.

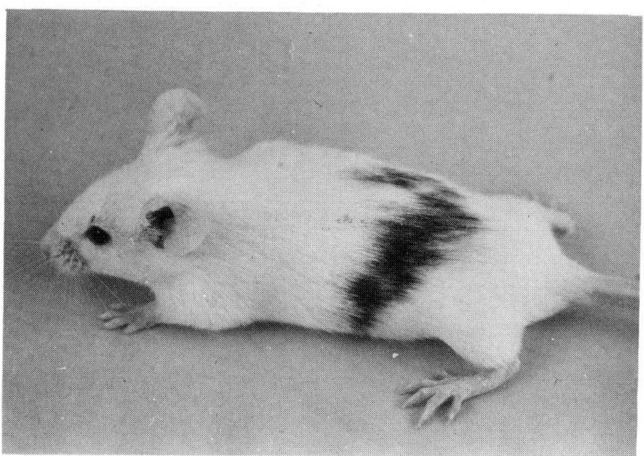

Fig. 1. Nine week old animal with embryonal carcinoma derived melanacytes producing black pigment.

*Progress in Differentiation Research, ed. N. Müller-Bérat et al.*



# IN VITRO DIFFERENTIATION OF NORMAL MOUSE EMBRYO AND TERATOCARCINOMA OF MOUSE

Y.-C. Hsu, J. Baskar, T. Matsuzawa and M. Müntener

Johns Hopkins University
Baltimore, Maryland 21205, U.S.A.

## 1. In vitro development of the mouse embryo from the two-cell egg stage to the early somite stage.

To gain a better understanding of the relationship between the cell differentiation and tumorigenicity of embryonal tumors of mice, an in vitro system to study the differentiation of both normal mouse embryo and embryonal carcinoma has been established[1].

Mouse embryos developed in vitro from the two-cell stage to the blastocyst stage in chemically defined medium. When these blastocysts were cultured in Eagle's minimum essential medium (MEM) plus 10 to 20% of fetal calf serum and human cord serum they differentiated normally to the early somite stage[2,3]. The blastocysts began to shed the zona pellucida on the first day of incubation in MEM. On the second day the denuded blastocysts lay on the surface. The mural trophoblast transformed to giant cells and spread out on the surface leaving the rounded inner cell mass protruding from the surface. Two germ layers of endoderm and ectoderm differentiated from the inner cell mass on the third day to form an egg cylinder. The egg cylinder consisted of a proximal extra-embryonic and distal embryonic region. The embryonic region of the egg cylinder developed into the embryo proper; the extra-embryonic region gave rise to the yolk sac. Blood islands containing primary erythroblasts appeared on the eighth day and became interconnected. As the heart began to beat, the erythroblasts circulated along the capillaries of the yolk sac. The neural plate, the neural tube and five to ten somites were formed in the embryo (fig. 1). Histologically and electron microscopically, the embryos appeared normal at this stage (equivalent to embryos of 8.5 days gestation).

2. Formation of the chorio-allantoic placenta; proliferation of trophoblastic giant cells and cells of the ectoplacental cone.

The trophoblastic giant cells fixing the embryo on the collagen increased their size and cell number and migrated outward after seven days of cultivation as small trophoblastic cells proliferated rapidly from the ectoplacental cone. The trophoblastic cells adjacent to the ectoplacental cone displayed a well organized three-dimensional discoid outgrowth pushing the surrounding giant cells outward. The ectoplacental disc ceased to increase its diameter of about 4 mm after 14 days. Some giant cells displayed nuclear fragmentation. Proloferation and outgrowth of the ectoplacental cone were observed also in the absence of the embryo proper suggesting that the development of the placenta is independent of the presence of an embryo proper.

3. *In vitro* differentiation of embryoid bodies derived from transplantable teratocarcinoma OTT6050.

Embryoid bodies of transplantable teratocarcinoma OTT6050 (derived from normal 6-day-old embryos[4]) proliferated and differentiated *in vitro* in fetal calf serum and/or human cord serum. This system made it possible to study the *in vitro* differentiation of teratocarcinoma in comparison with that of a normal mouse embryo beyond the implantation stage. The embryoid body differentiated *in vitro* with some similarity to the normal embryo of six days' gestation rather than to the morula stage (third gestational day). On the other side it has been shown that embryoid bodies OTT6050 possess antigens in common with morulae[5]. Embryoid bodies OTT6050 did not produce trophoblastic tissue. Disorganized embryonal carcinoma cells in the core of the embryoid body became well organized and formed rosettes. The embryoid body proliferated by budding, forming up to 30 cysts, each of which had an inner core of well-organized ectoderm and a one-cell layer of endoderm around it (fig. 2). Some embryoid bodies developed a yolk sac with blood islands containing primary erythroblasts whereas others developed sqamous epithelium.

4. Small molecular factor(s) in fetal calf serum which accelerates the growth and differentiation of embryoid bodies.

It is important to study the factors in fetal calf serum and human cord serum which are required for the _in vitro_ differentiation of normal mouse embryos. However, it is expensive and time-consuming to use the normal mouse embryo as the test material to detect factors in various serum fractions in a biological assay. Therefore as a preliminary study, the factors required for the growth and differentiation of embryoid bodies have been investigated. Fetal calf serum was separated with Diaflow membrane UM05 (AMINCON) into two fractions: a small molecular filterable (F) fraction and a large molecular non-filterable (L) fraction. It was found that both F and

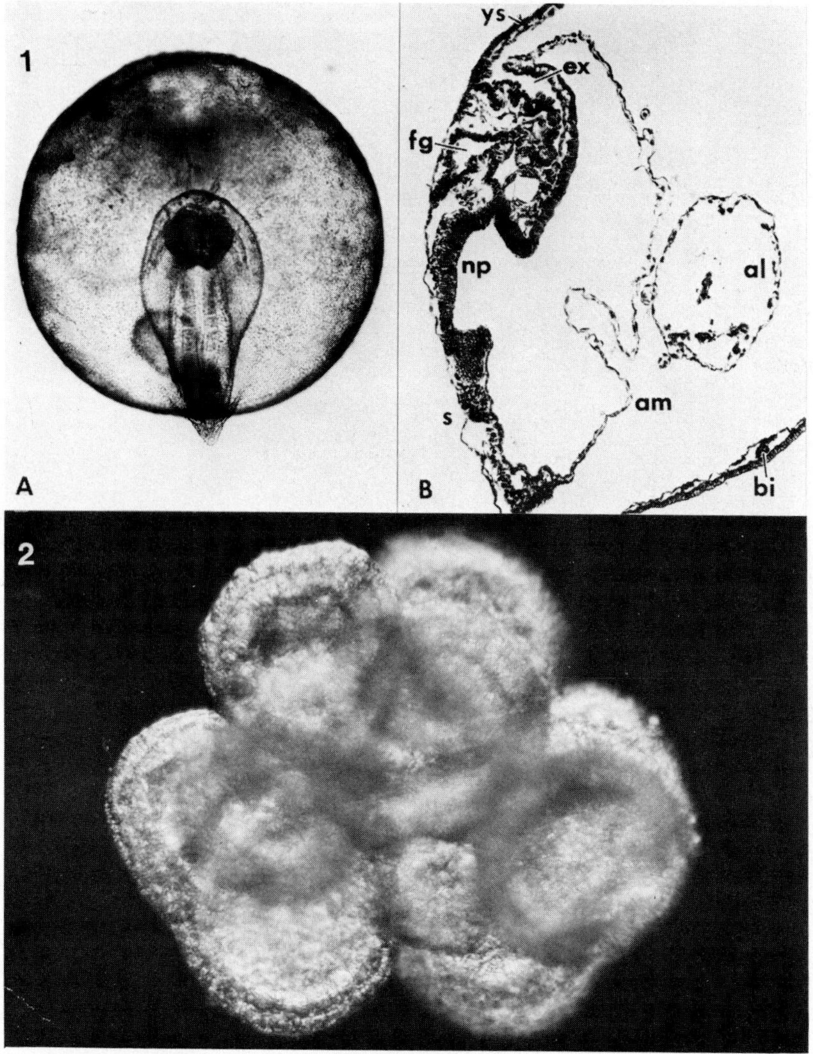


Fig. 1. Mouse embryo developed _in vitro_ ( about 8 days' gestation).

Fig. 2. Cyst formation by cultivating a single embryoid body for 6 days in Eagle's minimum essential medium plus 10% fetal calf serum.

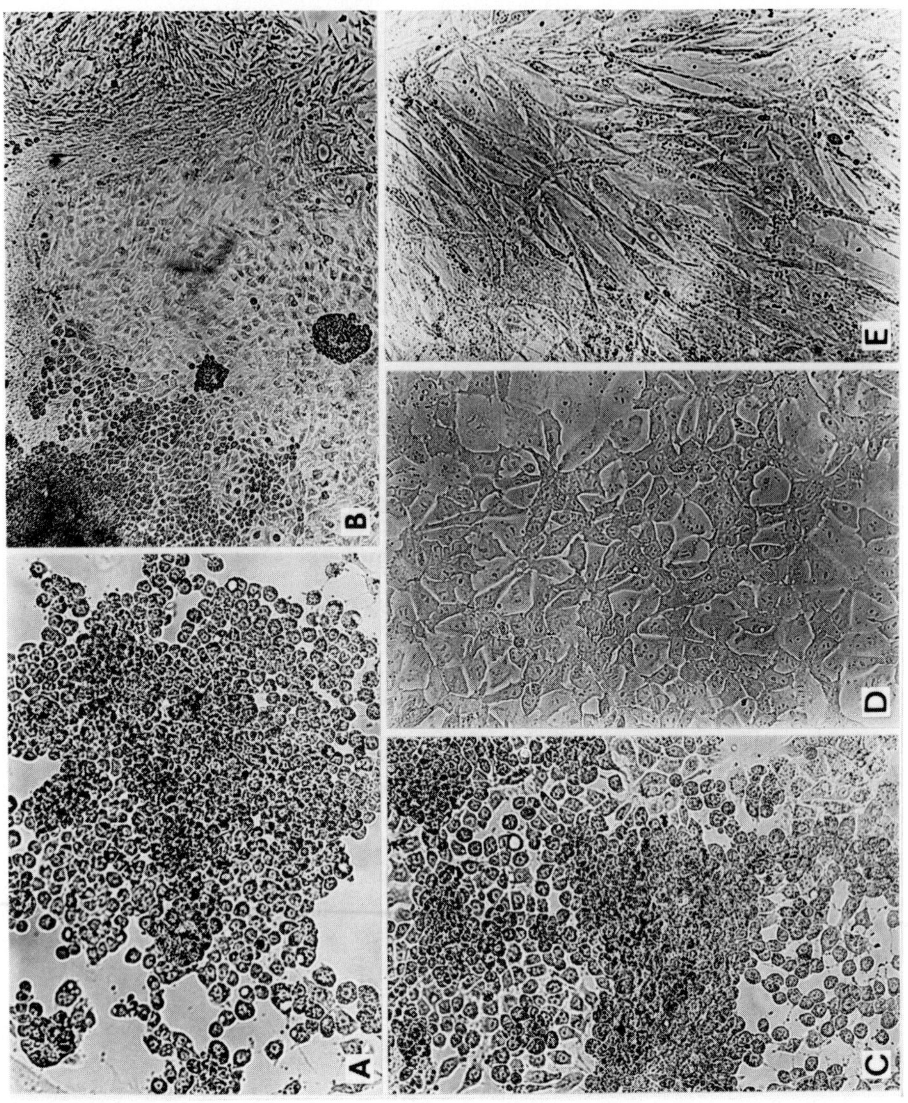


Fig. 3. Clonal isolation of teratocarcinoma stem cells.
(A) An early clone of round cells after 10 days in culture.
(B) The same clone after 21 days in culture. Note differentiation of polygonal cells and spindle cells from round cells.

L fractions were required for the growth and differentiation of embryoid bodies. The F fraction was heat stable (90°C for 30 min.) and could not be substituted with hydrocortisone (1 mg - 0.1 mg/ml), insulin ($10^{-4}$M - $10^{-8}$M) and adenosine 3':5'-cyclic monophosphoric acid (cyclic AMP, $10^{-5}$M - $10^{-6}$M), respectively. The L fraction could not be replaced by bovine plasma albumin, fetuin or transferrin, respectively.

5. Tumorigenicity of three cell types derived from an established cell line of embryonal carcinoma.

A teratocarcinoma cell line derived from transplantable embryoid bodies (OTT6050-C) was established in tissue culture following alternate passages in mouse and tissue culture (Baskar et al. under preparation). Even through clonal isolation, the established culture exhibited heterogeneity displaying at least three morphologically distinct cell types: round, polygonal (epithelial-like) and spindle. Clones of round cells (fig. 3A) tended to give rise to polygonal cells and spindle cells in a sequential fashion (fig. 3B). Round cells not sensitive to contact-inhibition of growth were tumorigenic. Polygonal cells which seemed to show contact-inhibition of growth were also tumorigenic to a lesser extent. The spindle cells sensitive to contact-inhibition of growth were not tumorigenic. Tumors produced by both round and polygonal cells were teratocarçinomas (table 1).

Table 1

Tumorigenicity of three cell types*

| Cell type | Subcutaneous inoculation of cells/mouse | Mice with tumor / Mice inoculated** | Appearance of tumors (days) | Type of tumor |
|---|---|---|---|---|
| Round cells | $10^7$ | 10/10 | 12-14 | Terato-carcinoma |
| Polygonal cells | $10^7$ | 4/10 | 21-23 | Terato-carcinoma |
| Spindle cells | $10^7$ | 0/10 | No tumor | |

* Combined results of two experiments.

** Development of tumor observed for 60 days.

Histochemically, both the round and polygonal cells had a high level of alkaline phosphatase (Gomori method), whereas only a trace amount of alkaline phosphatase could be detected in the spindle cells. Electron microscopy revealed that both round and polygonal cells had cytoplasmic characteristics similar to undifferentiated cells; they were devoid of organelles except for a few mitochondria, occasional strands of endoplasmic reticulum and a few small vacuoles. In contrast, the spindle cell ultrastructure revealed many organelles, a characteristic of differentiated cells. In a complement-dependent cytotoxicity test the antiserum against embryoid bodies was cytotoxic for the round cells. It was less cytotoxic for the polygonal cells, whereas the spindle cells showed no reaction with the antiserum (fig. 4).

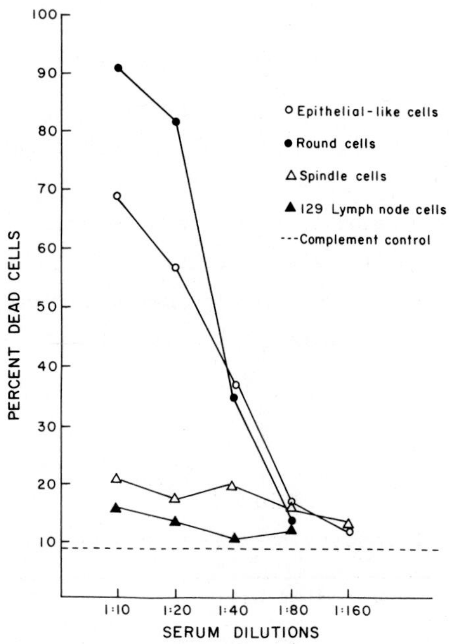


Fig. 4. Cytotoxicity test: anti-embryoid body antiserum against three cell types.

6. Aggregation and differentiation of established teratocarcinoma cell line.

The cell culture of the established teratocarcinoma cell line tended to aggregate and float in culture medium in human cord serum more than in fetal calf serum. Histologically, these aggregates appeared to contain only embryonal carcinoma stem cells without surrounding endoderm. By culturing these stem cell aggregates in Eagle's minimum essential medium plus fetal calf serum, the development of endoderm has been observed. These aggregates developed into a yolk sac-like cyst. The differentiation of these aggregates did not proceed as far as did in vitro differentiation of embryoid bodies from the peritoneal cavity. However, in some cysts, development of hemoglobin in primary hematoblasts was observed.

## 7. Discussion

Teratocarcinoma cell lines have been established either from solid tumors or from embryoid bodies[6-11]. The results and conclusions obtained from using these cell lines are similar. It is interesting to note that the stem cells of embryoid bodies retained their original properties to align as does the ectoderm of 6-day-old embryos. After reaching a certain cell mass by proliferation of stem cells, the embryoid bodies formed multiple cysts. Each cyst corresponded to an embryo. The simultaneous study of in vitro differentiation in both normal and aberrant embryos improves the probability that findings will relate to a better understanding of both cell differentiation and dysdifferentiation.

## References


1. Hsu, Y.-C. and Baskar, J., 1974, J. Natl. Cancer Inst. 53, 177.
2. Hsu, Y.-C., 1973, Develop. Biol., 33, 403.
3. Hsu, Y.-C., et al., 1974, J. Embryol. exp. Morph. 31, 235.
4. Stevens, L.C., 1970, Develop. Biol. 21, 364.
5. Artzt, K., et al., 1973, Proc. Nat. Acad. Sci., USA, 70, 2988.
6. Finch, B.W. and Ephrussi, B., 1967, Proc. Nat. Acad. Sci., USA, 57, 615.
7. Rosenthal, M.D., et al., 1970, J. Natl. Cancer Inst., 44, 1001.
8. Kahan, B.W. and Ephrussi, B., 1970, J. Natl. Cancer Inst., 44, 1015.
9. Jacob, H., et al., 1973, Ann. Microb. (Inst. Pasteur) 124B, 269.
10. Lehman, J.M., et al., 1974, J. Cell Physiol. 84, 13.
11. Martin, G.R. and Evans, M.J., 1974, Cell 2, 163.

*Progress in Differentiation Research, ed. N. Müller-Bérat et al.*



# TERATOCARCINOMA IN RATS

Halina SOBIS and M. VANDEPUTTE
Rega Institute for Medical Research,
University of Leuven, B-3000 Leuven, Belgium.

## INTRODUCTION

In a previous paper we described the induction of yolk-sac carcinomas which developed after inoculation of mouse sarcoma virus (MSV) in fetectomized rats[1,2]. Apart from the yolk-sac carcinomas, the morphology of which was described in detail, we also obtained in this experiment as well as in similar ones performed later, tumors consisting of poorly differentiated cells and of areas containing well differentiated tissues[2]. The fetal origin, as well as the histological characteristics of these tumors, indicates that they may belong to the group teratocarcinomas. In this paper we report in more detail the histological appearance of these tumors as well as their ultrastructural characteristics.

## MATERIALS AND METHODS

Since the methods used for the induction of tumors have been described previously[1] we shall give only a brief description here. Inbred R female rats were copulated with R males, and at day 12 of pregnancy fetectomy was performed in both uterine horns, leaving the placentas _in situ_. After fetectomy, all the placentas in one uterine horn were injected with 0.05 ml MSV [$10^{5.0}$ focus forming units, (FFU) per ml], the placentas in the contralateral horn receiving the same volume of control fluid. Control non-pregnant R female rats of the same age were injected in one uterine horn with MSV, the other horn receiving control fluid. A group of R female rats copulated with BN males was treated in a similar way to the first group.

The animals were killed 3 months after operation and a complete autopsy was performed. Fragments of tumors were used for subcutaneous transplantation into syngeneic young adult rats or were trypsinized and cultured _in vitro_. The organs were fixed for histological and ultrastructural examination and treated in the usual way. We tested for the presence of alkaline phosphatase on fresh frozen as well as on paraffin sections[3,4,5].

## RESULTS

Out of the total of 80 fetectomized R female rats copulated with a R male and inoculated with MSV, 19 developed tumors, 13 of which had the histological and ultrastructural characteristics of yolk-sac carcinoma[1,2]. However, 6 out of the 19 tumors although macroscopically similar to the yolk-sac carcinomas had quite different morphological characteristics. Like the yolk-sac carcinoma these tumors infiltrated the uterus and in most cases gave metastases in the peritoneum, in the lymph nodes and in the lungs. The uterine tumors were solid, but sometimes a cyst adhered to them. As previously described[1], no tumors were recorded in the fetectomized and inoculated R rats copulated with BN males and in the non-pregnant injected R female rats.

Histology and ultrastructure. As we mentioned previously, 6 out of the 19 invasive tumors showed a morphology different from that of the previously described yolk-sac carcinomas. Histological examination of primary tumors showed masses of round or fusiform, poorly differentiated cells, which infiltrated the uterine muscle (Figs. 1, 2). In one primary tumor, nests of yolk-sac carcinoma were set into areas of embryonal carcinoma (Fig. 3). In 3 cases, well differentiated tissues such as cartilage, bone, bone marrow (Fig. 4), striated muscle, intestinal cysts (Fig. 5) and skin were also observed.

The ultrastructural observation confirmed a low level of differentiation of most tumor cells. They had large nuclei containing 2-3 prominent nucleoli. Chromatin - not abundant - lay near a nuclear membrane (Fig. 6). Many free ribosomes and numerous mitochondria were present in the cytoplasm, but only a few other cellular organelles. Sometimes vesicles or small vacuoles were observed. In better differentiated cells, endoplasmic reticulum appeared and ribosomes formed polysomes (Fig. 7). Occasionally, microvilli were seen, but in general the cell membrane was rather regular, although not always very distinct. Often the cells formed finger-like protrusions which connected them. No desmosomes were observed. These cells showed the characteristics of embryonal carcinoma cells. Two kinds of cells - round and fusiform - showed a similar ultrastructure, but in the fusiform ones a better differentiated Golgi apparatus was sometimes present, and their cytoplasm contained delicate fibrils. In one case, virus-like particles were observed in the vacuoles of these cells (Fig. 8). Between the cells some collagen fibrils were seen but hyalin was absent.

The metastases mostly contained cells of embryonal carcinoma (Fig. 9), but in two cases areas of yolk-sac carcinoma were also found (Fig. 10).

---

Fig. 1. Embryonal carcinoma infiltrating uterine wall. H and E.
Fig. 2. An epithelial-like field of an embryonal carcinoma. H and E.
Fig. 3. Nests of yolk-sac carcinoma in an embryonal carcinoma. H and E.
Fig. 4. Bone and bone marrow in a primary tumor. H and E.
Fig. 5. Intestinal-like cyst in a primary tumor. P.A.S.
Fig. 6. A poorly differentiated cell of embryonal carcinoma. Chromatin not abundant, nucleolus prominent.
Fig. 7. Many polyribosomes and numerous mitochondria in an embryonal carcinoma cell.

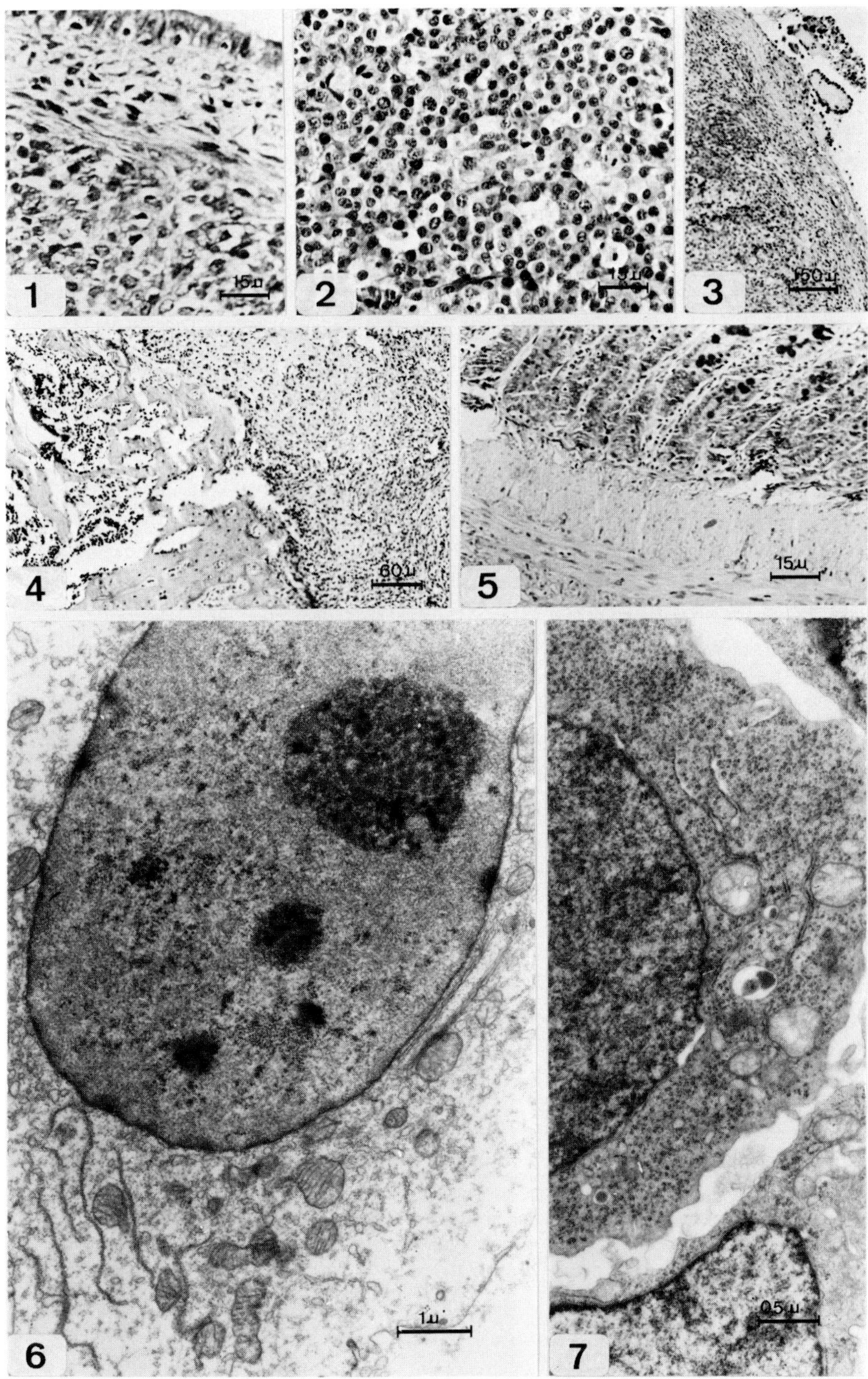
1
15μ
2
15μ
3
150μ
4
60μ
5
15μ
6
1μ
7
0.5μ

Tumor transplantation. In 3 cases we transplanted a primary *in utero* induced MSV tumor into young adult R rats by subcutaneous trocar implantation. All three tumors were successfully transplanted and have been kept as continuous transplantable lines in syngeneic adult rats. The morphology of these transplants retained the same characteristics as undifferentiated parts of the primary tumors, i.e. they were composed of embryonal carcinoma cells. However, in two tumors (lines F36 and F41) the formation of cartilage (Fig. 11) and glandular structures (Fig. 12) was observed among the undifferentiated cells.

One of the neoplasms was injected intraperitoneally to produce the ascitic form (line F3/1). The ascitic fluid was transplanted every week, and after 11 passages the growth slowed down for 2 months, after which it accelerated. The animals bearing the ascitic form of tumor also developed peritoneal neoplasms, the histology of which was similar to that of primary and subcutaneously transplanted tumors.

In the ascites fluid we observed single cells and groups of cells which were similar to embryoid bodies. The ultrastructural examination showed that they were not true embryoid bodies; the core in the middle was formed from only one cell, which was better differentiated than an embryonal carcinoma cell, but which was surrounded by epithelial cells (Fig. 13). After 11 passages, when the growth of the ascitic form became slower, histological examination showed that the tumors in the peritoneum were composed of embryonal and yolk-sac carcinoma cells. At this time, the embryoid-like bodies were composed of endodermal cells with well differentiated ergastoplasm which surrounded a hyalinic substance (Fig. 14). Many of the cells showed the presence of dilated endoplasmic reticulum with the formation of cisterns (Fig. 15). This indicated a secretory function of these cells.

When the growth of the ascitic form of the tumor became accelerated, the peritoneal tumors showed all the morphological features of embryonal carcinomas, which have been maintained until now.

Tissue culture. The cultures grow very well in monolayers and are trypsinized every 3-4 days. Two types of cells are always present : a round type, which does not adhere well to the plastic Falcon, and a fusiform type which does adhere and to which the former type of cells attach themselves. After a few days of cultivation the fusiform cells change into round cells and detach from the plastic. The ultrastructure of these cultured cells shows many free ribosomes, some mitochon-

---

Fig. 8. Virus-like particles in the vacuole of an embryonal carcinoma cell.
Fig. 9. Metastatic tumor-embryonal carcinoma. H and E.
Fig. 10. Metastatic tumor-yolk-sac carcinoma. P.A.S.
Fig. 11. Formation of cartilage in transplanted embryonal carcinoma. H and E.
Fig. 12. Formation of glandular structures in transplanted embryonal carcinoma. H and E.
Fig. 13. Embryoid body-like structure composed of an embryonal carcinoma-like cell surrounded by endodermal cells.

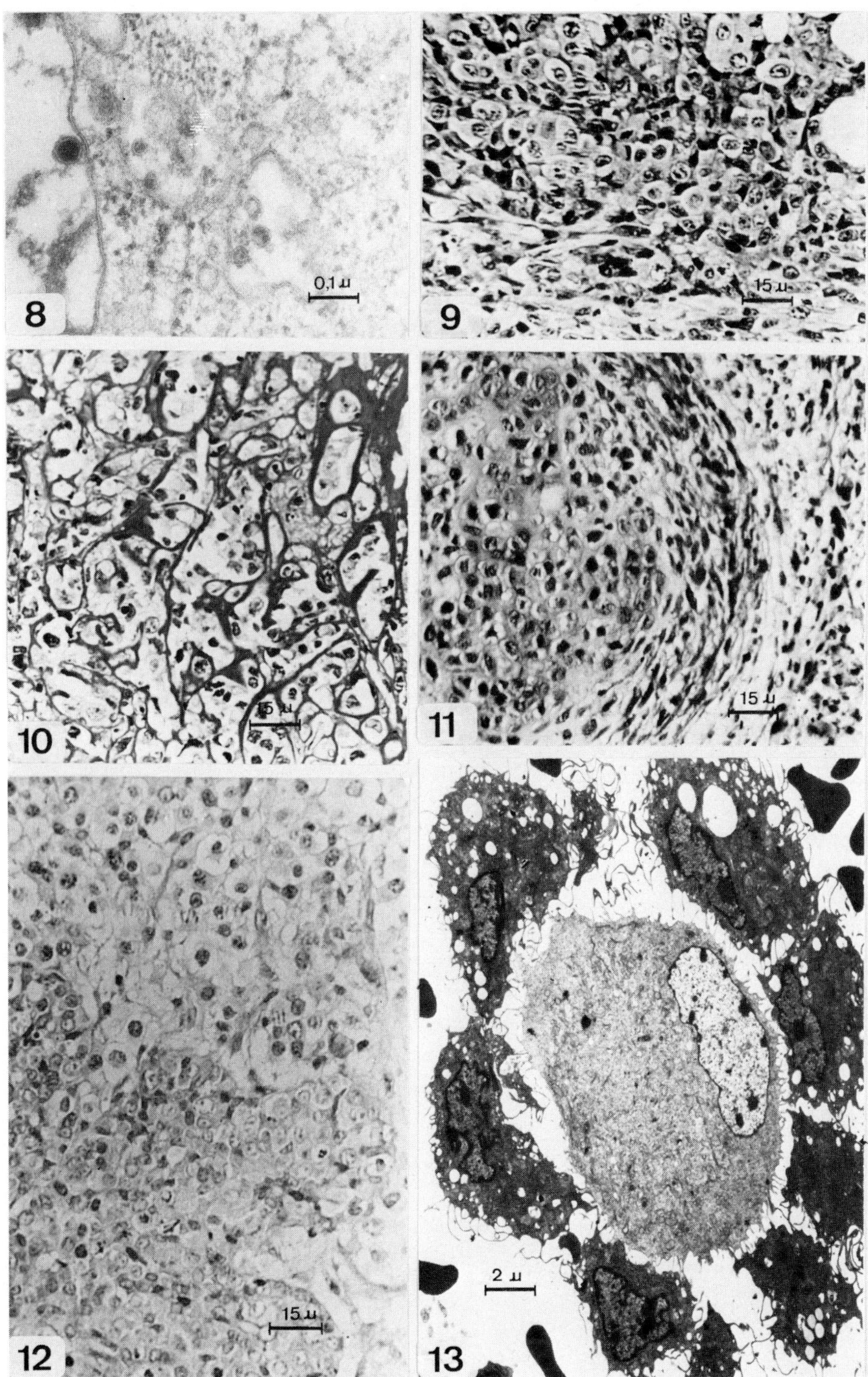
0,1 μ
8
15 μ
9
15 μ
10
15 μ
11
15 μ
12
2 μ
13

dria, and fragments of endoplasmic reticulum in the cytoplasm (Fig. 16). Very occasionally some lipid droplets are observed. The morphology of these cells is very similar to that of primary germ cells. The cells in the culture form protrusions, sometimes the borders of the cells adhere to one another, but they never form desmosomes. The trypsinized cells injected into adult R rats form embryonal carcinoma.

Alpha-fetoprotein. The large number of free ribosomes found in the cytoplasm of embryonal carcinoma cells makes them similar to other undifferentiated embryonic cells, which do not synthesize protein for "export" but only structural proteins for rapid growth and proliferation. However, some authors think that one plasma protein - alpha-fetoprotein - is synthesized in embryonal carcinoma cells and excreted into the serum[6].

We tested the level of alpha-fetoprotein in the sera of rats bearing embryonal carcinoma by using a radioimmunoassay[7]. The levels observed were never higher than 40 ng/ml which corresponds to that found in the normal rat.

Alkaline phosphatase. A known marker of embryonal carcinoma cells is alkaline phosphatase[8,9]. It was shown that in teratocarcinoma, which gives rise to 2 cell lines - tumorigenic and non-tumorigenic - only the former possesses high activity of this enzyme[9]. We observed the presence of alkaline phosphatase in the cell cultures *in vitro* and in the slides of all three transplanted tumors. The intensity of coloration differed, ranging from dispersed foci to near total slide coloration (Fig. 17).

## DISCUSSION

The histology and ultrastructure of cells in the described tumors as well as the presence of a high alkaline phosphatase activity qualifies them as embryonal carcinoma cells. Moreover, these tumors have a potential for differentiation as demonstrated by the formation of areas of yolk-sac carcinoma and of chondroid tissue and glandular structures. This differentiation potential is limited, however, probably by the overgrowth of poorly differentiated and very invasive cells. The presence in primary tumors of well differentiated tissues characteristic of benign teratomas, as well as embryonal carcinoma cells showing a potentiality for differentiation, indicates that all these neoplasms can be diagnosed as teratocarcinoma.

Fig. 14. Embryoid body-like structure composed of endodermal cells with well differentiated ergastoplasm surrounding a hyalinic substance.
Fig. 15. Cisterns in the cell of an ascitic form of embryonal carcinoma.
Fig. 16. Embryonal carcinoma cell from tissue culture. Many free ribosomes, mitochondria and fragments of endoplasmic reticulum are visible.
Fig. 17. Presence of alkaline phosphatase in embryonal carcinoma.

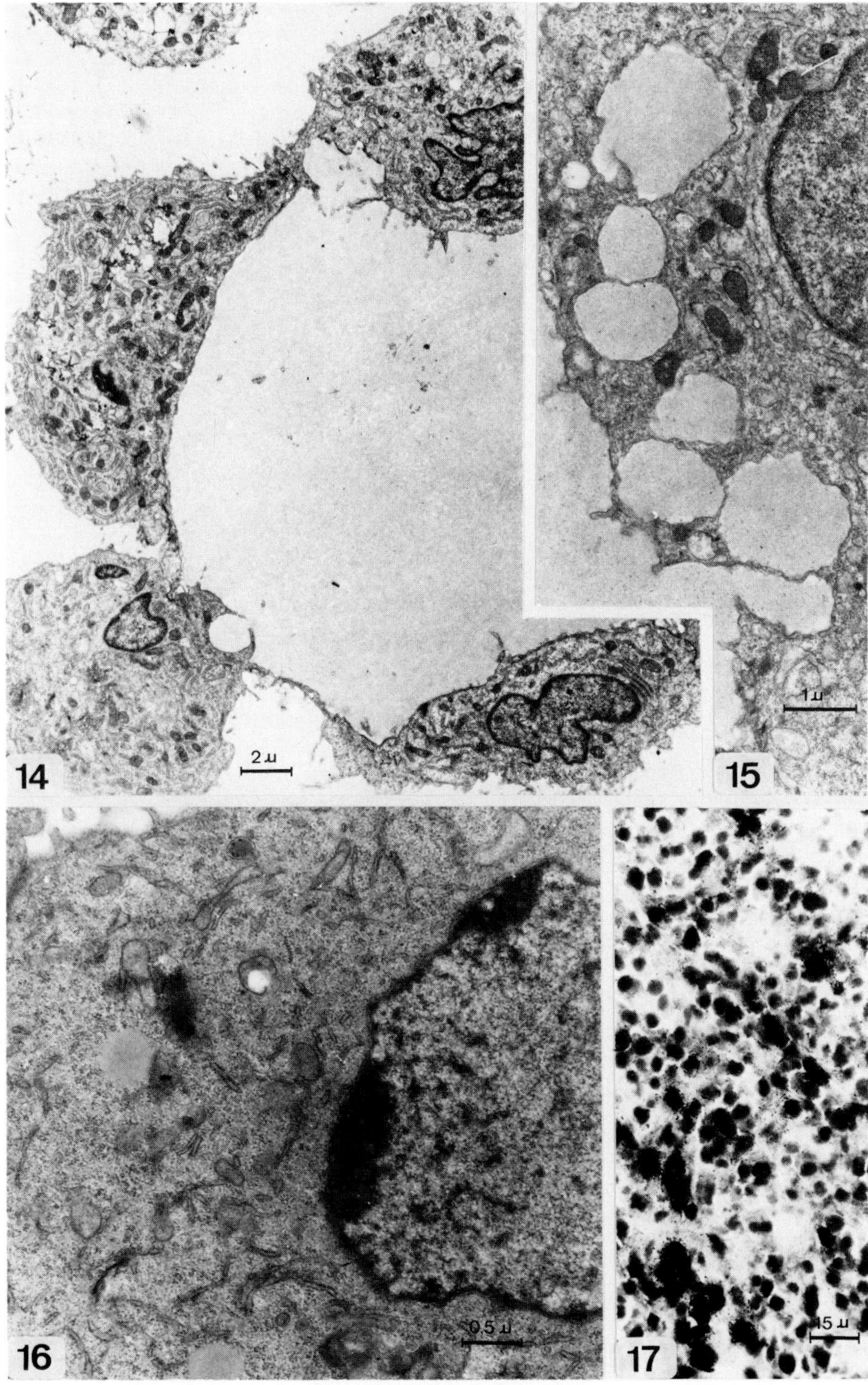
14
2 µ
15
1 µ
16
0.5 µ
17
15 µ

To our knowledge, teratocarcinomas in rats have never been described. These tumors, however, have been extensively studied in the mouse. In this species they either develop spontaneously in the testes of a particular strain (129)[10,11,12,13,14,15] or they can be experimentally induced by transplantation of genital ridges or 3 to 7 day old embryos into the testes or under the kidney capsule[16,17]. These mouse teratocarcinomas are composed of embryonal carcinoma cells; most of them also contain various well differentiated tissues and yolk-sac carcinoma. It was shown that after ascitic conversion of the tumor[18,19] single cells could be obtained from embryoid bodies floating in the ascitic fluid. After grafting, these single cells developed into embryonal carcinoma together with areas of teratocarcinomas containing well differentiated somatic tissue and occasionally trophoblast and yolk-sac. These experiments demonstrate the multipotentiality of single embryonal carcinoma cells. With in vitro cloning techniques using similar tumor material, Kahan and Ephrussi[20] came to the same conclusion. The clonal lines isolated by them from these embryonal carcinoma cells retained the potential to produce multidifferentiated tumors, containing mostly embryonal carcinoma and parietal yolk-sac tumors[21,22,23].

In the tumors which we induced in rats, after inversion into the ascitic form, we also observed embryonal carcinoma and occasionally yolk-sac carcinoma cells. Therefore we think that the yolk-sac carcinomas described earlier are probably derived from embryonal carcinoma and not from the parietal yolk-sac as previously stated[2]. Indeed, as we shall discuss later, some teratocarcinomas may have a limited capacity for differentiation.

The only cells known to have the ability to differentiate into various multipotential embryonal cells and into primordial tissues are germ cells. It is thought that in mice teratocarcinoma originates from germ cells[13,14,24]. A similar origin in humans is supported by the fact that teratocarcinomas and yolk-sac tumors are not only found in the gonads but also in the sacrococcygeal region, in the mediastinum, and in the entire midline area[25,26]. In this last case they probably originate from germinal elements misplaced during the embryonic migration of the primordial germ cells into the genital ridges.

There is some doubt as to whether the teratocarcinomas we induced in rats are of the same origin, because they develop from extraembryonic parts, the fetuses having been taken out during the operation. The only tissues left in the uterus are placenta and adhering parts of the visceral and parietal yolk-sac. On a purely morphological basis, the yolk-sac carcinomas previously described, may have developed from the parietal yolk-sac[2]. However, when implanted intraperitoneally in the rat, the parietal yolk-sac does not grow or differentiate, whereas the visceral yolk-sac does develop, and differentiates into various tissues[27]. Also, the parietal yolk-sac does not grow in tissue culture whereas the visceral yolk-sac can differentiate[28,29,30]. From these experiments, it can be concluded that the normal parietal yolk-sac has a limited growth and differentiation potential, but that the visceral yolk-sac can differentiate. Indeed, when this membrane is displaced outside the uterus it has the capacity to differentiate into various kinds of well differentiated tissues[31,32]. It is conceivable that cells having such a differentiating capacity are more susceptible to virus transformation than other tissues.

The evidence that MSV is responsible for development of the malignant tumors has already been discussed in a previous paper[1]. The fact that the teratocarcinomas described here appear only in the virus inoculated uterine horn, as well as the presence of C particles in one tumor, supports the theory that MSV is responsible for development of these neoplasms. Moreover, these tumors were not observed in yolk-sac-derived benign teratomas nor were they seen after polyoma virus inoculation[31,32,33].

It was stated previously that yolk-sac carcinoma is probably one of the furthest differentiated malignant forms of embryonal carcinoma. The latter tumor is characterized by a high alkaline phosphatase activity, which disappears in the yolk-sac carcinomas. In contrast we never found an abnormally high level of alpha-fetoprotein in the serum of animals bearing embryonal carcinomas, nor could we detect this protein in the cells of these tumors. However, this protein was found in the serum as well as in the cells of the yolk-sac carcinomas[7]. In mouse teratocarcinoma, Engelhart *et al.*[34] observed alpha-fetoprotein only in the epithelium of endodermal cysts and in the outer layers of embryoid bodies, which probably corresponds to visceral yolk-sac. The findings of Engelhart *et al.*[34] as well as our results are in agreement with observations in humans, in which alpha-fetoprotein is found in the serum of patients with embryonal carcinoma only when yolk-sac carcinoma structures are present in the tumor[35].

All the features described, such as the histology and ultrastructure, the potentiality of differentiation and high alkaline phosphatase activity, indicate a similarity between rat and mouse teratocarcinomas. The difference between MSV induced tumors in rats and those described in mice without virus treatment is, that teratocarcinomas can only be induced in some strains of mice, namely 129 S and C3H, and not in C57Black and originate from embryonal or germ cell, whereas in the rat these neoplasms develop only after inoculation of MSV and in both strains R and BN[36]. Moreover, in the rat they originate from the visceral yolk-sac and not from embryonal or primary germ cells.

Damjanov and Solter[17] think that two groups of factors control the development of teratocarcinomas : genetic, and environmental. It is possible that in mice the first are of greater importance and in rats the second, because of the action of virus on the fetal membrane which undergoes neoplasmic transformation in such privileged sites as the pregnant uterus.

This work was supported by the Belgian A.S.L.K. Cancer Fund and by the N.F.W.O. We are indebted to Mr. L. Bassi, Mr. J. Goebels, Mr. G. Hermans, Miss M. Naessens and Mr. C. Seghers for skillful technical assistance. The editorial assistance of Mrs. J. Edy and Mrs. J. Putzeys is appreciated.

REFERENCES

1. Vandeputte, M., Sobis, H., Billiau, A., Van de Maele, B. and Leyten, R. (1973) Int. J. Cancer, 11, 536-542.
2. Sobis, H. and Vandeputte, M. (1973) Int. J. Cancer, 11, 543-554.
3. Gomori, G. (1950) Stain Technology, 25, 81-85.
4. Wachstein, E. and Meisch, E. (1957) Am. J. Clin. Pathol., 27, 13-23.
5. Pearse, A.G.E. (1968) In Histochemistry, p. 711. Little Brown & Co., Publ.
6. Kahan, B. and Levine, L. (1971) Cancer Res., 31, 930-936.
7. Hooghe, R., Zeicher, M., Sobis, H. and Vandeputte, M. (1974) Proc. Intern. Conf. Alpha-Feto-Protein, INSERM, R. Masseyeff (ed.), pp. 271-273.
8. Damjanov, I., Solter, D., Belicza, M. and Skreb, N. (1971) J. Nat. Cancer Inst., 46, 471-480.
9. Berstine, E.G., Hooper, M.L., Grandchamp, S. and Ephrussi, B. (1973) Proc. Nat. Acad. Sci. USA, 70, 3899-3903.
10. Stevens, L.C. and Hummel, K.P. (1957) J. Nat. Cancer Inst., 18, 719-747.
11. Stevens, L.C. (1959) J. Nat. Cancer Inst., 23, 1249-1295.
12. Stevens, L.C. (1960) Developm. Biol., 2, 285-297.
13. Pierce Jr., G.B., Dixon Jr., F.J. and Verney, E.L. (1960) Labor. Invest., 9, 583-602.
14. Pierce Jr., G.B., Dixon Jr., F.J. and Verney, E.L. (1960) Cancer Res., 20, 106-111.
15. Pierce Jr., G.B. and Verney, E.L. (1961) Cancer, 14, 1017-1029.
16. Solter, D., Skreb, N. and Damjanov, I. (1970) Nature, 227, 503-504.
17. Damjanov, I. and Solter, D. (1974) Current Topics in Pathology, 59, 69-130.
18. Pierce, G.B. and Dixon Jr., F.J. (1959) Cancer, 12, 573-583.
19. Pierce, G.B. and Dixon Jr., F.J. (1959) Cancer, 12, 584-589.
20. Kahan, B.W. and Ephrussi, B. (1970) J. Nat. Cancer Inst., 44, 1015-1036.
21. Kleinschmidt, L.J. and Pierce Jr., G.B. (1964) Cancer Res., 24, 1544-1551.
22. Finch, B.W. and Ephrussi, B. (1967) Proc. Nat. Acad. Sci. USA, 57, 615-621.
23. Rosenthal, M.D., Wishnow, R.M. and Sato, G.H. (1970) J. Nat. Cancer Inst., 44, 1001-1014.
24. Stevens, L.C. and Bunker, M.C. (1964) J. Nat. Cancer Inst., 33, 65-78.
25. Huntington Jr., R.W. and Bullock, W.K. (1970) Cancer, 25, 1357-1367.

26. Huntington Jr., R.W. and Bullock, W.K. (1970) Cancer, 25, 1368-1376.
27. Payne, J.M. and Payne, S. (1961) J. Embryol. Exper. Morphol., 9, 196-116.
28. Sorokin, S.P. and Padykula, H.A. (1964) Am. J. Anat., 114, 457-478.
29. Padykula, H.A., Deren, J.J. and Wilson, T.H. (1966) Developm. Biol., 13, 311-348.
30. Pierce, G.B., Bullock, W.K. and Huntington Jr., R.W. (1970) Cancer, 25, 644-658.
31. Sobis, H. and Vandeputte, M. (1974) Int. J. Cancer, 13, 444-453.
32. Sobis, H. and Vandeputte, M. (1975) Developm. Biol., 45, 276-290.
33. Vandeputte, M., Meyer, G. and Sobis, H. (1975) J. Nat. Cancer Inst., in press.
34. Engelhart, N.V., Poltoranina, V.S. and Yazova, A.K. (1973) Int. J. Cancer, 11, 448-459.
35. Talerman, A. and Haije, W.G. (1974) Cancer, 34, 1722-1726.
36. Sobis, H. and Vandeputte, M. (1975) Not published observations.

Progress in Differentiation Research, ed. N. Müller-Bérat et al.
© 1976, North-Holland Publishing Company - Amsterdam, The Netherlands.



# CELL TYPE OF ORIGIN AND PRENEOPLASTIC STAGES IN FOREIGN BODY TUMORIGENESIS

K.G. Brand

University of Minnesota Medical School, Department of Microbiology

Minneapolis, Minnesota, USA, 55455

Foreign body (FB) tumorigenesis in mice or rats has the advantage of technical and interpretive simplicity as compared to other experimental cancer models. Recent investigations in our laboratory have led us to appreciate, in particular, aspects of differentiation and development in attempting to elucidate this tumorigenic process.

As has been established by early experimentors[1] the physical presence and nature of the FB material, not its chemical reactivity, are responsible for tumor development. Many investigators[1] studied the histology of tissue reactions in different animal species in response to FB's of different sizes, shapes, or surface properties. The results indicated that tumor incidence is strongly influenced by the type and course of the FB reaction. The degree of fibrosis in FB encapsulation and a chronic course of FB reaction with low cellular activity are positively correlated with FB tumor incidence.

In our laboratory we used the following basic method. Plastic films (unplasticized vinyl chloride acetate copolymer, 15 x 22 x 0.2 mm) were implanted in CBA/H or CBA/H-T6 mice. After various time periods the implants were excised and cut into 7- x 15-mm segments. These were transferred separately to recipient mice that were fully histocompatible yet distinguishable from the donor mouse on the basis of the T6 chromosomes. Sarcomas of donor origin developed in the recipients up to 2 years later, indicating that preneoplastic cells resided on the FB surface at the time of transfer. Tumors that arose from segments of the same original implant were often identical or closely related ("homologous") with regard to (a) tumor latency, in that neoplastic growth commenced in all recipients at closely spaced points in time; (b) specific chromosome aberrations, in terms of number and morphology; (c) histopathology, in terms of sarcoma type and degree of anaplasticity; and (d) growth characteristics and cell generation times _in vivo_ and _in vitro_.

These results prompted the following conclusions[2]. In the case of homology, tumors must have arisen from cells with the same neoplastic specificity, pointing to their clonal nature. Specific tumor properties must have been predetermined in these clonal cells long before they actually started neoplastic proliferation. In fact, specific clones must have been in existence before implants were cut and segments transferred. The demonstration of such clones implied prior existence of "parent cells" in which the initial event must have taken place, the event that not only created the neoplastic potential but moreover predetermined the specific tumor properties.

The nature of the tumor originator ("parent") cell.

Although millions of cells, mainly of the bone marrow-derived macrophage type, are mobilized in experimental tumorigenic FB reaction, the number of preneoplastic parent cells was found to be extremely small. By applying the maximum likelihood estimate, it was calculated that in CBA mice the most probable number of preneoplastic parent cells is only 1 in response to a single 7- x 15- x 0.2-mm plastic film and 3 in response to a single 15- x 22- x 0.2-mm implant[3]. These values are consistent with direct experimental counts[4]. It appears that the key tumorigenic event is a statistical chance event or else that a small constant proportion of the cell population has a primary neoplastic disposition or susceptibility.

Tumor incidence curves of numerous group experiments[5] were uniformly and consistently characterized by multiple distinct peaks which made it impossible to describe the tumor latency data in form of a single Normal Distribution Curve. Instead, the observed clustering of tumor appearances on the time scale suggested the existence of several subpopulations of tumors according to duration of latency. Since latency is one of the tumor characteristics which are predetermined already in the parent cell by the initial carcinogenic key event it follows that we are dealing not only with different latency-classes of tumors but also with different determination-classes of parent cells, and ultimately with different kinds of carcinogenic key events. Hence, the key events in FB-tumorigenesis may not be uniform; different molecular happenings may be involved and may occur at different loci or levels of the cellular regulation system.

Various experimental attempts were undertaken to determine the origin and identity of the preneoplastic parent cells[6]. The most conspicuous participants in FB reaction (i.e., monocytes, macrophages, and fibroblasts) were experimentally excluded as the likely progenitor cells. Instead, ultrastructural studies of tumor cells implicated a pluripotential mesenchymal cell type possessing morphological characteristics consistent with cell types of the microvasculature. First of all, tumor histopathology is varied and includes fibro-, myxo-, hemangio-, leiomyo-, and osteogenic sarcomas. These various histopathologic elements are found mixed in some tumors despite clonal origin, hence pointing to mesenchymal pluripotentiality of parent cells. Electron microscopic studies further revealed that despite histopathologic variability FB-sarcomas uniformly contain basal lamina and 60 Å microfilaments while collagen is generally sparse. These ultrastructural characteristics together with the preneoplastic pluripotentiality incriminate cell types of the microvasculature, especially pericytes and possibly endothelial cells, as the source.

The multifactorial complexity of preneoplastic "maturation".

Modified transfer experiments similar to those described above were carried out with segments of both implants and FB-reactive capsule tissue at various times during preneoplastic FB reaction[5]. One procedural step in these experiments was of particular significance beyond mere technical consideration. Demonstration of the preneoplastic cells in capsule tissue by transferring segments to recipients was achieved only when a new piece of FB material was inserted into the capsule segment to replace the original implant. Segments from these same capsules would usually not lead to tumors when transferred empty.

Upon implantation of the FB and through the 1st few weeks masses of monocytes and macrophage-type cells, variable numbers of polymorphonuclear leukocytes, as well as occasional fibroblasts infiltrate the FB environment. Capillary outgrowth is seen to take place as in wound healing. From the 12th day on, the surface of the FB is almost completely covered by a monolayer of macrophage-type cells. During the first 4 to 8 weeks postimplantation preneoplastic cells appear in the loose cellular FB-reactive tissue. However, preneoplastic cells were never demonstrated during this stage on the FB itself. It may be argued that preneoplastic cells nevertheless make transient or very loose contact with the FB at this time. However, this appears unlikely. When cells make contact with the FB surface they generally seem to adhere rather firmly, especially if the FB material is a plastic as in our experiments.

These findings indicate that the preneoplastic cells have acquired their neoplastic determination at a site distant from the FB and independent of direct physical or chemical reaction with it. Hence, the FB must be excluded from consideration as a direct inducer of cellular neoplastic potential and determination. Instead, if there is a factor that exerts an influence on cells of the local environment partaking in the FB reaction it is the dynamics of the FB reaction itself, which stimulates cellular mobilization, proliferation, and functional activity. The carcinogenic key event may then be assumed to constitute a spontaneous error within the regulation system of the cell, especially when it divides. The likelihood of such an event would increase during forced cell proliferation as it occurs in FB reaction. On the other hand, the event may take place also under normal conditions when cells proliferate at a regular pace. This would mean that preneoplastic cells may be residing in the tissue prior to the introduction of the FB. Both situations are consistent with available data and are presently subject to further experimentation.

Whenever such an error occurs in the cellular regulation system, loss of growth control with autonomous proliferation is apparently not an immediate outcome because there is always a long period of tumor latency. Therefore, we have to assume that the error remains unexpressed or that the cell is able to compensate for the defect for some time.

During the 2nd month following implantation, while the FB reaction shows the criteria of a subacute inflammation with continuing cellular activity, a grossly recognizable thin tissue capsule begins to develop around the FB fibroblast-like cells, present on the inner aspect of this capsule, are separated by a fine capillary space from the cell monolayer on the FB surface. Fibroblastic cells become more numerous and collagen fibers appear in the capsule wall, which gradually thickens and consolidates. At this time, preneoplastic cells are still not detectable on the FB surface itself. However, they can be demonstrated in some instances already as clones, in the consolidating capsule and in the loose FB-reactive tissue surrounding the capsule.

During the 3rd and the following months, fibrosis of the capsule with pronounced collagen formation continues until this process becomes stationary between the 4th and 6th month. The fibrotic element is consistently associated with the tumorigenic process as indicated by several observations and experiments. It has been shown that FB surface properties influence degree of fibrosis as well as tumor incidence and latency. For example, plastic film implants with sandpapered surfaces cause more cellularity, less fibrosis, fewer tumors, and longer latencies than do untreated smooth films[7]. Also, if FB reaction remains for an extended time period in the acute afibrotic form due to prolonged aseptic inflammation or mechanical irritation, tumor appearance is markedly delayed[2]. Furthermore, animal species differ in proneness to FB sarcomas and, again, this appears to be correlated with the degree of fibrosis in FB reactivity. In face of the association between fibrosis and FB tumorigenicity, the process of fibrosis itself or factors that stimulate fibrosis must be considered to be of critical significance in FB tumorigenesis.

While fibrosis develops,the macrophage-type cells, which still predominate on the FB surface, take on an appearance of inactivity and dormancy, signaled ultrastructurally by a decrease in the number of cellular organelles and the presence of electron-dense cell matrices[8]. From recent experiments on mice in which Millipore filters of different pore sizes were used as implants,a conspicuous correlation was found to exist between phagocytic inactivity of FB-attached macrophages, fibrosis of the FB reaction, and FB tumor incidence[9]. How this relates to preneoplastic maturation is difficult to assess at the present. The same factors that are responsible for stagnancy of tissue function and/or phagocytic inactivity of macrophages may in turn promote the maturation of preneoplastic cells; or phagocytic inactivity of macrophages by itself, in the continued presence of a FB, may create a functional stimulus for other cell types including collagen-producing fibroblasts and mesenchymal stem cells of the microvasculature possessing neoplastic determination.

Clonal preneoplastic cells are now regularly demonstrated in the capsule tissue and with increasing probability also on the FB surface itself as a component of the firmly attached cell monolayer. The earliest successful demonstration of preneoplastic cells on the FB surface was accomplished in 1 experiment out of several at 11 weeks postimplantation. After that time the success rate increased from about 30% during the 5th month to 70% and then nearly 100% during the 6th and 7th months, respectively. Although part of the preneoplastic clone establishes contact with the FB surface this is at first not a requirement for the maturation of preneoplastic cells to proceed. Evidence was obtained through the observation that tumor latencies are virtually of equal duration whether homologous preneoplastic cells were transferred with a FB segment or with a capsule segment plus a new FB insert. Hence, it must be assumed that the homologous cells remaining in the capsule tissue continue their preneoplastic maturation at the same pace as does the FB-attached part of the clone.

Only during the terminal phase of FB-tumorigenesis does direct FB-contact become a prerequisite for preneoplastic cells to mature beyond the point that can be reached by homologous cells in the capsule tissue. In fact, FB-contact is the prerequisite for finally attaining proliferative autonomy. Homologous preneoplastic cells may still be present in capsule tissue but will not give rise to tumors upon transfer unless a new FB is inserted. If that is done, however, preneoplastic cells will move out of the capsule tissue and soon settle on the new FB surface.

The requirement of FB contact for completing preneoplastic maturation is usually of short duration. This can be concluded from transfer experiments involving segments of advanced capsules plus new FB inserts. If such capsule segments contained preneoplastic cells it often took less than 2 months for palpable tumors to develop. But again, if segments of the same capsule were transferred without a new FB insert, the preneoplastic cells generally would not reach the state of proliferative autonomy and not develop into tumors.

This late phase then appears to be the only period of FB tumorigenesis during which physical contact between FB and preneoplastic cells becomes the essential feature. The nature of the interaction remains to be investigated. Presumably, the critical effect that leads to proliferative autonomy of the cell occurs primarily at the level of the cell membrane when it firmly adheres to the FB surface. However, these crucial events, representing the final step in the acquisition of neoplastic autonomy, must not distract from the fact that the specific neoplastic determination is a fixed property of the cells long before they come into direct contact with the FB surface.

## Conclusions, consequences, and projections.

It is generally agreed that the cellular mechanisms of growth control are central to every carcinogenic process. This system of growth control operates at various levels: genetic, epigenetic, and regulatory. Various functional subunits take part in it, e.g. those serving intercellular communication, reduplication and division mechanisms, relevant molecular biosynthesis and transport. These functional subunits are interlocked and correlated; they are subject to continuous checks and balances. A single error at any one of these sensitive points, (be it mutational defects, genetic interference of some sort, or epigenetic-regulatory failures), may lead to the ultimate breakdown of cellular growth control. Regarding FB tumorigenesis, indications are that the key defect may be the result of spontaneous chromosomal inbalance and, hence, excessive genetic load. In general, it appears that one of the keys to the understanding of carcinogenesis may well lie in the realization of the diversity of cellular carcinogenic events and/or target sites, an assumption which is supported by the observation of different latency-classes of FB-induced tumors and parent cells.

The phenomenon of tumor latency (which in FB tumorigenesis is secondary to the occurrence of determining carcinogenic key events at the earliest parent cell stage) indicates that there is no immediate expression of the key defect but temporary compensation. The compensatory capacity, however, depends in large measure on the tissue environment: pronounced macrophage activity but only faint fibrosis are seen in situations where tumor latency is significantly prolonged.

Then, there must be still another barrier guarding the cell against autonomous proliferation. This barrier breaks down only when the predetermined matured preneoplastic cell attaches to the FB surface. Here, and only here at this terminal preneoplastic maturation stage which immediately precedes overt tumor growth does direct interaction between FB surface and cell membrane become an essential tumorigenic requirement.

These conclusions direct further experimental attacks towards the following targets: cellular regulation and growth control; the roles of tissue-environmental factors; the regulatory involvement of the cell membrane. Admittedly, such a listing is almost commonplace considering that it encompasses virtually all the basic aims of modern experimental biology. Yet, the proposition of such a program can also be read as saying that the mysteries of cancer may remain unresolved until those basic biological aims have been reached. Furthermore it is implied, and this could be of great methodological significance, that the multifactorial complexity of carcinogenesis may be approachable only by multifactorial experimentation.

Analytical investigations along these lines at the subcellular level call for sufficiently large and uniform preparations of cells at various defined stages of preneoplastic maturation. We recently discovered[10] that these can be obtained through _in vitro_ culture of FB-reactive tissue and FB-attached cells. Aneuploid cell clones with neoplastic potential gradually outgrow the euploid cells which initially predominate. While the aneuploid cell clones are kept and expanded in culture their preneoplastic maturation is arrested. This is of practical importance and also shows that there are factors in the _in vivo_ environment, but not in culture, which are essential in promoting FB-induced preneoplastic cell maturation. Consequently, I would like to submit that for analyzing the complexities of carcinogenic processes a system combining developmental main tracks "in vivo" with investigative side tracks "in vitro" would be most appropriate: the *in vivo* periods to allow "natural" preneoplastic events and developments to take place; the *in vitro* periods to obtain and expand cell samples for analytical studies at different subcellular levels and at different stages of the carcinogenic process. The model of FB tumorigenesis would be especially suitable for such an approach. In particular, it would permit one to relate retrospectively specific tumor characteristics to specific features of the preneoplastic cells at defined maturation stages.

## References


1. Bischoff, F., 1972, Clin. Chem., 18,869.
2. Brand, K. G., Buoen, L. C., and Brand, I., 1967, J. Natl. Cancer Inst., 39, 663.
3. Brand, K. G., Buoen, L. C., and Brand, I., 1973, J. Natl. Cancer Inst., 51, 1071.
4. Thomassen, M. J., Buoen, L. C., and Brand, K. G., 1975, J. Natl. Cancer Inst., 54, 203.
5. Brand, K. G., Buoen, L. C., and Brand, I., 1971, J. Natl. Cancer Inst., 47, 829.
6. Johnson, K. H., Ghobrial, H. K., Buoen, L. C., Brand, I., and Brand, K. G., 1973, Cancer Res., 33, 3139.
7. Bates, R. B., and Klein, M., 1966, J. Natl. Cancer Inst., 37, 145.
8. Johnson, K. H., Ghobrial, H. K., Buoen, L. C., Brand, I., and Brand, K. G., 1972, J. Natl. Cancer Inst., 49, 1311.
9. Karp, R. D., Johnson, K. H., Buoen, L. C., Ghobrial, H. K., Brand, I., and Brand, K. G., 1973, J. Natl. Cancer Inst., 51, 1275.
10. Buoen, L. C., Brand, I., and Brand, K. G., 1975, J. Natl. Cancer Inst., 55, 721.


## Acknowledgment

Supported by Public Health Service research grant CA10712 from the National Cancer Institute.

Progress in Differentiation Research, ed. N. Müller-Bérat et al.



# INDUCTIVE EPITHELIO-MESENCHYMAL INTERACTION: IS IT INVOLVED IN THE DEVELOPMENT OF EPITHELIAL NEOPLASMS?

C.J. Dawe, W.D. Morgan, J.E. Williams, J.P. Summerour

Laboratory of Pathology, NCI, NIH, Bethesda, Md., U.S.A. 20014

## 1. Introduction

Much of the thought and experimental design in developmental biology is based on the concept that the processes of determination, morphogenesis, and cytodifferentiation, classed together as embryonic inductions, can be attributed to interactions between dissimilar cell populations.[1,2] The mechanisms by which these inductions occur are incompletely understood, but it is tentatively presumed that each of the three processes named above represents a stage in activation of specific gene expressions.[2]

It is also generally presumed that interactions between dissimilar cell populations--for example, between the mesenchyme and the epithelium of an exocrine gland--are needed for continued maintenance of the specific function of fully-developed organs as well as for controlled renewal and regeneration of the cell populations of these organs. This presumption is based in large part on repeated observations in vitro that dissociation of epithelium from its proper mesenchyme results in loss of specific functional properties of that epithelium. Often the dissociated epithelium also lacks ability to maintain itself at either a constant or an increasing population size. Moreover, it has been observed that epithelium from fully developed salivary gland retains the ability to interact with embryonic salivary mesenchyme in such a way that morphogenesis, comparable to that seen in rudiments in which both components are embryonic, is re-enacted.[3]

During development of epithelial neoplasms, epithelium remains for a time interactive with its supporting mesenchyme, but after a certain degree of independence, or autonomy, has been acquired it may display the ability to grow at sites remote from its proper mesenchyme. At the same time, neoplastic epithelium often shows a considerable ability to undergo cytodifferentiation.[4] Generally, it exhibits the phenotypic characteristics of the cell type from which it was derived, which is to say that neoplasms from squamous epithelium are squamous in character and often produce keratin; neoplasms from glandular epithelium form glands and secrete products similar to those secreted by the cells of the parent organ. It is on this principle that histopathologic identification and classification of neoplasms has classically been based. Exceptions occur, for example in neoplasms of non-endocrine organs that may produce endocrine products, and neoplasms of salivary and mammary glands that may produce cartilage. It would seem important to know how such aberrations come about.

If epithelio-mesenchymal interactions are controlling factors in determining organ type and cell type, they should also be influential 1) in determining the responsiveness to organ- and cell-specific carcinogens, and 2) in governing the phenotypic characteristics of neoplastic cells that evolve from the carcinogenic process. A way to test this hypothesis is to separate the components of a given inductive interaction system and then observe the responses of the isolated components to a carcinogen whose effects on the unperturbed, integrated system are reproducible and well characterized. An ideal inductive interaction system for application of this method is the submandibular salivary gland of the mouse, because 1) this epithelio-mesenchymal interaction system is among those most thoroughly analyzed by developmental biologists;[2] and 2) neoplasms of unique and well characterized histotype develop from the submandibular glands of mice infected with the polyoma virus.[5]

Work previously reported from this laboratory tended to support the concept that interaction between salivary epithelium and salivary mesenchyme is essential for salivary gland neoplasms to develop in response to infection by polyoma virus.[6] Thus: 1) Only those anlagen that had undergone a visibly demonstrable degree of morphogenesis were able to give rise to tumors;[7] 2) When epithelial and mesenchymal components were separated from each other, infected with polyoma virus, and transferred to syngeneic newborn mice, tumors failed to arise from either type of separated component. Recombination of the infected components before transplantation restored both morphogenetic capability and oncogenic capability;[7] 3) Phenotypic hybrid recombinations, e.g., salivary epithelium interacting with dental mesenchyme failed to undergo morphogenetic changes and also to give rise to tumors in response to virus;[8] 4) Tumors developing from allophenic recombinations of T6T6 chromosome-marked epithelium with unmarked mesenchyme were found to bear the karyotype of the epithelial component. Hence the tumors were of epithelial origin despite the fact that some function of mesenchyme was *evidently* required for epithelial transformation to a neoplastic state.[9]

In view of Grobstein's demonstration[10] that inductive interaction can take place between salivary epithelium and salivary mesenchyme separated from each other by thin (20 μ) Millipore membranes with a porosity as little as 0.45 μ, a most critical experiment would be to test the responsiveness of the two tissue components in the trans-filter system. We have attempted this experiment but have not performed it as initially intended because, with the technique used, morphogenesis did not occur in salivary epithelium trans-filter from salivary mesenchyme. We succeeded, however, in maintaining prolonged growth of the epithelium in such cultures and apparently this alone proved sufficient to permit epithelial transformation in the presence of polyoma virus infection. This outcome was fortuitous, as it has provided a means to induce neoplastic change in epithelium in an altered relationship to its proper mesenchyme and to compare the neoplasms thus induced with those derived from intact salivary glands.

## 2. Materials and Methods

For long-term experiments it was necessary to use a closed-system, double-chambered culture vessel in which the Millipore filter could be sealed around its periphery so as to prevent migration of cells from one side to the other. The chamber used is shown diagrammatically in Fig. 1, and was designed by one of the authors (W.D. Morgan).

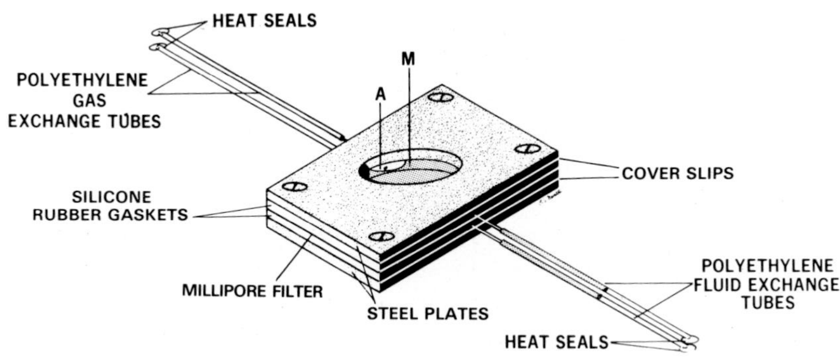


Fig. 1. DOUBLE VAULTED TRANSFILTER CHAMBER

With this chamber cultures were maintained for more than one year without microbial contamination.

Submandibular salivary gland rudiments were dissected from mouse embryos at 13 or 14 days of gestation, when rudiments were at Stages 2-5. Using aseptic technique throughout, we dissected epithelial components from mesenchyme using crude trypsin (Difco 1:250) in $Mg^{++}$-$Ca^{++}$-free Tyrode's solution. Rudiment components were then transferred to 50% fetal bovine serum in Hanks' BSS to inactivate residual enzyme. They were then placed in 1.0 ml. of a suspension of polyoma virus (passages 3 and 4 of strain 2 PTA, isolated by us from a salivary tumor) equivalent to 10 tumor-inducing doses by mouse-inoculation assay. After 2 hrs.' exposure to virus, the components were placed in the chambers on Millipore membranes in the various combinations indicated in Tables I to II. Control cultures were prepared similarly, omitting infection with virus. In most cultures a thin chicken plasma clot was applied to hold the rudiments on the membrane until cells attached. Culture medium was exchanged 3 times weekly through the polyethylene feeder tubes. The medium was Eagles' MEM in Hanks' BSS, supplemented with 10% fetal bovine serum and containing 100 µ Penicillin G per ml. Cultures were incubated at 36°C in a 5% $CO_2$ atmosphere. $CO_2$ diffusing through the polyethylene feeder tubes controlled pH.

At variable periods, filters were removed from the chambers and were either fixed in toto or cut into strips, some of which were transferred to Leighton tubes for further culture, while others were used to prepare fixed and stained whole-mount microscopic slides, or for electron microscopy.

Tumorigenicity tests of the various continuous cell lines obtained were done by inoculating 1-2 x $10^6$ cells subcutaneously into newborn syngeneic mice. Donors of the salivary rudiments were mainly of strain C3H/Bi from our own polyoma-free breeding colony. However, in 3 cultures (Table I) of epithelium opposite mesenchyme, from which continuous transformed lines were obtained, either the epithelium or the mesenchyme was taken from CBA/HaT6T6 embryos. This was done to permit karyotypic determination of the component that was transformed and gave rise to tumors. Thus when T6T6-marked epithelium and unmarked mesenchyme were used, mesenchymal tumors would lack the marker chromosomes, while epithelial tumors would have the T6 markers. $(CBA/HaT6T6 \times C3H/Bi)_{F1}$ hybrids were used as recipients of these 3 culture lines.

## 3. Results

Table I presents the results with cultures of salivary epithelium placed trans-filter from salivary mesenchyme. The duration of culture from time of initiation to time of removing the filter from the chamber varied from 25 to 375 days for the cultures represented in Table I, depending on the behavior of individual cultures. Those that did not thrive were fixed and examined when extinction appeared imminent. During the early phase of the work, some cultures that were growing well were fixed in order to assess morphological changes more accurately, since cytologic details cannot be seen in living cultures on the filters. The results given in Tables I-II, with respect to the number of continuous tumorigenic cell lines derived, therefore do not represent the maximal proportion of transformed culture lines that could have been developed.

Table I. Trans-filter Transformation
Experiments with Epithelium Opposite Mesenchyme

| Type of Culture on Millipore Memb. | | Fixed, to Date | | Morphologic Trans formation Pres. | | Continuous Growth on MM and Glass | | Tumorigenic in Syngeneic Mice | |
|---|---|---|---|---|---|---|---|---|---|
| | | 30d. | 30d. | Ep | Mes | Ep | Mes | Ep | Mes |
| PV- Infected | Ep/ Mes | 20 | 7 | 5 | 2 | 3* | 1 | 3 | 1 |
| Non- Infected | Ep/ Mes | 24 | 16 | 4(?) | 1 | NA | NA | NA | NA |

NA = Not attempted. * = Original cultures set up with either mesenchyme or epithelium marked by T6T6 chromosomes.

Four of the non-infected control cultures showed questionable transformation, morphologically. Pleomorphism was evident, and mitoses were present when the cultures were fixed after periods of 35, 45, 50 and 53 days respectively. None of these cultures showed evidence, such as intranuclear inclusion bodies or foci of lysis, that might indicate inadvertent infection with polyoma virus.

Two points were of key interest relevant to the experiment summarized in Table I. One was that in none of the cultures was evidence seen at any time of morphogenetic activity on the part of the epithelium, whether transformed or not (Figs. 2 & 3). Yet there was abundant proliferative activity in uninfected cultures where mesenchyme was present on the opposite side of the filter. In several examples most of the filter was covered by a sheet of epithelium after 30 days in culture.

The second point, probably related to the absence of morphogenetic activity in these cultures, concerns the morphologic and biologic characteristics of the tumors derived from the 3 transformed epithelial lines (Figs. 5-8). Twenty-seven neoplasms have been obtained by transfer of cells from these 3 lines back to syngeneic mice, and several of these have been passed through 3 to 7 serial mouse transfers. With only 2 exceptions all of the tumors so obtained were undifferentiated carcinomas (Fig. 5). They were unlike the salivary tumors induced by infecting newborn mice with polyoma virus (Fig. 4), in that their cells showed larger nuclei, greater nuclear hyperchromatism, greater pleomorphism, and no tendency to form gland- or duct-like structures. These tumors were also extremely invasive, extending deeply into muscle and causing ulceration of the skin by invasion and replacement of the dermis. We have not seen this in the salivary tumors induced in mice by virus infection in vivo.

One of the 2 exceptional tumors noted above arose from transformed epithelial line $TMI_4$, and was a poorly differentiated squamous carcinoma (Fig. 8). The other, from line $TMI_5$, was composed of a mixture of patterns. One of these patterns was highly anaplastic adenocarcinoma (Fig. 6), while adjacent to such areas were some foci in which tumor cells produced a cartilage-like histologic pattern (Fig. 7), and other foci where osteoid-like material was produced. Still other foci were completely undifferentiated. Histochemical studies showed that the cartilage-like material reacted as typical cartilage and the osteoid material reacted as typical osteoid.

---

Fig. 2. Uninfected salivary epithelium on filter opposite mesenchyme at 34th day. Note absence of morphogenesis.

Fig. 3. Polyoma-infected epithelium on filter opposite mesenchyme at 375th day. Cells show moderate atypia. This transformed culture ($TMI_5$) gave rise to tumors shown in Figs. 5-7.

Fig. 4. Polyoma salivary tumor induced in vivo. Note small, uniform cells forming gland-like structures as well as mesenchymoid pattern.

Fig. 5. Undifferentiated tumor derived from salivary epithelium transformed on filter opposite mesenchyme (line $TMI_5$). Note large, pleomorphic cells.

Fig. 6. Tumor derived from same transformed epithelial culture line as that in Fig. 4. Pattern is that of adenocarcinoma.

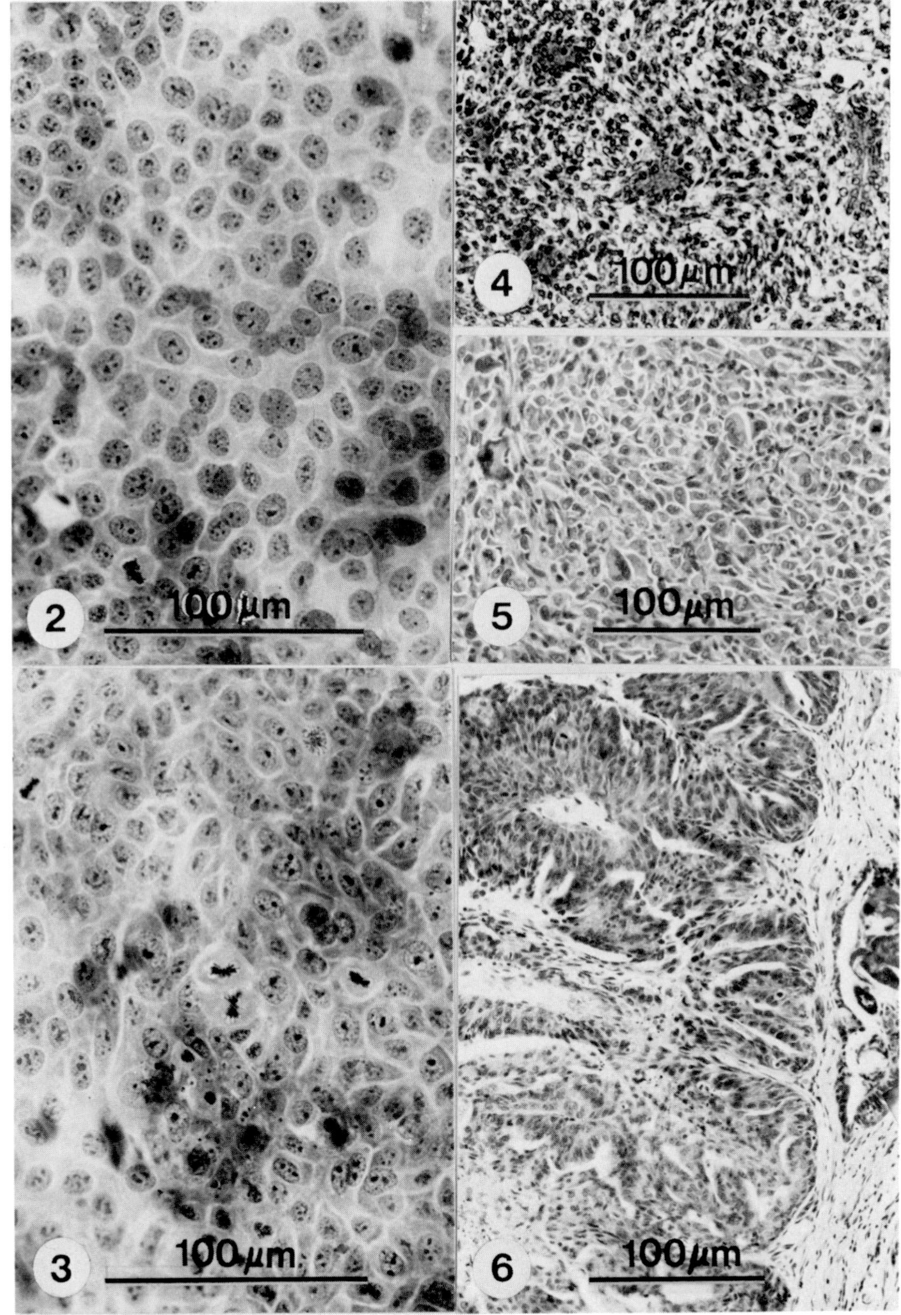
2
100 µm
3
100 µm
4
100 µm
5
100 µm
6
100 µm

One transformed mesenchyme line from Table I has been carried as a continuous culture line and transferred back to mice. In Figs. 9 & 10 the resulting tumor is compared with a subcutaneous fibrosarcoma induced in vivo by inoculation of a newborn mouse with polyoma virus. The tumor derived from the transformed culture line was the more highly cellular and more anaplastic of the two, but had the features to be expected in a poorly differentiated sarcoma. Little collagen was produced. Figs. 11 & 12 compare the appearances of transformed and untransformed mesenchyme growing on filters. Figs. 13 & 14 compare the appearances of transformed mesenchyme and transformed epithelium growing on glass.

Table II presents the preliminary results of trials to see if salivary epithelium can be transformed by polyoma virus when placed on one or both sides of the filter in the absence of mesenchyme. By purely morphological criteria, the positive results in 5 of 24 attempts suggest that it can. It will be necessary to determine whether the morphologically transformed cultures are tumorigenic before answering the question definitively. Interestingly, one non-infected culture (control) in Table II is presently growing as an apparently continuous line. Tissue culture infectivity tests show it contains no polyoma virus.

Morphologically, the transformed epithelium from the experiment in Table II was not distinguishable from epithelium transformed in cultures where it was grown opposite mesenchyme on the filter. However, back-transfer of these cultures into mice has not yet been done, and it is of course important to determine whether these cultures are tumorigenic, and if they are, to characterize the tumors that emerge.

---

Fig. 7. Section taken from same tumor shown in Fig. 6. In this area tumor cells produce cartilage-like material.

Fig. 8. Poorly differentiated squamous cell carcinoma derived from line $TMI_4$, transformed on filter opposite mesenchyme.

Fig. 9. Tumor derived from mesenchyme transformed on filter. Cellularity is greater, collagen deposition less, than in fibrosarcoma induced in vivo (Fig. 10).

Fig. 10. Fibrosarcoma induced by polyoma virus in vivo.

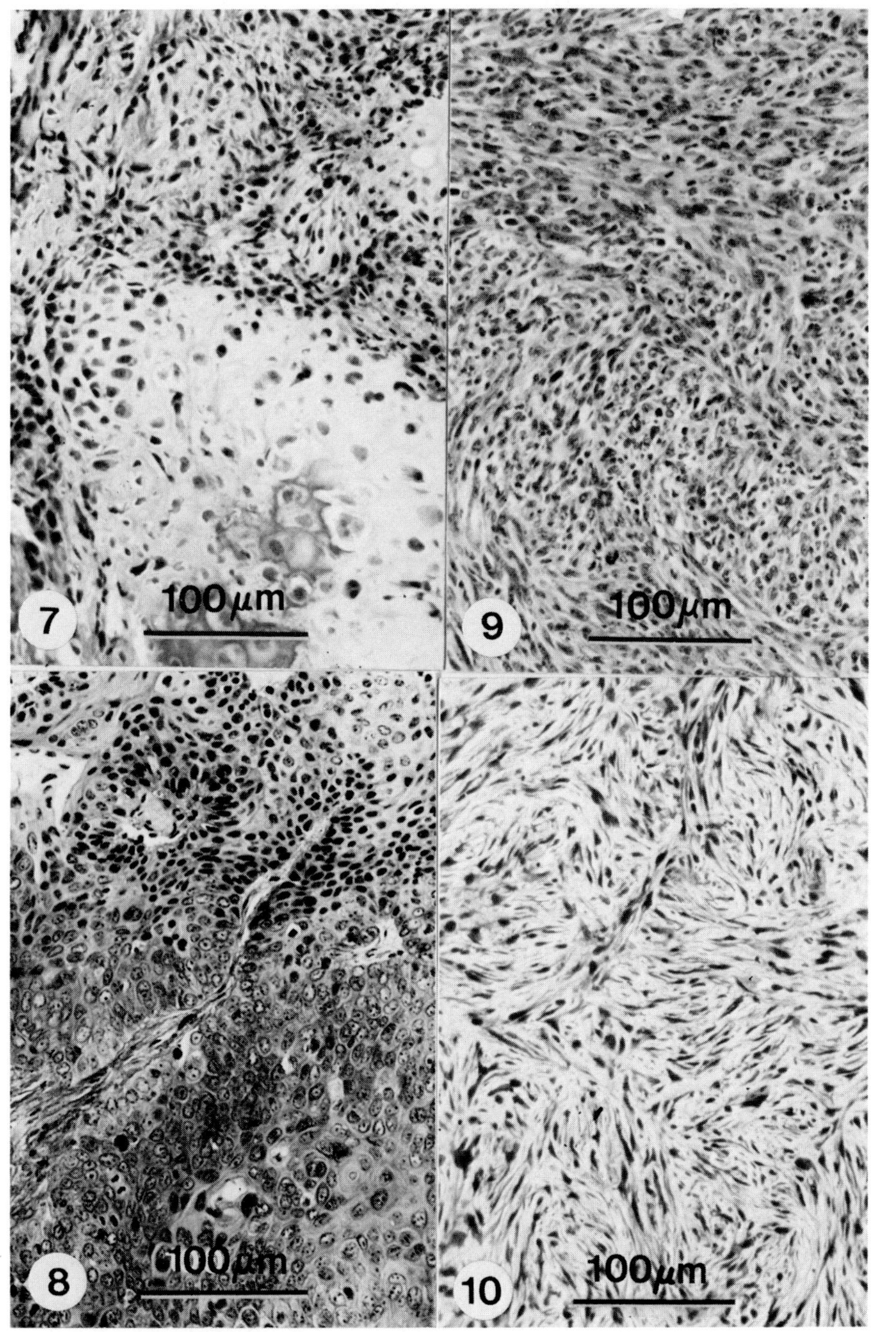
7
100 μm
9
100 μm
8
100 μm
10
100 μm

Table II. Trans-filter and Epi-filter Transformation Experiments with Epithelium Alone

| Type of Culture on Millipore Memb. | | Fixed, to Date | | Viable Cells Pres. | | Continuous Cultures Developed | |
|---|---|---|---|---|---|---|---|
| | | 30d. | 30d. | 30d. | 30d. | On Glass | On Memb. |
| PV-Infected | Ep/Ep | 19 | 12 | 11 | 0 | 2 (early) | 3 |
| | Ep/0 | 5 (+5 Inc.) | - | 2 | - | 1 (early) | 2 |
| Non-Infected | Ep/Ep | 13 | - | 4 | - | 1 | 1 |
| | Ep/0 | 14 (+2 Inc.) | - | 0 | - | 0 | 0 |
| Totals | | 51 | | | | 4 | 6 |

Inc. = Incomplete (filter cultures still in progress)

## 4. Discussion

The results of the trans-filter experiments leave little doubt that transformation of salivary epithelium by polyoma virus can occur in the absence of inductive epithelio-mesenchymal interaction accompanied by morphogenesis. Under the conditions used, no morphogenesis was observed, yet morphologic transformation occurred and tumors developed from transformed epithelium as well as from transformed mesenchyme. However, it cannot be assumed that no epithelio-mesenchymal interaction of any sort occurred in the experiments with epithelium opposite mesenchyme on the filter. In fact, the presence of mesenchyme opposite the epithelium had a decided effect on the growth of the epithelium in cultures uninfected by virus. When neither mesenchyme nor virus was present, salivary epithelium

Fig. 11. Transformed mesenchyme on filter, 126th day after infection. This culture line ($TMI_2$) gave rise to tumor in Fig. 10. Compare with untransformed mesenchyme in Fig. 12.

Fig. 12. Untransformed mesenchyme on filter, opposite epithelium shown in Fig. 2, on 34th day of culture.

Fig. 13. Phase contrast photo of transformed epithelial cell line, $TMI_5$, growing on glass. This line came from filter culture shown in Fig. 3, and gave rise to tumors shown in Figs. 5, 6, & 7. Compare morphology with that of transformed mesenchyme in Fig. 14.

Fig. 14. Phase contrast photo of transformed mesenchymal cell line, $TMI_2$, growing on glass. This line came from filter culture shown in Fig. 11, and gave rise to tumor in Fig. 9.

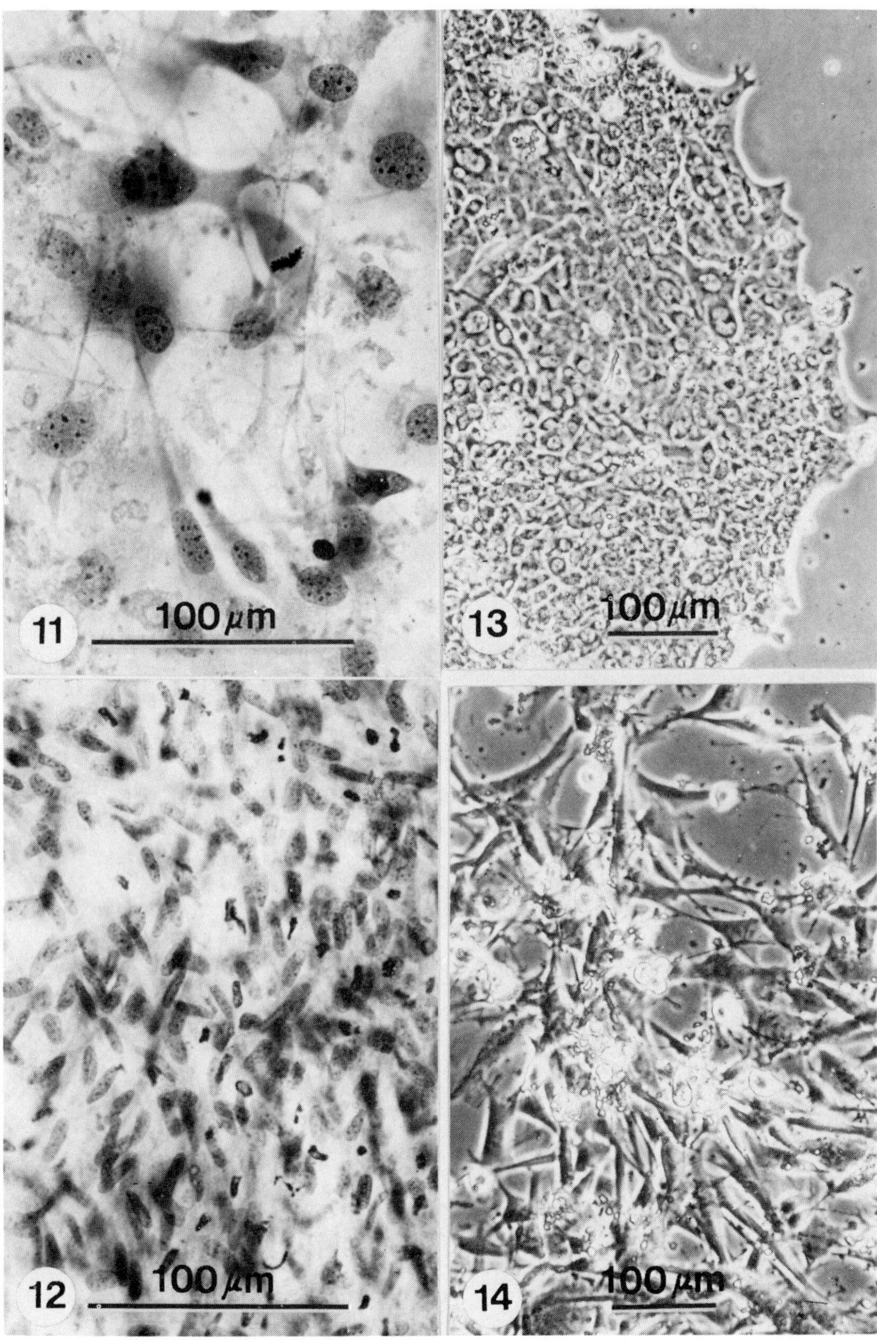
11
100 µm
13
100 µm
12
100 µm
14
100 µm

usually grew only transiently on the filters, whereas it regularly grew and covered the entire filter surface in about 30 days when mesenchyme was present.

At the same time, it must be noted that salivary epithelium alone survived and proliferated for a longer time and more vigorously on Millipore filter than it did on glass in our earlier experience. One control epithelial culture even went on to grow continuously on the filter and is presently growing on glass. This was never seen in a large number of attempts to grow epithelium alone, and to transform it with polyoma virus, on glass. This raises a question as to whether the Millipore filters themselves are carcinogenic or co-carcinogenic.

The salutory effect on epithelial growth of using Millipore filter as a physical substrate rather than glass may be an important factor in accounting for the difference in results obtained in this work as compared with the negative results obtained previously.[7] In studies of SV-40 transformation in vitro, Todaro and Green found that a prerequisite for transformation was mitotic cycling in the presence of virus.[11] Although we have seen mitoses during the first few days of culture of isolated epithelium on glass, these may have occurred in cells that had completed the S phase of the cycle prior to explanting, and hence also prior to infection with virus. The probability of an integrative transforming event may have been low under such conditions.

That the tumors resulting from transformation of epithelium in the absence of morphogenesis were virtually undifferentiated (with two exceptions) is not surprising. Since the cells at the time of transformation had undergone neither morphogenesis nor cytodifferentiation, they might well be expected to continue in such an undifferentiated condition after transformation.

However, the epithelium at the time of isolation had undergone determinative events. The specific restrictions and/or potentialities conferred upon the epithelium by these determinative events are unknown, but the occurrence of one tumor that showed squamous characteristics and another that showed glandular histologic patterns as well as formation of cartilage-like and bone-like material suggests that the "repertoire" of salivary cells at this stage may be considerable. This is of special interest in considering the pathogenesis of mixed tumors of salivary glands, mammary glands, and skin, which may display a similar melange of seemingly incongruous cell types.

The mechanism(s) through which culture on Millipore filter favors survival, growth, and transformation of salivary epithelium is unknown. It may be significant that preliminary EM examination of filters with epithelium on them show basal lamina laid down between epithelium and filter. In any case it is now evident that neoplastic transformation of embryonic salivary epithelium can occur in the absence of morphogenesis. Seemingly this would negate our earlier postulate that the fundamental system in salivary tumor development in response to polyoma virus is a 3-unit one composed of mesenchyme, epithelium, and virus.[6,7] However, the fact that tumors derived from the virus-infected epithelium in the absence of morphogenetic interaction with mesenchyme are strikingly different from those induced in vivo may equally well be interpreted to demonstrate that epithelio-mesenchymal interaction during tumor genesis is indeed effective in shaping the morphologic and biologic character of the tumor that emerges. How this interaction works and how it might be modified as a means of controlling the response to carcinogenic agents should be matters of much interest to those concerned with microenvironmental factors in carcinogenesis.

References

1. Grobstein, C., 1959, in: The Cell, vol. I, eds. J. Brachet and A.E. Mirsky (Academic Press, New York), p. 437.

2. Bernfield, M.R. and Wessells, N.K., 1970, Symp. Soc. Dev. Biol. 29, 195.

3. Auerbach, R., 1964, in: Retention of Functional Differentiation in Cultured Cells, ed. V. Defendi (Wistar Inst. Press, Phila.), p. 3.

4. Pierce, G.B. and Johnson, L.D., 1971, In Vitro 7, 140.

5. Law, L.W., Dunn, T.B., and Boyle, P.J., 1955, J. Natl. Cancer Inst. 16, 495.

6. Dawe, C.J., 1972, in: Tissue Interactions in Carcinogenesis, ed. D. Tarin (Academic Press, London & New York), p. 305.

7. Dawe, C.J., Morgan, W.D., and Slatick, M.S., 1966, Int. J. Cancer 1, 419.

8. Dawe, C.J., Main, J.H.P., Slatick, M.S., and Morgan, W.D., 1966, in: Recent Results in Cancer Research, vol. 6, ed. W.H. Kirsten (Springer, New York), p. 20.

9. Dawe, C.J., Whang-Peng, J., Morgan, W.D., Hearon, E.C., and Knutsen, T., 1971, Science 171, 394.

10. Grobstein, C., 1953, Nature, London 172, 869.

11. Todaro, G.J. and Green, H., 1966, Proc. Nat. Acad. Sci. 55, 302.

The authors gratefully acknowledge the photographic assistance of Mr. Ralph L. Isenburg, Laboratory of Pathology, DCBD, National Cancer Institute, and Mr. Robert W. Nye, Medical Arts and Photography Branch, Division of Research Services, National Institutes of Health. We are indebted to Mrs. Susan M. Hostler for her skill and patience in typing and assembling the script. Tissue sections and whole-mounts of filter cultures were prepared by the staff of the Pathological Technology Section, Laboratory of Pathology, DCBD, National Cancer Institute, under the supervision of the late Ms. Betty J. Sanders, who succumbed to pulmonary cancer while this Conference was in progress. Ms. Sanders' tenacious devotion to the cancer research effort will be a continuing inspiration to those who carry on.

*Progress in Differentiation Research, ed. N. Müller-Bérat et al.*



# FROG RENAL TUMORS ARE COMPOSED OF STROMA, VASCULAR ELEMENTS AND EPITHELIAL CELLS: WHAT TYPE NUCLEUS PROGRAMS FOR TADPOLES WITH THE CLONING PROCEDURE?[1]

R.G. McKinnell, L.M. Steven, Jr., and D.D. Labat
Department of Zoology
University of Minnesota, Minneapolis 55455, U.S.A.

## Introduction

Tadpoles result when renal tumor nuclei of North American leopard frogs, Rana pipiens, are inserted into activated and enucleated eggs[1,2,3,4]. The cloning studies of renal tumor nuclei may provide information concerning whether or not the genome of this particular neoplastic cell is entire, unaltered, and therefore equivalent to the genome of a zygote. They also may provide substantial information about the stability of the differentiated state of adult nuclei in general, and in particular, they witness to the potential reversibility of the malignant state.

The renal tumor of R. pipiens is a spontaneous adenocarcinoma of the mesonephros[5]. It has been long thought to have a viral etiology[6,7] and recently the putative etiological agent, the Lucké tumor herpesvirus, was shown to fulfill Koch-Henle postulates[8].

Because the tumor nuclear transfer studies have significance for those who study differentiation and for those who are concerned about the reversibility of the neoplastic state, we believed that it was exceedingly important to characterize the type nucleus that was being transplanted in these tumor cloning studies.

Parthenogenesis can be induced in Rana pipiens[9,10]. We sought to distinguish between parthenogenesis and development provoked by the inserted tumor nucleus, i.e., development of a parthenogenetically stimulated inadvertently retained maternal nucleus versus genuine embryonic development programmed by the transplanted nucleus. Lucké tumor herpesviruses were injected into triploid frog embryos produced by hydrostatic retention of the second polar body of inseminated eggs[11]. Frogs with triploid tumors ensued[12]. Nuclear transplantation of the triploid tumors resulted in triploid tadpoles[4]. These studies excluded the possibility of gynogenetic parthenogenesis; i.e., nuclei from an adult tumor were shown by cytogenetic analysis to program for embryonic development.

---

[1]Supported by Research Grant GB-40997 from the National Science Foundation.

An adenocarcinoma is an epithelial neoplasm of a gland. Accordingly, it is comprised of epithelial cells which are malignant, a connective tissue stroma which provides the foundation upon which the epithelium grows, and vascular elements. We report here for the first time, observations which lead us to reject the likelihood of connective tissue cells or vascular elements programming for the development of the recipient egg. These results lead us to affirm, in conjunction with our previous cytogenetic analysis, that development of embryos which in some cases become tadpoles follows nuclear transplantation of malignant epithelial nuclei obtained from the renal adenocarcinoma of R. pipiens.

Fluorescent microscopy of dissociated tumor cells

The fluorochrome acridine orange (AO) is useful in detection and identification of intracellular nucleic acids[13]. Nuclear DNA stains green and cytoplasmic RNA stains red[14]. Many tumor cell types, because of their increased rate of cytoplasmic RNA synthesis, may be distinguished with ease from their normal tissue of origin by fluorescent microscopy with AO[15,16]. Tweedell[17,18] reported that fluorochroming with AO distinguishes not only between normal and malignant renal epithelia of R. pipiens but the procedure also may be used to differentiate between tumor epithelium and stromal cells. The frog tumor epithelium, after staining with AO and viewed in the ultraviolet, has a light green nucleus with red nucleolus and the cytoplasm fluoresces orange-red to red. The stroma, lacking an abundance of red-fluorescing RNA, appears deep green in marked contrast with the epithelium. The observations of Tweedell are in harmony with the recent report of Harrison, et al[19]. We sought with the ultraviolet fluorescent AO technique to identify which cell type dissociated from a tumor fragment. Only dissociated cells are transplanted in a cloning experiment. Accordingly fluorescent microscopic examination of dissociated tumors reveals the cell type used for cloning.

Four sexually mature renal adenocarcinoma-bearing R. pipiens were obtained from Professor George W. Nace, The Amphibian Facility, University of Michigan, Ann Arbor, U.S.A. (Fig. 1). The tumor-bearing frogs were maintained at laboratory temperature (20-22°C) until killed by pithing which destroyed both the brain and spinal cord. A small sample of the untreated fresh tumor was fixed for histological study (Fig. 2a). Small fragments of freshly dissected tumor were placed into Calcium-free Steinberg's[20] electrolyte solution with or without

Fig. 1. Sexually mature Rana pipiens dissected to show large primary renal adenocarcinoma (RT) with several metastatic masses in the liver (M). Tumor-bearing frog courtesy of Professor George W. Nace.

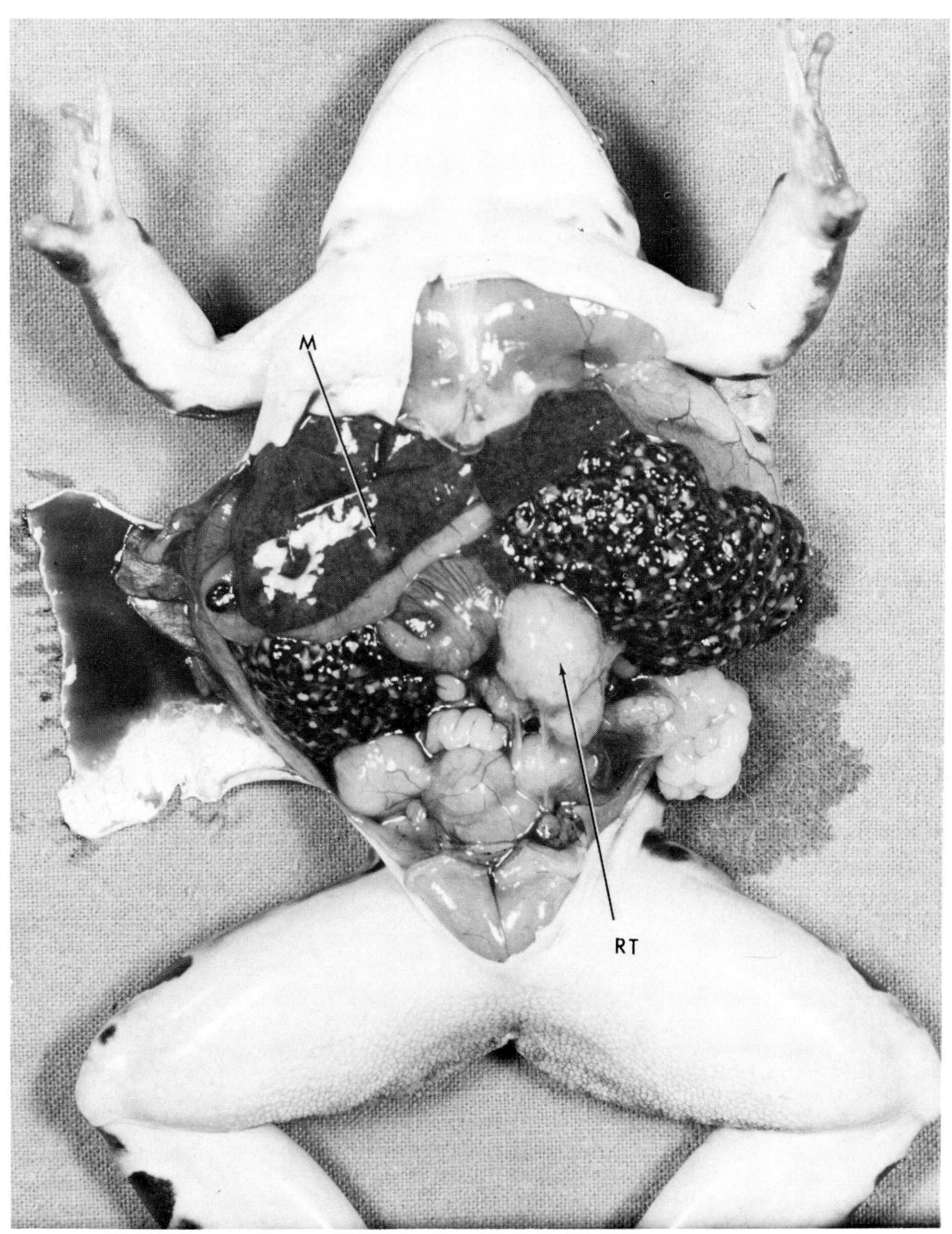
M
RT

the chaelating agent, disodium(ethylenedinitrilo)tetraacetate (Baker). Normal mesonephros, which is red-brown in color, was cut away and removed from the pale yellow tumor tissue. Fragments of tumor with blood vessels visible by observation with a dissecting microscope were discarded. The prepared tumor fragments were then placed in a clean dish containing fresh Calcium-free Steinberg's electrolyte solution.

Dissociation of free cells occurs within 15 to 30 minutes at laboratory temperature. Gentle shaking permits free cells to fall away from the tumor fragments which retain substantial structure; i.e., after the described treatment of cells has occurred, there are both free cells at the bottom of the dish and a tissue fragment refractory to the dissociation procedure. Fragments of tumor material assumed to be stroma persist even with 24 hours of the gentle dissociation treatment.

Ms. Louise A. Rollins, University of Minnesota, kindly provided an anterior eye chamber graft of a renal adenocarcinoma. After the grafted tissue was dissected from the adult host eye, it was treated in the same manner as primary tumor described above. There is substantial necrotic material in a large primary tumor. We examined a small graft growing in the eye of an adult frog because we desired to know if dissociation of tumor cells would be similar in a tumor graft versus a primary tumor.

When a nuclear transplantation experiment was completed (2 or more hours after the onset of treatment) dissociated cells were spread on 3 x 1 inch glass microscope slides. The tumor fragment refractory to dissociation was fixed for histological study.

Cells spread on slides were allowed to dry. They were washed 3X in 0.15 M phosphate buffer pH 5.0, stained for 4 minutes in $1 \times 10^{-4}$ M acridine orange NO C.I. 46005 (National Aniline) prepared in 0.15 M phosphate buffer pH 0.5. The slides were then rinsed 3X in 0.15 M phosphate buffer pH 5.0. The slides were dried and coverslips were mounted with a non-fluorescing medium or the slides were examined wet while still in buffer with a coverslip applied. The slides were examined with a Zeiss RA research microscope with an HBO 200 ultraviolet lamp. A non-fluorescent (Cargille, Type A) immersion oil was used.

---

Fig. 2a. Fluorescent microscopy of untreated primary renal adenocarcinoma stained with acridine orange. Stroma (S) and tumor epithelium (E) are indicated with arrows. Index marker = 100 μm.

Fig. 2b. Totally dissociated renal adenocarcinoma cell prepared by treatment with a chaelating agent and calcium-free electrolyte solution. The cytoplasm fluoresces a brilliant red revealing an abundance of RNA which is diagnotic for epithelium. Index marker = 10 μm.

Fig. 2c. Residual tissue fragment remaining after dissociation treatment. Note connective tissue stripped of its overlying epithelium (arrow). Index marker = 100 μm.

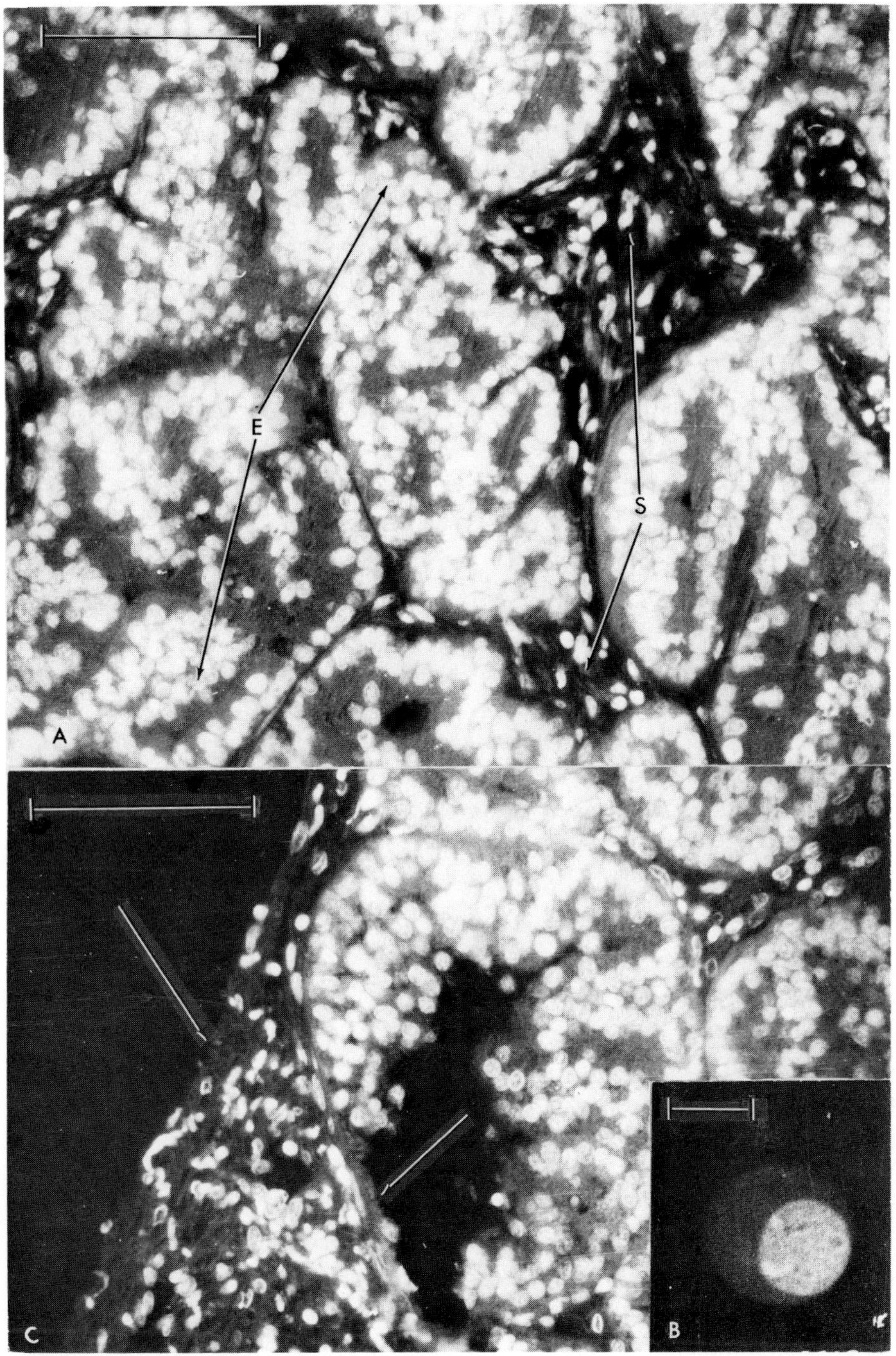
E
S
A
C
B

A total of 50 slides of dissociated cells were prepared from the 4 primary tumors and 10 slides were made of dissociated anterior eye chamber tumor cells. The impression gained by looking at random slides is that the overwhelming majority of totally dissociated and living cells were epithelial and therefore malignant. The diagnosis was made on the basis of red fluorescing cytoplasm (Fig. 2b). Despite the predominance of epithelial tumor cells, 1000 cells were individually diagnosed and counted: 98.5% were epithelial.

Nine hundred eighty-five dissociated cells were unequivocally of renal adenocarcinoma epithelia origin as ascertained by flame red cytoplasm. They had distinct nuclear and plasma membranes and were clearly alive at the time of slide preparation. Fifteen cells were counted which lacked red cytoplasm, had irregular nuclear and plasma membranes, and may or may not have been stromal cells. We believe that the cells which failed to stain the diagnostic red were dead and moribund and precise diagnosis was therefore not possible. In addition to intact cells, the slide contained extensive cellular detritus. To our surprise, the slides of anterior eye chamber grafts of renal tumor had many dead cells and much detritus. However, there was quantitatively more necrotic material in the slides prepared from primary tumors. The proportion of tumor epithelial cells to cells of undetermined origin was the same in both tumor types (1.5%).

What is the composition of the tissue fragment that was undissociated? Light microscope studies of hematoxylin and eosin sections reveal a grossly eroded epithelium on exposed surfaces of the fragment. The stroma was intact. Fluorescent AO observations, which more precisely discriminate between the epithelium and the stroma, confirmed this observation (Fig. 2c).

We conclude that living cells which provide donor nuclei for tumor nuclear transplantation therefore are entirely or overwhelmingly epithelial. Thus, development of the majority of recipient eggs to the tadpole stage, which may be as high as 7%, can only be attributed to transplanted nuclei from neoplastic cells.

We were not surprised with these results. Dispersal of adult living tissues to suspensions of isolated single cells generally required utilization of proteolytic and mucolytic enzymes coupled with mechanical agitation in addition to altering the ionic environment[21,22]. Our gentle dissociation procedure, lacking enzymes by design, seemed to be inadequate to disperse the connective tissue matrix of the frog tumor.

Eggs stimulated to develop by blood cell insertion: parthenogenesis or nuclear transfer?

It is exceedingly unlikely that stromal cells could provide nuclei that promote the development of tadpoles that we have described previously. All vascular material that can be discerned with the aid of a dissecting microscope is carefully removed prior to dissociation (see above). Yet, it might be argued that since capillaries in the connective tissue stroma are certainly not easily resolved with our methods, some of the dissociated cells in the operating dish may be blood cells. Erythrocytes as well as leukocytes are nucleated in *R. pipiens*. We therefore sought to ascertain if nucleated blood cells could program for embryonic development.

Conventional nuclear transfer seemed an unlikely method for analysis. Blood cells inserted into mature egg cytoplasm replicate DNA but do *not* enter into embryonic development[23,24]. Perhaps the elapse of time in the conventional methodology, from the moment of cell pickup to the time of nuclear insertion into recipient cytoplasm, permits damage to occur to fragile blood cells of *Xenopus*. Is there a quicker mode of inserting blood cells into recipient cytoplasm that might indicate whether or not blood cell nuclei could promote embryonic development?

Guyer[25], as a result of an early investigation of parthenogenesis in frogs, postulated that eggs which developed after pricking with a microneedle dipped in blood, were in some cases, attributable to proliferation of leucocyte nuclei that were thrust into the egg cytoplasm.

Could any of our tadpoles have developed as the result of blood cell nuclear transfer as Guyer believed his frogs did? In order to answer this question, we repeated Guyer's experiment using triploid blood, instead of diploid blood, to induce parthenogenesis. Our reasoning was as follows: If parthenogenesis is due in any part to leucocyte or erythrocyte proliferation, then a proportion of parthenogenetically developing embryos should have a ploidy corresponding to the ploidy of the erythrocytes and leucocytes used to promote virgin development. Haploid and diploid parthenogens clearly cannot descend from introduced triploid leucocyte or erythrocyte nuclei.

Triploid frogs, which served as blood cell donors, were produced by hydrostatic repression of the second polar body following insemination of eggs[11]. A total of 1995 freshly ovulated but *not* inseminated eggs were stimulated by pricking with triploid blood cells. Fifty-two blastulae and 43 gastrulae were obtained. Nineteen embryos developed as far as tadpole stages so that chromosome number could be determined from tail tip preparations. Fifteen were haploid (1n=13) and 4 were diploid (2n=26). None of the parthenogenetically developing tadpoles stimulated by triploid blood were triploid.

We believe that the parthenogenetic experiments with triploid blood cells provide important evidence that embryonic development of our tumor nuclear transfer tadpoles is not promoted by introduced blood cells. In our parthenogenesis experiment development was due to replication of the mature gamete nucleus (haploid) or it was due to the gamete nucleus after a delay in cleavage with fusion of the first 2 mitotic descendents of that nucleus (diploid). The authenticity of the tumor nuclear transfers is corroborated by these experiments with gynogenesis. We believe that although it is possible that an occasional blood cell could be inadvertently inserted at the time of tumor nuclear transplantation to the egg cytoplasm, it seems extremely unlikely that the blood cell is competent to participate in development.

The experiment with parthenogenesis affirms again the critical need for proper enucleation of the recipient egg. Gynogenesis permits development in some cases to metamorphosis. For this reason, we continue to seek a methodology that assures high enucleation success coupled with minimum cytoplasmic damage[26].

Nuclei transplanted from tumors are indeed derived from malignant cells.

Whether or not tumor nuclear transplant tadpoles are in fact derived from the genome of a tumor cell is a major question after a decade and a half of experimentation. Klein[27] stated: "MacKinnell's experiment is obviously extremely important, provided that it can be shown convincingly that the donor cell was a neoplastic cell, rather than a stromal cell." To this same issue, Gurdon[28] wrote: "there is no proof that the cells from which these nuclei were transplanted were tumour cells and not non-malignant host cells present in the tumour tissue."

Our studies with ultraviolet fluorescence of acridine orange stained dissociated tumor cells and with blood cell gynogenesis suggest the extent to which we agree with the importance of distinguishing non-malignant cells from malignant cells in a tumor. We believe that there is no valid reason to doubt the malignancy of the nuclear donor cells in our studies.

## Discussion

A euploid chromosome complement is essential for normal development[29,30,31]. Cells from normal early embryos may be presumed to be euploid. When blastula nuclei are used as donors for transplantation purposes, many of the experimental transplant embryos develop normally[32]. Nuclei from an aneuploid *in vitro* cultured adult liver cell line program for limited development[33]. Less than perfect development would be expected with imperfect chromosomes. The chromosome constitution of frog tumors is therefore of great significance to cloning experiments. Fortunately for this purpose, many of the tumor cells have euploid chromosomes[34,35]. Therefore, normal development of nuclear transplant embryos should not be precluded on the basis of light microscopic examination of renal tumor chromosomes. However, the Lucke tumor herpesvirus is associated with all frog renal tumors either in a latent state[36] or as a productive viral infection[37,38]. It is not known at the present time what effect, if any, the viral genome has on the capacity of the tumor genome to be reprogrammed for embryonic development.

There are now several adult cell types in addition to the renal adenocarcinoma that are known to program for extensive development with the nuclear transplantation procedure. Nuclei from mitotically active lens epithelial cells in culture are reported to produce normal frogs by cloning[39]. Cultured adult skin nuclei yield advanced embryos when cloned[40]. One tadpole which fed was produced by the transfer of an adult gut cell epithelial nucleus[41]. What permits the exceptional adult nucleus to enter into embryonic development after transplantation? Is it mitotic activity as proposed by Gurdon[28] or is it the relative state of differentiation of the donor cell that allows for development[41]?

The renal adenocarcinoma is mitotically active in its calid phase[42]. Transplantation of normal kidney nuclei, which are not mitotically active, promotes little embryonic development[2]. Exposure of frog kidney cells to the Lucke tumor herpesviruses results in transformation to the malignant state and a mitotically active renal adenocarcinoma[12]. The mitotically active renal tumor nuclei promote extensive embryonic development by cloning. Thus, it may be that mitosis is an important aspect of the capacity of adult somatic tumor cells to be reprogrammed for embryonic development. However, mitosis is not the only effect of the Lucke tumor herpesvirus on mesonephric cells. The tumorous state is characterized by varying degrees of anaplasia. Anaplasia refers to an apparent loss of the differentiated state. How anaplastic is the frog renal adenocarcinoma? The tumor retains a relatively high degree of epithelial structure (Fig. 2a). Certainly, the pseudostratified columnar cells are not dedifferentiated in a morphological sense. The tumor cells do resemble embryo cells in at least one biochemical aspect. Many rapidly dividing embryonic cells are characterized by cytoplasmic basophilia which is indicative of much cytoplasmic RNA[43]. Renal tumor cells are also characterized by cytoplasmic RNA. Indeed, the acridine orange fluorescent test for tumor cells is a test for cytoplasmic RNA. Thus, an actively synthetic genome, perhaps in some ways not unlike that of embryonic cells, may be essential for an adult nucleus to enter into embryonic development through the cloning process.

Certainly the other adult cell types which program for extensive development by means of nuclear transplantation are not only mitotically active as the tumor is but are less than terminally differentiated.

Nuclei from lens epithelium proliferate, elongate, become transparent, and are terminally "fully" differentiated when they lose their nuclei. Adult lens nuclei which are reprogrammed to form frogs are mitotically active cells which have not yet begun the elongation process, nor lost their nuclei, and are therefore less than terminally differentiated. Terminally differentiated skin epithelium consists of keratinized cells. Skin cells in culture which fill up with a birefringent material believed to be keratin are unsuited for cloning[28].

Differing from the above studies, McAvoy et al[41] have shown that crest cells of adult gut, mitotically inactive, yield the same results by cloning as trough cells from adult gut which are mitotically active. They conclude that mitotic activity and the stage of the cell cycle which characterize the donor nuclei are less important in cloning studies than the degree of differentiation of the donor cell. These studies, however, are just beginning and only 1 feeding tadpole has ensued.

Therefore, it seems to us that it is premature to conclude what biochemical or physiological aspect of an adult cell is requisite for it to promote embryonic development by cloning.

Comprehension of the mechanism which permits an adult cell to be reprogrammed, viz. mitotic activity, a synthetically active genome, or a less than fully differentiated state, is not essential to the conclusion of the present study which is that neoplastic epithelial nuclei may be reverted to a condition which mimics the embryological potentialities of a zygote nucleus.

References


1. King, T.J. and McKinnell, R.G., 1960, in: Cell physiology of neoplasia (Univ. Texas Press, Austin) p. 591.
2. King, T.J. and DiBerardino, M.A., 1965, Ann. New York Acad. Sci. 126, 115.
3. DiBerardino, M.A. and King, T.J., 1965, Develop. Biol. 11, 217.
4. McKinnell, R.G., Deggins, B.A., and Labat, D.D., 1969, Science 165, 394.
5. Lucké, B., 1934, Am. J. Cancer 20, 352.
6. Lucké, B., 1952, Ann. New York Acad. Sci. 54, 1093.
7. McKinnell, R.G., 1973, Am. Zool. 13, 97.
8. Naegele, R.F., Granoff, A., and Darlington, R.W., 1974, Proc. Nat. Acad. Sci. (USA) 71, 830.
9. Paramenter, C.L., 1933, J. Exp. Zool. 66, 409.
10. Fraser, L.R., 1971, J. Exp. Zool. 177, 153.
11. Dasgupta, S., 1962, J. Exp. Zool. 151, 105.
12. McKinnell, R.G., and Tweedell, K.S., 1970, J. Nat. Cancer Inst. 44, 1161.
13. Rigler, Jr., R., 1966, Acta Physiol. Scand. 67, Suppl. 267, 1.
14. Bertalanffy, L. von, and Bickis, I., 1956, J. Histochem. Cytochem. 4, 481.

15. Bertalanffy, L. von, Masin, F., and Masin, M., 1956, Science 124, 1024.

16. Bertalanffy, L. von and Bertalanffy, F.D., 1960, Ann. New York, Acad. Sci. 84,225.

17. Tweedell, K.S., 1960, J. Morph. 107, 1.

18. Tweedell, K.S., 1965, Ann. New York Acad. Sci. 126, 170.

19. Harrison, F.W., Zambernard, J., and Cowden, R.R., 1975, Acta Histochem. In press.

20. Steinberg, M., 1957, in: Carnegie Institution of Washington Yearbook 56, p. 347.

21. Moscona, A., Trowell, O.A., and Willmer, E.N., 1965, in: Cells and tissues in culture, vol. 1, ed. E.N. Willmer (Academic Press, London) p. 19.

22. Barofsky, A-L, Feinstein, M., and Halkerston, I.D.K., 1973, Exp. Cell Res. 79, 263.

23. Graham, C.F., Arms, K. and Gurdon, J.B., 1966, Develop. Biol. 14, 349.

24. Gurdon, J.B., 1967, Proc. Nat. Acad. Sci. (USA) 58, 545.

25. Guyer, M.F., 1907, Science 25, 910.

26. Ellinger, M.S., King, D.R., and McKinnell, R.G., 1975, Radiat. Res. 62, 117.

27. Klein, G., 1972, in: Oncogenesis and herpesviruses, eds. P.M. Biggs, G. de-Thé, and L.N. Payne (International Agency for Research on Cancer, Lyon) p. 501.

28. Gurdon, J.B., 1974, The control of gene expression in animal development (Oxford, London).

29. Boveri, T., 1902, Verh. Phys.-med. Ges. Wurzburg, N.F. 35, 67.

30. Fankhauser, G., 1934, J. Exp. Zool. 68, 1.

31. Fankhauser, G., 1955, in: Analysis of development, eds. B.H. Willier, P.A. Weiss, and V. Hamburger (Saunders, Philadelphia) p. 126.

32. McKinnell, R.G., 1972, in: Cell differentiation, eds. R. Harris, P. Allin, and D. Viza (Munksgaard, Copenhagen) p. 61.

33. Kobel, H.R., Brun, R.B., and Fischberg, M., 1973, J. Embryol. Exp. Morph. 29, 539.

34. DiBerardino, M.A., King, T.J., and McKinnell, R.G., 1963, J. Nat. Cancer Inst. 31, 769.

35. DiBerardino, M.A. and Hoffner, N., 1969, in: Biology of amphibian tumors (Springer-Verlag, New York) p. 261.

36. Zambernard, J. and McKinnell, R.G., 1969, Cancer Res. 29, 653.

37. McKinnell, R.G. and Ellis, V.L., 1972, Cancer Res. 32, 1154.

38. McKinnell, R.G., Ellis, V.L., Dapkus, D.C., and Steven, Jr., L.M., 1972, Cancer Res. 32, 1729.

39. Muggleton-Harris, A.L. and Pezzella, K., 1972, Exp. Geront. 7:427.

40. Laskey, R.A. and Gurdon, J.B., 1970, Nature 228, 1332.

41. McAvoy, J.W., Dixon, K.E., and Marshall, J.A., 1975, Develop. Biol. 45, 330.

42. McKinnell, R.G. and Ellis, V.L., 1972, in: Oncogenesis and herpesviruses, eds. P.M. Biggs, G. de-Thé, and L.N. Payne (International Agency for Research on Cancer, Lyon) p. 183.

43. Brachet, J., 1957, Biochemical cytology (Academic Press, New York).

Note added after preparation of manuscript:

M. R. Wabl, R. B. Brun, and L. Du Pasquier, in a poster display at the Second International Conference on Differentiation, described development by serial nuclear transfer of *X. laevis* tadpoles to stage 43-44 of Nieuwkoop and Faber. The donor nuclei were derived from splenic lymphocytes. How does this new information affect the interpretation of our blood cell study? It is of crucial importance to note that their primary transfers resulted in partial blastulae. We reported no development programmed by mixed white blood cell injection. Thus, our results and those of Wabl and his colleagues are in essential harmony in that no tadpoles resulted from primary leucocyte transfer. We conclude, therefore, that the only nuclear type that programs for swimming tadpoles in a primary transfer of dissociated tumor cells must be nuclei of tumor epithelium.

*Progress in Differentiation Research, ed. N. Müller-Bérat et al.*



# CONTRIBUTION OF THE IMMUNE RESPONSE TO THE DEVELOPMENT OF TUMOURS: TRANSFORMATION OF REACTING LYMPHORETICULAR CELLS INTO TUMOUR CELLS

J. V. Spärck
Immunological Laboratory
Statens Seruminstitut
Copenhagen, Denmark

It is still a common opinion that a malignant tumour is a kind of a parasitic organism growing in spite of the host reaction against it, and that the animal organism is equipped with an immune surveillance of neoplasia which, for some reason, a tumour only occasionally escapes.

I would like to discuss some evidence which has accumulated recently, and to suggest some alternative interpretations. I would venture to suggest a completely different way of looking at the relationship between host response and neoplasia.

Interest in the stromal response to neoplasia is by no means new. As long ago as 1863 Virchow considered that the frequent presence of lymphoreticular cells in tumours reflected the origin of cancer at sites of chronic inflammation, the "chronic irritation theory". Waldeyer (1872) supported this idea and suggested that local disturbance of connective tissue was of pathogenetic importance in the origin of tumours, including epithelial tumours. Later there was a change in the general opinion, and many workers considered round cell infiltration in tumours to be a "defensive process", such infiltration being considered a good prognostic sign.

However, many conflicting accounts have accumulated concerning the effect of lymphoreticular infiltration in tumours, and it is not possible at this time to establish a causal relationship between such infiltration and a favourable prognosis (Underwood 1973). There is no convincing evidence available to show that the close spatial relationship between tumour cells and infiltrating lymphoreticular cells has a toxic effect likely to retard growth in the spontaneous tumour situation. On the other hand, there are several experimental reports concerning an increased tumour growth in connexion with intensivated local inflammation.

The question is then: what is the role of lymphoreticular tissue in the development of tumours? What is the relation of the immune response to neoplasia? The "immune surveillance" hypothesis, which has prevailed for a long time now, implies that it is a main function of the immune apparatus to survey and destroy neoplastic cells. The hypothesis implies that there is a correlation between immune competence and tumour frequency. It assumes that somatic mutations continously produce neoplastic cells, the majority of which have specific antigens sufficiently immunogenic to cause their destruction by the thymus-derived lymphocytes of the host. Where there is a reduced immune responsiveness, such neoplastic cells have a possibility of avoiding the immune surveillance and of forming tumours.

But is this the case in nature? - To-day there are many reports of observations and experimental findings which are in direct contrast to what would be expected if immune surveillance of neoplasia was operating. On the contrary, much of the data shows that a direct positive relation exists between immune reactivity and tumour growth which forces us to consider a different basic concept.

Indeed, a tumour-specific immunogenicity has now been established in a number of experimental and transplantable neoplasias, and it has been shown that even the organism in which a tumour has arisen may in some experiemntal systems react specifically to its own tumour. However, most cases in which such immunogenic potential of tumours has been demonstrated have been concerned with chemically or virally induced tumours which may prove to be laboratory artifacts not representative of spontaneous tumours. If an immunological defence against neoplasia exists in the animal organism, it is paradoxical that tumours so seldom regress. Apparently the immune surveillance against the incipient neoplasia does not function!

What is then the function of the immune response in tumour growth? In a number of tumour systems it has now been shown that the immune reaction stimulates the tumour, i.e. an acceleration of tumour growth is obtained when the host immune response is activated. It has also been shown in several cases of immunosuppression and growth of spontaneous mouse carcinomas that there is a direct positive correlation between tumour growth and the immune reactivity of the host. An illustration of this is provided by the results of experiments published some years ago on growth of mammary carcinomas in mice with changed reaction capacity (Spärck 1962, 1969). These were the first experiments in which a well-defined model was used to show that host responsiveness and tumour growth may be positively correlated. In order to have a close approximation to the natural tumour phenomenon, the transplantation system used was limited to the transfer of primary spontaneous tumours to syngeneic recipients (see fig. 1 - 5).

From such studies of immunosuppression in an immunogenetically well-defined system, it is apparent that tumour growth in the syngeneic situation is correlated with the capacity of the host to react, i.e. in the situation which compares best with the natural situation. Where there is a genetical difference between donor and host and a strong antigenic incompatibility, the opposite relationship is found.

The important point is the difference between the two situations with a different degree of antigenic stimulus. When immuno-genetical relations are such that tumour development is possible, then the growth does not depend on a defect immune mechanism of the host as postulated by the immune-surveillance hypothesis.

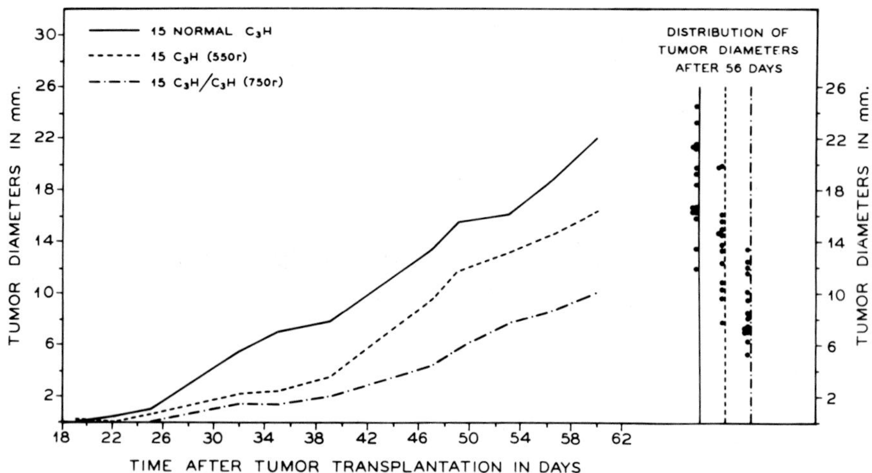


Fig. 1. Effect of whole-body irratiation on the tumour growth following syngeneic transplantation. To the right is shown the distribution of the individual tumour diameters in each group 56 days after tumour grafting.

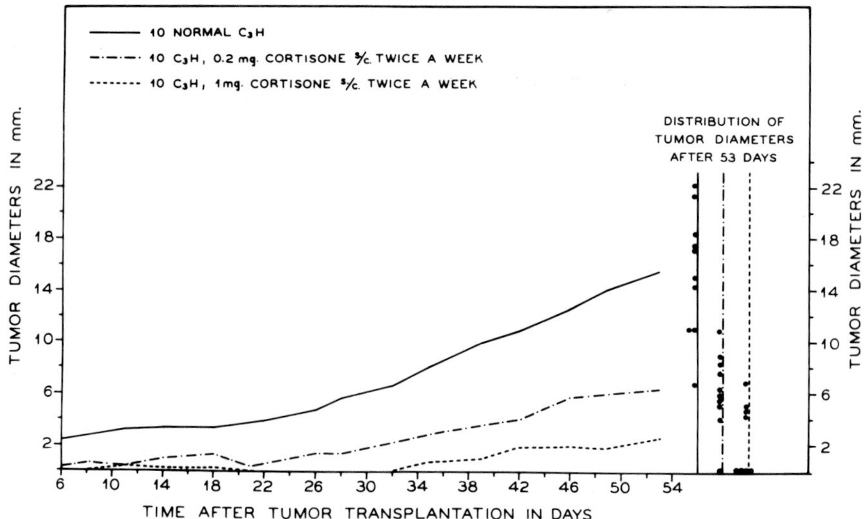


Fig. 2. Effect of cortisone on the progressive growth of tumours following the implantation of a syngeneic carcinoma. To the right are shown the average tumour diameters.

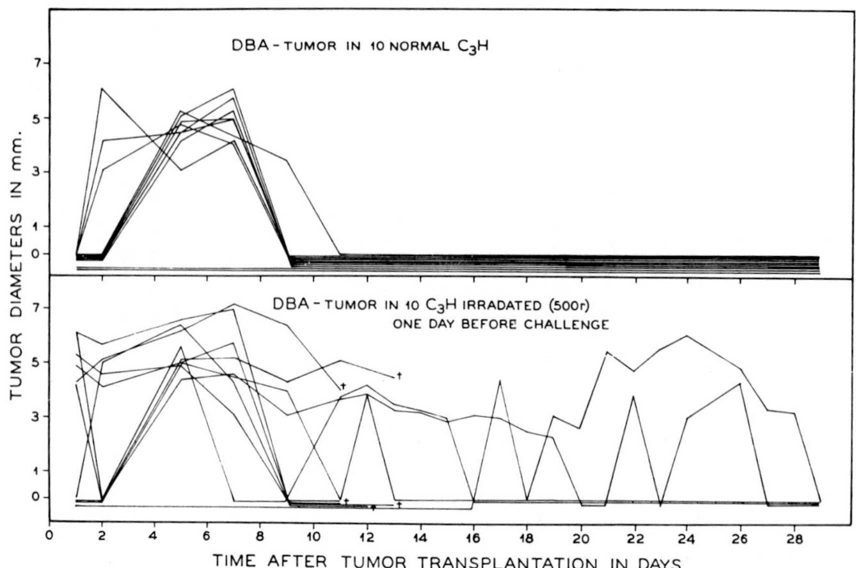


Fig. 3. Effect of whole-body X-irradiation on allogeneic tumour transplantation. The growth curve of each individual tumour is shown.

On the basis of these findings I suggested the hypothesis that the immune reaction might not only be beneficial to but indeed necessary for tumour growth. It is suggested that the weak immune stimulus which the ultra weak tumour antigens represent to the antochtonous or syngeneic host releases a reaction supporting tumour growth, while a strong immune response released in the host results in regression of the tumour. This hypothesis implies that the lymphoreticular system is not exclusively a defensive organ but that it has growth stimulating and regulating function in relation to normal and malignant tissues.

Later Prehn (1969) obtained similar results in experiments on immunosuppression and tumour growth, namely reduced growth of murine carcinoma in syngeneic hosts given immunosuppression. He was furthermore able to demonstrate(1972) an enhancement of tumour growth in syngeneic hosts mediated by in vivo sensitized spleen cells. He thymectomized the recipient mice and then gave them 450 r irradiation. When spleen cells from mice with the same tumour were added to the tumour inoculum, a significant and reproducible increase of tumour growth was obtained in the recipients mentioned. Prehn also found that the lymphocyte/tumour cell ratio was essential: very large number of spleen cells admixed resulted in tumour inhibition. On the basis of this, he formulated a similar hypothesis of immune stimulation also stressing that a weak immune reaction promotes tumour growth, while a strong immune response inhibits the tumour.

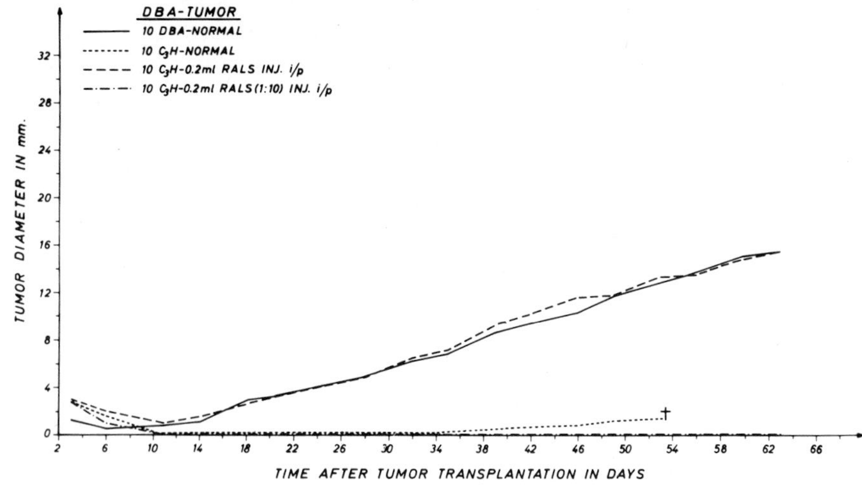


Fig. 4. Treatment of C3H mice with rabbit anti mouse lymphocytic serum. It is seen that in recipient mice given undiluted ALS the effect is a progressive tumour growth from the allogeneic graft which is normally rejected.

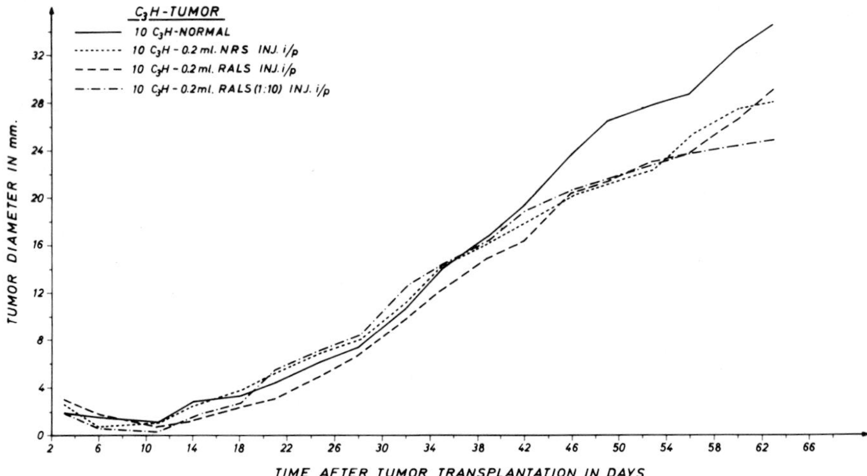


Fig. 5. Effect of rabbit anti mouse lymphocytic serum on tumour growth in C3H mice. The same treatment which in allogeneic tumour grafting causes an enhanced growth results in the syngeneic situation in a tumour growth which is reduced compared to that in untreated controls.

Attempts have been made to explain the failure of immune surveillance by the complexity of the reaction against the tumour, including a complicated interaction of cellular and humoral immunity. The assumption is that the rejection is mediated by immune T-lymphocytes while humoral antibodies or circulating antigen-antibody complexes block the cytotoxic cellular immunity or it is being suppressed by a population of suppressor T-lymphycytes. Much experimental evidence is available for the existence of such mechanisme. However, this is hardly the whole explanation of the positive relation of tumour growth to immune response. Blocking agents or cellular suppressors of the immune response cannot explain my findings mentioned before. They are proof that the tumour growth is immediately and directly dependent on the functional level of the lymphoreticular system: growth cannot take place in the total absence of the host response. This was shown by treatment with very large doses of immunosuppression(see fig. 6).

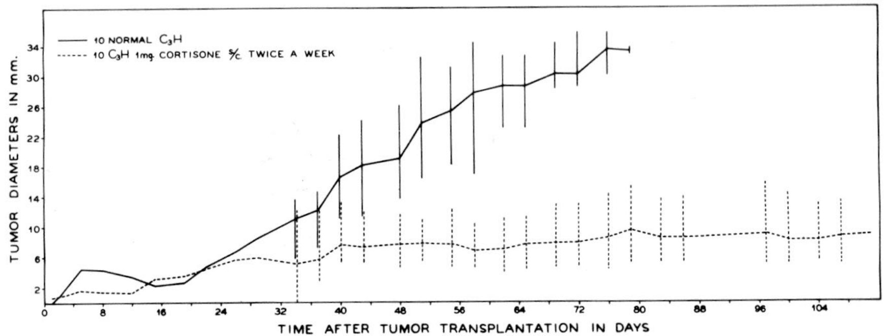


Fig. 6. Effect of cortisone injections on the tumour growth following the grafting of a primary C3H carcinoma to syngeneic recipients. Each point on the curve represents the average tumour diameter in the group; the range is indicated by a vertical line.

In situations like the one just referred to (fig. 6) it would rather be expected that the tumour growth would be increased if it was actually counteracted by the immune reaction and that it only escaped because of blocking or suppressing factors.

It has been demonstrated recently by Treves et al.(1974)that lymphoid cells sensitized in vivo or in vitro against the 3LL mouse tumour enhance the growth and metastasis of this tumour when injected into the tumour recipients. Evidence was presented suggesting that this effect is caused by a suppression of the immune response and not by a direct influence on the tumour. Thus, in heavily irradiated or thymectomized mice there was no enhancement of tumour by lymphoid cells, presumably because there was no immune response to suppress. However, this observation might be interpreted in a different way. The drastic tumour reduction following thymectomy could also be explained by assuming that the treatment had removed an essential condition for tumour development, viz. an active immune response of the host.

There are several reports on similar effects of thymectomy on the growth of tumours, but there are also conflicting reports: some on increased growth, some on reduced growth. For instance, Heppner, Wood and Weis (1968) found that neonatal thymectomy of BalbC mice resulted in reduction and a marked delay in the development of spontaneous mamma carcinomas. Balner and Dersjant (1966) report similar effects of thymectomy on chemical carcinogenesis. When considering the "immune surveillance" hypothesis it is also most important that the nude mouse, which is born without a thymus and is an immunological cripple, does not show any increased frequency of cancer.

When I suggested that a tumour is a reaction phenomenon I carried out some experiments to examine to what extent the tissue of a transplantation tumour is of host origin and to what extent of donor origin (Spärck 1969). The experimental system consisted in transplanting spontaneous mammary tumours from C3H-donors to recipients which were hybrids of C3H and DBA mice. In this system cell surface antigens could be used as markers by exposing the tumour cells to specific antisera produced by cross-immunizing the two mouse strains. The test was the cytotoxic antibody test using the dye exclusion criterion. By exposing cell suspensions prepared from tumours of the hybrid recipients to both anti-DBA and anti-C3H antibodies, it was found that almost all the cells recovered were of host type(see table 1).

Evidence of the decisive role of the connective tissue in carcinogenesis was also given by the experiemnts of Billingham,Orr and Woodhouse (1951). They treated mouse skin with methylcolanthrene for 12 weeks, and then left them untreated for 2 weeks to allow the methylcolanthrene to disappear. The carcinogen-treated epidermis did not yield tumours when transplanted to untreated body sites, while autografts of untreated epidermis to the treated area gave tumours.

On the basis of the data I have presented, I would like to put forward the hypothesis that tumours may be formed by the host reaction by a particular pathological type of inflammatory proliferation. I would suggest from evidence available that reacting lymphoreticular cells are transformed into tumour cells and I propose an immunological mechanism as basis for this phenomenon. A weak autoimmune response is the primary carcinogenic event. It is a characteristic of the multicellular animal organism that it is capable of a self-non self discrimination. Foreign matter is rejected, self material is catabolized by enzymatic mechanism. It is only reasonable to anticipate the occasional occurrence of ultraweak antigens, only slightly changed tissue constituents, which will stimulate

Table 1.

Cytotoxic action of anti-DBA and anti-C3H sera on tumour cells from C3HxDBA $F_1$-hybrids with implants of C3H-tumour.

| Cell suspensions | Days after implantation of C3H-carcinoma | Per cent unstained (surviving) cells after incubation with dilutions of antiserum | | |
|---|---|---|---|---|
| | | anti-DBA (serum 2) | anti-C3H (serum 3) | Control (saline) |
| Spont DBA carcin. | - | 3.5 | 53.5 | 51.5 |
| Norm .DBA spleen | - | 7.5 | 95.5 | 94.5 |
| Spont.C3H carcin. | - | 49.o | 8.5 | 54.o |
| Norm.C3H spleen | - | 92.o | 7.o | 91.5 |
| C3HxDBA-$F_1$No.1o carcin.32,C3H | 44 | 5.o | 2.o | 52.o |
| C3HxDBA-$F_1$No 4 carcin.32 | 46 | 6.o | 7.o | 61.o |
| C3HxDBA-$F_1$No 2 carcin. | 47 | 8.5 | 6.o | 47.5 |
| C3HxDBA-$F_1$No 2 spleen | 47 | 5.o | 12.o | 91.5 |
| C3HxDBA-$F_1$No 8 carcin.32 | 47 | 9.5 | 11.5 | 51.o |
| C3HxDBA-$F_1$No 8 spleen | 47 | 4.5 | 11.o | 92.o |

the lymphoreticular system to a reaction - an inflammatory proliferation - but which, on the other hand, lack the sufficient antigenecity - foreignness - to release a rejection and thereby a termination of the process. The result is thus, for immunological reasons, a continuous reaction, a particular form of pathological inflammation which eventually causes the lymphoreticular tissue to transform into tumour tissue. This might be an explanation of the immune stimulation of tumour growth.

References

1. Virchow,R.(1863) Krankhaften Geschwülste. Berlin.

2. Waldeyer,H.W.G.(1872) Virchows Arch.path.Anat. 55,67.

3. Underwood,J.C.E.(1974) Brit.J.Cancer 3o, 538.

4. Spärck,J.V.(1962) Immunity and host response in the growth of transplanted tumors.Munksgaard,Copenhagen.

5. Spärck,J.V. (1969) Acta path.microbiol.scand.77,1 .

6. Prehn,R.T.(1969) J.nat.Cancer Inst. 43,1215.

7. Prehn,R.T. (1972 ) Science 176,17o.

8. Treves,A.J.et al. (1974) Eur.J.Immunol. 4,722.

9. Heppner,G.H. et al.(1968) Israel J.Med . Sci. 4,1195.

1o. Balner,H.and Dersjant,H. (1966) J.nat.Cancer Inst. 36,513.

11. Billingham,R.E. et al.(1951) Brit.J.Cancer 5,417.

Progress in Differentiation Research, ed. N. Müller-Bérat et al. 


# CHANGES IN DIFFERENTIATION OF THE URINARY BLADDER DURING BENIGN AND NEOPLASTIC HYPERPLASIA

R.M.Hicks

School of Pathology, Middlesex Hospital
Medical School,London W1P 7LD

## 1. Introduction

The mammalian urinary bladder is lined by a 3 cell thick highly differentiated epithelium, the urothelium, in which morphogenesis and cytodifferentiation is complex but orderly. The cells increase in size and ploidy from basal to superficial layers. The small basal cells of the germinal layer are diploid, the intermediate cells tetraploid and the nuclei of the frequently multinucleate very large superficial cells are octaploid or higher[1]. It appears that cells from the germinal layer elongate, project above the basal layer and then fuse to produce the intermediate cells[2]. After cell fusion, the nuclei may remain separate or combine to form a tetraploid nucleus. The process is repeated, thus producing the large polyploid cells of the superficial layer. Since these cells are formed primarily by a process of fusion, the initial ratio of cytoplasm to nucleus is maintained. In the normal urothelium, the final stage of gene expression which is revealed in the cytodifferentiation of the superficial cells results in the production of a remarkable plasma membrane on the free luminal surface, which acts as a permeability barrier between the blood and the urine[3,4,5]. This membrane has a unique substructure composed of plaques of hexagonally arranged subunits separated by thinner, unstructured hinge regions[6-9]. The metabolism of these cells is orientated round the turnover of this membrane[10], and its presence is a convenient marker of normal differentiation. The superficial cells have an exceptionally long life span, estimated at 200 days or longer[11], and it takes more than 11 weeks for a cell to migrate from the basal to the superficial cell layer[2]. There is thus plenty of time for progressive and orderly changes in gene expression and for the maturation of the barrier membrane.

After cytotoxic or physical damage to the urothelium, there is a simple, compensatory hyperplasia. The basal stem cells proliferate rapidly resulting in a multilayered epithelium of small, diploid cells in which there has been insufficient time for the normal orderly pattern of morphogenesis and cytodifferentiation to develop. The determination of the tissue is still normal, and after a few weeks, the usual, polyploid pattern is re-established as surplus cells die and desquamate, and the remaining cells fuse and mature. But during the few weeks it takes for this to happen, the epithelium appears to be undifferentiated. The individual cells, however, are not de-differentiated but are more correctly regarded as immature.

The urothelial stem cells also have the potential for an alternative form of development. The morphogenesis and cytodifferentiation of the epithelium can be modified as the result of a change in cell determination brought about by epigenetic factors, for example by the Vitamin A status of the animal. In common with other, normally non-cornifying mucosae, including those of the oesophagus, mouth, salivary gland ducts, trachea and vagina, the urothelium in animals maintained on a Vitamin A-deficient diet undergoes squamous metaplasia with cornification[12,13,14]. The effect on the bladder epithelium of reducing the Vitamin A level in the animal is to switch morphogenesis from the pattern of normal urothelium to that of normal epidermis. Cell determination is still for differentiation then death, but the tissue type produced is inappropriate for the anatomical situation.

During the development of bladder cancer, cell differentiation in the early stages of neoplastic hyperplasia is difficult to distinguish from that of benign hyperplasia. In this paper, some of the similarities and differences are illustrated and discussed. Squamous metaplasia is also a feature of many human and experimentally induced animal bladder tumours, and its development in tumours is considered here in terms of cell differentiation.

## 2. Materials and methods

Specific pathogen free Wistar rats were used throughout.

Benign hyperplasia of the urothelium was induced either by a single, intravesicular dose of 2 mg N-methyl-N-nitrosourea (MNU)[15], or by a single i.p. injection of 200 mg/Kg.b.wt. of cyclophosphamide.

Neoplastic hyperplasia of the urothelium was induced either by 4 intravesicular doses of 1.5 mg MNU[15], or by a single dose of MNU followed by a saccharin-containing or cyclamate-containing diet[16], or by weekly s.c. injections of 200 mg/Kg.b.wt. of dibutylnitrosamine or by including nitrofurylthiazolyl formamide as 0.18% of the diet.

For comparison, untreated rats or animals maintained on a Vitamin A-deficient diet[14] were used.

Bladders were fixed in osmium tetroxide and embedded in Epon. 2μ sections stained with toluidine blue were used for histology and thin sections, contrast stained with lead and uranyl salts, were examined by electron microscopy.

## 3. Results and Discussion

### a. Squamous metaplasia

Epidermalisation of the urothelium induced by Vitamin A-deficiency results in an increase in the number of cell layers, formation of a characteristic "prickle cell" layer with interdigitating processes and desmosome attachments, and development of a stratum granulosum and stratum corneum[14]. The keratohyalin granules formed have identical staining properties to those of normal epidermis and oral mucosa and the stratum corneum is composed of birefringent fibrils[17]. In fact, in every way except for the absence of secondary appendages such as hair and sweat glands, it appears to have transformed from urothelium to epidermis. Like other Vitamin A-sensitive epithelia, the urothelium will also cornify in response to irritation. Thus a urinary calculus or persistent vesicular bacterial infection will produce leukoplakia, even when the Vitamin A supply of the animal is normal.

Squamous metaplasia is also frequently observed in the neoplastic bladder epithelium. In man, squamous metaplasia in bladder cancer is relatively rare in Europe and America, but common in the Middle East where it is usually superimposed on bilharzial cystitis. In the latter situation, local irritation of the epithelium by the bilharzial ova is probably responsible for the squamous metaplasia and cornification of the epithelium, and this condition may persist for many years before malignant growth patterns develop. In experimentally induced animal tumours, squamous metaplasia almost always develops as the tumours increase in size and complexity[18,19]. In such tumours, the foci of cornification start either at the free surface of an area of transitional cell hyperplasia or in the centre of the hyperplastic nodule of transitional cells (fig.1). Always, the cornifying area is at some distance from the blood capillaries of the underlying connective tissue, and probably the process starts

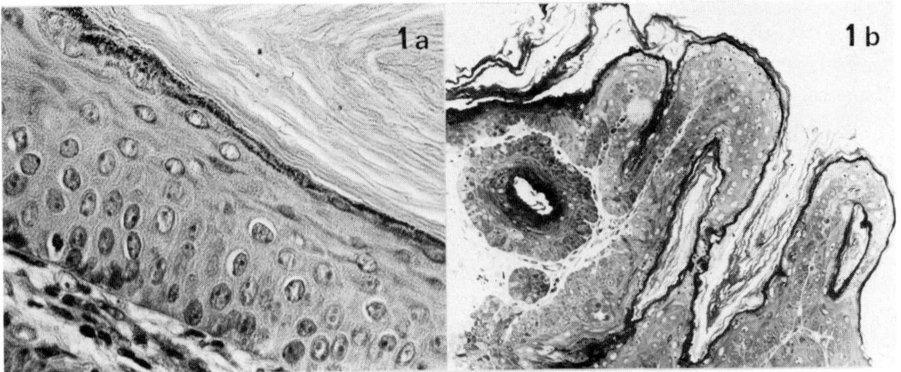


Fig. 1a. Bladder epithelium from rat maintained for 37 weeks on a Vitamin A-deficient diet. The epithelium has the characteristics of normal epidermis, including a stratum granulosum and stratum corneum.

Fig. 1b. Squamous metaplasia in bladder tumour of rat maintained 20 weeks on a cyclamate-containing diet after receiving a single dose of MNU. Keratin plaques have formed on the surface of the epithelium and within an epithelial nodule.

as a consequence of nutritional imbalance in the cell rather than as a consequence of the neoplastic transformation *per se*. Clinical observation confirms that tumours associated with leukoplakia have a relatively poor blood supply while transitional cell tumours of the bladder tend to be very well vascularised.

The subcellular changes which occur during squamous metaplasia in neoplastic epithelia are the same as in the Vitamin A-deficient urothelium. The two main features of the keratinisation process are first, the synthesis of $\alpha$-keratin filaments and their aggregation into tonofibrils (fig.2), and second, the formation of keratohyalin granules (fig.3). Although these two processes are normally concurrent in keratinising epithelia, in the cornifying bladder epithelium both processes can be observed and may proceed concurrently or independently[14]. The normal urothelial cells also contain fine, approximately 8nm diameter filaments, which appear to be identical to those found in the basal cells of skin and other keratinising tissues such as the oesophagus and oral mucosa. In skin these cytoplasmic filaments are believed to be the direct precursors of keratin fibrils of the stratum corneum. The cells of the cornifying bladder also contain cytoplasmic filaments with the same

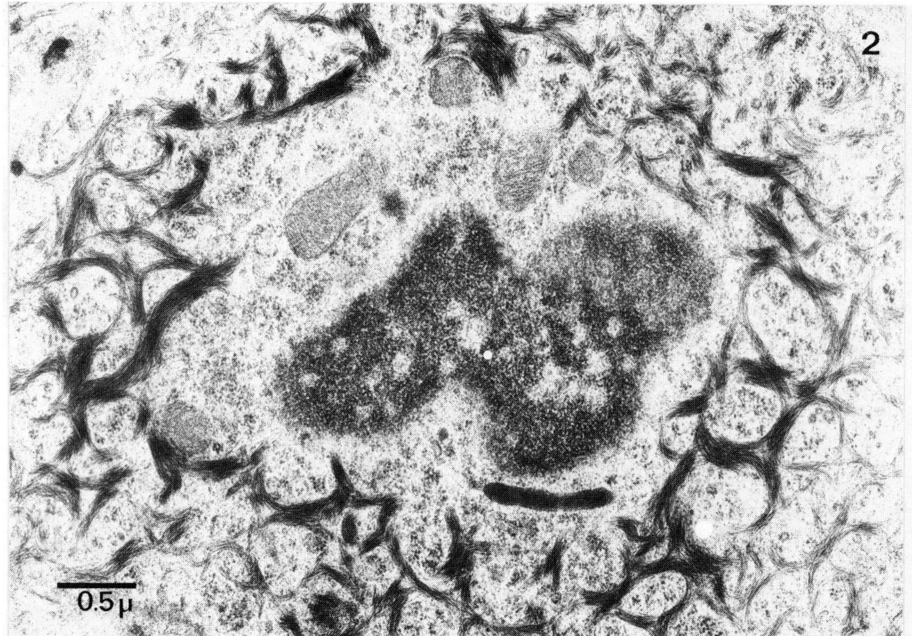


Fig. 2. Tonofibrils forming an interlaced mesh around the nucleus of a cell from an area of squamous metaplasia in a bladder tumour induced by 23 weeks treatment with dibutyl nitrosamine.

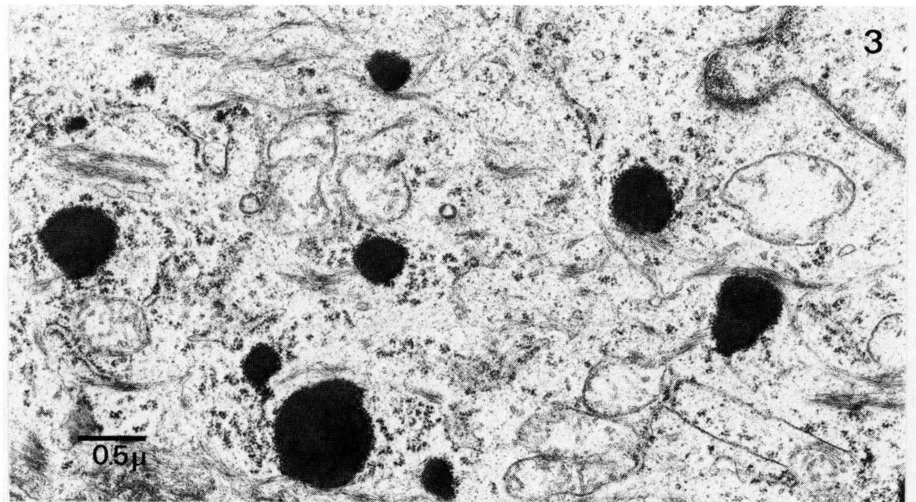


Fig. 3. Keratohyalin granules in the stratum corneum of a keratinising bladder tumour induced by 4 doses of MNU.

appearance as those in the normal cells, but present in much larger numbers. These filaments aggregate to form the tonofibrils of the "prickle cell" layer, and subsequently the birefringent fibrils of the stratum corneum, in an exactly analogous manner to the formation of the stratum corneum in skin. There is no reason to suppose, however, that the chemical nature of the cytoplasmic filaments in the cornifying epithelium differs from that of the filaments in the normal urothelium[14]. Similarly, the keratohyalin granules which are formed in the cornifying epithelium, although they appear to be a new morphological structure, in fact derive from aggregates of ribosomes which become progressively more condensed until their morphology changes from a granular to a dense amorphous form[14]. Again there is no reason to propose a de novo synthesis of new molecules in the cornifying epithelium.

Whether squamous metaplasia represents a complete switch in cytodifferentiation or whether it is simply an expression of hyperactivity of existing cell mechanisms is an interesting theoretical consideration. If the cytodifferentiation has completely changed, then the appearance of keratohyalin granules and the stratum corneum presumably should reflect the production of new molecular species as a result of an altered reading of the genetic code. If, on the other hand, it represents hyperactivity of existing synthetic pathways, no new molecular species needs to be produced which is not also present in the normal urothelium, but the quantitative production of certain molecules may be increased. In my opinion, this change in morphogenesis is more logically regarded as a quantitative than a qualitative phenomenom; it appears to be a non-specific response to stress involving normal cell synthetic pathways rather than true metaplasia, or a qualitative change in gene expression. Nevertheless squamous metaplasia in the bladder is still regarded by many pathologists as an integral part of the neoplastic process. More logically, it would seem that a percentage of cells opt out of the mitotic cycle to differentiate and die, and that although the course of that differentiation is determined, it may be influenced by epigenetic factors such as the position of the cells in a nutritional gradient.

b). Differentiation in benign hyperplasia and preneoplastic hyperplasia

In the compensatory response to chemical or physical traumatisation of the urothelium, the remaining basal, or stem cells, divide

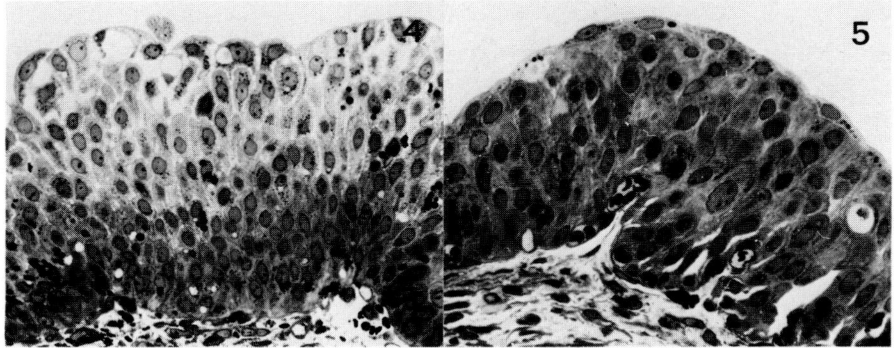


Fig. 4. Benign hyperplasia of the urothelium 2 weeks after a single dose of MNU.

Fig. 5. Premalignant hyperplasia of the bladder epithelium after 20 weeks treatment with dibutyl nitrosamine.

rapidly to produce a multilayered hyperplastic epithelium of immature, diploid, undifferentiated cells (fig.4). The hyperplastic epithelium produced in response to treatment with chemical carcinogens is also composed of immature, undifferentiated, diploid cells (fig.5), which suggests that the effective target cells for the carcinogens are the diploid, basal cells of the normal urothelium. In theory, even the highly differentiated superficial cells of the urothelium could be target cells, for they retain their ability to divide even though they do so rarely under normal circumstances. Indeed, clones of cells producing barrier membrane are occasionally found deep within tumours (fig.6), and these may derive from transformed superficial cells in which the synthetic mechanisms for specialisation have not been completely suppressed in favour of growth and division. In general, however, the preneoplastic epithelium is composed of undifferentiated cells and there is no way of predicting by histological examination that after some 30 weeks, the benign hyperplasia of Fig.4 will have regressed and been replaced by a normally differentiated urothelium (fig.7)while the preneoplastic epithelium of Fig.5, will have progressed to form a tumour (fig.8).

At first sight examination of the subcellular changes is also unrewarding, for in both conditions similar changes in cytoplasmic structure can be observed. These include increase in numbers of cytoplasmic filaments and ribosomes, changes in the numbers of lysosomes, changes in nuclear morphology and projections of basal

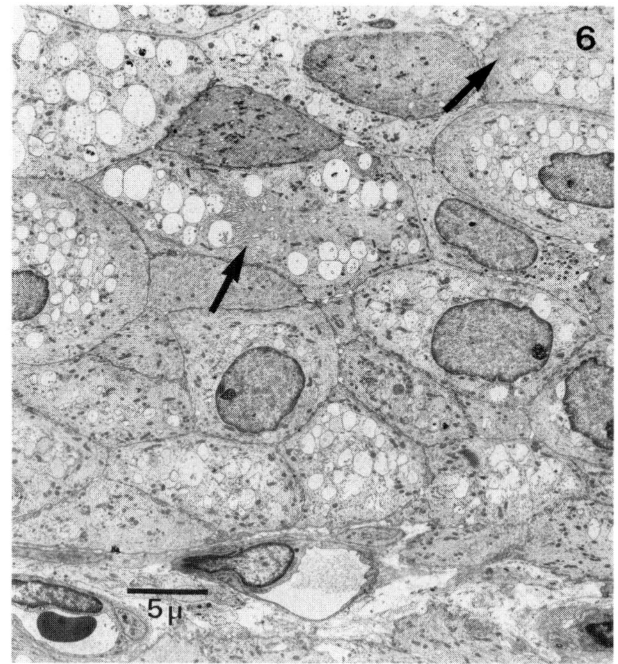


Fig. 6. Cells producing vesicles of differentiated barrier membrane (arrows) deep within a bladder tumour after 34 weeks treatment with dibutyl nitrosamine.

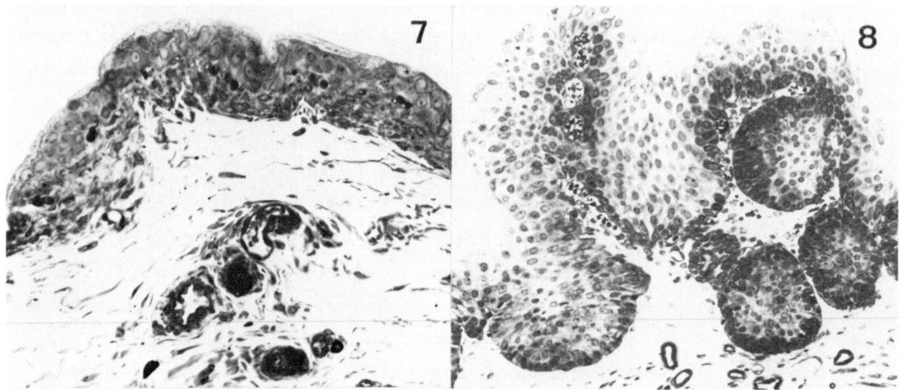


Fig. 7. Normally differentiated urothelium, re-established 30 weeks after a single dose of MNU cf. Fig.4.

Fig. 8. Ingrowths of epithelium in early stage of tumour development after 30 weeks treatment with dibutyl nitrosamine cf. Fig.5.

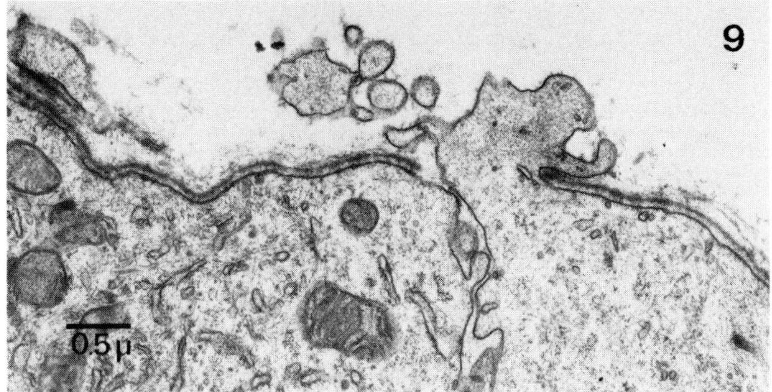


Fig. 9. "Microinvasion" i.e. extension of a process from a basal cell through a break in the basal lamina, in the epithelium of a bladder four days after treatment of the rat with the non-carcinogenic, cytotoxic agent, cyclophosphamide.

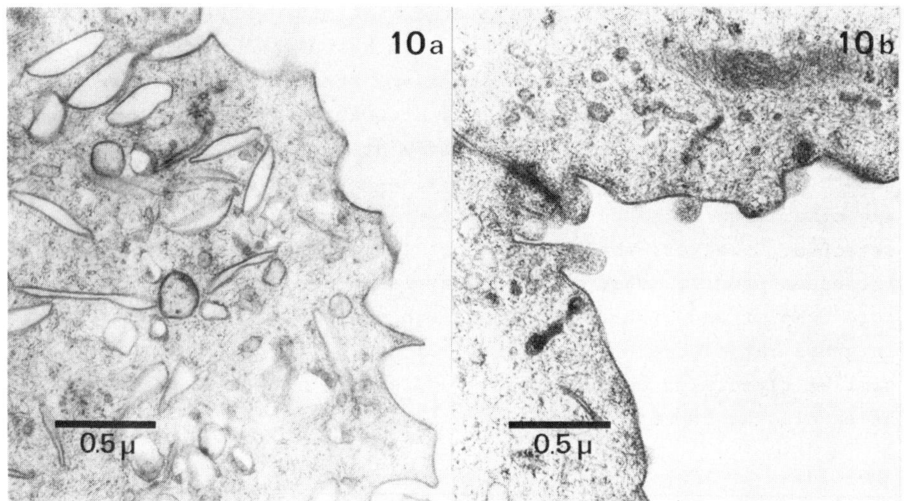


Fig.10a. Part of the free edge of a normal superficial cell from rat urothelium showing the normal, differentiated, angular barrier membrane which limits the luminal face of the cell and also the cytoplasmic fusiform vesicles.

Fig.10b. Undifferentiated membrane limiting the luminal face of the most superficial layer of cells in reversible, benign hyperplasia of the urothelium induced by a single dose of MNU.

cell cytoplasm through breaks in the basal lamina into the mesenchyme[20]. The last, sometimes termed microinvasion, was most frequently seen in the reversible benign hyperplasia which followed cyclophosphamide treatment (Fig.9). However, detailed examination of the most superficial layer of cells, revealed a regular morphological change in the structure of the cell membrane adjacent to the urine in the preneoplastic epithelium, which has not, so far, been observed in benign hyperplasia[5,21]. The normal specialised barrier membrane of the luminal face of the epithelium is not produced in either benign or preneoplastic hyperplasia. In benign hyperplasia, the luminal face is limited by a thinner, more flexible membrane with a thin, amorphous glycocalyx (Fig.10). This membrane appears morphologically the same as that surrounding the basal and intermediate cells of the normal urothelium, which, since the hyperplastic epithelium is composed of essentially normal undifferentiated cells derived from the normal basal layer, is only to be expected. By contrast some, but by no means all, of the surface cells in the various preneoplastic epithelia studied show an entirely new pattern of differentiation of the surface membrane. The surface of these cells is covered with microvilli (Fig.11) and is limited by a flexible membrane which has a characteristic glycocalyx (Fig.12) which is more complex and structured than that seen in simple, benign hyperplasia. This structured glycocalyx can be found on a few cells in the very early stages of preneoplastic growth, in one case within 8 weeks of starting treatment, before any other sign of neoplastic transformation could be histologically detected. Moreover, the number of cells carrying the glycocalyx increases progressively as the hyperplasia progresses to become more complex and bizarre in its growth pattern. It is also present in gross rat bladder tumours (Fig.13) and we have also observed a similar glycocalyx on the surface cells of a number of biopsies of human bladder tumours.

---

Fig.11a. Appearance of the luminal face of superficial cells in a normal, dilated rat bladder as observed with a scanning electron microscope.

Fig.11b. The luminal face of the superficial cells of a transitional cell tumour. The cells are smaller than normal, and covered with microvilli.

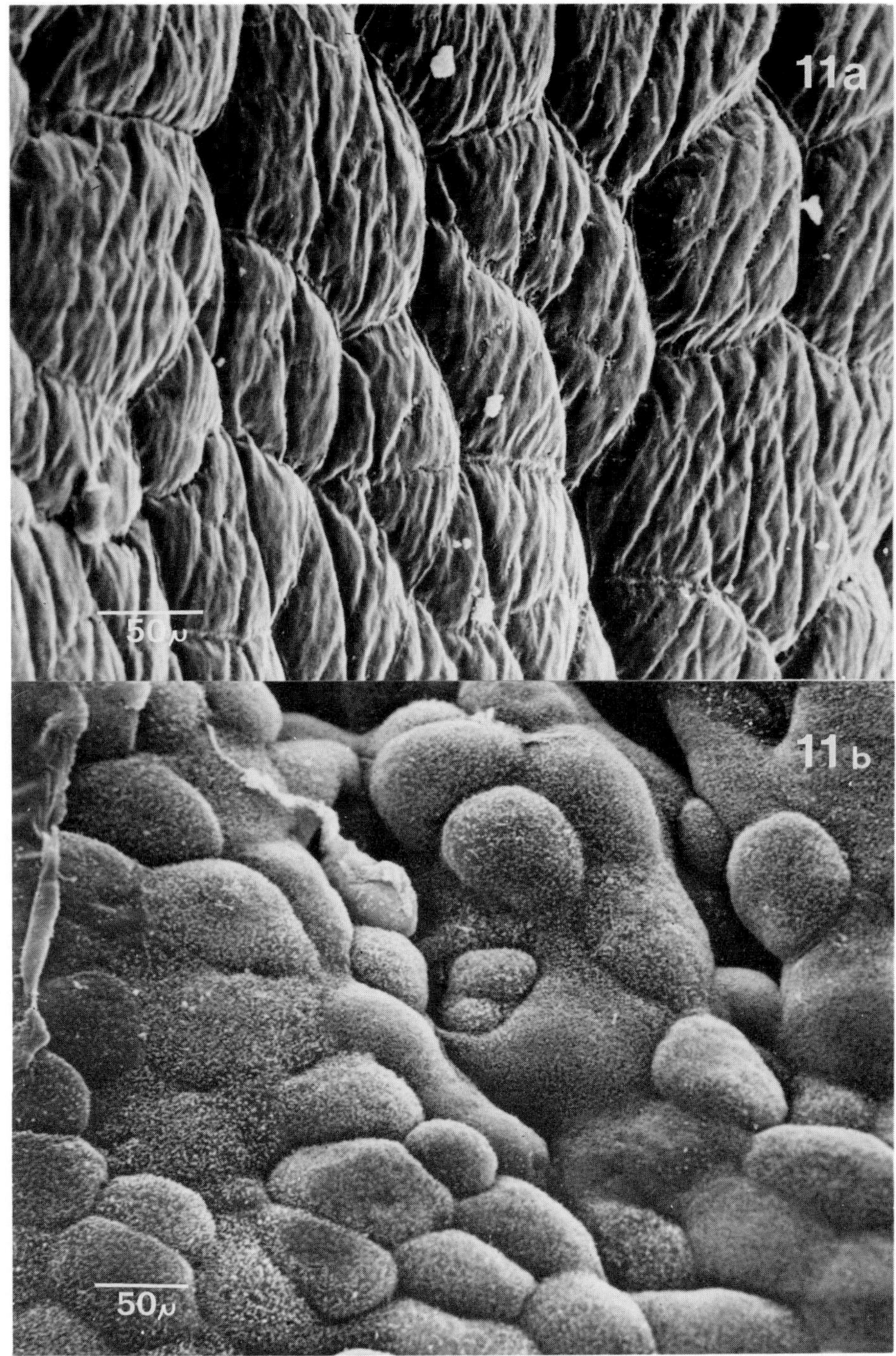
11a
50μ
11b
50μ

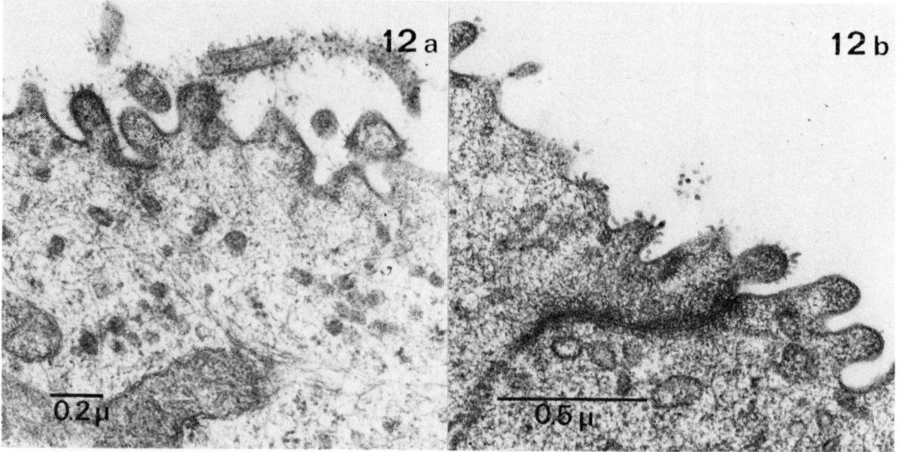


Fig.12a. Structured glycocalyx on the membrane limiting the microvilli on the free edge of a transitional cell rat bladder tumour, induced by multiple doses of MNU.

Fig.12b. Structured glycocalyx on the luminal membrane of cell in a preneoplastic epithelium, only 6 weeks after starting treatment with MNU.

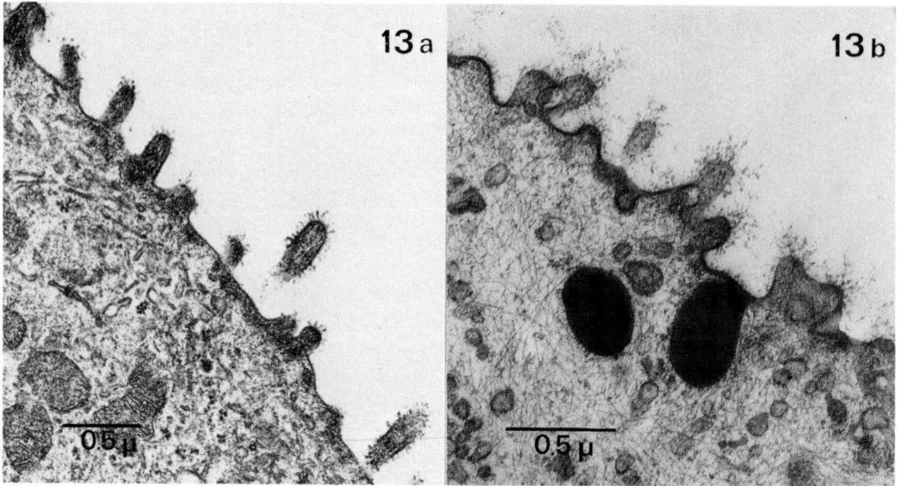


Fig.13a. Structured glycocalyx on cell from a gross transitional cell tumour of the rat bladder induced by 40 weeks treatment with dibutyl nitrosamine.

Fig.13b. Structured glycocalyx on superficial cell membrane of human transitional cell carcinoma of the bladder. Surgical specimen.

This glycocalyx has never been observed in the normal urothelium, or in benign hyperplasia, and thus represents an entirely new pattern of gene expression. It is more likely that this is achieved by derepression of a normally repressed area of the code, rather than by a fundamental alteration in the genetic information, since the same type of glycocalyx is produced in animal and human tumours, irrespective of the carcinogenic stimulus involved. Furthermore, the work of the Hellströms[22] indicates that, unlike other animal tumours, induced rodent bladder tumours produce cross reactive tissue type antigens. This suggests an alteration in surface glycolipids and/or glycoproteins which is common to all bladder tumours, which we may be observing as the production of this unusual glycocalyx.

The ultimate distinguishing mark of the neoplastic cell is not its structure, but its determination for unchecked growth and division. For a limited period of time this is also true of the urothelial cells involved in the compensatory hyperplasia following physical or chemical traumatisation of the epithelium. For cell growth and division, different metabolic products are required from those produced during specialised differentiation. It can be assumed that in benign hyperplasia there is merely a temporary switch in the pattern of gene expression, for the hyperplasia is completely reversible and after a period of days or weeks the epithelium again responds to growth regulating factors and the metabolism leading to normal specialisation of the urothelium is resumed. In theory, neoplastic growth could be considered in the same way, as a change in gene expression due to persistent, instead of transitory, activation of the biosynthetic systems required for growth and division which are present in the normal cell but are usually repressed by growth regulating factors. If this were the only cause of neoplasia, then the cells of early neoplastic hyperplasia would be indistinguishable from those of benign hyperplasia. However, this cannot be so, for the neoplastic cells are unable to respond to growth regulating factors which may be chalones[23], hormones[24], or proteins[25]. At least some of these growth regulating factors need to react with specific receptor sites on the cell membrane before they can be effective[26]. The most likely receptors in the cell membrane are the complex glycoproteins and glycolipids with their carbohydrate moieties exposed on the external face of the membrane. There is thus good reason to believe that the unusual glycocalyx seen on the surface of the transformed cells of the preneoplastic and neoplastic bladder epithelium represents a significant change in cell structure and that it may be a marker for irreversible neoplastic transformation in this tissue.

Clearly, much work is needed to confirm this hypothesis. It is tempting to assume that this glycocalyx is the carrier or morphological counterpart of the tissue specific, tumour antigen produced by both human and animal tumours. This possibility is being investigated further and an attempt is also being made to compare the chemical nature of the carbohydrate-containing, lectin binding sites in the normal urothelium with those in preneoplastic hyper-plasia and in fully developed tumours.

Acknowledgments

This work has been done over a period of many years, during which time I have been generously supported by the Cancer Research Campaign. I am also indebted to J.St.J.Wakefield and J.Chowaniec for their collaboration in the work on experimental carcinogenesis.

References

1. Levi, P.E., Cowan,D.M., & Cooper, E.H., 1969, Cell Tissue Kinet. 2, 249.
2. Martin, B.F., 1972, J.Anat.Lond., 112,433.
3. Hicks, R.M., 1965, J.Cell Biol., 26,25.
4, Hicks,R.M., 1966, J.Cell Biol., 28,21.
5. Hicks,R.M., Ketterer, B., & Warren,R.C., 1974., Phil.Trans. R.Soc.Lond. B., 268, 23.
6. Hicks, R.M. & Ketterer, B., 1969, Nature, Lond., 224,1034.
7. Vergara, J., Longley, W., & Robertson,J.D., 1969,J.Molec. Biol., 46,593.
8. Warren, R.C., & Hicks, R.M., 1970, Nature, Lond., 227, 280.
9. Warren, R.C., & Hicks, R.M., 1973, Micron, 4, 257.
10. Hicks, R.M., 1966, J.Cell Biol., 30,623.
11. Cooper E.H.,1972, Ann.Roy.Coll.Surg.Eng., 51,1.
12. Wolbach, G.B., & Howe, P.R., 1925, J.Exptl.Med., 47, 753.
13. Capurro, P., Angrist, A., Black, J., & Moumgis, B., 1960, Cancer Res., 20, 563.
14. Hicks, R.M., 1968, J.Ultrastruct. Res., 22, 206.
15. Hicks, R.M., & Wakefield, J.St.J., 1972, Chem.-Biol.Interactions, 5, 139.
16. Hicks, R.M., Wakefield,J.St.J., & Chowaniec, J., 1975, Chem.-Biol. Interactions, 11,225.

17. Hicks, R.M., 1969, J.Anat. Lond., 104,327.

18. Hicks, R.M., & Wakefield, J.St.J., 1975, in: Scientific Foundations of Urology, eds. D.Innes Williams & G.D.Chisholm (W.Heinemann Medical Books Ltd) in press.

19. Hicks, R.M., Wakefield, J.St.J., Vlasov, N.N., & Pliss, G.B., 1975, in: Pathology of Tumours in Laboratory Animals, Vol.1, part.2, ed. V.Turusov (Lyon: International Agency for Research on Cancer) in press.

20. Wakefield, J.St.J., & Hicks, R.M., 1973, Chem.-Biol.Interactions, 7, 165.

21. Hicks, R.M., 1973 Il Cancro, 26, 193.

22. Taranger, L.A., Chapman, W.H., Hellström, I., & Hellström, K.E., 1972, Science, 176, 1337.

23. Bullough, W.S., & Laurence, E.B., 1960, Proc.Roy. Soc.Lond.B., 151, 517.

24. Tata, J.R., 1970, in : Biochemical Actions of Hormones, Vol.1, ed. G.Litwack, (Academic Press, New York) p.89.

25. Benbow, R., & Ford, C., 1975, Proc.Natl.Acad.Sci. U.S.A., 72, 2437.

26. Brachet, J., Baltus, E., Hanocq-Quertier, J., Hubert, E., Pays, A., Steinert, G., & Wiblet, M., 1975, Abstracts of 2nd Int.Conf. on Differentiation, Copenhagen, Sept. 8-12,1975, p.77.

Progress in Differentiation Research, ed. N. Müller-Bérat et al.



# DEDIFFERENTIATION ASSOCIATED WITH CELL-TYPE CONVERSION IN THE NEWT LENS REGENERATING SYSTEM : A REVIEW

Tuneo Yamada
Institut Suisse de Recherches Expérimentales sur le Cancer,
Lausanne, Switzerland

## 1. Introduction

Cell dedifferentiation is an area which has remained obscure, inspite of its potential importance in understanding the control mechanism of cell differentiation and carcinogenesis. Here I would propose to define cell dedifferentiation as a process by which cells in the differentiated state are released from the differentiative controls. From the view point that the process of differentiation is a bipartite developmental phenomenon in which determination precedes realization, a definite dedifferentiation should release cells from developmental determination as well as from structural and physiological control of the differentiated state. Under the influence of the generally accepted idea that the differentiated state is stable, there is a strong tendency to deny the possibility of dedifferentiation altogether. It is true that in most of metaplasia we are dealing with alteration of the future pathway of specific stem cells destined to produce one particular cell-type, and not with dedifferentiation above defined. The same holds for transdetermination. However, there is some experimental evidence available for reality of dedifferentiation by demonstrating that cells which have attained the differentiated state can lose the state and engage in a new differentiation pathway.[1,2,3,4,5] This article reviews recent data on dedifferentiation of iris epithelial cells, which provides the base of lens regeneration in adult newts.

## 2. General characterization of the system

When the lens is removed from the eye of adult newts, a new lens is formed by iris epithelial cells (I E C)[5,6,7,8]. The I E C are fully differentiated melanocytes and non-dividing in the normal adult. No stem cells are present for I E C[9]. Lentectomy puts I E C in the cell cycle. In a fraction of the proliferating I E C, melanosomes which characterize the normal condition of I E C completely disappear. Subsequently lens specific proteins accumulate in the cytoplasm, and differentiation of lens fiber proceeds as the cells retreat from the cell cycle. The whole process starting after lentectomy and terminating with formation of lens fiber cells can be called cell-type conversion. This self explanatory expression is preferred to the older expression, cellular metaplasia, because of ambiguities attached to the concept of metaplasia. Since overt differentiation is lost simultaneously with disappearance of melanosomes, and since lens differentiation by progenies of I E C during lens regeneration in situ or under various experimental conditions only occurs after complete loss of melanosomes, it is proposed that loss of melanosomes correlates with dedifferentiation. One important aspect of this system is that formation of the regenerated lens is a recapitulation of ontogenesis of the lens from its start in the tail-bud embryo. This would suggest that after dedifferentiation the adult I E C recover an embryonic condition, and rechannelling of cells in the new differentiation pathway requires retrogenesis of cells.

3. Pathways of dedifferentiation and the cell cycle time.

In lens regeneration in situ some of I E C enter into the cell cycle subsequent to lentectomy. Only a part of proliferating I E C participate in cell - type conversion after complete loss of melanosomes (the pathway of conversion). The rest of I E C in proliferation lose melanosomes partially, retreat from the cell cycle, resynthesize melanosomes, and revert to the original condition of I E C (the pathway of retrieval). A study of cell cycle parameters of those two pathways in situ has been done utilizing a computerized mathematical treatment[10]. In the pathway of conversion, the estimated means of S-, G2-, and G1-phases are 27 hr, 8 hr, and 11 hr respectively with the total cell cycle time of 46 hr. In this computation, the M-phase is divided into G1 and G2. In the pathway of retrieval, the corresponding values are 40 hr, 8 hr, and 30 hr respectively with the total cell cycle time of 79 hr. Thus there is a considerable difference in the total cell cycle time between the two pathways, which is significant at the 95% confidence limits. By combining those data with the known time sequence of cellular events in lens regeneration, it is suggested that the minimum number of cell cycles required for definite dedifferentiation in the pathway of conversion is 4, and that two more cell cycles are passed in the minimum before the cells start lens fiber differentiation. The cell cycle parameters of cultured I E C are similar to those of I E C in situ in the pathway of retrieval[11].

The above data open up the possibility that dedifferentiation of I E C is controlled by the cell cycle time. The notion is in accordance with the studies of the effect of X-radiation of the system in situ and in vitro, which indicates that inhibition of lens regeneration by irradiation is due to suppression of replication of I E C which is accompanied by inhibition of depigmentation[12].

4. Discharge of melanosomes.

The close association of loss of melanosomes by I E C with their dedifferentiation suggests that the basic mechanism of dedifferentiation is related to disappearance of melanosomes or connected with the process of their disappearance. Whether one subscribes to the hypothesis or not, it appears meaningful to understand how I E C become depigmented. The available data show that melanosomes are not synthesized in the pathway of conversion, and that dilution of melanosomes from I E C by cell replication and discharge of melanosomes from I E C are both instrumental in depigmentation. That melanosomes are discharged from I E C during lens regeneration has been observed already by Wolff in his classical paper[13], and later confirmed by Eguchi with electron microscopy[14]. A later electron microscopic investigation shows that the main mode of melanosome discharge is by a membrane-bound complex, in which melanosomes are accompanied by other organelles like mitochondria and ribosomes[15]. Extending the observation to the lens regenerating system in vitro, it has been shown that the membrane-bound complex is originated in the cytoplasm as a field of condensation of melanosomes which become invested with membrane. The membrane-bound unit moves to the cell periphery and is discharged[16]. There is a tendency that mitochondria and other organelles inclosed in the complex undergo degeneration during the transport. In the _in situ_ condition, the discharged complex is taken up by macrophages which infiltrate into the iris epithelium in a large number[17]. Enhancement of the activity of glucosaminidase, a lysosomal enzyme, in the I E C in dedifferentiation in situ and in the ocular fluid of the regenerating eyes is in accord with the notion that lysosomes are involved in the dedifferentiative process[18].

The formation of melanosome concentrates within the cytoplasm and subsequent discharge, concluded from electron microscopic observations of thin sections, has been later confirmed by the observations on the living I E C in primary culture[19].

5. The possible role of the cyclic AMP level and its control by adenosine.

In the sparce primary culture of I E C, the majority of cells assume flat discoidal shape with more or less extended ondulating membrane. When 0.01 - 5.00 mM adenosine is added to culture medium, in some of those cells the membranous cytoplasm is transformed into an array of branching strands emanating from the central nuclear area[20]. Such condition called the stellate configuration is expressed maximally around 80 min. after the start of the treatment. Subsequently the cells go back to the original configuration in the presence of adenosine. 5'-AMP has the same effect, but adenine, guanosine, uridine, and cytidine are ineffective. Cyclic AMP, dibutyryl and monobutyryl cyclic AMP, and theophylline are effective but only in the concentration range higher than that of adenosine. On the other hand, the derivatives of cyclic GMP, cyclic UMP or cyclic IMP do not have the effect. From those results it has been suggested that formation of the stellate configuration is due to a rise in the intracellular level of cyclic AMP, which can be caused by exogenous cyclic AMP, its derivatives, and theophylline or more effectively by exogenous adenosine and 5'-AMP. This idea has been now supported by synergistic effects of adenosine combined with cyclic AMP or with theophylline[21]. As to be expected the combination of cyclic AMP and theophylline or that of dibutyryl cyclic AMP and theophylline shows synergism. A more direct support for the idea that the intracellular level of cyclic AMP is the decisive factor comes from the preliminary results of micro-injection of cyclic AMP into individual I E C in culture which causes formation of cell projections similar to those of the stellate configuration[22]. The morphogenesis after injection is completed within 20 to 30 min, while it takes 80 min after the adenosine treatment. Since the stellate configuration corresponds to the cell surface condition of I E C in dedifferentiation in situ[15], the possibility has been suggested that during dedifferentiation a variation of the cyclic AMP level occurs in individual I E C and causes the specific structural alteration. There are a number of data available suggesting that the alteration is associated with discharge of melanosomes and other cytoplasmic organelles[23]. Whether extracellular adenosine or 5'-AMP controls the level of cyclic AMP in I E C during lens regeneration in situ remains to be studied.

## 6. Concluding remarks.

Other important parameters of dedifferentiation in the present system appears to be physical and chemical properties of cell surface. Since this is covered by a report by Dr. S.E. Zalik in this Conference, it would suffice to mention that sequential alterations in physical and chemical properties of cell surface occur in close temporal correlation with the cell cycle, structural alteration of cell surface, and depigmentation. How those events are causally related with each other and what are their roles in cell dedifferentiation are the questions posed for future research.

References

1. Eguchi, G. and Okada, T.S., 1973, Proc. Nat. Acad. Sci. U. S. 70, 1495.

2. Eguchi, G., Abe, S., and Watanabe, K., 1974, Proc. Nat. Acad. Sci. U. S. 71, 5052.

3. Connelly, T.G., Ortiz, J.R. and Yamada, T., 1973, Develop. Biol. 31, 301.

4. Yamada, T., Reese, D.H. and McDevitt, D.S., 1973, Differentiation 1, 65.

5. Yamada, T. and McDevitt, D.S., 1974, Develop. Biol. 38, 104.

6. Reyer, R.W., 1954, Quart. Rev. Biol. 29, 1.

7. Yamada, T., 1967, Cur. Top. Develop. Biol. vol. 2, eds. A. Monroy and A.A. Moscona (Academic Press, New York) p. 247.

8. Yamada, T., 1972, Proceedings of the 1st International Conference on Cell Differentiation, eds. R. Harris, P. Allin, and D. Viza (Munksgaard, Copenhagen) p. 56.

9. Yamada, T. and Roesel, M.E., 1971, J. Exp. Zool. 171, 119.

10. Yamada, T., Roesel, M.E. and Beauchamp, J.J., in press, J. Embryol. Exp. Morphol.

11. Horstman, L.P. and Zalik, S.E., 1974, Exp. Cell Res. 84, 1.

12. Michel, C. and Yamada, T., 1974, Differentiation 2, 193.

13. Wolff, G., 1895, Arch. Entwicklungsmech. Organismen 1, 380.

14. Eguchi, G., 1963, Embryologia 8, 47.

15. Dumont, J.N. and Yamada, T., 1972, Develop. Biol. 29, 385.

16. Dumont, J.N. and Yamada, T. unpublished.

17. Yamada, T. and Dumont, J.N., 1972, J. Morphol. 136, 367.

18. Idoyaga-Vargas, V. and Yamada, T., 1974, Differentiation 2, 91.

19. Yamada, T., Ortiz, J.R. and Michel, C., unpublished.

20. Ortiz, J.R., Yamada, T. and Hsie, A.W., 1973, Proc. Nat. Acad. Sci. US. 70, 2286.

21. Ortiz, J.R. and Yamada, T., in press, Differentiation.

22. Yamada, T., unpublished.

23. Ortiz, J.R. and Yamada, T., unpublished.

*Progress in Differentiation Research, ed. N. Müller-Bérat et al.*



# CHANGES AT THE CELL SURFACE DURING *IN VIVO* AND *IN VITRO* DEDIFFERENTIATION IN CELLULAR METAPLASIA

Sara E. Zalik, Vi Scott and Eva Dimitrov

Department of Zoology
University of Alberta
Edmonton, Alberta, Canada T6G 2E9

In some species of urodeles, the phenomenon of lens regeneration occurs. In this process, after removal of the lens from the eye, the pigmented cells of the dorsal iris give rise to a new lens. Lens regeneration involves a metaplastic change in which the dorsal iris melanocytes dedifferentiate to give rise to a non-pigmented cell population, which then redifferentiates and forms a new lens. The question as to the origin of the lens from the pigmented iris cells has been supported by experimental evidence and has been reviewed by Yamada[1,2].

We have investigated whether or not changes at the cell surface can be detected during dedifferentiation of the pigmented iris cell. The surface markers in this system are not yet very specific; however, evidence collected so far supports the notion that changes at the cell surface occur during the dedifferentiation of the pigmented iris melanocyte.

The first approach in this study was to investigate changes in the cell surface charge density, as determined by cell electrophoresis, at different stages of lens regeneration. As is shown in fig. 1, the mean electrophoretic mobility of the dorsal iris cells decreases significantly at 10 days after lens removal. Regenerates removed at this time are composed of a cell population which has a predominance of dedifferentiated cells in which pigment granules are either absent or present in very small numbers in the cytoplasm[1].

Electrophoretic mobility (EPM) determinations give only estimates of the average cell surface charge density. In an effort to further characterize the observed changes in EPM we decided to study the effect of various enzymes which have been reported to induce a decrease in the EPM by removing certain charged groups from the cell periphery[3,4]. A series of experiments was performed in order to determine the effect of neuraminidase (from vibrio cholera), ribonuclease (pancreatic and $T_1$), and chondroitinase ABC, on the EPM of normal pigmented iris melanocytes. The results of these experiments are presented in table 1. It can be observed that the EPM of the normal pigmented iris cells is reduced significantly by prior treatment with neuraminidase, ribonuclease and chondroitinase. The effects of neuraminidase and ribonuclease on the EPM of normal iris cells are additive and this leads us to assume that these enzymes are removing different groups from the cell periphery[5]. The effects of

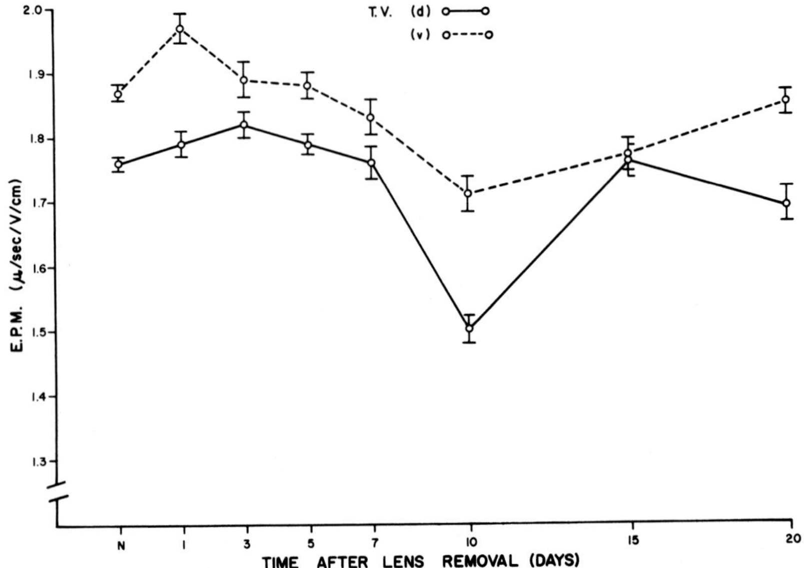


Fig. 1. Mean electrophoretic mobilities of dorsal and ventral iris cells from the normal eye (N) of *Notophthalmus viridescens* and regenerates at various time intervals after lens removal. Each point represents mean EPM and standard error. Average EPM values were obtained from 150 or more measurements.

ribonuclease and chondroitinase, however, are not additive, suggesting that these two enzymes may be removing similar groups from the cell periphery. We have interpreted these results as suggesting that ribonuclease-sensitive groups may be associated with glycosaminoglycan-like molecules[6].

It can also be observed in table 1 that cells obtained from regenerates 10 days after lentectomy are no longer affected by the above mentioned enzymes, a situation that is also present in cells from 15 day regenerates. These results lead us to suggest that neuraminidase, ribonuclease and chondroitinase-sensitive groups are either present in a masked form or absent from the cell periphery of the dedifferentiated iris cells.

A series of experiments was designed in order to determine if the above mentioned enzyme susceptible groups disappeared at random or in a sequential manner during the process of dedifferentiation. To elucidate this problem we examined enzyme effects on the EPM of cells from dorsal irises obtained at closer time intervals after lentectomy, when cells are either at the onset or involved in early dedifferentiation. The results of these experiments are presented in table 2 and in fig. 2. A decrease in EPM after treatment with any of the enzymes

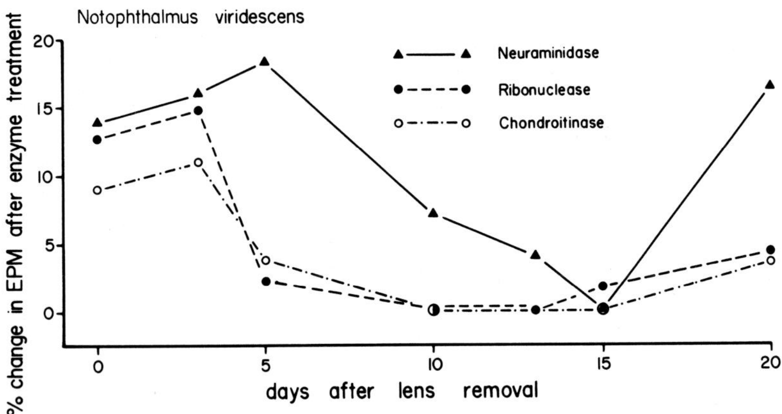


Fig. 2. Effect of neuraminidase, ribonuclease and chondroitinase on the electrophoretic mobilities of cells from dorsal iris and lens regenerates. The data show decrease (%) induced by these enzymes at different stages of regeneration. The values shown at 0 days represent the normal iris. In the majority of the experiments decreases in EPM of less than 8% are not significant as assessed by 't'-tests.

previously mentioned is observed in cells from regenerates at 3 and 4 days after lens removal. However, the mobilities of cells from 5 day regenerates are no longer affected by ribonuclease and chondroitinase, and a similar result is also obtained when cells are treated with testicular hyaluronidase[6]. The lack of effect of these enzymes on the cells' EPM continues until dedifferentiation has been completed (10 to 15 days after lens removal). Neuraminidase, however, decreases the EPM of cells obtained from regenerates at 7 to 10 days after lentectomy. These results suggest that the neuraminidase-sensitive groups become non detectable at the cell surface after the ribonuclease and chondroitinase-sensitive groups. An interpretation of these results is that ribonuclease and/or chondroitinase-sensitive groups become non detectable at the cell surface at the onset of dedifferentiation, when expansion of the intercellular space becomes evident[7]. On the other hand, the neuraminidase-sensitive component is absent from the surface ionic layer once the pigmented cell is completely dedifferentiated. It is interesting to note that melanosomes are removed from the cell surface of the dedifferentiating cell, together with fragments of the cell membrane with cortical cytoplasm[7]. The absence of certain charged groups from the periphery of the depigmented cell is supported by ultrastructural studies using colloidal iron (Yamada, personal communication), and lend support to the notion of selective shedding of the cell membrane during dedifferentiation[2] (Yamada, this symposium).

Table 1

Electrophoretic mobility of dorsal iris cells after different enzyme treatments

| Control | Neuraminidase | Ribonuclease | Chondroitinase |
|---|---|---|---|
| Normal Iris | | | |
| 1.660 | 1.472† | | |
| 1.809 | 1.480† | | 1.570* |
| 1.729 | | 1.495 (p)† | |
| 1.907 | 1.638†† | 1.58 (p)†† | |
| 1.857 | | 1.645 (p)† | |
| 1.803 | | | 1.632* |
| 1.806 | | | 1.566†† |
| 1.804 | 1.499 | 1.579 ($T_1$)* | |
| 1.809 | | 1.556 ($T_1$) | |
| 10 Day Regenerates | | | |
| 1.667 | 1.657 | | |
| 1.814 | 1.826 | | |
| 1.605 | | 1.618 (p) | |
| 1.827 | | 1.881 (p) | |
| 1.763 | 1.602 | 1.757 (p) | |
| 1.617 | | 1.507 ($T_1$) | |
| 1.728 | | | 1.709 |
| 1.812 | | | 1.823 |

Cells were obtained by mild trypsinization. Controls were placed in 0.118 M NaCl, pH 7.2, for the time interval of enzyme treatment. Each mean electrophoretic mobility ($\mu m\ sec^{-1}\ V^{-1}\ cm^{-1}$) represents an average of 30 to 40 determinations. Enzyme concentrations used were: neuraminidase, 20 units $ml^{-1}$; ribonuclease pancreatic (p) and $T_1$, 180 units $ml^{-1}$; chondroitinase, 0.5 units $ml^{-1}$. Probability values refer to the differences between control and a particular enzyme treatment. Where probability values are missing, the differences are not significant.

* $P < 0.05$ or less
† $P < 0.01$ or less
†† $P < 0.001$ or less

Changes at the cell surface have also been detected during depigmentation in culture. Under conditions of cell culture, one of the first observable events is the attachment of the iris melanocyte to the substrate, which occurs concurrently with the onset of depigmentation[8]. This process is followed by cell division and completion of depigmentation. Evidence from our laboratory (Zalik and Dimitrov, in preparation) and other investigators[9] have shown that this induced dedifferentiation can be followed by redifferentiation into lens protein producing cells *in vitro*. In order to investigate whether cell surface changes occur during *in vitro* dedifferentiation, cell suspensions of normal iris melanocytes were obtained and cultured for several time intervals. The effects of ribonuclease and neuraminidase on the EPM of normal and cultured cells were determined at subsequent time intervals. The data presented in table 3 show that the decrease in EPM induced by ribonuclease disappears in cells cultured for 5 days, when cells have attached to the substratum and onset of depigmentation is evident (fig. 3). The effect of neuraminidase on the cells' EPM is always present in cultured cells, reflecting perhaps the synthesis of new glycoproteins as an adaptation to culture conditions.

Table 2

Electrophoretic mobilities of cells from early stage regenerates after different enzyme treatments

| Stage | Control | Neuraminidase | Ribonuclease | Chondroitinase |
|---|---|---|---|---|
| 3 d | 1.725 | 1.299† | 1.43 ($T_1$)* | 1.458* |
| | 1.832 | | 1.541 (p)†† | |
| | 2.034 | 1.607†† | 1.525 (p)†† | |
| | 1.911 | | | 1.706* |
| 5 d | 1.528 | 1.230† | | 1.407 |
| | 1.833 | 1.497†† | 1.751 (p) | |
| | 1.661 | 1.267* | 1.633 (p) | |
| | 1.625 | | 1.70 ($T_1$) | 1.631 |
| 7 d | 1.837 | 1.704* | 1.771 (p) | |
| | 1.807 | 1.506†† | 1.736 (p) | |
| 10 d | 1.617 | 1.395* | 1.507 ($T_1$) | |
| | 1.763 | 1.602 | 1.757 (p) | |
| | 1.667 | 1.657 | | |
| | 1.728 | | | 1.709 |
| | 1.812 | | | 1.803 |
| | 1.814 | 1.826 | | |
| 15 d | 1.867 | 1.891 | | |
| | 1.724 | 1.716 | | |
| | 1.724 | | 1.716 (p) | |
| | 1.888 | | 1.923 (p) | |

d - days after lens removal.

Average EPM values (μm $sec^{-1}$ $V^{-1}$ $cm^{-1}$) represent an average of 30 to 40 determinations. For further explanation see table 1.

The presence of RNA at the cell surface has been reported by several investigators[10,11,12,13]. Glycosyl-amino-glycans have also been detected at the cell surface[4,14,15,16]. An interpretation of the results presented in this paper is that removal of an RNAse-sensitive component at the cell surface, presumably associated with a carbohydrate, is the first surface event that occurs during dedifferentiation. The disappearance of neuraminidase-sensitive groups may be associated with the appearance of cell surface sites involved in cell adhesion at the onset of lens fiber elongation[17].

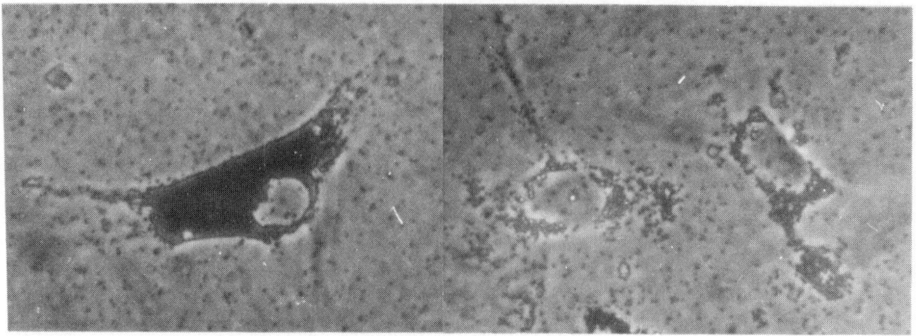

Fig. 3. Pigmented cells from the dorsal iris after 5 days in culture. Cells have attached to the substrate and signs of depigmentation are visible. X300.

Table 3

Effect of neuraminidase and ribonuclease on the electrophoretic mobility of cultured iris cells

| Days in Culture | Control | Ribonuclease | Neuraminidase |
|---|---|---|---|
| 0 | 1.45 | 1.27† | 1.21† |
| 5 | 1.47 | 1.48 | 1.05* |
| 8 | 1.41 | 1.54 | 1.06† |
| 14 | 1.46 | 1.39 | 1.10† |
| 21 | 1.56 | 1.48 | 1.00†† |
| 35 | 1.25 | 1.29 | 0.94†† |

Cells were obtained by mild trypsinization, cultured in L-15 medium diluted with distilled water 1:1 and cultured at 24°C. See table 1 for further explanation.

We have used a variety of lectins in an attempt to further characterize cell surface changes during dedifferentiation. Wheat germ agglutinin agglutinated normal as well as dedifferentiated cells from 13 day regenerates. A slightly lower incidence in agglutination was found when normal iris cells were studied. Similar results were obtained when soybean agglutinin was used. Normal and dedifferentiated cells did not agglutinate with *Ricinus communis* agglutinin. These data suggest that EPM determinations and lectin mediated agglutination may reflect different cell surface properties.

References

1. Yamada, T., 1967, in: Current topics in developmental biology, vol. 2, eds. A.A. Moscona and A. Monroy (Academic Press, New York) p. 247.

2. Yamada, T., 1972, in: Cell differentiation, eds. R. Harris, P. Allin and D. Viza (Munksgaard, Copenhagen) p. 56.

3. Weiss, L., 1970, in: Permeability and function of biological membranes, eds. A. Katchalsky, R.D. Keynes and W.R. Lowenstein (North-Holland, Amsterdam) p. 94.

4. Kojima, K. and Yamagata, T., 1971, Exp. Cell Res. 67, 142.

5. Zalik, S.E. and Scott, V., 1972, J. Cell. Biol. 55, 134.

6. Zalik, S.E. and Scott, V., 1973, Nature, New Biol. 244, 212.

7. Dumont, J.N. and Yamada, T., 1972, Develop. Biol. 29, 385.

8. Horstman, L.P. and Zalik, S.E., 1974, Exp. Cell Res. 84, 1.

9. Eguchi, G., Abe, S.-I. and Watanabe, K., 1974, Proc. Nat. Acad. Sci. U.S.A. 71, 5052.

10. Bierle, J.W., 1968, Science 161, 798.

11. Weiss, L. and Mayhew, E., 1969, J. Cancer 4, 626.

12. Davidova, S.Y. and Shapot, V.S., 1970, Febs Letters 6, 349.

13. Rieber, M. and Bacalao, J., 1974, Proc. Nat. Acad. Sci. U.S.A. 71, 4960.

14. Suzuki, S., Kujohide, K. and Utsumi, K.R., 1970, Biochim. Biophys. Acta 222, 240.

15. Kramer, P.M., 1971, Biochemistry 10, 1437.

16. Kramer, P.M., 1972, J. Cell. Biol. 55, 713.

17. Lloyd, C.W. and Cook, G.M.W., 1974, J. Cell Sci. 15, 575.

We thank the National Cancer Institute of Canada for their continuous support. Portions of this work were also supported by the National Research Council of Canada.

*Progress in Differentiation Research, ed. N. Müller-Bérat et al.*



# STUDIES ON THE CONTROL OF GENE EXPRESSION, ENZYME LEVELS AND CATALYTIC EFFICIENCY IN NORMAL AND NEOPLASTIC TISSUES.

Philip Feigelson
The Institute of Cancer Research and the Department of Biochemistry
College of Physicians and Surgeons, Columbia University
New York, New York, U.S.A.

The biochemical lesion(s) responsible for the normal cell - neoplastic cell transformation has (have) thus far eluded precise determination. What is recognized are functional differences in the constraints on cellular multiplication, which *in vivo*, are manifested as unrestricted growth with or without invasiveness, and *in vitro* as unlimited cellular multiplication and subculturability with absence of contact inhibition. In addition to alterations in the cell surface which are detectable with the scanning electron microscope, a frequent cytologic alteration in neoplastic cells is the appearance of prominent nucleoli reflecting high rates of ribosomal RNA synthesis. From the biochemical viewpoint, tumors frequently manifest altered metabolic patterns and enzymic profiles when compared with their normal tissue counterparts. Quantitative shifts occur in the relative proportions of certain enzymes and alterations in the ratio of isozymic species are detectable. The responsiveness of tumors to hormonal regulators is also often aberrant. Furthermore, in addition to these quantitative shifts in enzymic patterns and their regulatability, qualitative changes are also evident; protein species present in normal cells are sometimes deleted from their corresponding tumors and reciprocally in certain instances, new protein species appear within tumor cells, are sometimes designated as carcinofoetal antigens, which are absent from the corresponding, non-neoplastic cell.

This laboratory has focused, for some years, in a series of studies designed to provide some insight into the processes by which normal and neoplastic cells determine the presence of and regulate the levels of specific intracellular enzymes. The primary biosystems employed have been rat liver and Morris hepatoma. We have investigated the events by which glucocorticoids regulate hepatic and tumor levels of tryptophan oxygenase. This enzyme was chosen as it had been established that its catalytic level in liver was subject to both substrate and hormonal induction. This enzyme also manifests complex positive and negative allosteric control and the levels of this enzyme in hepatomas and its control therein were aberrant.

We have continued our attempts to gain insight into the processes by which hormones and other biochemical regulators determine the catalytic efficiency, enzyme levels, and gene expression in normal and neoplastic tissues. In persuit of these objectives, our studies have carried us into various aspects of the control of gene transcription by RNA polymerases I and II, considerations of certain aspects of the processing of the gene transcript, and measurement of the functional levels of specific species of mRNA under varying conditions of control in normal and neoplastic tissues.

We have continued to purify the glucocorticoid receptor from normal liver and hepatomas and to study the chemical transformations which are responsible for its activation which enables it to interact with target tissue nuclei, chromatin, and DNA, and furthermore are attempting to evaluate the changes in genomic function which result from this interaction (1, 2, 3, 4, 5, 6, 7, 8).

A rapid, simple purification procedure has recently been developed which takes advantage of the principle that as the glucocorticoid-receptor complex becomes activated, it develops the ability to bind to DNA and phosphocellulose. Hence, passage of unactivated glucocorticoid-receptor complex through a phosphocellulose column, followed by thermal activation and rechromatography on phosphocellulose enables its 1,000-fold purification. Sucrose gradient centrifugation and exclusion chromatography indicate that no dramatic change in molecular weight accompanies its activation. Millimolar levels of calcium chloride markedly accelerate the rate of conversion of inactive glucocorticoid-receptor complex to the form which is able to interact with the genome. Time and dose response studies indicate that both the the degree of saturation of glucocorticoid receptor with the hormone *in vivo* and the extent of its migration from the cytoplasm into the nucleus are correlated with hepatic enzyme induction. All findings are compatible with the belief that the steroid hormone after entering its target tissues combines with its receptor which then undergoes an activation process accelerated by calcium ions. This steriod-receptor complex then migrates into and binds to nuclear components and presumably regulates transcriptional events (1, 2, 3, 4, 5, 6, 7, 8).

To determine whether hormonal modulation in rates of synthesis of inducible enzymes was via transcriptional or translational control we performed a series of studies, upon normal liver and hepatoma, measuring the effects of glucocorticoids on the tissue levels of the specific messenger RNA which codes for tryptophan oxygenase. Employing a messenger RNA-dependent Krebs ascites cell - free ribosomal protein synthesizing system, followed by immunoprecipitation with monospecific antibodies and SDS acrylamide gel electrophoresis, it was determined that this system carried out a quantitative and faithful translation of the mRNA for tryptophan oxygenase synthesizing the 40,000 Dalton protomeric units of tryptophan oxygenase. Employing this system, it was found that, in liver, in both hormonal dose response and time course studies, a close parallelism between the functional amount of the specific mRNA for tryptophan oxygenase and the hepatic level of the catalytic activity of this enzyme was observed. Thus, the hormonally elevated mRNA for tryptophan oxygenase is responsible for the rise in the rate of synthesis of this enzyme during hormonal enzyme induction (8, 9, 10, 11, 12).

The hepatic mRNA for tryptophan oxygenase was also determined when the level of this enzyme was "superinduced" by actinomycin D. We have found that when this enzyme is superinduced by actinomycin D, the level of the mRNA for tryptophan oxygenase does not rise, but rather corresponds to that of control animals. Thus, factors other than the mRNA levels must be responsible for this "anomolous" effect of actinomycin D. These findings do not support the prevalent hypothesis proposing the existence of an actinomycin D sensitive repressor which determines the rate of degradation of messenger RNA species (13).

We have studied the expression of the genes for tryptophan oxygenase and alpha 2u globulin and their regulation in host liver and hepatoma cells from Morris hepatoma bearing rats. The tryptophan oxygenase levels from host livers of rats bearing several different strains of Morris hepatoma are similar to those of normal liver whereas the hepatoma cells from all strains, we have studied, contain undetectably low levels of tryptophan oxygenase. These hepatomas were found to be devoid of detectable levels of the mRNA coding for this enzyme whereas the host livers did contain the mRNA capable of coding for tryptophan oxygenase (14).

Alpha 2u globulin is a protein synthesized in the liver of male rats and not in female rats. We have found that a direct correlation exists between the functional level of the mRNA which codes for alpha 2u globulin and the <u>in vivo</u> rate of synthesis of this protein in the livers of animals under various endocrine states. This correlation exists throughout regulation by androgens, glucocorticoids and thyroid hormone. These studies are the first direct evidence indicat-dicating that thyroid hormone exerts its effect upon specific protein synthesis by modulating the tissue level of the gene product, i.e., the mRNA which codes for that protein. One of the hepatomas studied (7793), is incapable of synthesizing either alpha 2u globulin or another hepatic marker, tryptophan oxygenase. Under all endocrine conditions studied, it has been found that neither <u>in vivo</u> synthesis of alpha 2u globulin nor of tryptophan oxygenase occurs nor can detect-detectable levels of the mRNA which codes for either of these proteins be detected in such hepatomas. We have found that in hepatomas the gluococorticoid receptor protein is present and apparently interacts with nuclei normally; nevertheless these hepatomas are unable to respond to glucocorticoidal induction of either tryptophan oxygenase or alpha 2u globulin. These data support the view that deleted synthesis of these hepatic proteins following the neoplastic transformation are due to the absence of their respective mRNA species and may be consequences of a normal hepatic gene being deleted, or more probably, becoming repressed during this dedifferentiation process (15, 16).

Thus during hormonal modulation of hepatic levels of tryptophan oxygenase and alpha 2u globulin the rates of their intracellular synthesis are consequences of hormonal modulation of the tissue levels of the specific mRNA species coding for these proteins. Furthermore, the neoplastic transformation which modifies specific enzyme patterns is also reflecting changes in the levels of the specific mRNA species coding for these proteins. It is therefore probable that hormonal regulation by steroids and aspects of the dedifferentiation which characterizes the neoplastic process are reflecting modification in the transcription of specific genes.

1. Beato, M., Kalimi, M., and Feigelson, P. (1972) Correlation between glucocorticoid binding to specific liver cytosol receptors and enzyme induction *in vivo*, Biochim., Biophys. Res. Commun., Vol. 47, pg. 1464.

2. Beato, M., and Feigelson, P. (1972) Glucocorticoid binding proteins of rat liver cytosol, I. Separation and identification of the binding protein, J. Biol. Chem., Vol. 247, pg. 7890.

3. Koblinsky, M., Beato, M., Kalimi, M., and Feigelson, P. (1972) Glucocorticoid binding proteins of rat liver cytosol, II. Physical characterization and properties of the binding proteins., J. Biol. Chem., Vol. 247, pg. 7897.

4. Kalimi, M., Beato, M., and Feigelson, P. (1973) Interaction of glucocorticoids with rat liver nuclei I. Role of the cytosol proteins, Biochemistry, Vol. 12, pg. 3365.

5. Beato, M., Kalimi, M., Konstam, M. and Feigelson, P. (1973) Interaction of glucocorticoids with rat liver nuclei II. Studies on the nature of the cytosol transfer factor and the nuclear acceptor site, Biochemistry, Vol. 12, pg. 3372.

6. Beato, M., Kalimi, M., Beato, W., and Feigelson, P. (1974) Effect of adrenalectomy and cortisol administration, Endocrinology, Vol. 94, pg. 377.

7. Kalimi, M., Colman, P., and Feigelson, P. (1975) The "activated" hepatic glucocorticoid receptor complex: Its generation and properties, J. Biol. Chem., Vol. 250, pg. 1080.

8. Feigelson, P., Beato, M., Colman, P., Kalimi, M., Killewich, L., and Schutz, G. (1975) Studies on the hepatic glucocorticoid receptor and on the hormonal modulation of specific mRNA levels during enzyme induction, Recent Progress in Hormone Research, Vol. 31, pg. 213.

9. Schutz, G., Beato, M., and Feigelson, P. (1972) Isolation of eukaryotic messenger RNA on cellulose and its translation in vitro, Biochim, Biophys. Res. Commun., Vol. 49, pg. 680.

10. Schutz, G., Beato, M., and Feigelson, P. (1973) Messenger RNA for hepatic tryptophan oxygenase: Its partial purification, its translation in an heterologeous cell-free system and its control by glucocorticoid hormones, Proc. Nat. Acad, Sci. USA, Vol 70, pg. 1218.

11. Schutz, G., Beato, M., and Feigelson, P. (1974) Isolation on cellulose of ovalbumin and globin m-RNA and their translation in an ascites cell-free system, Methods of Enzymol., Vol. 30, pg. 701.

12. Schutz, G., Killewich, L., Chen, G., and Feigelson, P. (1975) Control of the mRNA for hepatic tryptophan oxygenase during hormonal and substrate induction, Proc. Nat'l Acad. Sci., U.S.A. Vol. 72, pg. 1017.

13. Killewich, L., Schutz, G., and Feigelson, P. (1975) Level of rat liver tryptophan oxygenase messenger RNA during superinduction of the enzyme with actinomycin D., Proc. Nat. Acad, Sci., USA, In Press.

14. Feigelson, P., Murthy, L.R., Sippel, A.E., Morris, H.P. (1975) On the absence of specific mRNA species in hepatoma cells, Proc. of the Amer. Assoc. for Cancer Res., Vol. 16, pg. 26.

15. Sippel, A.E., Feigelson, P., and Roy, A.K. (1975) Hormonal regulation of the hepatic messenger RNA levels for α2u globulin, Biochemistry, Vol. 14, pg. 825.

16. Murthy, L., Colman, P.D., Morris, H., and Feigelson, P., Studies upon the mRNA for tryptophan oxygenase in host livers and Morris hepatomas under basal and hormonally induced states. Manuscript in preparation.

*Progress in Differentiation Research, ed. N. Müller-Bérat et al.*



# STEM CELLS IN ACUTE MYELOID LEUKEMIA

Sven-Aage Killmann

Department of Medicine A
Rigshospitalet
University Hospital of Copenhagen

Normal blood contains five major cell components, i.e. red cells, neutrophilic granulocytes, monocytes, platelets, and lymphocytes of various subclasses. Monocytes and lymphocytes will not be discussed here.

The lifespan of red cells, platelets and neutrophilic granulocytes is limited although highly variable according to cell type. Because of this obligatory cell loss due to cell senescence, a continuous cell formation is needed. This goal is achieved by a rather complex system of cells characterized by various degrees of division and maturation capacity. Diagrammatically, the system can be illustrated as follows:

Pluripotent hemopoietic stem cells
↓
division
↙ ↓ ↘
committed stem cells
1. red cell line
2. platelet line
3. granulocyte + monocyte line
↓ ↓ ↓
division
↓
observable precursors
(divide and mature)
↓ ↓ ↓
mature
↓ ↓ ↓
egress to blood

The bone marrow and some other tissues, depending on the species, contain pluripotent stem cells. They are characterized by a high potential for consecutive division, by self-maintenance and a certain choice of which cell line they will ultimately feed into, i.e. the stem cells committed to red cell, platelet, or granulocyte/monocyte formation. The best way to demonstrate the pluripotent stem cells is to inject them into lethally irradiated mice. Here, they protect against radiation-induced bone marrow failure and also form colonies in the spleen of the recipient. These colonies are most often erythroid or granulocytic, less often megakaryocytic, i.e. committed to platelet production. Finally, colonies of mixed lines are also observed.

All these colonies have two characteristics: 1) they are clonal in origin and 2) they have the capacity of self-replication. In other words, irrespective of their morphologic appearance, a single colony can create new colonies with the characteristics mentioned when injected into a new lethally irradiated mouse.

The pluripotential stem cells are in rather slow cell cycle, whereas the committed stem cells are rapidly cycling. How long a cell line, e.g. the erythroid, can be maintained by the committed stem cell variety without input from the pluripotent stem cell pool is limited. At this meeting, Dr. Lajtha mentioned that it may be in the order of one week. Perhaps it should be noted that the rigid separation between pluripotent stem cells and committed stem cells maybe is too simplistic although it is conceptually handy. To refine our concepts at this point we need additional markers which we do not have as yet.

The stem cells (pluripotent or committed) mentioned so far are not recognizable in the light microscope. In contrast, the progeny formed by the committed stem cells can be recognized. They have different names which may be of interest for hematologists but otherwise are not exciting.

The first stages in the recognizable hemic precursor compartments are characterized by cell divisions. In man, statistically there are four consecutive cell divisions in the red cell line and the granulocytic line. Thus, the earliest recognizable precursor cell gives rise to 16 descendant cells. These divisions are accompanied by considerable cell maturation. The divisions are well synchronized with maturation; in fact, it is this synchronized behaviour that allows identification of the various cell types and stages. After cell division has definitely ceased, maturation in the bone marrow continues for some time until the point where the cells egress to the blood stream.

These systems are controlled by various feed back systems which except for erythropoietin are poorly understood as yet. Now, after this sketchy outline of normal hemopoiesis[1,2,3], what is the difference between hemopoiesis in acute myeloid leukemia (AML) and in the normal situation?

AML is characterized by 1) a vast accumulation of blastic cells, primarily in the bone marrow with little or no maturation and no demonstrable function, and 2) a deficit of mature functioning blood cells. It is this latter component which mainly determines the clinical picture of AML, i.e. anemia, granulocytopenia with increased susceptibility to serious infections, and thrombocytopenia which results in severe bleeding tendency in all organs.

Until about 15 years ago it was believed that leukemic blast cells proliferate very quickly and thus "outcrowded" the normal cells of the marrow. All seemed simple although it was vague. Mainly based on the advent of $^{3}H$-thymidine, it can now be said with considerable certainty that leukemic blast cells proliferate at a slower rate than those marrow cells with which we can compare them[4] Their large accumulation must therefore be a matter of rather long "life span" coupled with a fair degree of "stemness". By stemness is meant the potential for further consecutive divisions.

Still, present day chemotherapy of AML is based on these old concepts of rapid proliferation and outcrowding. Nevertheless, it must be admitted that empiricism has shown that the treatment in use today works to some extent although nobody really understands why. This being said, it must be added that leukemia is not an entity. Real progress in treatment has been made practically only in childhood leukemia (acute lymphocytic leukemia). Otherwise, the outlook for other types of leukemia, acute myeloid, chronic myeloid and chronic lymphocytic leukemia is still sad with little or no progress over the past 25 years.

What then about the remaining apparently normal hemic cells in florid AML? Customarily, they have been considered to be the remainder of normal hemopoiesis. However, by looking at things from another point of view, the alternative has been raised[5,6,7] that they could well be of leukemic origin but be less blocked in their maturation than the "standard" leukemic blast cells and thus achieve a status of functional usefulness. These apparently normal cells therefore deserve considerable scrutiny. In the last analysis the problem is whether in AML, all blood cells are leukemic, albeit with differences with respect to maturational and functional capacity (single population theory) or whether in the individual with AML normal stem cells are left which under appropriate conditions can express themselves and give rise to a normal progeny (double population theory).

Hence, let us look at the supposedly normal cells in AML, keeping in mind that in AML, one rarely gets a clear-cut answer to any question posed which can be answered by laboratory methods. The answer is often like those from the Oracle in Delphi, i.e. interpretable. In the case of AML, the interpretation depends to some degree on temperament. The optimist hopes for the existence of the double population which offers possibilities of cure of the disease by present day therapeutic methods which aim at killing cells. The pessimist may focus on the parameters supporting the single population theory that requires a complete re-thinking with respect to therapeutic approach.

In the following, some of the abnormalities observed in the apparently normal cells of AML will be listed: most of these abnormalities are cellular in origin and consistent with but surely not proving the single population theory.

In neutrophilic granulocytes there usually is a lack of primary and/or secondary granules[8,9], which contain enzymes that are important for the bactericidal potential of the cells. Lack of myeloperoxidase, which normally is linked with the primary granules is also often encountered in AML, even when the primary granules with which the enzyme is associated are demonstrable by transmission electron microscopy[9]. The so-called Auer-bodies, that are an abnormality of the primary granules, and which so far have not been seen in any other condition outside of myeloid leukemia, may occasionally be found in mature neutrophils.

The cell amplification factor in the recognizable granulocytopoiesis, which normally is about 1:16[3], is severely reduced in AML, indicating either less divisions than usual in granulocytopoiesis or intramedullary cell death of precursor cells in the marrow[10]. I would like to point out that this latter phenomenon is not per se a sign of malignancy. Thus it is also found in deficiency states such as lack of vitamin $B_{12}$ or folic acid.

A much more specific point of disturbed neutropoiesis in AML come from agar culture studies. Here, granulocytic colonies, normal with regard to growth pattern, colony morphology, and cytology is almost absent in AML marrow.

Regarding the red cell there is now convincing evidence that abnormal karyotypes in AML are present not only in leukemic blast cells but also in red cell precursors[11]. This is important because it clearly demonstrates that AML is not a "white cell neoplasia" and strongly suggests that AML (like chronic myeloid leukemia[12]) probably as its target cell has the pluripotential hemopoietic stem cell. Formal proof that the erythroid line is involved in AML comes from a combination of radio-iron-labeling and karyotypes: in this way, abnormal karyotypes have been observed in iron-incorporating cells, that means red cell precursors[13]. In red cells from AML, many other abnormalities are present such as an excess of iron in red cell precursors[8], changes in red cell antigens[14], and sensitivity of red cells to cold antibodies.

Platelets in AML display an abnormal ultrastructure[15] and abnormal behaviour as measured by functional tests in vitro[16]. Hence, there are good reasons to assume that AML is a disease that resides in the pluripotent hemopoietic stem cell, involving directly neutrophil precursors, red cell precursors and platelet precursors.

Not all AML cases are so dominated by blast cells as indicated in the beginning. Some cases show only minor abnormalities which easily can be mis-diagnosed even by experienced workers. Such cases often reveal their true nature when the marrow is cultured in agar where they grow as abnormally as florid cases do. It is true that the so-called CSA (colony stimulating activity) is a growth promoting factor in many situations but one thing appears quite certain: CSA is not able to induce full differentiation in a culture of human leukemic cells [17].

At this point it is perhaps appropriate to insert that in the so-called preleukemic conditions (i.e. high risk groups with respect to developing AML), most of the aberrations which I have briefly mentioned have also been observed. From my point of view, a preleukemia is probably nothing but an AML that starts in a partial remission phase which lasts for a certain time until finally, full blown AML develops.

I have now introduced the term remission. Remission means that normal marrow function is restored, blood counts normalized and the patient feels fit. In other words, at this point one cannot make the diagnosis of AML. Things appear normal. To obtain this remission state from the state of florid AML, intensive chemotherapy is required which eradicates the majority of the leukemic cells and leaves the patient in temporary bone marrow aplasia during which the blood counts of normal-appearing cells decline even further than during untreated disease. At this point three things may happen: 1) the patient may die in bone marrow aplasia, 2) the leukemia may recur (at times even more "malignant" as judged by the agar growth pattern) without intervening remission, or 3) remission may ensue. Unfortunately, in this latter case relapse sooner or later takes place. This usually happens within a few months. Nevertheless, the main point of interest in this context is the remission phase. That remission takes place appears to be the best argument for the validity of double population theory: the vacuum left in the marrow allows normal stem cells to take over hemopoiesis for a while and accordingly, morphology becomes normal. Also when the cells are cultured in agar, they apparently grow normally but closer observations are needed. Moreover, karyotypes which are abnormal in ca. 40 - 50 per cent of untreated cases of AML usually but not invariable [18] return to normal. So, there are several lines of evidence that hemopoiesis returns to normality. However, none of the parameters are final academic proof of normality. It is disturbing namely that some parameters remain abnormal in remission. This goes for red cell sensitivity to cold antibodies[19], abnormalities of serine incorporation into granulocytes and most importantly: the acquired virus sequences which can be demonstrated in AML cells in relapse remain in the cells of AML patients during remission[20,21,22]. It must also be mentioned in this context that neutrophilic granulocytes from AML patients in "remission" do not behave normally as far as CSA is concerned.

It is conceivable therefore, that hemopoiesis has not been replaced by completely normal cells but by basically leukemic cells which however are sufficient in numbers and function to mimic normality. The change from the relapse to the remission and back to the relapse state is certainly one of a succession of cell clones induced by the cytostatic drugs but in addition, direct differentiating effect by the drugs used remains a possibility. The drugs now used are classified as being cytostatic or cytocidal agents but to my knowledge it has not been studied whether they also might have differentiating effects. It should be noted that even "primitive" blast cells in AML cytochemically show some signs of "intended" maturation which suggests that the maturation block in the leukemic blast cells is manifest only at a rather late stage[23]. So, if one wants to study human neoplasia further in terms of differentiation, material from patients with AML (blood and/or marrow) containing large numbers of blast cells are readily available for study.

In my opinion the present strategy of killing leukemic cells does not give much promise for the future irrespective of whether the single or double population theory holds true. The reason for this pessimism is that remissions are obtained in only about half of the cases, they are usually short lasting, and the treatment needed is taxing on the patients with many highly unpleasant side-effects Moreover, when relapse takes place a new remission is usually more difficult to obtain than the first one was, maybe because of resistance of leukemic cells. This, parenthetically, raises another interesting question: why do normal cells not become resistent to chemotherapy? This lack of non-resistance of normal cells is well known from chemotherapy of non-leukemic neoplastic conditions.

Rather than the present therapeutic approach, efforts should be made to facilitate or force the non-maturing blast cells to divide and mature normally. From a clinician's point of view, it should be explored whether an artificial RNA or DNA sequence could (some time in the future) be introduced into the diseased stem cells to negate the effect of the "leukemic sequences" already present in the cells of AML or in other ways make those genes express themselves which are needed to make the "leukemic" cells functional and hence useful rather than destructive members of the human body. This, admittedly, is at the present time wishful thinking but once one has achieved the present goal to map the functions of cells in general and the control mechanisms involved, why should it not be possible to repair defective cells? At least it has been demonstrated in principle that certain virus infections may have antineoplastic effects. Thus, bovine enterovirus may cause regression of solid and ascites tumors in rodents. The killing effect on neoplastic cells was also seen in vitro when it was further shown that untransformed lines were resistant to the virus[24].

References

1. Wolstenholme, G.E.W. and O'Connor, M., eds.Ciba Foundation Symposium on Haemopoiesis, 1960, (Churchill, London).
2. Stohlman, F.,jr., ed., 1970, Hemopoietic Cellular Proliferation (Grune &Stratton, New York).
3. Cronkite, E.P. and Vincent, P.C., 1969, Ser. Haemat. 2(4), 3.

4. Killmann,S.-Aa., 1968, Ser.Haemat. I(3), 38.

5. Killmann,S.-Aa., 1968, Ser.Haemat. I(3), 103

6. Killmann,S.-Aa., 1970, in: Hemopoietic Cellular Proliferation, ed. F.Stohlman,F.jr., (Grune & Stratton, New York), p.267.

7. Killmann,S.-Aa., 1972, Proc.Int.Cancer Conf.,Sydney,Australia, p. 205.

8. Bessis,M.,and Breton-Gorius,J., 1969, Nouv.Rev.franc.Hémat. 9,245.

9. Bainton,D.F., 1975, Blood Cells 1, 191.

10. Fliedner,T.M., Hoelzer,D.,and Steinbach,K.H., 1975, 3rd.Meeting Europ.Afr.Div., Int.Soc.Haemat., Abstract, p.24.

11. Krogh Jensen,M.,and Killmann,S.-Aa., 1971, Acta med.Scand. 189, 97.

12. Whang,J., Frei,E.III, Tijo,J.H., Carbone,P.P.,and Brecher,G., 1963, Blood 22, 664.

13. Blackstock,A.M.,and Garson,O.M., 1974, Lancet ii, 1178.

14. Salmon,Ch., 1969, Ser.Haemat. II(1), 3

15. Maldonado,J.E., 1975, Ser.Haemat. 8(1), 102.

16. Cowan,D.H.,and Graham,R.C., 1975, Ser.Haemat. 8(1), 68.

17. Muller-Bérat,C.N., 1975, IInd.Int.Conf.Differentiation,Copenhagen, abstracts, p.165.

18. Galbraith,P.R., 1974, Can.Med.Assoc.J. 110, 1147.

19. Catovsky,D., Lewis,S.M.,and Sherman,D., 1971, Brit.J.Haemat. 21, 541.

20. Mak,T.W., Aye,M.T., Messner,H., Sheinen,R., Till,J.E.,and McCulloch,E.A., 1974, Brit.J.Cancer 29, 433.

21. Spiegelman,S., 1974, Cancer Chemother.Rep. 58, 595.

22. Spiegelman,S., 1975, Proc.Symp., Lüneburger Heide, in press.

23. Castoldi,G.L., Grusovin,G.D.,and Scapoli,G.L. 1975, Biomed. 23, 12.

24. Taylor,M.W., Cordell,B., Souhrada,M.,and Prather,S., 1971, Proc.Nat.Acad.Sci.USA, 68, 836

Progress in Differentiation Research, ed. N. Müller-Bérat et al.



# PLASMA CELLS AND BLAST PROLIFERATION IN COLONIES OF LEUKEMIC HUMAN HEMOPOIETIC CELLS IN VITRO

## Cytology, Cytochemistry and Immunocytochemistry of Colony Cells in Agar Gel Culture *

Hertha Beckmann[1], Rolf Neth[1], Helga Soltau[1], Jörn Ritter[3], Kurt Winkler[2], Manfred Garbrecht[4], Kurt Hausmann[5] and Gabriel Rutter[6]

1 Molekularbiologisch-hämatologische Arbeitsgruppe,
2 Abteilung für Gerinnungsforschung und Onkologie,
3 Abteilung für Klinische Immunpathologie, Universitäts-Kinderklinik,
4 II. Medizinische Universitätsklinik,
5 Hämatologische Abteilung, Allgemeines Krankenhaus St.Georg,
6 Heinrich-Pette-Institut für Experimentelle Virologie und Immunologie an der Universität Hamburg, West Germany.

## 1. Abstract

The colony forming ability of blood or bone marrow cells of 43 hematologically normal persons and of 36 patients with hemoblastosis were studied. Stimulated by feeder layer of peripheral leukocytes, between 66 and 154 colonies proliferated out of 100,000 mononuclear bone marrow cells, and between 2 and 14 colonies out of 500,000 blood cells. As shown by cytological and cytochemical differentiation, the colonies consisted of monocytes, eosinophils, macrophages, and only rarely, neutrophils. Monocytes were the most frequent cell type observed.
Cultures from patients with different types of acute leukemia in relapse, or in the acute transformation phase of chronic myeloid leukemia were charakterized by a lack of normal colony formation, with the exception of patients with acute myeloid-promyeloid leukemia, who had greatly increased numbers of colony-forming cells, as did patients with CML. As shown by cytological and cytochemical differentiation the majority of colonies from bone marrow and blood cells of leukemic patients contained the same cell type as normal cells, i.e. predominantly monocytes and macrophages. But in five colonies there were blasts and three of them contained plasma cells as well as blasts.

* Supported by a grant from the Deutsche Forschungsgemeinschaft.

Eight further colonies contained plasma cells without any blasts.
The in vitro proliferated plasma cells showed typical cytological and cytochemical characteristics and also a positive immunocytological reaction as proof for immunoglobulin production. Only those colonies which contained blasts and/or plasma cells were assumed to originate from leukemic cells. It might be possible that in mixed colonies of blasts and plasma cells, or of macrophages and plasma cells, there is a cellular interaction between immunocompetent cells and tumor cells in vitro.

## 1. Introduction

Under appropriate stimulation hemopoietic cells from patients with acute myeloid, myelomonocytic and chronic myeloid leukemia proliferate as colonies in the semi-solid agar culture system (1-11). Results obtained by various laboratories generally correspond to each other with regard to patients with chronic myeloid leukemia. In opposition to this greater variation was found in reports describing culture findings in AML. The reason for this discrepancy might be the use of different culture techniques and a different cytologic interpretation (10). Certain cytological and cytochemical differentiation of single cells needs technically perfect preparation and staining of the colonies. This is the only reasonable basis for comparative investigations which - in addition to quantitative counting of the colonies and identification of their size - require qualitative analysis of the composition of the colonies. In this study colony cells of normal and leukemic blood cells were demonstrated and classified ny means of cytological, cytochemical and immunocytochemical techniques. In contrast to cytologic findings in cultures from normal persons mixed colonies of blasts and plasma cells were found in colonies of leukemic patients.

## 2. Materials and Methods

### a) Source of material

Normal bone marrow was obtained by aspiration from iliac crest or sternum, taken during the hematologic examination of 13 children and two adults. Six of these patients were defined as normal, and the other nine showed no evidence of hemoblastosis or granulo-

poietic abnormalities.
Peripheral blood samples taken from 27 healthy volunteers were used as source for blood leukocytes. These control persons had peripheral leukocyte counts within the normal range, and with the normal white cell differential counts.
Marrow or peripheral blood samples were taken once or several times from 36 patients. Of these, six had undifferentiated leukemia, three acute lymphoblastic leukemia (ALL), two acute acid phosphatase-positive leukemia (T-cell leukemia), 19 either acute myeloid leukemia (AML), acute myeloid-promyeloid, or acute monocytic leukemia, and four chronic myeloid leukemia (CML) in various clinical stages. One patient had preleukemia and another one with the diagnosis of reticulosis became leukemic. For the cytological and cytochemical classification of bone marrow cells the usual criteria were used (12-15).
The ALL therapy concept was proposed by PINKEL and co-workers (16). All patients of this group received the same combination of induction chemotherapy including vincristine (VCR) and prednisone (PRED), followd by central nervous system (CNS) leukemia prophylaxis with CNS radiation and intrathecal methotrexate application (MTX). For consolidation, a combination of endoxan, MTX, and 6-mercaptopurine (6-MP) was given. - Patients with AML were treated with a combination of 6-thioguanine and cytosine arabinoside.

b) Cell separation

For the separation of leukocytes the method of BÖYUM (17) was used. The resulting buffy coat was suspended in culture medium. The nucleated cells were counted with a hemocytometer. For the preparation of the feeder layer, granulocytes and monocytes were counted. The cell layer was prepared by suspending two or three drops of aspirated bone marrow in the culture medium, by a shock with hypotonic treatment repeated twice for lysis of erythrocytes, and by counting the number of mononucleated bone marrow cells omitting non-deviding cells as metamyelocytes and polymorphs.

c) Culture technique

The agar cultures were prepared using the double-layer-agar-technique of PIKE and ROBINSON (18). McCoy's 5A medium containing 15% fetal calf serum and supplemented with amino acids and vitamins, was mixed in a 9:1 ratio with boiled 5% agar (Difco). After addition of the appropriate number of leukocytes (1,5-1,8 x $10^6$), 1 ml of this agar-cell-medium-mixture was pipetted into 35 mm plastic petri dishes (Falcon Plastics). These prepared feeder layers were stored at 37°C in a humidified incubator continously flushed with 8% $CO_2$. The washed peripheral blood leukocytes or bone marrow cells were mixed with culture medium and boiled 3% agar, in a 9:1 ratio; of this, 1 ml aliquots were thenpipetted onto the feeder layer. The final concentration for the cell layer was 5x $10^5$ mononucleated peripheral blood leukocytes per ml, or 1x $10^5$ mononucleated bone marrow cells per ml. Following preparation no aggregates, clumps, or tissue fragments were found in bone marrow suspensions or in agar. The dishes were incubated for three weeks. During this time, at intervals of 12 to 14 days, the number of colonies was counted with an inverted microscope (Diavert) at 40x magnification. Only colonies containing 50 or more cells were counted. For each experiment, at least four plates with, and two plates without the feeder layer were examined; the counts were expressed as the mean result of these plates. The number of colonies in two plates never varied more than 10% for bone marrow cells, and maximally 100% for peripheral leukocytes.

d) Source of colony stimulating factor

A feeder layer of peripheral blood leukocytes was used as source of colony stimulating factor (CSF). Since the introduction of the proliferation of colonies depends on the age of the feeder layer (19), the feeder layer was used during the day of preparation or only a few days later. After seven days of incubation, the feeder layer had lost about 50% of its original stimulating activity.

Mature granulocytes inhibit the proliferation of colonies (20). Therefore the number of granulocytes in the feeder layer was not allowed to exceed 1x $10^6$ cells per ml (19). At this concentration of granulocytes the feeder layer contained nearly 1-2x $10^5$ monocytes per ml.

e) Cytologic analysis of colonies

For cytologic analysis, colonies were picked out of the agar under the inverted microscope with a fine Pasteur pipette. They were put on slides and incubated for ten minutes in a humidified chamber with a 1% solution of agarase (Calbiochem., Los Angeles). The colonies were prepared according to the method of TESTA and LORD (22). After incubation, the agarase was drawn off carefully, and the colonies were fixed according to the cytochemical reaction necessary (10% formalin alcohol for peroxidase reaction, 60% cold acetone for acid phosphatase reaction). The fixation solution was dropped onto the slide. Then a coverslip was placed on top of the drop. This in turn was covered with a piece of filter paper and gently pressed. The slide was frozen on dry ice for 10 minutes, the cover slip quickly removed, and the slide immediately dried. The cells of the colonies were stained with May-Grünwald-Giemsa for peroxidase (12), or acid phosphatase (23).

f) Demonstration of immunoglobulins

To demonstrate immunoglobulin within plasma cells, some colonies were fixed as described above, stained for 30 minutes with FITC conjugated anti-human immunoglobulin from a goat (Behringwerke AG, Marburg), and then washed three times with PBS.
The colonies were examined with a Leitz ortholux microscope equipped with an Opak-Fluor vertical immuminator, a 4mmBG exciting filter, and two 495mm barrier filters: K 495 and AL 530.

## 4. Results

Cultures were analyzed from the peripheral blood or bone marrow of 43 normal volunteers and of 36 patients with leukemia. Corresponding to the type of leukemia, the plated cells showed differences in their colony forming ability, in spite of the stimulating activity of the feeder layer. The number of proliferated colonies of patients with cytochemically different types of leukemic blasts was very diminished (table 1, Nrs. 2-5), as has been observed before (1, 3-11, 24). With plated blood and bone marrow cells of patients with myeloid-promyeloid leukemia and chronic myeloid leukemia the number of colonies was elevated (table 1, Nrs. 6, 8) as described previously by several authors for CML (1, 7, 10, 24, 25). For three cases of monocytic leukemia (table 1, Nr. 7), and one with reticulosis in the leukemic phase (table 1, Nr. 10) the number of colonies was diminished when compared to normal.

<u>Table 1</u> - Colony forming ability of blood and bone marrow cells from leukemic patients in relapse

| diagnosis | CFC from | number | % atypical cells[1] plated | number of colonies[2] | | cytologic composition of colonies |
|---|---|---|---|---|---|---|
| | | | | stimulated[3] | unstimul. | |
| 1. preleukemia | bone marrow | 1 | 15 | 7 | 2 | monocytes, esosinophils, neutrophils, macrophages, <u>plasma cells</u> |
| 2. undifferentiated leukemia | bone marrow | 6 | 10-94 | 1-11 | 0 | monocytes, eosinophils, macrophages, <u>blasts</u> (1)[4] |
| 3. ALL | bone marrow | 2 | 98-100 | 0-0,5 | 0 | monocytes, eosinophils, macrophages |
| | blood | 1 | 94 | 0,2 | 0 | |
| 4. acute acid phosphatase positive leukemia | bone marrow | 2 | 44-100 | 0-13 | 0-1 | monocytes, eosinophils, macrophages |
| 5. ALL | bone marrow | 3 | 19-75 | 0,8-5 | 0-3 | monocytes, eosinophils, neutrophils, macrophages, <u>blasts</u> (3), <u>plasma cells</u> (3) |
| | blood | 4 | 87-97 | 0-1 | 0 | |
| 6. acute myeloid promyeloid leukemia | bone marrow | 7 | 22-92 | 2-352 | 0-0,3 | monocytes, eosinophils, neutrophils, macrophages, <u>plasma cells</u> (3) |
| | blood | 2 | 13-86 | 0-80 | 0-47 | |

Table 1 - continuation

| diagnosis | CFC from | number | % atypical cells plated | number of colonies | | cytologic composition of colonies |
|---|---|---|---|---|---|---|
| | | | | stimulated | unstimul. | |
| 7. acute monocytic leukemia | bone marrow | 3 | 27-95 | 1-3 | 0-2 | monocytes, eosinophils, neutrophils, macrophages, plasma cells (1) |
| 8. CML | bone marrow | 1 | 50 | 387 | 0 | monocytes, eosinophils, macrophages |
| | blood | 2 | 0-14 | 32-33 | 0-15 | |
| 9. CML, blastic transformation | blood | 1 | 83 | 0 | 0 | --- |
| 10. reticulosis, leukemic development | bone marrow | 1 | 97 | 12 | 5 | monocytes, eosinophils, neutrophils, macrophages, blasts, plasma cells |
| 11. normal | bone marrow | 15 | -- | 66-154 | 0-24 | monocytes, eosinophils, neutrophils, macrophages |
| normal | blood[5] | 16 | -- | 0,4-2,2 [6] | 0,1-1,4 [6] | |
| normal | blood[5] | 12 | -- | 1,1-2,8 | 0,1-2,1 | |

1 - atypical blasts, promyelocytes and monocytic cells
2 - per $10^5$ mononuclear cells plated after 12-14 days of culture
3 - stimulated by feeder layer of leukocytes
4 - in () number of patients with blasts and/or plasma cells in colonies
5 - plated as $5x10^5$ mononuclear blood cells
6 - McCoy not supplemented as culture medium

No colonies proliferated in the agar culture of plated blood cells from a patient with CML in the blastic transformation phase (table1 Nr. 9). Cultures of patients without CSF hardly contained any colonies (table 1, Nrs. 1, 4-8, 10).

In stimulated and unstimulated cultures the colonies proliferating between the eighth and fourteenth day of incubation only contained a small number of premature cells of granulopoiesis (up to segmented neutrophils and eosinophils), whereas monocytes and macrophages predominated. In general, the cytologic composition did not differ from that of those colonies which proliferated from normal mononuclear blood and bone marrow cells under our culture conditions (table 1, Nr. 11).

But in contrast to this pattern, blasts were detected in colonies of five patients (table 1, figure 1a).

In only a few cases these blasts showed peroxidase positive reaction. Besides these blasts, the colonies consisted of cells which usually proliferate (monocytes, macrophages, eosinophils and rarely neutrophils), or macrophages exclusively. These colonies were of extraordinary size (maximum 3,400 cells). A proliferation of blasts in colonies of plated bone marrow cells from leukemic patients has

---

Figure 1

Cells of colonies derived from bone marrow cells of patients with acute myeloid leukemia

a: 8 days after seeding
b-g: 14 days after seeding of bone marrow cells
a-d: peroxidase reaction
e: acid phosphatase reaction
f-g: immunocytofluorescence

a - blasts, peroxidase negative (artificial deformation of nuclei by pressing), at the left top one plasma cell, in the middle right two plasma cells

b - different cell types of a colony: blasts, peroxidase positive granulopoietic cells (black reaction in the cytoplasm), monocytes, macrophages and plasma cells

c - a group of plasma cells and macrophages

d - four plasma cells, two macrophages, in the middle left one peroxidase positive granulopoietic cell (promyelocyte ?)

e - three plasma cells with a granular acid phosphatase reaction in the cytoplasm, no fixation

f - plasma cells of a colony with positive flourescence in the cytoplasm, spread between several negative cells

g - two plasma cells with positive fluorescence in the cytoplasm, on the right hand a negative macrophage

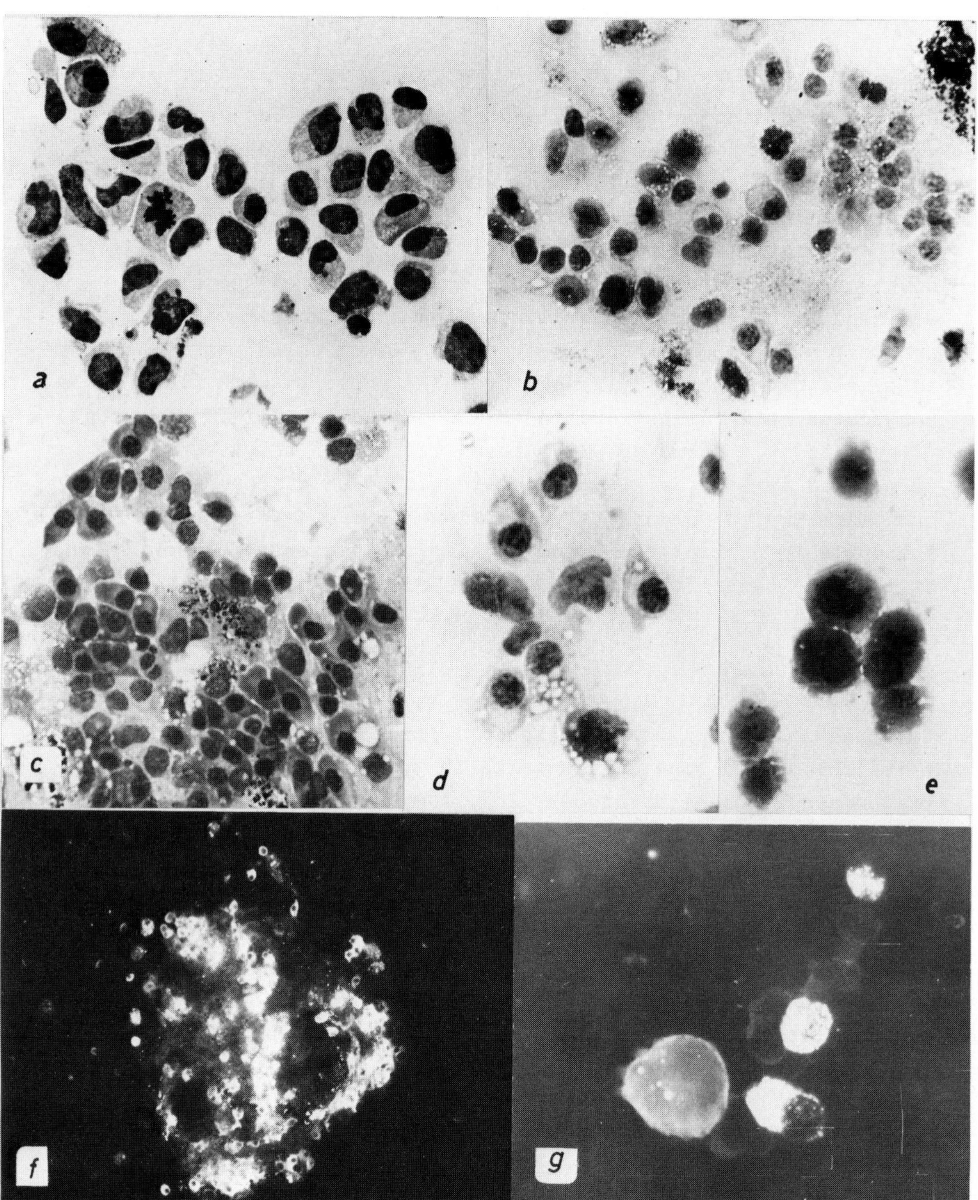
a
b
c
d
e
f
g

not yet been described. Moreover, colonies of three of these patients contained a great number of plasma cells besides blasts. Colonies of eight patients contained plasma cells without any blasts. According to the cytological characteristics of plasma cells (figure 1a-c), and the typical localization of their acid phosphatase reaction (figure 1e), there cannot by any doubt as to their classification. Plasma cells were spread throughout the whole colonies; they amounted to at least 10-35% of the cell number per colony. When the colonies were isolated, no plasma cells could be found in the surrounding agar. The in vitro production of immunoglobulins by these plasma cells was demonstrated by staining colony cells of three AML patients with fluorescein conjugated anti-immunoglobulin (FITC conjugated anti-IG of a goat) (figure 2f, g). Colonies with plasma cells were only seen in stimulated as well as in unstimulated cultures of leukemic patients, predominantly in relapse (seven patients - one before chemotherapy), in partial remission of AML (two patients), in remission of AML (two patients), and in preleukemia (one patient).

## 5. Discussion

Colonies consisting of granulopoietic cells proliferate out of normal blood and bone marrow cells in the agar culture system predominantly using an inducer (26). A myeloid committed stem cell, the so-called colony forming cell (CFC) can be assumed to be the stem cell of these colonies (27). The question as to whether leukemic cells in colonies can be the origin of normal myeloid elements cannot be answered precisely even now, in spite of many results (10, 11). MOORE and METCALF showed with cytogenetic markers that leukemic cells proliferate in agar culture (28). The density of CFC isolated in BSA gradient of bone marrow from CML, and the density of cluster (  50 cells) forming cells isolated in a BSA gradient of bone marrow from acute leukemias, is different from the density of normal CFC (7). But there is no final proof that these leukemic CFC develop further in vitro; as blasts (as shown in table 1 and figure 1), or along the normally occurring maturation process towards normal granulocytes or monocytes.
Colonies from normal persons, and the major part of colonies which proliferate by plating leukemic bone marrow and blood, did not differ in their cytologic composition. Therefore it cannot be ex-

cluded that a remaining population of normal stem cells, indistinguishable from leukemic cells by cytologic criteria, is responsible for the proliferation of colonies.
Only those colonies with an abnormal cytological composition, for example those containing blasts, presumably develop out of leukemic cells.
But in blast colonies we saw all cell types of normal colonies, or only macrophages, which proliferated as well. If we assume that all colonies are of clonal origin we must conclude that some of the leukemic cells develop further into mature cells.

So far, the proliferation of plasma cells in colonies of normal hemopoietic cells has not been described. In our investigations, too, we could not find this cell type in cell cultures derived from 43 normal persons (table 1). The proliferation of plasma cells exclusively in colonies from leukemic bone marrow supports the assumption of some connection between tumor cells and the immunologic system.
There are two possible explanations for the origin of plasma cells in agar culture: In fact plasma cells develop out of B-lymphocytes, which must have been programmed before in the organism. But supposing that each colony is of clonal origin (2) granulopoietic cells as well as plasma cells might proliferate out of a common stem cell. On the other hand the stem cell in agar culture is assumed to be myeloid committed. Therefore the origin from a common stem cell seems to be improbable - for both, granulopoietic cells and plasma cells. The second explanation: Plasma cells originate from B-lymphocytes, which are contained in the plated mononuclear cell suspension of bone marrow cells. B-lymphocytes, accidentally laying in the agar where the colony is forming, could be stimulated by contact with leukemic cells or macrophages of the colony, and the resulting plasma cells could become integrated into the colony.
We would like to speculate about a cellular interaction between immunocompetent cells and tumor cells in these mixed colonies of blasts and plasma cells. This interaction might happen in vitro or could have already worked in vivo. Macrophages are apparently able to stimulate B-lymphocytes to become plasma cells in vitro.
Further investigations will show whether the immunocytologically demonstrated product of plasma cells is a specific immunoglobulin acting against tumor cells.

References


1. Paran, M., L. Sachs, Y. Barak, R. Resnitzky: In vitro induction of granulocyte differentiation in hematopoietic cells from leukemic and non-leukemic patients. - Proc.nat.Acad.Sci., USA 67, 1542 (1970).
2. Chervenick, P.A., L.D. Ellis, S.F. Pan, A.L. Lawson: Human leukemic cells: in vitro growth of colonies containing the Philadelphia (Ph[1]) chromosome. - Science 174, 1143 (1971).
3. Iscove, N.N., J.S. Senn, J.E. Till, E.A. McCulloch: Colony formation by normal and leukemic human marrow cells in culture: effect of conditioned medium from human leukocytes. - Blood 37, 1 (1971).
4. Robinson, W.A., J.E. Kurnick, B.L. Pike: Colony growth of human leukemic peripheral blood cells in vitro. - Blood 38, 500 (1971)
5. Aye, M.T., J.E. Till, E.A. McCulloch: Growth of leukemic cells in culture. - Blood 40, 806 (1972).
6. Moore, M.A.S., D. Metcalf, N. Williams: Human leukemic colony formation in agar culture. - in: D.W. van Bekkum and K.A. Dicke (Eds.), In vitro culture of hemopoietic cells. Radiobiological Institute TNO, Rijswijk 1972, p. 334.
7. Moore, M.A.S., N. Williams, D. Metcalf: In vitro colony formation by normal and leukemic human hematopoietic cells: characterization of the colony-forming cells. - J.nat.Canc.Inst. 50, 603 (1973).
8. Bull, J.M., M.J. Duttera, E.D. Stashick, J. Northup, E. Henderson P.P. Carbone: Serial in vitro marrow culture in acute myelocytic leukemia. - Blood 42, 679 (1973).
9. Duttera, M.J., J.M. Bull, J.D. Northup, E.S. Henderson, E.D. Stashick, P.P. Carbone: Serial in vitro bone marrow culture in acute lymphocytic leukemia. - Blood 42, 687 (1973).
10. McCulloch, E.A., T.W. Mak, G.B. Price, J.E. Till: Organization and communication in populations of normal and leukemic hemopoietic cells. - Biochim.biophys.Acta 355, 260 (1974).
11. Stohlman, F., Jr.: The leukemic cell in vivo and in vitro. - in: R. Neth, R.C. Gallo, S. Spiegelman, F. Stohlman, Jr. (Eds.), Modern trends in human leukemia. J.F. Lehmanns Verlag, München and Grune & Stratton, New York 1974, p. 50.
12. Undritz, E.: Hämatologische Tafeln Sandoz. - II.Auflage, Sandoz 1972.
13. Hayhoe, F.G.J., D. Quaglino, R. Doll: The cytology and cytochemistry of acute leukaemias. - London, Her Majesty's Stationery Office, 1964.
14. Löffler, H.: Cytochemische Klassifizierung der akuten Leukosen.- in: A. Stacher (Ed.), Chemo- und Immuntherapie der Leukosen und malignen Lymphome. Wien, Bohmann 1969, p. 120.
15. Beckmann, H., R. Neth, G. Gaedicke, G. Landbeck, U. Wiegers, K. Winkler: Cytology and cytochemistry of the leukemic cell. - in: R.Neth, R.C. Gallo, S. Spiegelman, F. Stohlman, Jr. (Eds.), Modern trends in human leukemia. J.F. Lehmanns Verlag, München and Grune & Stratton, New York 1974, p. 26.
16. Aur, R.J.A., J. Simone, H.O. Husta, T. Walters, L. Borella, C. Pratt, D. Pinkel: Central nervous system therapy and combination chemotherapy of childhood lymphocytic leukemia. - Blood 37, 272 (1971).
17. Böyum, A.: Separation of leukocytes from blood and bone marrow.- Scand.J.clin.Lab.Invest. (Suppl.97) 21, 1 (1968).
18. Pike, B.L., W.A. Robinson: Human bone marrow colony growth in agar-gel. - J.cell.Physiol. 76, 77 (1970).

19.Beckmann, H., H. Soltau, M. Garbrecht, K. Winkler: Plasmazell-kolonien aus leukämischem Knochenmark in vitro. - Klin.Wschr. 52, 603 (1974).
20.Bruch, Ch.: personal communication
21.Mintz, U., L. Sachs: Differences in inducing activity for human bone marrow colonies in normal serum and serum from patients with leukemia. - Blood 42, 331 (1973).
22.Testa, N.G., B.J. Lord: A technique for the morphological examination of hemopoietic cells grown in agar. - Blood 36, 586 (1970).
23.Leder, L.D.: Der Blutmonocyt. - Springer-Verlag, Berlin - Heidelberg - New York 1967.
24.Craddock, C.G., E.F. Hays, N.R. Forsen, D. Rodensky: Granulocyte-monocyte colony-forming capacity of human marrow: a clinical study. - Blood 42, 711 (1973).
25.Moore, M.A.S., E. Ekert, M.G. Fitzgerald, A. Carmichael: Evidence for the clonal origin of chronic myeloid leukemia from a sex chromosome mosaic: clinical, cytogenetic, and marrow culture studies. - Blood 43, 1 (1974).
26.Chervenick, P.A., D.R. Boggs: In vitro growth of granulocytic and mononuclear cell colonies from blood of normal individuals.- Blood 37, 131 (1971).
27.Haskill, J.S., T.A.NcNeill, M.A.S. Moore: Density distribution analysis of "in vivo" and "in vitro" colony forming cells in bone marrow. - J.cell.Physiol. 75, 167 (1970).
28.Moore, M.A.S., D. Metcalf: Cytogenetic analysis of human acute and chronic myeloid leukemic cells cloned in agar culture. - Internat.J.Cancer 11, 143 (1973).

---

Acknowledgement:

The authors are indebted to Professor H. Busch, director of the blood bank of the University Hospital Hamburg-Eppendorf, who kindly supplied research material, and to Miss H. Stührk for her excellent technical assistance.

*Progress in Differentiation Research, ed. N. Müller-Bérat et al.*



# TURNOVER AND SHEDDING OF CELL-SURFACE CONSTITUENTS IN NORMAL AND NEOPLASTIC CHICKEN CELLS

M. Kapeller*, Y.M. Plesser, N. Kapeller and F. Doljanski

Department of Experimental Medicine and Cancer Research
Hebrew University-Hadassah Medical School, Jerusalem, Israel

## 1. Introduction

The surface membrane is a cell organelle undergoing rapid turnover (1-5). Much has been learned in recent years about the biochemical and morphological pathways leading to the biogenesis of membrane components, but as to the elimination process, knowledge is rudimentary.

We attempted to approach this problem by labelling cell-surface components specifically in order to study their metabolic fate. It was found (6) that when confluent monolayers of chicken embryo cells (CEC) are labelled with $[^3H]$ or $[^{14}C]$ glucosamine, it is the cell-surface carbohydrate-containing macromolecules that are predominantly labelled, confirming earlier observations of Onodera and Sheinin on 3T3 cells (7). These surface components were found to be released into the culture medium. The DEAE-cellulose profiles of these components were identical to the profiles of macromolecules released from intact cells in sister culture by mild trypsin treatment. In the case of mouse (C3H) fibroblasts, it was found that the molecules in the medium carried an antigenic surface determinant - the H-2 specificity. It was concluded that, as a consequence of surface membrane turnover, certain membrane components are continuously shed from the cell into the environment, and the term shedding was proposed for this process. In the present communication the kinetics of the biosynthesis of these surface components and their shedding are analysed. Experiments designed to gain insight into the mechanism of shedding and its possible significance are also reported.

## 2. Materials and methods

Cells: Normal cells: Secondary cultures of chicken embryo cells (normal CEC) grown in medium 199 (Gibco) supplemented with 10% tryptose phosphate, 5% heat inactivated calf serum and antibiotics, were grown in 35 or 50mm plastic petri dishes (Nunc or Falcon), coated with 0.2% gelatin solution. Transformed cells: CEC were infected with Schmidt-Ruppin Rous Sarcoma Virus.

Labelling procedure: The medium of confluent monolayer CEC culture was replaced by medium containing insulin (400 units/Liter), in place of serum (insulin medium). 24-48 hours later radioactive glucosamine was added to this conditioned medium. In some experiments the labelling was done in fresh 199-insulin medium. D-$[6-^3H]$

* Died in the Yom Kippur War, October, 1973.

glucosamine HCl (Amersham) (sp. act. 10-25 mCi/mmol) 1.5-3.5 μCi/ml, or D-[u-$^{14}$C] glucosamine HCl in 3% ethanol, (sp. act. 318 mCi/mmol) 0.1-0.5 μCi/ml were used.

Preparation of "trypsinisate": Labelled cultures were incubated for 10 min at 37°C with 0.1 mg/ml trypsin (Difco 1-250 or Worthington x2 Cryst.) in Ca++ and Mg++ free phosphate-buffered saline. This enzyme concentration was the minimal amount needed for maximal release of macromolecules. The trypsin solution containing the radioactive macromolecules liberated from the cell was centrifuged in cold for 10 min at 2000g. The supernatant termed "trypsinisate", free of cells, was assumed to contain only cell-surface components on the following grounds:

1. Cell number and viability in the monolayer culture remained practically unaltered after trypsinisation. This excluded the possibility that macromolecules in the trypsinisate were due to hydrolysis of moribund or dead cells.
2. When cells were treated with $I^{125}$ labelled trypsin (2.5 mg/ml) for 15 min, followed by thorough washing in order to determine the amount of trypsin remaining with the cells, it was found that only 0.013% of radioactivity remained within the cell pellet. This represents 0.3 μgr/ml trypsin, which was shown not to liberate glucosamine-labelled macromolecules from the cells.
3. When cells were prelabelled for 24 hours with $^{3}$H-leucine, only minute amounts of label were found to be released by trypsin, indicating that the cells do not become "leaky".
4. Sepharose-bound trypsin released more than 50% of the amount of radioactivity released by the standard 0.1 mg/ml solution.

These results strongly indicate that most of the material released by trypsin from intact cell is derived from components outside the permeability barrier and is not of intracellular origin. This is in line with the observations of Snow and Allen (8).

Shedding: At the end of the labelling period, the monolayers were washed 3 times, and fresh medium or conditioned-medium, containing insulin in place of serum, was added to the cultures. At specified times, medium was collected and centrifuged for 10 min at 2000g, to precipitate cells that might have become detached from the monolayer. The radioactivity of samples from the supernatant - called shed - was determined. Each determination was done in duplicate.

Scintillation counting: Samples of 0.5-1 ml from the shed or the trypsinisate material were transferred to vials containing 4-10 ml of scintillation fluid [4 gr PPO and 0.1 gr POPOP/Liter toluene (Mallinckrodt-Scintill AR) with Triton x-100, in the ratio 2:1 respectively]. Counting was carried out in a liquid scintillation counter (Tri Carb, Packard) at 4°C, with correcting for the quenching by external standard (A.E.S.) and channel ratio.

3. Results

Biosynthesis of cell-surface molecules: The biosynthesis of the surface membrane can be studied by labelling cells with metabolic precursors and following their rate of incorporation into isolated whole surface membrane or into a particular membrane constituent. We chose to examine the most peripheral part of the plasma membrane, the trypsin-removable ("trypsinisate") surface moiety, which we assumed to be involved in cell-environment interactions. In order to determine the kinetics of their synthesis, confluent monolayers of CEC were labelled with radioactive glucosamine, and the incorporation of the isotope into the trypsinisate (as >85% nondialysable molecules) was examined (Fig. 1).

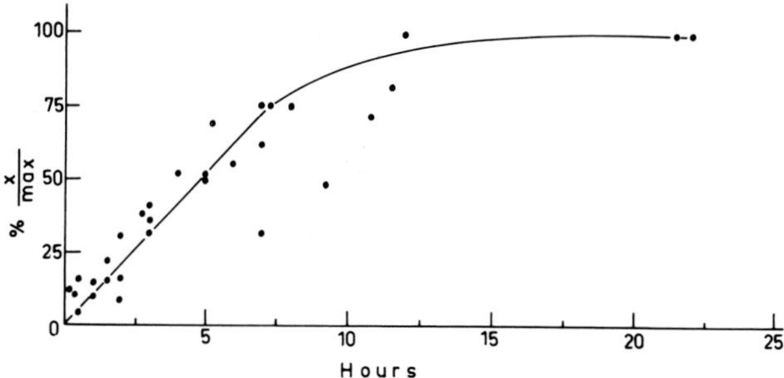


Fig. 1 Incorporation of glucosamine-$^3$H into "trypsinisate" of chick embryo cells. $\frac{X}{Max}$ = radioactivity at each time point/maximum radioactivity attained in the experiment. Summary of 5 experiments.

During the first 8-10 hours a rapid rate of incorporation was observed with no initial lag period, followed by a much slower rate of increase, which reached a plateau at 25-30 hours, indicating that these membrane components are actively synthesized. The plateau represents the steady state where incorporation equals elimination. The kinetics of the incorporation process, however, should be interpreted with caution (1), since the curve will only reflect true rate of synthesis if a number of conditions are fulfilled. The label, for example, must be present in excess and should not be reutillized, and the rate of entry of the isotope into the cell, as well as pool size, must not alter during the labelling period. It is therefore much more accurate to determine turnover rates from decay curves, which in a steady state situation must be the mirror-image of the real incorporation curves. For this purpose, cells were labelled for 24 hours in order to label most of trypsin-removable surface components, and then their rate of elimination was examined.

Elimination of cell-surface molecules: After labelling, the radioactive medium was removed and a non-radioactive medium added. Thereafter, at different times, the amount of radioactivity in the trypsinisate was determined and compared to the amount that was present at the end of the labelling period (trypsinisate $t_o$). It was observed that the specific activity dilution from the cell-surface exhibited a biphasic pattern (Fig. 2), with a fast component ($t_{1/2}$ 4-5 hours) and a slow component ($t_{1/2}$ of the order of 2,5-3,5 days). This slow component comprised about 50% of the total radioactivity in the trypsin-removable surface moiety-"trypsinisate $t_o$".

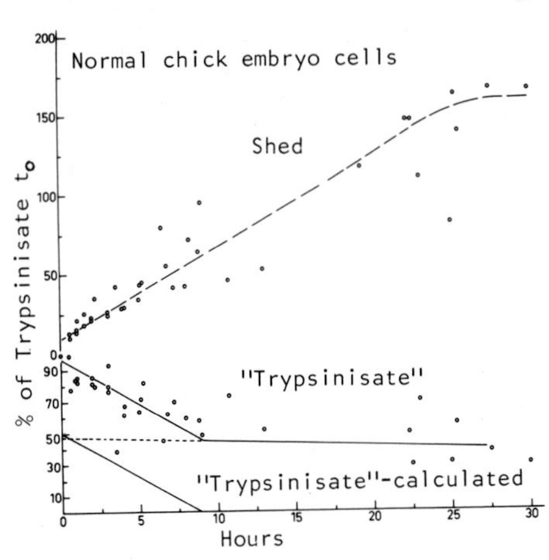


Fig. 2 Kinetics of elimination of radioactivity from the cell surface ("trypsinisate"), and accumulation of radioactive macromolecules in the medium ("shed") of normal chick embryo cells, as percentage of the total radioactivity at the end of labelling ("trypsinisate $t_o$").

The rate of elimination of the rapid components was calculated by subtracting the values of the slow rate of elimination from the total rate. The results summarize five experiments.

Shedding: If the metabolic fate of these molecules is to be shed into the medium, as we previously suggested (6), then the rate of their accumulation in the medium must parallel their rate of elimination from the cell-surface.

Fig. 2 depicts the kinetics of their appearance in the medium. The rate of appearance of the rapidly shed macromolecules was identical to the rate of their elimination from the cell. The rate of accumulation of the slower components was higher than their rate of elimination. It should be recalled that these molecules are identical in charge and size to the molecules in the trypsinisate (6, 9).

Effect of inhibitors on the turnover process: In order to get some insight into the mechanism of the biosynthesis and shedding of these macromolecules, the effect of different metabolic inhibitors was examined. Shedding was markedly affected by temperature (Fig. 3). At 4°C the process was almost completely inhibited. The rate of loss of radioactivity from the trypsinisate also decreased correspondingly, as expected. Sodium azid had only a slight inhibitory effect on shedding at non-toxic concentrations ($10^{-5}$-$10^{-2}$M).

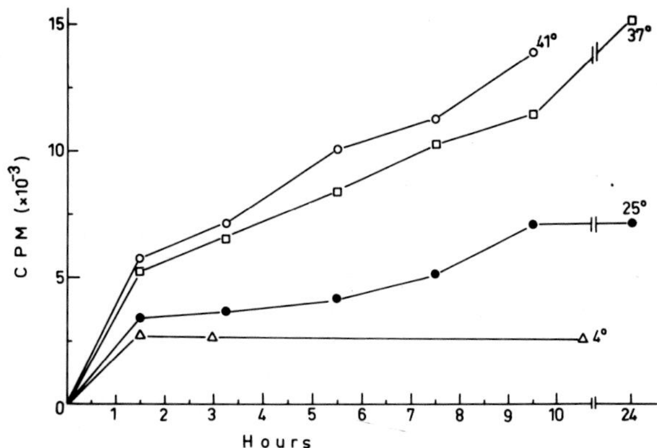


Fig. 3 Effect of temperature on shed accumulation. Cells were prelabelled with glucosamine $^{14}C$ for 24 hours in 37°C, and then incubated with non-radioactive insulin medium. Accumulation of radioactivity in the medium was determined as cpm/sample.

Cycloheximide (10 μg/ml) was found to inhibit biosynthesis by 30-50% after a lag period of 2 hours, whereas protein synthesis was almost immediately inhibited to >95% of the control values. The inhibitory effect of this drug on shedding was already detected after 1 hour, and reached 30-50% after 2 hours.

Preliminary results on the effect of soybean trypsin inhibitor, at non-toxic concentration, in the range of 5-100 μg/ml indicated that shed was inhibited by about 40% and less, with doses lower than 100 μg/ml. Trasylol (50 units/ml) also inhibited shedding by approximately 30%.

Biosynthesis and shedding of SR-RSV transformed CEC: The kinetics of loss of radioactivity from the trypsinisate and its accumulation in the medium also displayed a biphasic pattern in the transformed cell. Two differences could be detected in comparing these cells to normal CEC: the rapidly turning over molecular population exhibited a shorter half life ($t_{1/2} = \sim 2.5$ hours), and the accumulation of the slower component in the medium was twice that of normal cells.

Do shed molecules affect the shedding process? Already in 1965, Rubin (10) proposed a dynamic surface membrane model, where membrane sub-units are released into the medium and can be reinserted into the membrane. He suggested that this interchange is important to maintain the stability of the membrane. To test this model, we examined the effect of shed molecules and cell population density on the shedding process. A 24 hour medium of confluent CEC cultures was filtered by Daiflo membranes, (Amicon, XM 10, PM 30), in order to increase the concentration of the shed molecules 10-40 times, and these were added to cultures at the end of the 24 hour glucosamine labelling period in order to determine their effect on shed accumulation kinetics. It can be seen (Fig. 4) that concentrated shed in the medium slowed the shedding process.

Fig. 4 Effect of shed molecules on shedding and trypsinisate. Cells prelabelled for 18 hours with glucosamine-$^3$H, and then incubated with either non-radioactive insulin medium (control) or with 28 hours conditioned medium from sister cultures concentrated 10 times by PM 30 Diaflo membrane (+ shed). Control medium was also filtered. ----"trypsinisate".——shed.

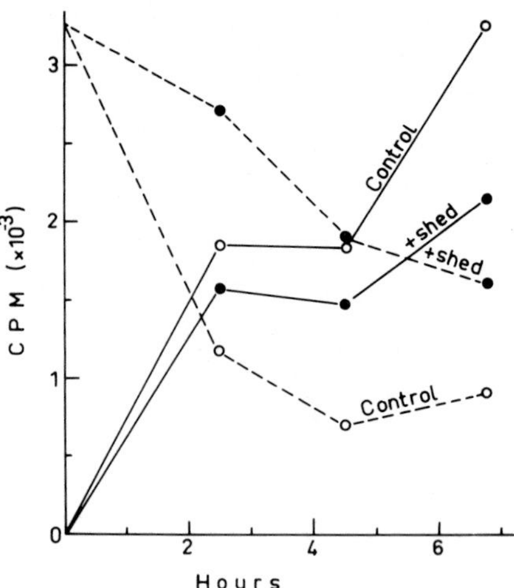


Similarly, it was found that the higher the cell population density in the dish, the slower the shedding process.

Surface membrane biosynthesis after trypsinisation: In many systems it was shown that surface moieties removed by trypsin are regenerated within several hours (11). Regeneration could be brought about either by the normal turnover process or by a biosynthetic burst in response to a deficit. It was found that there is no increase, but rather a decrease, for 20 hours, in the rate of biosynthesis after trypsinisation. This decrease could be due to residual trypsin adherent on the cells after washings.

## 4. Discussion

We are dealing here with certain cell-surface components that can be metabolically labelled with glucosamine, a precursor known to label glycoprotein and glycosaminoglycans (11), which are part of the external face of the plasma membrane. It is reasonable to suspect that molecules located in a cell-site of this type are involved in cell-cell and cell-molecular interactions. Our results clearly demonstrate that these molecular entities are actively synthesized and the kinetics of their elimination from the cell surface show that there are at least two molecular populations involved - one with a short life span of about 8-10 hours and a second population with a much longer life span of 5-7 days. (The possibility that these represent two different cell populations cannot be excluded, however.) It is of interest that in most systems studied, the time required for full recovery of different surface molecules, such as transplantation antigens (12, 13, 14), sialic acid (15) or lectin receptors (6), after their removal with proteolytic enzymes, was also found to be several hours. Thus, it seems that the rapidly turning over molecular species may be carriers of many of the surface recognition structures. The rate of synthesis of "cell coats" or microexudate (16) is also similar to that of the rapidly turning over components. Cell coats, dish-bound material (17), "carpets" (18), and trypsin-removable components, all probably represent the same type of cell-surface components.

The life span of the longer-lived components is in the range observed by Glick & Warren (2) and Kaplan & Moskowitz (4) for metabolically labelled surface membranes. Recently Hubbard & Cohn (3), studying the turnover and fate of iodine labelled surface proteins of Hela cells grown as a suspension culture, and Kaplan & Moskowitz (4), examining the turnover of plasma membranes of monkey epithelial kidney cell grown as monolayer cultures, also observed biphasic rates of elimination of labelled membrane components. These observations corroborate our conclusion that at least two short-lived (hours) and long-lived (days) molecular population comprise the surface membrane.

The fact that the curves depicting rate of surface membrane biosynthesis exhibit no lag period after initiation of labelling indicates that there are no prominent intracellular pools of formed membrane component. The partial inhibition of biosynthesis caused by cycloheximide may indicate that about half of the trypsin-removable cell-surface components are glycoproteins. The glucosamine labelled macromolecules not inhibited by cycloheximide may be glycosaminoglycans such as heparane sulphate (19) or chondroitin sulphate (20) - all shown to be cell-surface constituents in at least certain cell types. The 2 hour lag period observed in the inhibitory effect of cycloheximide probably indicates that certain reserve pools of non-glycosilated membrane protein are present in the cells.

The fate of these surface molecules can be deduced from the kinetics of the appearance of labelled macromolecules in the medium. The fact that their rate of appearance during the first 10 hours after labelling was identical to the rate of the elimination of radioactivity in the trypsinisate, strongly suggests that the molecules in the medium are the molecules released from the cell-surface. The identity of the profiles of the shed molecules and the trypsinisate when chromatographed on DEAE Cellulose (6) or SDS-poly-acrylamide gels (9) lends further support to our previous proposition that the fate of glucosamine-labelled, trypsin-removable macromolecules is to be shed from the cell-surface into the medium. These results corroborate the findings of Kraemer (21) on hamster ovary cell line. Glick & Warren (2), in their pioneering work, mentioned the possibility that surface components are shed into the medium. They believed, however, that it is a characteristic only of the non-dividing cell in vitro. We assume that it is a continuous process taking place in dividing cells both in vitro and in vivo. Kaplan & Moskowitz (4) and Hubbard and Cohn (3) came to the conclusion that most membrane proteins are eliminated by internal proteolysis. The latter authors, however, found that a large surface glycoprotein is continuously released into the medium.

The shedding of surface components was recently described in several other systems. The B-lymphocyte is the most thoroughly studied cell type. The biosynthesis and shedding of a B-lymphocyte surface protein - the IgM - has been described in great detail (22-24) and shedding of several other lymphoid cell-surface markers has also been reported (25). Other cell types, such as tumor cells, have been reported to release surface glycoproteins (26).

If shedding of surface components is indeed the normal physiological consequence of surface membrane turnover, then presence of cell-surface determinants in the biological fluid of the organism is to be expected. In this context, the experiments of Ruoslahti et al. (27) are especially interesting. These authors found that antisera against a fibroblast specific surface antigen give a precipitation reaction of complete immunological identity with chicken serum. The same was demonstrated with human fibroblasts and serum (28).

In line with these observations are the findings on the release of certain surface glycoproteins from tumors into serum in vivo (26). The presence of CEA in serum of patients with gastrointestinal tumors can be interpreted in the same way (29).

These observations lead to the question of the role, if any, of these shed macromolecules. Are they just waste products of the turnover process, or do they play some physiological role? The fact that the concentration of shed molecules in the medium affects rate of shedding lends support to Rubin's hypothesis (10) that there is a dynamic interchange between membrane components within the membrane and those released from the cells.

References

1. Schimke, R.T., 1975, in: Methods in Membrane Biology, vol. 3, ed. E.D. Korn (Plenum Press, N.Y.) p. 201.
2. Warren, L. and Glick, M.C., 1968, J. Cell Biol. 37, 729.
3. Hubbard, A.L. and Cohn, Z.A., 1975, J. Cell Biol. 64, 461.
4. Kaplan, J. and Moskowitz, M., 1975, Biochim. Biophys. Acta 389, 290.
5. Leblond, C.P. and Bennett, G., 1974, in: The Cell Surface in Development, ed. A.A. Moscona (John Wiley & Sons, N.Y.) p. 29.
6. Kapeller, M., Gal-Oz, R., Grover, N.B. and Doljanski, F., 1973, Exp. Cell. Res. 79, 152.
7. Onodera, K. and Sheinin, R., 1970, J. Cell Sci. 7, 337.
8. Snow, C. and Allen, A., 1970, Biochem. J. 119, 707.
9. Kapeller, M., Plesser, Y.M. and Doljanski, F., 1975, in preparation.
10. Rubin, H., 1966, in: Major Problems in Developmental Biology, ed. M. Locke (Academic Press, N.Y.) p. 317.

11. Kraemer, P.M., 1971, in: Biomembranes, vol. 1, ed. L.A. Manson (Plenum Press, N.Y.) p. 67.

12. Turner, M.J., Strominger, J.L. and Sanderson, A.R., 1972, Proc. Nat. Acad. Sci. U.S.A. 69, 200.

13. Schwartz, B.D., Wickner, S., Rajan, T.V. and Nathenson, S.G., 1973, Transplant.Proc. V, 439.

14. Bhandari, S.C. and Singal, D.P., 1973, Tissue Antigens 3, 140.

15. Marcus, P.I. and Schwartz, V.G., 1968, in: Biological Properties of the Mamalian Surface Membrane, ed. L.A. Manson (Wistar Inst. Press, Phil.) p. 143.

16. Poste, G., Greenham, L.W., Mallucci, L. and Reeve, P., 1973, Exp. Cell Res. 78, 303.

17. Culp, L.A., Terry, A.H. and Buniel, J.F., 1975, Biochem. 14, 406.

18. Yaoi, Y. and Kanaseki, T., 1972, Nature 237, 283.

19. Kraemer, P.M. and Smith, D.A., 1974, Biochem. Biophys. Res. Commun. 56, 423.

20. Suzuki, S., Kojima, K. and Utsumi, K.R., 1970, Biochim. Biophys. Acta 222, 240.

21. Kraemer, P.M. and Tobey, R.A., 1972, J. Cell Biol. 55, 713.

22. Melchers, F. and Cone, R.E., 1975, Europ. J. Immunol. 5, 234.

23. Vitetta, E.S. and Uhr, J.W., 1972, J. Exp. Med. 136, 676.

24. Nossal, G.J.V., Warner, N.L., Lewis, H. and Sprent, J., 1972, J. Exp. Med. 135, 405.

25. Ramsier, H., 1975, Europ, J. Immunol. 5, 23.

26. Cooper, A.G., Codington, J.F. and Brown, M.C., 1974, Proc. Nat. Acad. Sci. U.S.A. 71, 1224.

27. Ruoslahti, E., Vaheri, A., Kuusela, P. and Linder, E., 1973, Biochim. Biophys. Acta 322, 352.

28. Ruoslahti, E. and Vaheri, A., 1974, Nature 248, 789.

29. Laurence, D.J.R. and Neville, A.M., 1972, Br. J. Cancer 26, 335.

# Section 5.
# HORMONAL INDUCTION OF CELL DIFFERENTIATION

*Progress in Differentiation Research, ed. N. Müller-Bérat et al.*



# THE CELL BIOLOGY OF ACTION OF GROWTH AND DEVELOPMENTAL HORMONES

J.R. Tata

National Institute for Medical Research,
Mill Hill, London NW7 1AA, England

## 1. Introduction

Most growth and late embryonic developmental processes at one stage or another are regulated by hormones. Table 1 lists a few growth and developmental hormones which act on cells in a variety of plant and animal tissues.

Table 1

Initiation or maintenance of functions obligatorily dependent on hormones

| Hormone | Function |
|---|---|
| Ethylene | Fruit ripening |
| Ecdysone | Insect metamorphosis |
| Thyroxine | Amphibian metamorphosis; CNS maturation |
| Oestradiol | Growth and maturation of uterus; egg protein synthesis in birds and amphibia |
| Testosterone | Growth and maturation of prostate, seminal vesicles |
| Prolactin | Milk protein synthesis |

There are two important points to be retained from table 1: (a) that there is no one class of substance that acts as growth and developmental hormones; (b) that the same hormone has multiple actions in the same or different target cells.

## 2. Hormones and the regulation of protein synthesis

It is now well known that growth and developmental hormones exert their action by a regulation of protein synthesis at both the transcriptional and translational levels. It is also known that the same hormones provoke multiple responses, some of which are manifested more rapidly than those directly concerned with protein synthesis (permeability to ions, amino acids, sugars; intracellular levels of cyclic nucleotides). What is poorly understood from a purely biochemical point of view is how the multiple actions of the hormones are co-ordinated to give a well integrated expression of growth and development. A wider cell biological approach may prove to be more fruitful in exploring the co-ordination. In this presentation, therefore, particular emphasis will be laid on the following two aspects of the action of developmental hormones:

(a) The role of the structural organization of the nucleus in influencing transcription, modification and processing of RNA.

(b) The possibility of segregation on membranes of the endoplasmic reticulum of different populations of polysomes synthesizing different classes of proteins involved in the developmental process.

These two aspects will be illustrated by results of experiments on oestrogen-induced synthesis of egg proteins in birds and amphibia and on thyroid hormone-induced amphibian metamorphosis. However, the mechanisms discussed are applicable to all growth and late embryonic developmental processes, whether or not regulated by hormones.

(a) The nucleus

Most growth and developmental processes which are triggered off by hormones, such as in metamorphosis or maturation of eggs, are dependent on a rapid stimulation of synthesis of all species of RNA and their processing. Since transcription of different classes of RNA and their processing and turnover occur in different nuclear compartments (euchromatin, nucleoli, nucleoplasm, etc.), a consideration of the sequential events in the nucleus as a function of the different compartments in which they occur may be helpful in elucidating the co-ordination between separate mechanisms of gene transcription.

An example of sequence of events at the level of transcription in the nucleus, preceding the stimulation of growth by a hormone, is illustrated in fig. 1 for growth hormone acting on rat liver. Similar sequences have been recorded for other hormone-induced growth and developmental systems also, in particular for steroid hormones.

The induction of ovalbumin in birds and of vitellogenin in amphibia by the hormone oestradiol-17β has further revealed the following sequence of nuclear events:

(i) A subtle and specific change in the composition of acidic (non-histone) chromosomal proteins of the target cell accompanied by a small and transient increase in template capacity of chromatin and euchromatin bound RNA polymerase B (II) activity very soon after hormone administration.

(ii) A lag period during which the preferential synthesis of a few minor proteins in the cytoplasm is critical for further nuclear response.

(iii) A massive and sustained increase in the rate of synthesis of all species of

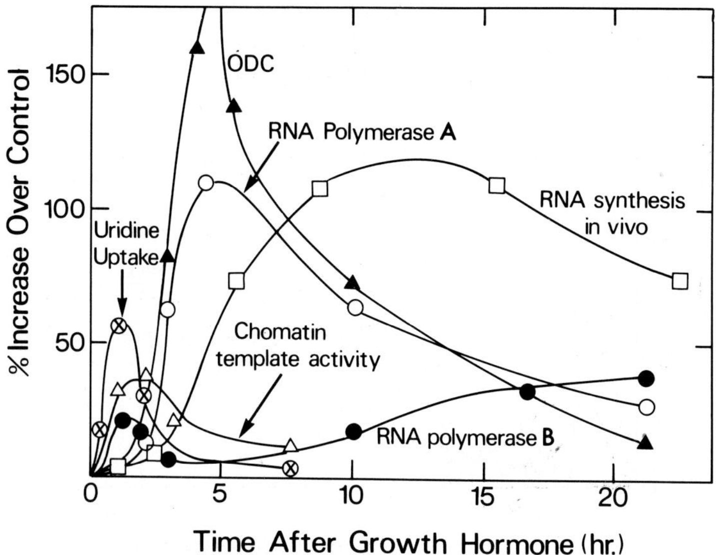


Fig. 1. An idealized set of curves illustrating the response of the liver of a hypophysectomized rat following the administration of growth hormone (somatotrophin). ODC = ornithine decarboxylase.

RNA, particularly that of ribosomal RNA and RNA polymerase A (I) activity in nucleoli.

(iv) An almost simultaneous enhancement of polyadenylation of nuclear RNA, increased content of high molecular weight poly(A)-containing RNA associated with euchromatin and an accelerated breakdown of newly-synthesized RNA recovered in nuclear sap.

(v) A relatively late but sustained elevation of RNA polymerase B activity and increase in nuclear ribonucleoprotein particles containing ribosomal and non-ribosomal RNA (presumably messenger RNA's for induced proteins) followed by their accelerated accumulation in cytoplasmic polyribosomes.

(b) The cytoplasm

A hormone-induced acceleration of nuclear RNA synthesis is soon followed by an increased rate of protein synthesis. As illustrated in fig. 2 for growth hormone, this acceleration of the rate of protein synthesis is achieved by a combination of enhanced translational capacity of the ribosome as well as by the overall increase in the number of ribosomes per cell.

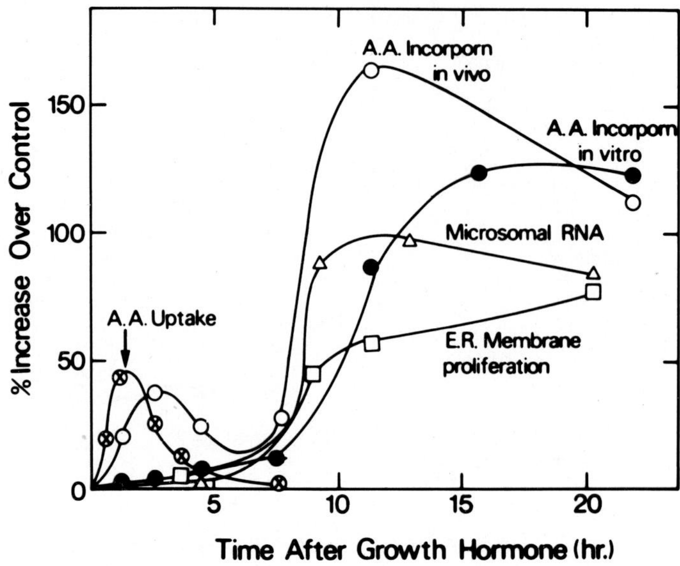


Fig. 2. An idealized set of curves depicting the incorporation of amino acids (a.a.) into protein by liver following the administration of growth hormone to a hypophysectomized rat.

Even when growth is not the primary action of a hormone but the initiation of development, based on the induction or preferential synthesis of some proteins, there occurs a massive increase in the number of ribosomes and a proliferation of the rough endoplasmic reticulum. This type of change is illustrated for the thyroid hormone-induced synthesis of urea cycle enzymes in frog tadpole liver (fig. 3) and for the synthesis of the egg-yolk protein precursor, vitellogenin, by oestrogen in male or female adult frog liver (fig. 4).

The detection of induced protein or the accelerated rate of protein synthesis is particularly characterized by the following changes occurring almost simultaneously in the translational machinery.

(i) An increase in the size, amount and stability of polyribosomes of the target cell accompanied by an enhancement of the amino acid incorporating activity per ribosome.

(ii) A greater capacity of soluble translational factors and an altered pattern of tRNA's to support polypeptide chain initiation, elongation and release.

(iii) A proliferation of membranes of the endoplasmic reticulum to which increasing number of newly-synthesized ribosomes are preferentially attached, whether the developmental change involves enhanced secretory or intracellular proteins.

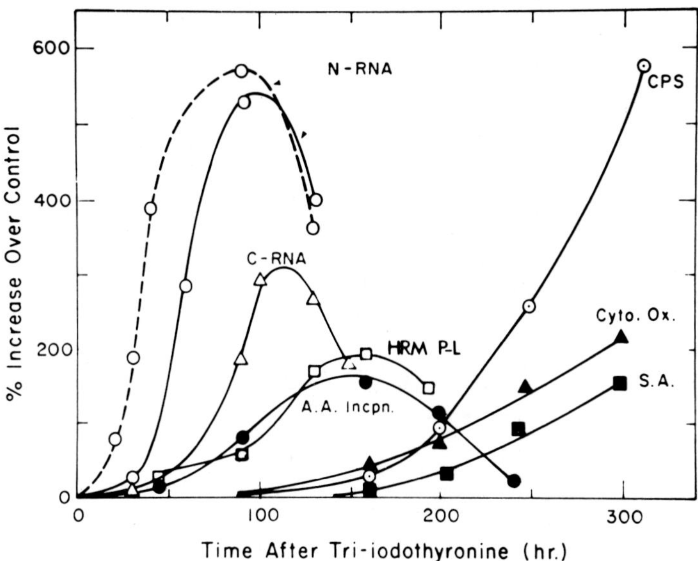


Fig. 3. Schematic representation of the responses of the hepatocyte of tadpole (Rana catesbeiana) liver to the administration of tri-iodothyronine ($T_3$) to induce precocious metamorphosis. An increase in the rates of synthesis of nuclear (N-RNA) and cytoplasmic (C-RNA) RNA's is followed by the proliferation of rough endoplasmic reticulum (HRM-PL) and the capacity of ribosomes to synthesize proteins (A.A. incpn.). Some hours later, one notices the appearance of carbamyl phosphate synthetase (CPS), proliferation of mitochondria (Cyto. ox.) and the secretion of serum albumin (S.A.).

The last point raises the question of the importance during development of a possible segregation of different populations of polyribosomes synthesizing different classes of protein. The role of membranes of the endoplasmic reticulum in post-translational modification of proteins, such as phosphorylation, glycosylation, lipidation and activation of enzymes, may also determine the requirement for the attachment of ribosomes to membranes.

3. Integration of early and late responses

It can be seen from figs. 1 and 2 that growth hormone caused a very rapid increase in the uptake by the liver of precursors for the synthesis of RNA and

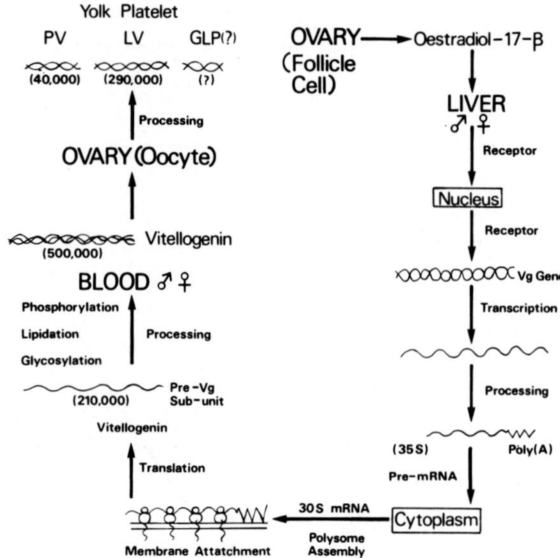


Fig. 4. Sequence of events leading to the induction of vitellogenin by oestradiol-17β in male or female frog, Xenopus laevis.

and protein. At the same time as the above modification of the target cell's transcriptional and translational machinery at multiple levels, very rapid responses are particularly observed in the following properties of the developing cell:

(a) The altered permeability of the cell towards ions (particularly $Ca^{++}$ and $Na^{++}$), amino acids, sugars and nucleosides.

(b) Re-adjustment of the intracellular levels of key metabolites such as cyclic AMP and GMP.

Some of the multiple responses of a target cell to the appropriate growth and developmental hormones are summarized in table 2 as early and late responses. These responses are not restricted to hormones but are common to all animal cells which respond to a variety of other stimuli, such as nutrition, drugs, etc., which act on their protein synthetic apparatus.

The difficulty of explaining the events associated with regulation of RNA and protein synthesis on the basis of these rapid metabolic responses, as a cause-effect relationship, raises the possibility that the multiple hormonal effects

Table 2

Early and late events in hormonal control of protein synthesis

| Early events | Late events |
|---|---|
| Enhanced uptake of ions, amino acids, nucleosides | Accelerated rate of synthesis of all RNA's |
| Alteration of cAMP and cGMP levels | Stimulation of RNA polymerases A and B |
| Redistribution or modification of acidic chromatin proteins | Increase in synthesis of ornithine decarboxylase and nuclear polyamines |
| Small increase in RNA polymerase B activity | Higher content of polysomes and enhanced amino acid incorporation |
| Small increase in amino acid incorporation | Redistribution of free and membrane-bound ribosomes |
| | Accelerated assembly of membranes (endoplasmic reticulum) |

may be initiated at more than one site in the target cell. A model integrating the rapid metabolic responses with the changes in protein synthetic functions is presented in fig. 5 in order to conclude that the ultimate expression of growth and development results from the co-operative interplay between the rapid and slow responses generated by the hormonal trigger.

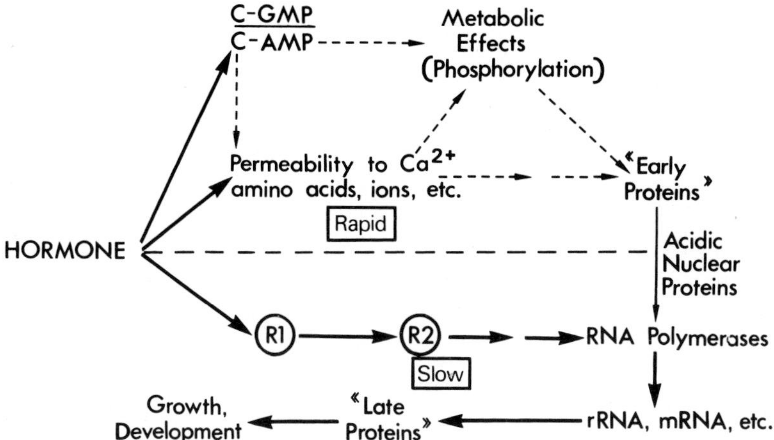


Fig. 5. Scheme summarizing the major "early" and "late" sequential events that follow the hormonal induction of growth or development as based on the regulation of "early" and "late" protein synthesis.

Further reading

1. Tata, J.R.: Regulation of protein synthesis by growth and developmental hormones. In "Biochemical Actions of Hormones", ed. by Litwack, G., vol. 1, (Academic Press, New York), pp. 89-134, 1970.

2. Tata, J.R.: Ribosome-membrane interaction and protein synthesis. 6th Karolinska Institute Symposium, pp. 192-224, 1973.

Progress in Differentiation Research, ed. N. Müller-Bérat et al.



# Induction of maturation (meiosis) in amphibian oocytes by steroid hormones and organomercurials.

by J. Brachet, E. Baltus[°], J. Hanocq, E. Hubert, A. Pays and G. Steinert.

Department of Molecular Biology
University of Brussels
and
Laboratorio di Embriologia molecolare,
Arco Felice (CNR) Naples.

## I. Introduction.

Oogenesis is characterized by intense synthetic activity : the oocyte accumulates enormous quantities of glycogen, lipid, protein (those of the yolk are always of exogenous origin), ribosomes, mRNA, etc. This high activity is seen, from a morphological point of view, as decondensation of the chromosomes (lampbrush chromosomes) and exceptional development of the nucleoli (fig. 1). The ovarian oocyte is in meiotic prophase (diplotene) and is consequently tetraploid; the lampbrush chromosomes are fully active synthetically and produce a considerable number of mRNA molecules; these are stored in the cytoplasm of the oocyte. The huge number of nucleoli (more than 1.000 in Amphibia) underlines the great importance of rRNA synthesis : there are 1 million copies of rRNA genes in the oocyte, while a somatic cell only contains 1.000 in Amphibia; this is called the amplification of ribosomal genes. More information on this very interesting period of development is found in the books by Denis (1974) and Brachet (1974 a,b).

After the oocyte attains its adult size, this synthetic activity declines by a variable degree, depending on the species : while that of the virgin sea urchin egg is nearly completely suppressed, the oocyte of *Xenopus* (which is of special interest to us) is still synthesizing significant quantities of mRNA and protein up until the moment that oogenesis finishes and maturation begins : this process brings about the continuation of meiosis, leading to the expulsion of the first polar body. The latter is diploid, just as oocytes of the 2nd order : it is at this stage that maturation ceases in Amphibia; for the oocyte to become haploid following expulsion of the 2nd polar vesicle, a new stimulus is necessary : fertilization or parthenogenetic activation (by pricking, for example).

(°) Maître de Recherches au Fonds National de la Recherche Scientifique.

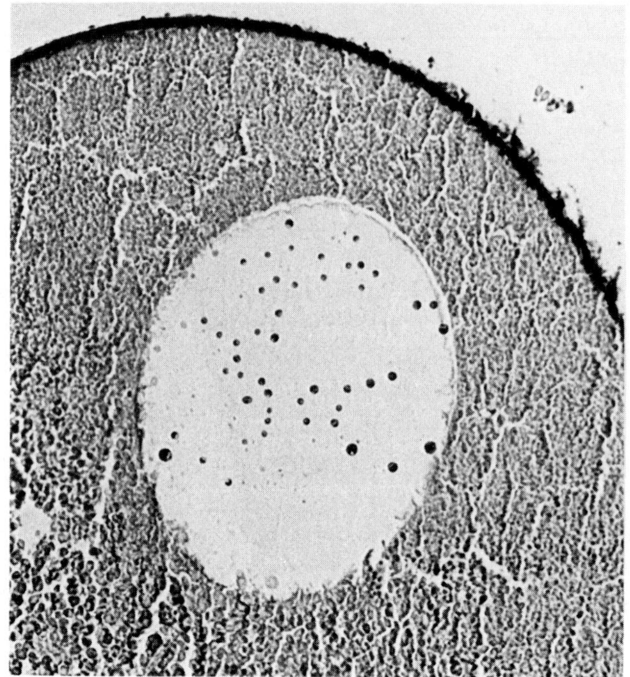

Figure 1.
Ovarian oocyte of Xenopus having reached the stage of growth. The germinal vesicle contains hundreds of nucleoli.

Under physiological conditions, maturation results from increased production of gonadotrophic pituitary hormones; these act on the follicular cells which surround the oocyte and induce there the synthesis of a steroid hormone which is very probably progesterone.

It is possible to induce maturation *in vitro* in Amphibia by isolating the oocytes by dissection, and placing them in Ringer containing a pituitary extract of the same species or progesterone. Numerous experiments have shown that, in such systems *in vitro*, a pituitary extract stimulates follicular cells, while progesterone acts directly on the oocyte; in effect, pituitary hormones become inactive if the follicular cells are removed by treatment with pronase or EDTA, the same effect is produced by inhibition of progesterone synthesis by a specific inhibitor, cyanoketone.

The morphological changes (fig. 2) undergone by the oocyte are identical whether maturation is induced *in vivo* or *in vitro* : breakdown of the nuclear membrane (germinal vesicle : G.V.) at its basal pole, disappearance of the

Figure 2.
Start of maturation. The basal part of the nuclear membrane is in the process of ruptuting.

nucleoli of which only the nucleolar organizers (ribosomal genes) remain condensation of the lampbrush chromosomes (diakinesis) formation of a maturation spindle, migration of the latter to the animal pole, expulsion of the first polar body by reduction division of the chromosomes.

The amphibian oocyte, isolated from the ovary by dissection and placed in a physiological solution containing progesterone is thus ideal material for studying the effects of a hormone on a single target cell. As we shall see, the mechanism of maturation is different, from a molecular viewpoint, from that generally considered when steroid hormones (progesterone, oestrogen) act on their target tissues such as the uterus or the oviduct of mammals and birds. As shown in fig.3, the accepted idea is that the hormone, having crossed the cellular membrane unmodified, attaches to a proteinacous receptor present in the soluble cytoplasm (cytosol); this enzyme-receptor complex induces modifications in the cytoplasm penetrates the nucleus, attaches to the chromatin (directly onto DNA or its associated proteins) and derepresses certain genes : the transcription of new molecules of mRNA results, followed by synthesis of new proteins in the cytoplasm. An essential feature of this mechanism is thus <u>transcription</u> of DNA into mRNA, a step which can be inhibited by actinomycin D.

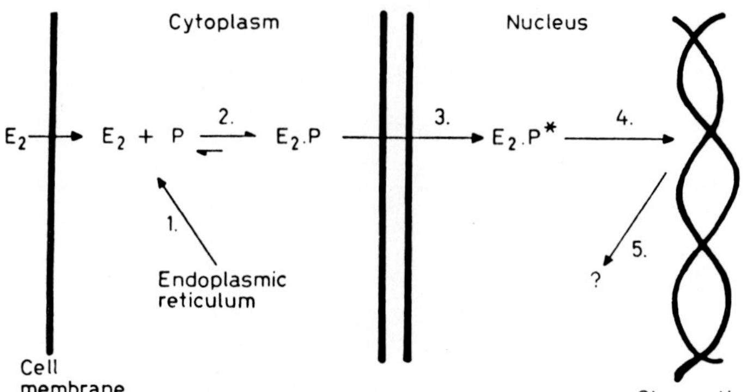


Fig. 3.: Schematic representation of action of an oestrogen P2 on target cell.

P : cytoplasmic receptor
P': modified cytoplasmic receptor in the nuclei.

After these general remarks, let us review the principal characteristics of maturation induced <u>in vitro</u> by steroid hormones, progesterone in particular.

## 2. Main characteristics of the in vivo induction of maturation by progesterone.

We can be brief as the question has been recently the object of several reviews, where many more details can be found (Smith and Ecker, 1970; Schuetz, 1972, 1974; Brachet, 1974; Brachet et al. 1974).

### a) Specificity of the hormone.

Either progesterone, testosterone, hydrocortisone and or aldosterone induce maturation <u>in vitro</u> in <u>Xenopus</u> oocytes (Iacobelli et al., 1974). On the other hand, oestrogens, prostaglandins and cyclic nucleotides are ineffective. The presence of calcium ions in the medium is necessary for progesterone to act.

### b) Role of the plasma membrane, nature of the receptor.

Progesterone is only active if present in the medium; it is ineffective when injected into oocytes. On the other hand, hydrocortisone induces maturation after micro-injection. The reasons why these two hormones behave differently remain unknown; but the fact that progesterone must interact with the membrane to become active indicates that it probably brings about a transformation there before attaching to the cytoplasmic receptor : no similar situation has been found when one studies the effects of steroid hormones on classical target organs.

The time for which progesterone has to be in contact with the oocyte to provoke maturation varies with its concentration in the medium; it is generally short (5-10 min. at 5-10 µg progesterone $ml^{-1}$).

The hormonal receptor is cytoplasmic; but it is not, as in classical cases, in solution in the cytosol: this receptor is bound to the pigment granules (melanosomes). It binds to all hormones which induce maturation with high affinitiy; there is only a weak affinity for other steroids (Ozon and Bellé, 1973; Iacobelli et al., 1974). Attachment of the hormone to the receptor does not require energy production or protein synthesis. (Brachet et al., 1975b).

c) Metabolic changes during maturation induced by progesterone. Effects of various inhibitors.

Addition of progesterone to Amphibian oocytes raises, just before G.V. breakdown, their oxygen consumption. Inhibitors of energy production (KCN, dinitrophenol) reversibly prevent the onset of maturation (Brachet et al., 1975b).

Conversely, inhibitors of nucleic acid synthesis (actinomycin, -amanitin for RNA synthesis, hydroxyurea, 2'deoxyadenosine, ethidium bromide for DNA synthesis) have no effect on the induction of maturation *in vitro* by progesterone. There is only a minimal synthesis of DNA during this period of development, and we have been able to show that this is strictly mitochondrial (Hanocq et al., 1974). The situation concerning RNA is less clear : a recent paper (Morrill et al., 1975) reports that progesterone induces a transitory increase, followed by a strong decrease, in total RNA synthesis; however this increase probably plays no role in maturation itself. In effect, as we have seen, hydrocortisone, which is a good inducer of maturation, produces a marked decrease (and not an increase) in RNA synthesis by oocytes.

Progesterone stimulates, after several hours, protein synthesis in *Xenopus* and frog oocytes (Smith and Ecker, 1970; Baltus et al., 1973). Synthesis of many proteins is stimulated to a similar degree; however, new enzymes always appear in oocytes treated with progesterone notably a new species of DNA polymerase (Grippo and Lo Scavo, 1972; Benbow et al. 1975) and a new protein kinase which phosphorylates histones (Wiblet, 1974). Experiments still in progress (Wiblet and Brachet) suggest that the histone kinase, which appears during maturation could play a role in chromosome condensation and even in maturation itself. Injection of histonekinase isolated from *Xenopus* oocytes produces maturation in axolotl oocytes and a very strong condensation of the lampbrush chromosomes; in other experiments, injection of histone kinase extracted from ascites tumour cells also induced complete maturation after injection in axolotl oocytes.

The effects of a temporary or permanent arrest of protein synthesis by cycloheximide have been the object of several investigations (Baltus et al., 1973; Brachet et al., 1974, 1956, Morrill et al., 1975). It has been demonstrated that maturation remains possible when the level of protein synthesis has fallen by 50%. A large inhibition (95%) blocks maturation so long as cycloheximide is applied during the 3 hours that follow progesterone treatment. If addition of cycloheximide is delayed more than this, G.V. breakdown is no longer inhibited, but maturation is abnormal, abortive : expulsion of the first polar body never occurs. Finally, if oocytes are treated continuously with a mixture of cycloheximide and progesterone, they undergo a particular type of degeneration which we have called pseudomaturation (fig. 4). : the membrane of the G.V. ruptures, but at its apical, instead of basal, pole ; the lampbrush chromosomes do not condense and the nucleoli remain intact; the morphology of the cortex and endoplasm is altered and the oocyte proceeds to cytolysis (Baltus et al., 1973; Brachet et al., 1974; Steinert et al., 1974).

d) Maturation factor (MPF), pseudomaturation factor (PIF) and cytostatic factor (CSF).

If the cytoplasm from a frog oocyte which has been treated for several hours with progesterone is injected into a normal recipient oocyte, the latter undergoes maturation (Smith and Ecker, 1970; Masui and Markert, 1971, Masui, 1972). The same occurs with *Xenopus* (Schorderet-Slatkine, 1972;

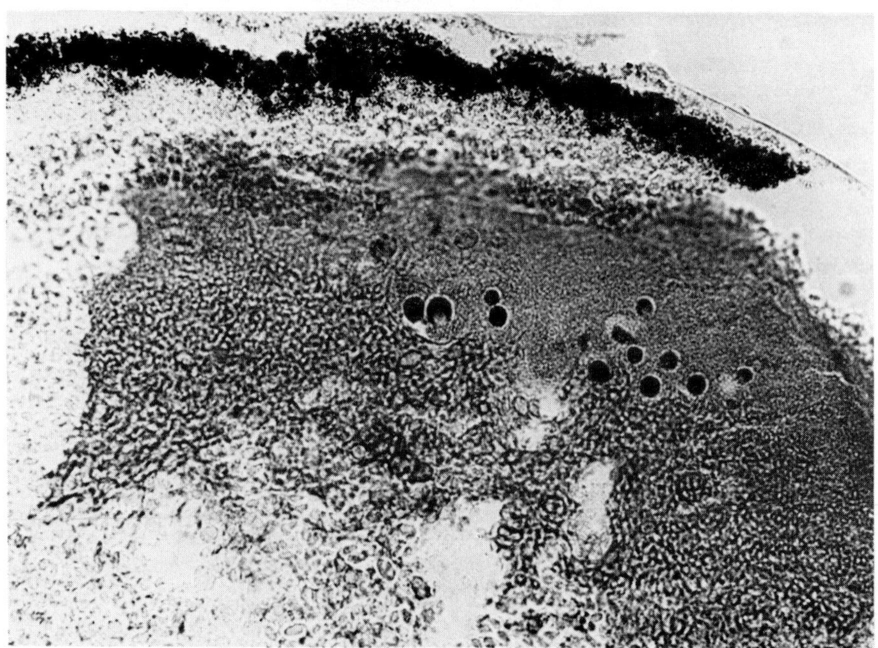

Figure 4.

Pseudomaturation in Xenopus oocytes.

Schorderet-Slatkine and Drury, 1973; Baltus et al., 1973). Hormonal stimulation thus induces the appearance of a factor capable of provoking maturation (MPF) in the oocyte. One remarkable feature is that this factor displays "autocatalytic" properties : the cytoplasm of an oocyte, in which maturation has been induced by injection of MPF, synthesizes its own MPF. More than 10 serial transfers can be effected in this way of the cytoplasm derived from an oocyte having received an initial injection of MPF, without any loss of biological activity. Only little is yet known of MPF; its production is induced by progesterone, and its serial propagation require protein synthesis, MPF is probably itself a protein (Reynhout and Smith, 1974; Drury and Schorderet-Slatkine, 1974, Masui, 1975).

During experiments to isolate MPF from Xenopus oocytes treated with progesterone, we have surprisingly found that injection of a centrifuged homogenate of these oocytes into normal recipient oocytes induces pseudomaturation, and not maturation (Baltus et al., 1973; Brachet et al., 1974). This homogenate contains therefore, instead of MPF, a factor inducing pseudomaturation (PIF). This factor, which has been partially purified and which is a protein, strongly inhibits (by 90%) protein synthesis in the recipient oocyte. While MPF occurs in the endoplasm of oocytes treated with progesterone, PIF accumulates in their cortex or membrane. PIF probably pre-exists in normal oocytes in an inactive form; progesterone would release PIF from such an inactive pre-existing complex, or activate a precursor. If melanosomes are isolated from normal oocytes, then treated in vitro with progesterone, they acquire the capacity to induce pseudomaturation after injection into recipient oocytes (Iacobelli et al. 1974).

2nd order oocytes, which have eliminated their polar vesicle, contain a cytostatic factor (Masui and Markert, 1971; Masui, 1974) : if a little of their cytoplasm is injected into one of the blastomeres of a 2 cell-stage egg, it stops dividing and any mitosis that was in progress is blocked at metaphase : this cytostatic factor (CSF) is still present in fertilized eggs, but in an inactive form; its inactivation would be linked to the elimination of $Ca^{++}$ ions at the moment of fertilization (Masui, 1974).

We still do not know whether PIF and CSF are identical or two distinct factors. What seems clear is that hormonal stimulation of the oocyte triggers a double control mechanism : positive control by MPF and negative control by PIF and CSF.

e) Role of the G.V. in maturation.

G.V. Breakdown is obviously the most spectacular biological event during maturation. One would expect to observe important changes at the moment when the nuclear sap (in which genes have functioned and been immersed for several years) mixes with the cytoplasm. Actually, little happens.

It is easy to remove the G.V. and treat the enucleated oocyte with progesterone : the synthesis of mitochondrial DNA is not affected by the absence of a nucleus and the enucleate oocyte is capable of increasing its rate of protein synthesis under the influence of progesterone (Smith and Ecker, 1970). Enucleate oocytes also respond to progesterone by producing MPF in a normal manner.

Thus, it can be concluded, that in the case of amphibian oocytes, progesterone does not act on gene transcription, as it does in the uterus or oviduct. Positive and negative controls exerted by MPF, PIF and CSF act on the translation of mRNA's of maternal origin which have accumulated during oogenesis. The uneffectiveness of actinomycin already suggested this; the enucleation experiments provide formal proof of control at the level of translation.

The presence of the G.V. is neither necessary for the important morphological changes occurring in the oocyte, in particular the disappearance of the microvilli which border the surface of the egg under the action of progesterone: this coincides with the end of the absorption by pinocytosis, of the serum proteins which give birth to the yolk (Schuetz et al., 1974). It is at the moment of maturation that special granules, ribonucleoprotein aceous in nature, accumulate at the vegetative pole of the egg; these granules characterize the germinal plasma, which is essential for differentiation of the gonads. This characteristic migration of the specific granules of the germinal plasma occurs in a normal manner in enucleated oocytes, after progesterone treatment.

If the G.V. plays no direct role in maturation, this does not necessary mean that the nuclear sap which is abundantly present, is totally devoid of biological interest : it is known that injection of this nuclear sap into fertilized frog eggs induces hyper-development of the brain and cement gland (Malacinski, 1971, 1974). Remarkably, such injections suppress a lethal mutation (mutation 0) of axolotl : mutants 0 becomes blocked at the blastula stage of development, when homozygous (0/0) : they reach an advanced larval stage if injected with nuclear sap from the G.V. from a normal individual (+/+), while injection of cytoplasm from the same eggs is only slightly effective : the product of the normal allele of gene 0 thus accumulates in the GV (Briggs, 1972). It seems that the nuclear sap of the G.V. is necessary for assembly of molecules the constitutive of microtubules present in the asters and mitotic spindles after fertilization. If nuclei isolated from the brain of adult *Xenopus* are injected into an oocyte, their DNA is only replicated if this oocyte has been stimulated to by progesterone (Gurdon, 1970 ). For this nucleus to divide, the chromosomes must condense eand form a mitotic apparatus; for this, cooperation between the cytoplasm and nuclear sap from the G.V. is indispensable : this has been shown by experiments where the nuclei of adult cells have been introduced into enucleated oocytes, stimulated by hormones and injected (or not) with nuclear sap from the G.V. (Dettlaff et al., Ziegler and Masui, 1973).

In conclusion, the G.V. is not indispensable to the correct functioning of maturation : the latter is, therefore, controlled at the level of translation; but the nuclear sap of the G.V. certainly plays a role in several events which follow fertilization of the egg.

## 3. Induction of maturation by organomercurials.

Surprisingly, we have recently found that maturation of *Xenopus* oocytes can be induced with a high efficiency (20% - 100%) by organomercurial derivatives which do not penetrate (or poorly penetrate) the cell (Brachet et al., 1975a). The aim of these experiments was to determine whether penetration of progesterone into the oocytes, and its attachment to the receptor present in the melanosomes requires the integrity of -SH groups (sulphydryls, thiols). These experiments gave a negative answer, but it quickly became obvious that many oocytes treated by -SH reagents in the absence of progesterone, underwent maturation, provided that these substances did not, or poorly entered the cell.

### a) Which -SH reagents induce maturation ?

Fig. 5 shows the formulae of substances which have proved very active in

Cl-Hg-$C_6H_4$-COOH
PCMB

Cl-Hg-$C_6H_4$-$SO_3H$
PCMBS

$C_6H_4$(CO-NH-$CH_2$-CH($OCH_3$)-$CH_2$-HgCl)($OCH_2COOH$)
MERSALYL

CO NH $CH_2$ CH($OCH_3$) $CH_2$ HgCl
NH-CO-$CH_2$-$CH_2$-COOH
MERALLURIDE

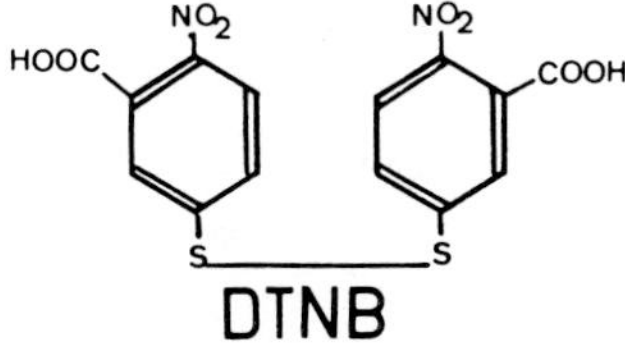


Figure 5.
Formulae of organomercurials and dithiols inducing maturation.

inducing maturation in Xenopus oocytes: these are p-chloromercuribenzoate (PCMB), p-chloromercuribenzene sulphonate (PCMBS) and two diuretic mercurials (Salyrgan or Mersalyl and meralluride). Weak activity (about 10% induction of maturation) has been observed with dithiónitrosobenzene, which oxidizes -SH groups at the cell surface; another -SH reagent, which is used to inhibit trypsin, l-l-tosylamide - 2 - phenyl - ethyl chloromethylcetone (TPCK) has sporadic activity.

Organomercuriels which easily enter the oocytes (mercurochrome, fluoresceine mercuriacetate) alter the structure of the G.V., but do not, induce maturation. Mercuric chloride ($Mg\ Cl_2$), iodoacetamide, N-ethylmaleinide (which inhibit -SH groups and easily penetrate the cell) are inactive; this also applies for organic derivatives of PCMB lacking in Hg (p-chlorobenzoic acid, and p-aminobenzoic acid, for example) as well as phenylarsinoxide.

b) Characteristics of the induction of maturation by organomercurials.

A notable difference exists between organomercurials and steroids concerning induction of maturation : while the latter are already fully active after a few minutes treatment, oocytes must be submitted to constant treatment with organomercurials for several hours (5-6 h. at least) to obtain G.V. breakdown. The concentrations required (0.1 - 0.3 mM) for organomercurials are nearly 100X those generally used for progesterone. In routine experiments, we use continuous treatment with PCMBS (0.1mM).

From a cytological point of view (fig. 6) rupture of the G.V. often occurs in a perfectly normal manner; in other cases the G.V. swells considerably, appearing as a clear spot at the animal pole and bursts from all sides. In both cases, subsequent maturation is normal; expulsion of the 1st polar body can even occur quicker in oocytes trated with PCMBS than in those stimulated with progesterone. In 50% of cases, the pigmentation of the egg takes on a mottled aspect; in these cases, the basophilic cytoplasm is distributed in a heterogeneous fashion and tends to form cytasters.

It may be wondered whether organomercurials act, like progesterone, on the oocyte itself or whether they induce, like pituitary hormones, synthesis or liberation of protesterone by follicle cells. Our experiments show that removal of these cells by various treatments (pronase, collagenase, EDTA) does not reduce the efficiency of PCMB or PCMBS treatment; on the contrary, organomercurials continue to induce maturation when progesterone synthesis by follicular cells is inhibited by cyanoketone. It must be concluded from these experiments that the target of organomercurials is the oocyte itself.

Like progesterone, organomercurials are inactive on injection (50 nl of PCMB or PCMBS at 1-10 mM) : no case of maturation has been observed in 52 oocytes that have been micro-injected, then analysed cytologically.

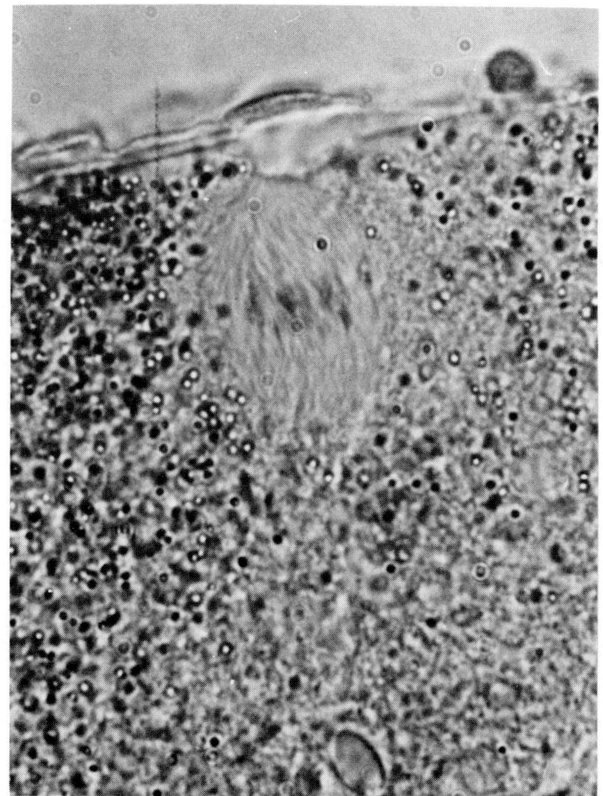

Figure 6.
Maturation induced by PCMBS 0.1 mM

Organomercurials act, therefore, on the plasma membrane or the adjacent cortex. As for progesterone, the presence of $Ca^{++}$ in the incubation medium is necessary for the efficiency of organomercurials.

As already mentioned, pretreatment with PCMBS does not inhibit induction of maturation by progesterone or pituitary hormones; but, as shown in table 1, progesterone and PCMBS have additive synergisnic effects, PCMBS at a concentration too low to induce maturation greatly apotentiates the effect of threshold doses of progesterone.

The effect of organomercurials is suppressed by addition of thiols (cysteine, dithiothreitol, mercaptoethanol) which do not induce maturation by themselves. This inhibition still takes place if the thiols are added 1-2 hours before G.V. breakdown to oocytes treated with PCMBS. In preliminary experiments, we found that dithiothreitol (1-10 mM) also inhibits progesterone induced maturation; like cycloheximide, it becomes inactive at the time when

Table 1.

Percentage of maturations obtained by treatments combining progesterone (Pg) and para-chloromercuribenzene-sulfonate (PCMBS).

| PCMBS \ Pg | 0 µg/ml | 0.005 µg/ml | 0.01 µg/ml | 0.05 µg/ml |
|---|---|---|---|---|
| 0 | 0 | 2.5 | 26 | 97 |
| 5 $10^{-7}$ M | 0 | 5.2 | 18 | 100 |
| $10^{-6}$ M | 0 | 32 | 67 | 100 |
| 5 $10^{-6}$ M | 0 | 42 | 82 | 100 |
| $10^{-5}$ M | 0 | 59 | 68 | 100 |
| $10^{-4}$ M | 100 | | | |

MPF is produced. Dithiothreitol (which is inactive as injection) inhibits the production of MPF, but does not inactivate this factor once it has been formed.

Maturation induced by organomercurials also resembles that provoked by progesterone from another viewpoint : it is inhibited by cycloheximide but not actinomycin D. Thus organomercurials cannot be effective when protein synthesis is completely inhibited, but they do not require RNA synthesis. Maturation induced by organomercurials would be therefore (like that induced by steroids) controlled at the level of translation and not transcription.

We have been able to demonstrate that treatment of oocytes with organomercurials is followed by MPF and PIF production : injection of endoplasm isolated from oocytes, which had been treated for several hours with PCMBS, into normal recipient oocytes, induces maturation (fig. 7); on the other hand, injection of a homogenate of these same oocytes provokes pseudomaturation.

Oocytes treated continuously with PCMBS are naturally more fragile than those briefly stimulated with progesterone. However, they can react to a parthenogenetic stimulus (pricking, treatment with the $Ca^{++}$ ionophore A23187 (Steinhardt et al., 1974, Hanocq et al., 1974)) with a typical activation reaction : retraction of microvilli, breakdown of cortical granules upheaval of the fertilization membrane, rotation of orientation, appearance of asters, multiplication of nuclei (implying synthesis of chromosomal DNA). Activated oocytes never divide and proceed rapidly to cytolysis when maturation has been induced with organomercurials.

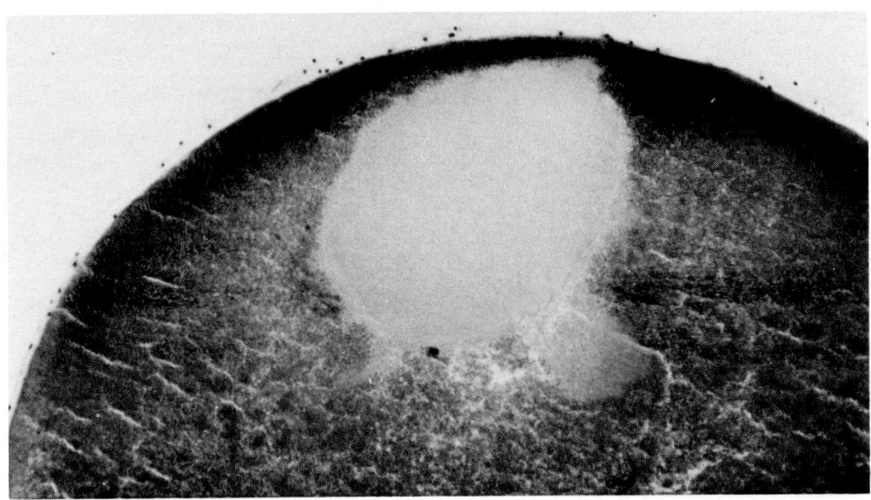

Figure 7.
Xenopus oocyte in the process of maturation after injection of MPF from a PCMBS (0.1 mM).

c) Biochemical changes.

Biochemical analysis of PCMBS treated oocytes has only just begun and needs following up. Here is a summary of the first results obtained (Brachet et al., 1975a and unpublished observations).

As with progesterone, PCMBS raises oxygen consumption and protein synthesis (fig. 8a,b). A complication arises in the latter case : PCMBS causes an almost total inhibition of amino acid penetration into the oocyte (table 2); this demonstrates that it has an important effect on the permeability of the cellular membrane. The graph shown in fig. 8b shows that the stimulation of protein synthesis in oocytes treated with PCMBS, which can only be observed when the radioactive precursor ($^{3}$H phenylalanine) is injected into the oocytes.

Fig. 9 shows the kinetics of the binding of Hg-PCMBS by oocytes : maturation is induced when about 40 M of organomercurials is fixed per oocyte. This means that, if the distribution of PCMBS is homogeneous over the surface of the oocyte, one atom of radioactive Hg would be fixed per 18.84 sq. Å. This high concentration suggests to us that PCMBS probably penetrates a short distance (several um) into the cortex of the oocyte. Saturation is reached when the oocyte has fixed 100 - 200 pM of organomercurial; a homogenized oocyte can bind up to 10.000 pm of PCMBS, about 250 times more than is necessary to induce maturation. The latter is thus induced when 0.4% of the total -SH sites have been blocked by Mg. This confirms that the -SH reagents inducing maturating are localized in the superficial layers of the oocyte.

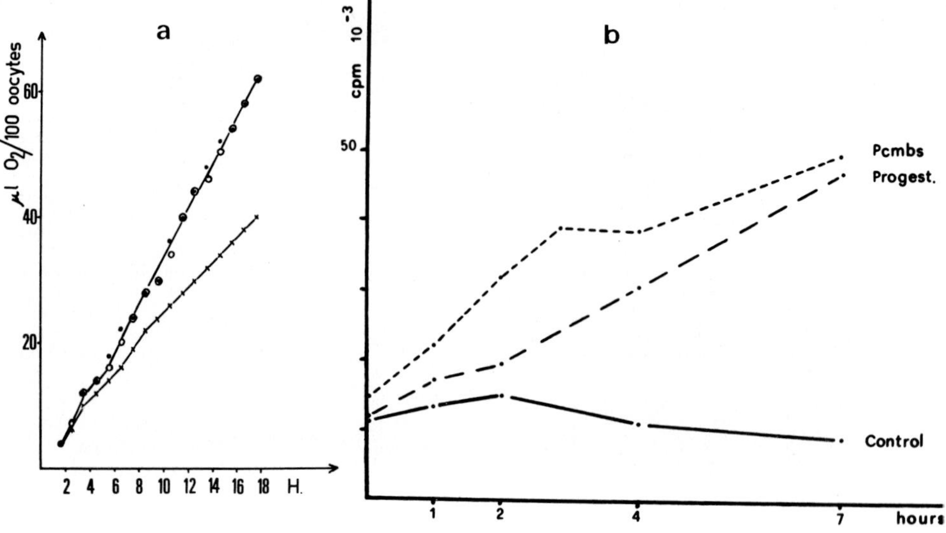


Figure 8a.

Oxygen consumption of Xenopus oocytes.

. Progesterone treated (2 µg/ml)

o PCMBS treated (0.1 mM)

x Control.

Figure 8b.

Incorporation of phenylalanine $^3$H into the TCA-insoluble fraction of batches of 8 oocytes treated with progesterone (1 µg/ml) or with PCMBS (0.1 mM). Each oocyte received an injection of 50 nl of phenylalanine $^3$H (1 mC/ml) at various times after the beginning of the treatment and were fixed 30 minutes after injection.

d) Is the induction of maturation by PCMBS a general phenomenum ?

Our experiments are yet too few to enable us to give a full answer to this question. It is certain that organomercurials induce maturation in diverse species of Amphibia, such as the axolotl (Ambystoma mexicanum) and the frog (Rana temporaria). Some attempts made with toad and trout eggs have yielded negative results up to the present. It is very possible that positive results could be obtained with all species whose oocytes respond well to steroid hormones but the question needs closer examination.

Conversely, PCMBS does not easily induce maturation of starfish oocytes (J.B. - personal observations). In Asterias, the normal inducer of maturation is not a steroid, but a purine : 1-methyladenine (Kanatani, 1973).

Table 2.

$^{3}$H Phenylalanine (1 µCi/ml) penetration into oocytes treated with progesterone (1 µg/ml) or PCMBS (0.1 mM). Penetration was measured after 60 minutes incubation in the isotope for various times after the beginning of the treatment.

| Times after the beginning of the treatment | | Total cpm per oocyte | % of control |
|---|---|---|---|
| | Controls | 739 | 100 |
| 1 hour | Progesterone | 794 | 107 |
| | PCMBS | 442 | 59 |
| | Controls | 382 | 100 |
| 2 hours | Progesterone | 407 | 106 |
| | PCMBS | 297 | 77 |
| | Controls | 554 | 100 |
| 3 hours | Progesterone | 509 | 91 |
| | PCMBS | 182 | 32 |
| | Controls | 604 | 100 |
| 5 hours | Progesterone | 607 | 100 |
| | PCMBS | 91 | 15 |
| | Controls | 611 | 100 |
| 7 hours | Progesterone | 591 | 96 |
| | PCMBS | 42 | 6 |
| | Controls | 480 | 100 |
| 19 hours | Progesterone | 442 | 91 |
| | PCMBS | 18 | 3 |

Maturation can also be produced by treating oocytes with a di-thiol for example dithiothreitol; as for PCMB, this di-thiol inhibits maturation induced by 1-methyladenine (Kinoshito and Kanatani, 1973). Our own observations confirm that, as in the Japanese species studied by Kanatani, dithiothreitol induces maturation in Asterias glacialis and Astropecter aurianticus; continuous treatment of oocytes in these two species with PCMBS (0.1 mM) somewhat enhances the percentage of oocytes entering maturation, which increases from 10-15% to 25-40%; but this is far from the 95% maturation obtained with 1-methyladenine and dithiothreitol. These experiments show that, in starfish as well, -SH groups play an important role in maturation; but it will not be possible to say whether this role is the same, despite appearances, in starfish and Amphibians until the molecular mechanism conditioning G.V. breakdown has been characterized.*

*In axolotl, chromosomes destined to give birth to the polar vesicle undergo pycnosis at anaphase as opposed to those remaining in the oocyte; this indicates that PCMBS does not need to penetrate the egg deeply.

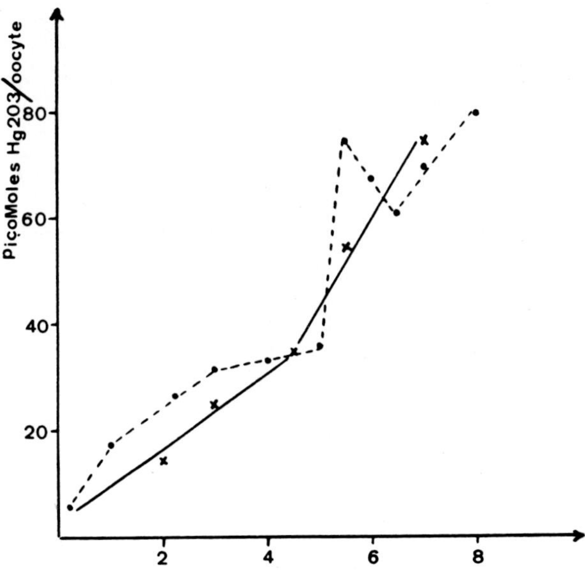


Figure 9.
Measure of $^{203}$Hg fixation of Xenopus oocytes during incubation in solutions of radioactive 0,1 mM PCMB----- and PCMBS ______ (average of 4 experiments).

## 4. Conclusions.

Maturation constitutes, from the two points of view of Genetics and Embryology, a crucial stage in ontogenesis. It can be seen as a rejuvenation of the egg cell (Eaves, 1973); the oocyte has the age of the maternal organism and is hence an old cell; the unfertilized egg, resulting from maturation is by definition a very young cell. Without maturation, the fertilized egg would be triploid and thus of a genetically abnormal constitution.

From a molecular viewpoint, maturation is a particularly important stage: it is the moment when RNA synthesis, which predominates during all oogenesis and which produces all the machinery necessary for the rapid DNA replication characterizing segmentation of the fertilized egg, stops : this is shown by the nuclear transplantation experiments of Gurdon (1970). We still do not know all the factors which regulate RNA synthesis and allow DNA synthesis : their identification is an important problem for the future; substantial progress in the right direction has recently been made by Benbow and Ford (1975) who succeeded in isolating a protein factor which promotes DNA synthesis in nuclei from adult liver; this factor is present in large amounts in unfertilized _Xenopus_ eggs, but is almost absent in oocytes.

We have seen that maturation is controlled, at the level of translation of mRNA's which accumulate during oocytenesis, by the production of antagonistic factors (MPF, PIF, CSF): their true chemical nature is still unknown and perhaps their identification will bring surprises. It is, anyway, logical that the maturing egg is under a double, negative and positive, control: stimulation by progesterone is ephemeral and the second order oocyte falls back to a stage of inertia from which only fertilization (or parthenogenesis) releases it.

There is no doubt that the primary stages of maturation take place at the level of the plasma membrane and the underlying melanosomes: it is from there that the signal driving synthesis of MPF and hence the machinery necessary for DNA replication starts. This signal can be given by substances as different as steroid hormones and organomercurials; both are inactive when injected into the oocyte and must threfore react with the plasma membrane. It is possible that steroids and organomercurials modify the activity of one of the enzymes regulating at the level of the plasma membrane, the ionic composition of the oocyte. In support of this hypothesis, it can be noted that:the presence of $Ca^{++}$ ions in the medium is necessary for progesterone and organomercurials to act[(1)]; onabain, which inhibits the sodium pump (ATPase,($Na^{+}K^{+}$), accelerates maturation with organomercurials as well as with progesterone (Vitto and Wallace, 1974; personal observations); this ATPase ($Na^{+}$, $K^{+}$) is inhibited by progesterone in frog oocytes and replaced by a $Ca^{++}$ dependent enzymatic activity (Morrill et al., 1971). During progesterone induced maturation, $Ca^{++}$ bound to the yolk platelets and the melanosomes is released. (Merriam, 1971).

Inhibition of amino acid transport, which we have observed in PCMBS treated oocytes, has also been observed in oocytes treated with Pregnyl, then progesterone (Pennequin et al., 1975). Identification of the key membrane enzyme whose activity is affected by maturation is an urgent problem; the fact that the membrane target (ATPase, cyclase, phosphodiesterase ?) is sensitive to both steroid hormones and organomercurials, should facilitate its identification.

## Acknowledgements.

We wish to thank Dr. Christopher Evans for considerable help in the preparation of the english manuscript.

---

(1) Maturation is not induced by addition of the $Ca^{++}$ ionophore A23187 or procaine (which inhibits the binding of $Ca^{++}$ ions to the membrane). Procaine provokes the appearance of many cytasters in the cytoplasm of eggs treated simultaneously with PCMBS (unpublished results).

References.

Baltus, E., Brachet, J., Hanocq, J. & Hubert, E. Differentiation 1, 127, 1973.

Baltus, E., Hanocq, F., Hanocq-Quertier, J., Hubert, E., Iacobelli, S. & Steinert, G. 1974, Mol. Cell Biochem. 3, 189.

Baltus, E., De Schutter-Pays, A., Hanocq-Quertier, J., Hubert E. & Steinert, G. Proc. natl. Acad. Sci. US , 1975, 72, 1574.

Benbow, R.M. & Ford, C.C. Proc. natl. Acad. Sci. US 72, 2437, 1975.

Benbow, R.M., Pestell, R.Q. & Ford, C. Developm. Biol. 43, 159, 1975.

Brachet, J. Année Biologique 13, 404, 1974.

Brachet, J. Introduction to molecular Biology. Ed. Springer Verlag, 1974.

Brachet, J. Introduction à l'Embryologie moléculaire. Ed. Masson, Paris, 1974.

Brachet, J. De Schutter-Pays, A. & Hubert, E. Differentiation3, 3, 1975.

Briggs, R. J. Exp. Zool. 181, 271, 1972.

Denis, H. Précis d'Embryologie moléculaire. Ed. Hermann Paris, 1974.

Dettlaff, T.A., Nikitina, L.A. & Stroeva, O.G. J. Embryol. exp. Morphol.12, 851, 1971.

Drury, K. & Schorderet-Slatkine, S. Cell 4, 269, 1974.

Eaves, G. Mechan. Ageing Development 2, 19, 1973.

Grippo, P. ° Lo Scavo, A. Bioch. biophys. Res. Comm. 48, 280, 1972.

Gurdon, J.B. Current Topics Developm. Biol. 5, 39, 1970.

Hanocq, J., Baltus, E. & Steinert, G. C.R. Acad. Sci. Paris 279, 2111, 1974.

Hanocq, F., De Schutter-Pays, A., Hubert, E. & Brachet, J. Differentiation 2, 75, 1974.

Iacobelli, S., Hanocq, J., Baltus, E. & Brachet, J. Differentiation 2, 129, 1974.

Kanatani, H. Int. Rev. Cytol. 35, 253, 1973.

Kinoshito, T. & Kantani, H. Exptl. Cell Res. 83, 296, 1973.

Malacinski, G.M. Cell Differentiation 1, 253, 1971.

Malacinski, G.M. Cell Differentiation 3, 31, 1974.

Masui, Y. & Markert, C.L. J. exp. Zool. 177, 129, 1971.

Masui, Y. J. exp. Zool. 179, 365, 1972.

Masui, Y. J. exp. Zool. 187, 141, 1974.

Masui, Y. Communication at the Congress on Differentiation. Copenhagen, 1975.

Merriam, R.W. Exptl. Cell Res. 68, 81, 1971.

Morrill, G.A., Kostellow, A. & Murphy, J.B. Exptl. Cell Res. 66, 289, 1971.

Morrill, C.A. & al. in press.

Ozon, R. & Bellé, R. Bioch. biophys. Acta 330, 588, 1973.

Pennequin, P., Schorderet-Slatkine, S., Drury, K.C. & Baulieu, E.E. FEBS Lett. 51, 156, 1975.

Schorderet-Slatkine, S. Cell Differentiation 1, 179, 1972.

Schorderet-Slatkine, S. & Drury, K. Cell Differentiation 2, 247, 1973.

Schuetz, A.W. Oogenesis. Ed. J.D. Biggers & A.W. Schuetz, Univers. Press Baltimore 479, 1972.

Schuetz, A.W. Biology of Reproduction 10, 150, 1974.

Schuetz, A.W., Wallace, R.A. & Dumont, J.N. J. Cell Biol. 61, 26, 1974.

Smith, L.D. & Ecker, R.E. Current Topics Developm. Biol. 5, 1, 1970.

Smith, L.D. & Reynhout, J.K. Developm. Biol. 38, 394, 1974.

Steinert, G., Baltus, E., Hanocq-Quertier, J. & Brachet, J. J. ultrastr. Res. 40, 188, 1974.

Steinhardt, R.E., Epel, D., Carroll, E.J. & Yanagimachi, R. Nature 252, 41,1974.

Vitto, Jr. A. & Wallace, R.A. J. Cell Physiol. 84, 360a, 1974.

Wiblet, M. Bioch. biophys. Res. Comm. 60, 926, 1974.

Ziegler, D. & Masui, Y. Dev. Biol. 35, 283-292, 1973.

*Progress in Differentiation Research, ed. N. Müller-Bérat et al.*



# REGULATION OF OVALBUMIN GENE EXPRESSION BY STEROID HORMONES

G. Stanley McKnight
Institut de Chimie Biologique
Faculté de Médecine, 11, Rue Humann
Strasbourg - France

## INTRODUCTION

The magnum portion of the oviduct in immature female chicks responds to the administration of estradiol with the proliferation and differentiation of specific cell types. The predominant cell type, tubular gland cell, synthesizes the four egg-white proteins (ovalbumin, conalbumin, ovomucoid, and lysozyme) in large quantities and the major product, ovalbumin, constitutes approximately 50-60% of the total protein synthesis in these cells after continued hormonal stimulation. The initial response to estradiol is slow and requires DNA synthesis. Ovalbumin synthesis begins after about 20 hours of stimulation and reaches a maximal value of 50% of total protein synthesis after 10 days of continual treatment with estradiol. If the hormone is removed (withdrawal)the rate of ovalbumin synthesis declines below detection but some of the tubular gland cells remain and can be restimulated with estradiol (secondary stimulation) even in the absence of DNA synthesis. For a more complete review of the system see either Schimke et al.[1] or Palmiter[2].

We have studied the early events in secondary stimulation using complementary DNA hybridization to detect ovalbumin messenger RNA (OVmRNA) sequences with the following questions in mind.

1. Is the rapid reappearance of ovalbumin synthesis (3-4 hours after estradiol injection) a result of the accumulation of newly made OVmRNA or is OVmRNA present in the withdrawn oviduct which can be utilized for translation only after some estrogen-dependent event?
2. Why does one observe a 3-4 hour lag between estradiol injection and ovalbumin synthesis since Palmiter (manuscript in preparation) has shown that the injected estradiol reaches and saturates the chromatin with receptors after only 30 minutes? One can consider that either the binding of estradiol-receptor complex to chromatin does not directly allow the ovalbumin gene to be transcribed or that OVmRNA transcription begins immediately and 3 hours are required for processing, transport, and incorporation into polysomes.

3. Is the OvmRNA synthesized as part of a larger precursor as has been suggested for most animal mRNAs?

## MATERIALS AND METHODS

A detailed description of materials and methods used in this study has been published.[3,4] Briefly, the detection of OVmRNA sequences was done by hybridization to a complementary DNA (cDNA) prepared from highly purified OVmRNA with reverse transcriptase isolated from RSV. Hybridizations were done at 68°C in 0.3 M NaCl, 10 mM Tris-Cl (pH 7), 2 mM EDTA, and 0.1% SDS. Complementary DNA was incubated with oviduct RNA from 0-3 days and the percent hybridization determined by digestion with $S_1$ nuclease. The concentration of OVmRNA sequences in the oviduct RNA samples was calculated by comparing either the Crt 1/2 or the slope of a double-reciprocal plot to the standard which was pure OVmRNA ($Crt^{1/2} = 1.4 \times 10^{-3}$)[5]. Size determinations of the OVmRNA were done by centrifugation in sucrose-SDS gradients after heating the RNA to 65°C for 10 minutes which eliminates aggregation. Individual gradient fractions were hybridized with cDNA for 24 hours and the percent hybridization converted to nanograms of OVmRNA using a standard curve generated under the same conditions with pure OVmRNA.

## RESULTS AND DISCUSSION

Ovalbumin gene repression - Although we could not detect ovalbumin synthesis in withdrawn oviduct we were able to demonstrate a low level of OVmRNA (60 molecules per tubular gland cell) in this tissue.

This prompted us to examine several other non-ovalbumin synthesizing tissues from the chick for the presence of trace amounts of OVmRNA. In both unstimulated chick oviduct (no previous estradiol treatment) and hen liver we find approximately 1 molecule per cell of OVmRNA (see Fig. 1). By comparison the fully-stimulated hen oviduct contains 78,000 OVmRNA molecules per cell. As shown in Fig. 1, RNA from either rabbit reticulocyte or rat liver gives no hybridization with cDNA ruling out a non-specific effect of high RNA concentrations. The small increase in the background at Crt $10^4$ is seen with this cDNA preparation in the absence of added RNA. In a more thorough search for OVmRNA sequences in non-ovalbumin synthesizing

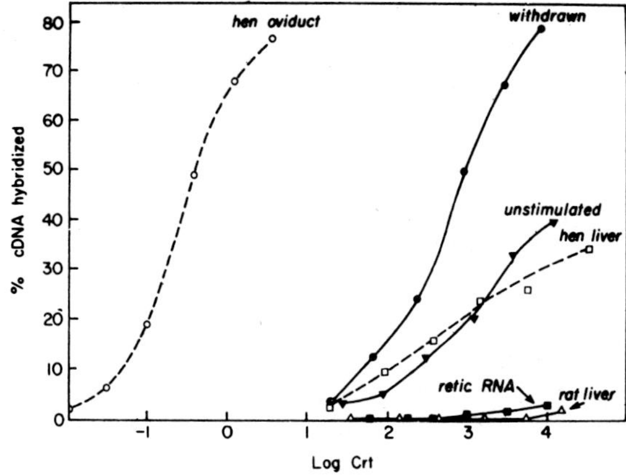


Fig. 1. Hybridization to ovalbumin specific cDNA with total cellular RNA prepared from laying hen oviduct, oviduct from unstimulated immature chicks or chicks withdrawn from estradiol treatment for 6 weeks, laying hen liver, rabbit reticulocyte, or rat liver. (From McKnight et al [4])

tissues from the chick J.P. LePennec et al (manuscript in preparation) have found sequences which hybridize to ovalbumin cDNA in hen or rooster liver as well as brain. These sequences are not due to DNA contamination since they are not destroyed by extensive DNase digestion and the majority remain at approximately 16S, the size of authentic OVmRNA.

Since we would not expect gene repression to be absolute (for a discussion of some of the theoretical considerations see Lin and Riggs [6]) the level of OVmRNA in the liver, for example, is surprisingly low. In fact, by comparison with bacterial genes this level of "gene leakage" is much lower than is seen for the repressed lactose [6] operon and suggests that animal cells have acquired a more efficient mechanism for repressing unwanted gene expression. Such a general repression mechanism is widely postulated to involve histones, a class of basic DNA-binding proteins found only in eucaryotes. Recent evidence from the *in vitro* transcription of adenovirus DNA reconstituted with histones to form "nucleosomes" demonstrates that these histone complexes are capable of both inhibiting initiation and blocking elongation by RNA polymerase indicating their possible role *in vivo* (P. Chambon et al, manuscript in preparation).

OVmRNA induction by estradiol and progesterone - The OVmRNA present in withdrawn oviduct is found predominantly in the cytoplasm and is identical in size to hen polysomal OVmRNA[4]. Fig. 2 shows that after estradiol administration there is a 3 hour lag during which there is little change in the amount of OVmRNA followed by a rapid increase in OVmRNA sequences per cell and, at the same time, induction of ovalbumin synthesis. Since the estradiol receptor saturates the chromatin at 30 minutes (see introduction) the delay before the accumulation of OVmRNA indicates that some intervening regulatory event must take place to allow the accumulation to begin. We can

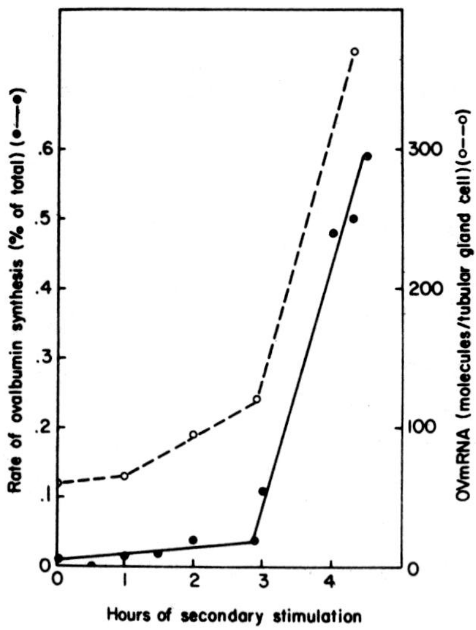


Fig. 2. The response of withdrawn oviduct to a single injection of estradiol at 0 time. Ovalbumin synthesis was measured by precipitation with a specific antibody after a 30 min. incubation of oviduct fragments *in vitro* with $^{3}$Hleucine. OVmRNA was measured in oviduct RNA from the same animals by hybridization to cDNA and expressed as molecules per tubular gland cell assuming 15% tubular gland cells. (From McKnight et al[4]).

conclude that the induction of ovalbumin synthesis is not caused by the translation of pre-existing OVmRNA but is instead the result of a large increase in the amount of OVmRNA per cell. Since there is little delay between the increase in OVmRNA and ovalbumin synthesis (see Fig. 2) we can also conclude that the times necessary for processing and transport are short. Our results are in good agreement with those reported by R. Cox et al[7] but differ somewhat from the results of Rosen and O'Malley[8] which suggested that OVmRNA accumulation begins more rapidly after estrogen administration. The difference may be due to difficulties in quantitating hybridization results when RNA is not in sequence excess and the hybridization does not go to completion as is common when measuring very low OVmRNA concentrations in total cellular RNA. In the experiments summarized in Fig. 2 we have therefore used enough RNA to drive all the hybridizations to completion after a 3 day incubation.

In another series of experiments we have examined the induction of OVmRNA by progesterone and find a more rapid response beginning after 1.5 hours with a dramatic increase in both OVmRNA and ovalbumin synthesis[4]. This difference between estrogen and progesterone has been confirmed by Palmiter (manuscript in preparation) and may indicate that there is more than one mechanism for inducing ovalbumin gene activity. A model for gene regulation including the possibility that multiple control elements might influence the expression of a single gene has been proposed by Britten and Davidson[9].

Absence of OVmRNA precursor - We have taken advantage of the OVmRNA induction results shown in Fig. 2 to examine the size of the newly-synthesized OVmRNA sequences by hybridization to detect a possible high-molecular weight precursor to polysomal OVmRNA of the type reported by T. Imaizumi et al[10] for hemoglobin mRNA. We isolated oviduct total cellular RNA from chicks withdrawn from estradiol and then restimulated by estradiol for 4.5 hours. At this time only approximately 300 new OVmRNA molecules have appeared in the cell and if we assume that a possible short-lived precursor has reached a steady state level we would expect the precursor to represent a significant percentage of the total OVmRNA (if the precursor half-life is 10 minutes we estimate that 16% of the OVmRNA should be in the precursor form)[3]. Fig. 3 shows the size distribution of the OVmRNA sequences measured by hybridization to cDNA of this 4.5 hour RNA under conditions which eliminate OVmRNA aggregation. The homogenous peak at 16S is identical to the results obtained with hen polysomal RNA containing large amounts of translatable OVmRNA indicating that

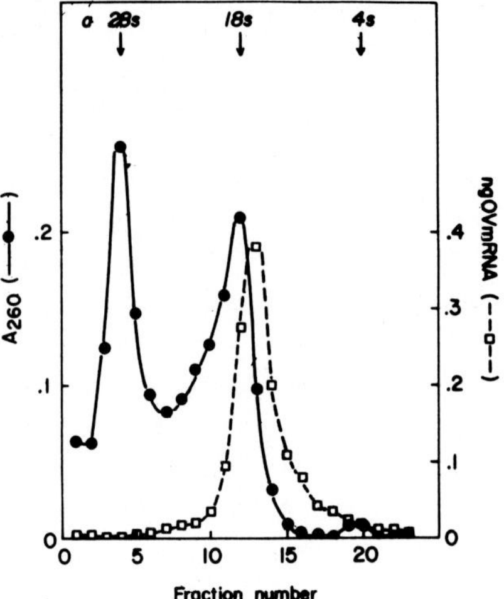


Fig. 3. Hybridization to newly synthesized OVmRNA. 1.2 $A_{260}$ units of total RNA from chicks given a secondary stimulation with estrogen for 4.5 hr was centrifuged under conditions which eliminate aggregation. The $A_{260}$ was measured and after ethanol precipitation, the total RNA in each fraction hybridized to cDNA for 24 hrs. (From McKnight and Schimke[3])

this newly synthesized OVmRNA does not contain any large precursor molecules.

In a separate series of experiments we examined high-molecular weight RNA (greater than 28S) from the laying hen oviduct and could find no evidence for a large precursor. A preparation of high-molecular weight RNA from the laying hen oviduct and the resulting hybridization results are shown in Fig. 4. We can calculate that 1 molecule per cell of precursor would have represented more than 0.2 ng of OVmRNA in the RNA preparation shown in Fig. 4. We conclude that there are no OVmRNA sequences larger than polysomal OVmRNA involved in the normal production of the messenger. Since the polysomal-associated OVmRNA is already nearly 1000 nucleotides longer than necessary to code for a polypeptide the size of ovalbumin[5]

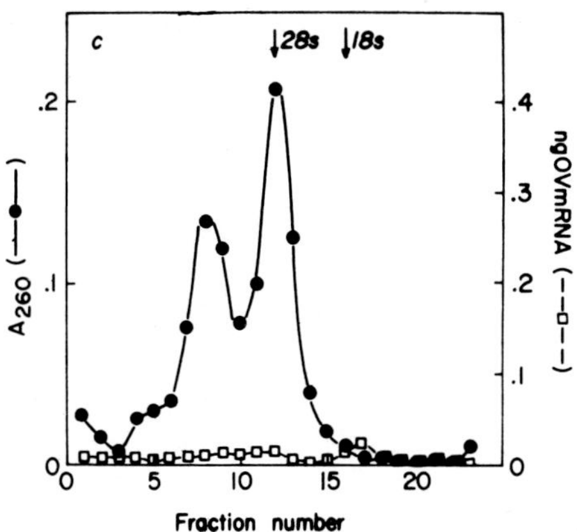


Fig. 4. Hybridization to greater than 28S RNA from hen oviduct. 0.6 $A_{260}$ units (isolated from 70 $A_{260}$ units of total RNA) of large RNA was centrifuged and hybridized as in Fig. 3. (From McKnight and Schimke[3])

it may already contain the necessary information for its post-transcriptional regulation.

## ACKNOWLEDGMENTS

I am grateful to Brigitte Chambon and Viviane Friederici for help in preparing this manuscript. This work was completed in the laboratory of Dr. Robert T. Schimke and supported in part by Research Grant GM 14931 from the National Institute of General Medical Sciences.

## REFERENCES

1. Schimke, R.T., McKnight, G.S. and Shapiro, D.J. (1975) Biochemical Actions of Hormones Vol. III (ed. G. Litwack) Academic Press, p. 245.
2. Palmiter, R.D. (1975) Cell 4, 189.
3. McKnight, G.S. and Schimke, R.T. (1974) Proc. Nat. Acad. Sci.

USA 71, 4327.
4. McKnight, G.S., Pennequin, P. and Schimke, R.T. (1975) J. Biol. Chem. 250, 8105.
5. Shapiro, D.J. and Schimke, R.T. (1975) J. Biol. Chem. 250, 1759.
6. Lin, S. and Riggs, A.D. (1975) Cell 4, 107.
7. Cox, R.F., Haines, M.E. and Emitage, J.S. (1974) Eur. J. Biochem. 49, 225.
8. Rosen, J.M. and O'Malley, B.W. (1975) Biochemical Actions of Hormones Vol. III (ed. G. Litwack) Academic Press, p. 271.
9. Britten, R.J. and Davidson, E.H. (1969) Science 165, 349.
10. Imaizumi, T., Diggelmann, H. and Scherrer, K. (1973) Proc. Nat. Acad. Sci. USA 70, 1122.

*Progress in Differentiation Research, ed. N. Müller-Bérat et al.*



# POSTNATAL DIFFERENTIATION OF HEPATIC STEROID METABOLISM

Jan-Åke Gustafsson, Paul Skett and Åke Stenberg

Department of Chemistry and Department of Germfree Research,
Karolinska Institutet, S-104 01 Stockholm 60, Sweden

## Introduction

During the neonatal period the neural mechanisms which regulate gonadotrophin secretion and sexual behaviour in the adult rat are differentiated. In the neonatal male rat, it is thought that testicular androgens are responsible for the suppression of cyclicity, whereas, in the female rat, the absence of gonadal activity during this period allows the hypothalamus to remain in the undifferentiated or cyclic state[1,2]. Evidence is accumulating that also several other physiological processes are "imprinted" or irreversibly programmed at birth by testicular androgens. We have been interested in hepatic metabolism of steroids since this system offers unique possibilities for exact quantitation of differentiation in terms of enzyme activities.

## Neonatal programming of hepatic metabolism

Hepatic metabolism of steroid hormones in the rat is characterized by large sexual differences[3,4]. Part of these differences are predetermined at birth by imprinting by testicular androgens[5-7]. We have studied this process using 4-androstene-3,17-dione, 5α-androstane-3α,17β-diol and 5α-androstane-3α,17β-diol 3,17-disulphate as substrates for enzymes in the microsomal and cytosol fractions. These studies have shown that the hepatic steroid-metabolizing enzyme activities may be grouped into three classes with regard to the mechanisms regulating their activity:

I. Enzymes irreversibly programmed by androgens neonatally and reversibly influenced by sex hormones postpubertally (the 2α-hydroxylase active on 5α-androstane-3α,17β-diol 3,17-disulphate and the 16α-hydroxylase, the 5α- and 5β-reductases and the 3β- and 17α-hydroxysteroid reductase active on 4-androstene-3,17-dione).

II. Enzymes reversibly influenced by sex hormones (the 2β-, 7β- and 18-hydroxylases active on 5α-androstane-3α,17β-diol and the 6β-hydroxylase active on 4-androstene-3,17-dione).

III. Enzymes almost or completely sex hormone-independent (the 7α-hydroxylase active on 5α-androstane-3α,17β-diol and the 7α-hydroxylase and the 3α- and 17β-hydroxysteroid reductases active on 4-androstene-3,17-dione).

In another series of experiments it has been shown that neonatal androgenic programming also sets the level for the degree of androgen responsiveness in adult life so that a female rat or a neonatally castrated male rat responds less well to androgen than a normal male rat[8].

## Pituitary control of hepatic metabolism

Hypophysectomy has been found to lead to an over-all masculinization of hepatic steroid metabolism in female rats (decreased activity of the 15β-hydroxylase system active on 5α-androstane-3α,17β-diol 3,17-disulphate, increased activities of the 2α-, 2β-, 7β- and 18-hydroxylase systems active on 5α-androstane-3α,17β-diol and of the 6β- and 16α-hydroxylase systems, 3β- and 17α-hydroxysteroid reductases and 5β-reductase active on 4-androstene-3,17-dione and decreased activity of the 5α-reductase active on 4-androstene-3,17-dione)[9]. These results indicate the existence of a pituitary "feminizing factor" (FF) that feminizes a basic masculine level of the hepatic sex-dependent enzyme activities. The masculinizing and feminizing effects on liver enzyme activities observed following treatment with testosterone propionate and estradiol benzoate, respectively, are not seen in hypophysectomized rats. The estrogen unresponsiveness characterizing these rats may be explained by the lost capacity to induce secretion of FF. It may be suggested that the androgen unresponsiveness in hypophysectomized rats is due to the loss of a pituitary factor necessary for the action of androgens in the liver cell.

In order to investigate the nature of the hypophyseal control of hepatic steroid hormone metabolism we have castrated and hypophysectomized adult male and female rats and transplanted a pituitary from a male rat under the kidney capsule[10]. Control rats that were only castrated and hypophysectomized displayed the expected masculinization of hepatic steroid metabolism whereas transplanted rats of both sexes showed a feminine type of hepatic metabolism. These experiments indicate that the pituitary _in situ_ in both male and female rats has a basic FF secreting tonus which is modulated by central influences. As a further means to strengthen the validity of this interpretation we have studied the effects of the presence of a transplanted pituitary tumor on liver metabolism of steroids in rats of both sexes[11]. Transplantation of $MtT/F_4$ pituitary tumor cells to male rats of the Fisher strain resulted in a change of the masculine type of hepatic steroid metabolism into a feminine pattern of enzyme activities. Liver metabolism of steroid hormones in female rats was relatively unaffected following transplantation of pituitary tumor cells. These experiments indicate that, similar to the transplanted pituitary, the autonomous $MtT/F_4$ pituitary tumor secretes FF.

## Hypothalamo-pituitary control of hepatic metabolism

In order to investigate the role of the hypothalamus in regulating hypophyseal secretion of FF, experiments have been carried out to relieve the male pituitary from the postulated inhibiting control of the secretion of FF. Male and female rats were subjected to electrolytic lesions with a steel electrode placed in the median eminence[12]. Histologic examination revealed lesions in principally the entire hypothalamus. Following electrocoagulation, hepatic steroid metabolism in male rats was generally feminized (increased 5α-reduction and decreased 6β- and 16α-hydroxylation of 4-androstene-3,17-dione, decreased 2α-, 2β-, 18- and 7β-hydroxylation of 5α-androstane-3α,17β-diol and induced 15β-hydroxylation of 5α-androstane-3α,17β-diol 3,17-disulphate) whereas hepatic metabolism in female rats remained essentially unchanged. These findings indicate that the release of the pituitary FF is controlled by means of a release-inhibiting factor (FFIF = FF inhibiting factor) from the hypothalamus. This factor is not secreted in female rats; it seems reasonable to believe that its secretion in male rats is turned on as a result of neonatal imprinting by testicular androgens (Fig. 1). Several experiments have been carried out in order to understand the nature of neonatal imprinting of the FFIF-secreting centre.

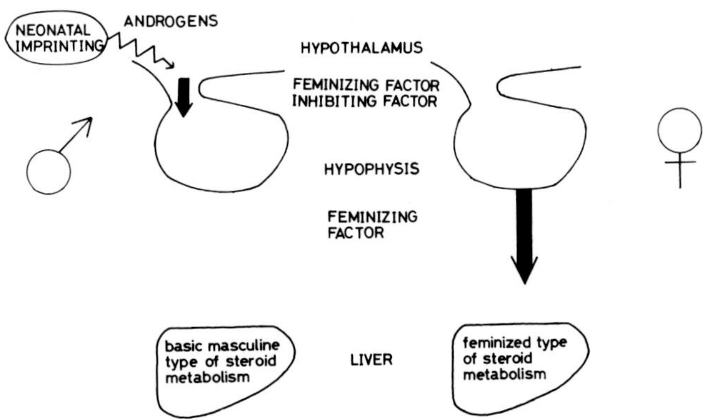


Fig. 1. Current concept of neonatal programming of sex-dependent hepatic enzyme levels.

A hereditary type of male pseudohermaphroditism in rats was described in 1964 by Stanley and Gumbreck[13]. The affected rats were phenotypic females with an XY-karyotype and bilateral inguinal testes[14]. Bardin _et al._ have suggested that the androgen insensitivity that characterizes the male pseudohermaphroditic rat is due to an inherited abnormality of a steroid receptor protein which makes the nucleus of the cell incapable of concentrating androgens at their proposed intracellular site of action[15]. The hepatic metabolism of these animals was found to be identical to that of female but different from that of male littermate rats[16]. The female character of the liver metabolism in male pseudohermaphroditic rats is probably due to the lack in these rats of neonatal androgen imprinting of the hypothalamic FFIF-secreting centre.

An interesting experimental model of male pseudohermaphroditism in rats is obtained by pre- and postpubertal treatment with cyproterone acetate. These animals possess outer genitalia with female characteristics including a vagina and also display a partially feminized sexual behaviour. These rats were found to have partially feminized enzyme levels, whereas the liver enzyme activities in treated female rats generally were not affected[17]. These results may be interpreted in the following way: treatment with the anti-androgen during the neonatal period results in less efficient imprinting of the hypothalamic FFIF-secreting centre leading to less pronounced masculine setting of sex-dependent enzyme levels and also to a relative androgen unresponsiveness.

Experiments have also been carried out to imprint the FFIF-secreting centre with other agents than testosterone. The results showed that neonatally administered 5α-dihydrotestosterone propionate and estradiol benzoate were almost as efficient as testosterone propionate in programming hepatic steroid metabolism[18]. The specific nature of this imprinting phenomenon was obvious from the fact that epitestosterone propionate and etiocholanolone propionate were almost without effect. The fact that the non-aromatizable androgen 5α-dihydrotestosterone is active as an imprinting agent indicates that the mechanism involved is different from that behind the androgen-induced development of persistent estrus syndrome and acyclic gonadotrophin secretion. The involvement of different mechanisms is further supported by the fact that op'-DDT, administered in doses similar to those inducing persistent vaginal estrus[19], did not have a masculinizing effect on hepatic steroid metabolism.

## On the nature of the pituitary feminizing factor (FF)

In order to investigate the nature of FF, an assay system has been developed using the steroid-metabolizing capacity of a hepatoma cell line in culture (HTC cells). This cell line was originally isolated from a chemically-induced hepatoma (no. 7288 C) produced in a rat of the Buffalo strain and has been shown to carry at least one liver-specific marker enzyme, a glucocorticoid-inducible tyrosine aminotransferase[20].

The cells were grown in minimum essential medium (with Earle's salts) supplemented with 5% calf serum, 5% fetal calf serum, penicillin (200 U/cm$^3$) and streptomycin (125 μg/cm$^3$) in 60 mm petri dishes. Each dish, containing 5 cm$^3$ of medium, was seeded with 5 x 10$^4$ cells. After 4 days the medium was changed and the extracts to be tested were added. The cells were maintained for a further 3 days in the new medium and then harvested, washed free of medium with phosphate-buffered saline (pH 7.4) and immediately analyzed for enzyme activity. Homogenates were prepared in a modified Bucher medium[21], pH 7.4, using an all-glas Potter-Elvehjem homogenizer. The steroid substrate used in the incubations was 4-androstene-3,17-dione. Duplicate incubations were performed in all cases and the results expressed as pmoles of product formed per min per mg of protein.

The major enzyme activity in the HTC cells was found to be the 5α-reductase which, in the intact rat liver, is a sex-dependent enzyme[6,7]. The high molecular weight fraction of homogenized female pituitary glands (separated using Sephadex G-25) was shown to increase the 5α-reductase activity of HTC cell homogenates at subsaturation concentrations of the substrate whereas the corresponding male extract was without effect. The response of the cells was linear with respect to log dose of pituitary extract over a wide range of concentrations of female pituitary extract. The change in the activity of the 5α-reductase was thus taken as a measure of the activity of FF. Studies on the kinetics of the 5α-reductase before and after addition of female pituitary extract, showed that the apparent increase in the activity of the enzyme at subsaturation concentrations of substrate was due to a decrease in the apparent $K_m$ and not to an increase in apparent $V_{max}$. This is similar to the differences observed between the 5α-reductases in gonadectomized male and female rats (the apparent, increased 5α-reductase in the female being due to a lower $K_m$)[22].

A series of purified pituitary hormones have been tested for activity in this assay (including LH, FSH, prolactin, GH, TSH, ACTH, oxytocin and vasopressin) but only FSH showed any activity and this was only 15% of the activity of the female pituitary extract (based on the concentration of FSH in fresh pituitary tissue).

Isoelectric focusing of female pituitary extracts indicated that FF is a single protein with a pI of 8.3. This value further distinguishes it from the other pituitary hormones which, in general, have a pI below 8.0 (except LH which showed a pI of 9.0) (Fig. 2). Separation of the secretory granules of the female pituitary gland using sucrose-gradient centrifugation (60 min at 105,000 x g, continuous gradient, d = 1.11-1.26) showed that FF is stored in granules with a density of 1.12-1.15 gm/cm$^3$, again separating it from LH and FSH which are stored in much denser granules. The distribution pattern of the FF granules were similar to that of prolactin granules.

## Summary and concluding remarks

This work indicates that there is a factor in the pituitary that can influence 5α-reductase activity in HTC cells. This factor also seems to be the effector of hypophyseal regulation of liver metabolism in the intact rat. Secretion of the

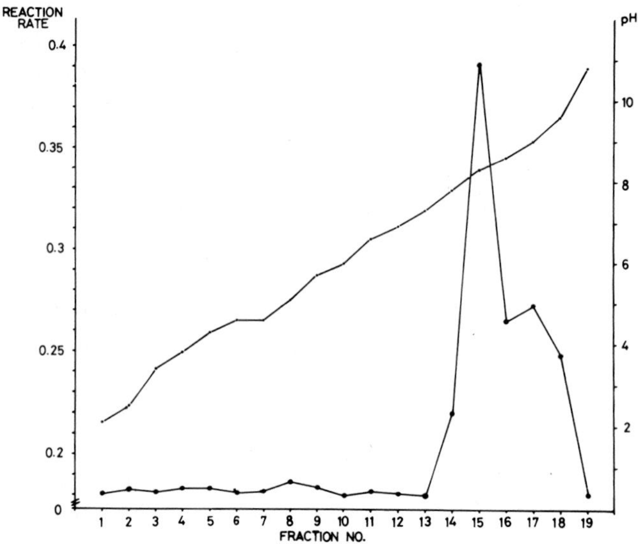


Fig. 2. Distribution of FF on an Ampholine column (pH 3-10) stabilized with sucrose. The pH gradient in the column is indicated.

factor does not occur from the male pituitary due to inhibitory hypothalamic influence. The postnatal sexual differentiation of hepatic metabolism in the rat therefore does not seem to occur as a result of direct androgenic programming of the liver cells but rather as a secondary phenomenon regulated by the hypothalamo-pituitary axis and related to irreversible changes elicited in the hypothalamus by androgens during the neonatal period.

Estrogens and androgens can only exert their normally highly significant effects on adult liver enzyme activities in the presence of an intact pituitary. The fact that sex steroids act on a target tissue *via* an indirect hypothalamo-pituitary pathway, or at least are completely dependent upon the hypothalamo-pituitary axis for exerting their effects, is not in agreement with the generally accepted view of a direct effect of sex hormones on target tissues. Much more information is needed about mechanisms involved in pituitary-mediated action of sex steroids but the question may be asked whether the indirect type of sex hormonal action on the liver represents an exceptional type of regulation or whether it in fact represents a common pathway for mediation of estrogenic and androgenic effects on target tissues.

## References


1. Gorski, R.A., 1966, J. Reprod. Fert. Suppl. 1, 67.
2. Gorski, R.A., 1971, in: Frontiers in neuroendocrinology, eds. L. Martini and W.F. Ganong (Oxford University Press, New York) p. 237.
3. Forchielli, E. and Dorfman, R.I., 1956, J. Biol. Chem. 223, 443.
4. Yates, F.E., Herbst, A.L. and Urquhart, J., 1958, Endocrinology 63, 887.
5. DeMoor, P. and Denef, C., 1968, Endocrinology 82, 480.
6. Einarsson, K., Gustafsson, J.-Å. and Stenberg, Å., 1973, J. Biol. Chem. 248, 4987.
7. Gustafsson, J.-Å. and Stenberg, Å., 1974, J. Biol. Chem. 249, 711.
8. Gustafsson, J.-Å. and Stenberg, Å., 1974, J. Biol. Chem. 249, 719.
9. Gustafsson, J.-Å. and Stenberg, Å., 1974, Endocrinology 95, 891.
10. Gustafsson, J.-Å. and Stenberg, Å., 1975, Proc. Natl. Acad. Sci., in press.
11. Eneroth, P., Gustafsson, J.-Å., Larsson, A., Skett, P., Stenberg, Å. and Sonnenschein, C., Proc. Natl. Acad. Sci., submitted.
12. Gustafsson, J.-Å., Ingelman-Sundberg, M., Stenberg, Å. and Hökfelt, T., Endocrinology, submitted.
13. Stanley, A.J. and Gumbreck, L.G,, 1964, in: Progr. 46th Meet. Endocrine Soc., p. 40.
14. Allison, J.E., Stanley, A.J. and Gumbreck, L.G., 1965, Anat. Rec. 153, 85.
15. Bardin, C.W., Bullock, L., Blackburn, W.R., Sherins, R.J. and Vanha-Perttula, T., 1965, in: Birth Defects: Original Article Series 7, 185.
16. Einarsson, K., Gustafsson, J.-Å. and Goldman, A.S., 1972, Eur. J. Biochem. 31, 345.
17. Gustafsson, J.-Å., Ingelman-Sundberg, M., Stenberg, Å. and Neumann, F., 1975, J. Endocr. 64, 267.
18. Gustafsson, J.-Å. and Stenberg, Å., 1975, Science, in press.
19. Heinrichs, W.L., Gellert, R.J., Bakke, J.L. and Lawrence, N.L., 1971, Science 173, 642.
20. Thompson, E.B., Tomkins, G.M. and Curran, J.F., 1966, Proc. Natl. Acad. Sci. 56, 296.
21. Bergström, S. and Gloor, U., 1955, Acta Chem. Scand. 9, 34.
22. Gustafsson, J.-Å., Larsson, A., Skett, P. and Stenberg, Å., 1975, Proc. Natl. Acad. Sci., in press.


## Acknowledgements

This investigation was supported by grants from the Swedish Medical Research Council (grant no. 03X-2819), from Harald and Greta Jeanssons Stiftelse and from the World Health Organization. One of us (P.S.) is grateful to the CIBA-GEIGY Fellowship Trust for a fellowship.

*Progress in Differentiation Research, ed. N. Müller-Bérat et al.*



# THE ROLE OF GERM CELLS IN THE MORPHOGENESIS AND CYTODIFFERENTIATION OF THE RAT OVARY

H. Merchant-Larios

Instituto de Investigaciones Biomédicas, U.N.A.M.

Apdo. Postal 70228, México 20, D.F.

## 1. Introduction

Current information implies that cytodifferentiation and morphogenesis in vertebrate organ differentiation are coupled events. Epitheliomesodermal tissue interactions have been shown to be necessary for differentiation of several organs, however the interactions between somatic and germ cells and the consequences of such interactions in terms of cytodifferentiation in the mammalian ovary have not been considered as yet.

In a previous work[1] busulphan, a drug which selectively destroys the primordial germ cells (PGCs), was used in order to determine the role of these cells in the development of the fetal ovary. In this study it was shown that PGCs are not necessary for the establishment of an undifferentiated gonad and that the fetal morphogenesis of the rat ovary deprived of PGCs follows a pattern similar to that of the controls. As the busulphan treated rats seem to develop normally after birth, and recent lines of evidence[2,3,4] suggest that germ cells may play an important role in the regulation of the endocrine activity in mammalian adult ovaries, the present study was directed toward describing the structure of busulphan sterilized ovaries in terms of the role of the germ cells in the ultrastructural differentiation of steroid-synthesizing cells.

## 2. Materials and Methods

Pregnant Wistar rats were injected intraperitoneally on the 11th or 12th day of gestation with single doses of busulphan (1 mg/100 g body wt). Female offspring were sacrificed on days 7, 15, 21, 32 60 and 90 after birth. Ovaries were processed for high resolution light and electron microscopy according to a technique which has been described elsewhere[1]. Two groups of eight busulphan sterilized animals aged two and three months were treated with 25IU of PMS and sacrificed 12, 24, 48, and 60 hours later. Their ovaries were processed in the same way as the others. The size of the genital tract, vaginal openining or vaginal smear in adult treated rats was recorded.

## 3. Results

The structure of those ovaries where no germ cells survived the busulphan-treatment is slightly modified during the first month after birth. These ovaries appear identical to the ovaries of full term treated fetuses[1]. They possess a cuboidal surface epithelium which is continuous with the inner epithelial cords. The amount of stromal tissue is increased and there is a moderate growth of the ovary at the end of this period. From the first week on, abundant dark cells, which were identified by their ultrastructure as mast cells, appear, especially in the epithelial cords. (Fig. 1).

Under the electron microscope the epithelial cells, also present a fine morphology similar to that observed in the fetus. They have numerous small mitochondria, abundant polyribosomes and a well developed rough endoplasmic reticulum. The epithelial "cords" are enveloped by a continous basal lamina and the cells are tightly associated. (Fig. 2). The stromal tissue is formed essentially by fibroblast-like cells, collagen fibers and dispersed smooth muscle cells. From the third week on, a cluster of smooth muscle fibers appears beside the hilus (Fig. 1). Distinct from normal ovaries where ultrastructurally differentiated steroid-forming cells are present from the 10th day on, no cells with these characteristics were found.

In a high proportion of busulphan sterilized rats, a variable number of germ cells survived the treatment. In such cases normal-looking follicles seem to be organized for these cells and they were present simultaneously with sterile epithelial cords. (Fig. 3). Moreover, from the third week on abundant and highly differentiated interstitial tissue is found between the surviving follicles. (Fig. 4).

At the second month, in ovaries which were fully depleted of germ cells, some changes appear on the epithelial cords. Some of them develop a lumen (Fig. 5) and the epithelial cells bordering it have abundant microvilli and are tightly associated by extended gap junctions. These features and the presence of numerous folds of the plasma membrane could suggest a certain secretory activity (Fig. 6). However they do not reflect the characteristic fine morphology of steroid synthezising tissue which is very abundant in control rats of the same age. On the stromal tissue, besides on increase in the amount of collagen, there is no major modification in the quality of its cells (Fig. 6).

In ovaries taken on the third month, there is a slight reduction in the diameter of the epithelial cords (Fig. 7) and some degenerative changes appear among the epithelial cells. These degenerations consist in the presence of big vacuoles in the cytoplasm of certain cells which eventually lead to their total degeneration. (Fig. 8).

In ovaries where some follicles were formed during the first weeks after birth, they become atretic at different times before the second month (Fig. 9). This happens in such a way that by the third month no follicles can be found and in their place, prominent masses of "luteal" tissue may be found. (Fig. 10). Under the electron microscope the structure of these cells suggests a high steroid-synthezising activity (Fig. 11).

In general there is a good correlation between the amount of steroid-forming tissue in the ovaries and the degree of response in the genital tract. Animals with fully undifferentiated ovaries have an infantile genital tract and continue through the third month to have a closed vagina. Very often only one ovary was completely devoid of "luteal" tissue and the other contained different amounts.

Animals treated with PMS will also respond, depending on the amount of "luteal" tissue contained in their ovaries. When their two ovaries were undifferentiated, no response was found in terms of their ultrastructural differentiation. In cases where "luteal" tissue was present, this was somewhat hypertrophic in comparison with the untreated ones.

---

Fig. 1. Light microscope picture of a 20 day ovary devoid of oocytes. The epithelial cords contain several dark cells which correspond to mast cells. In the stromal tissue no steroid synthezising cells were found. A cluster of smooth muscle fibers is localized on the dorsal region near the hilus. (arrow).

Fig. 2. Electron micrograph of the same ovary of figure 1. Parts of two epithelial cords (EC) and the stromal tissue (ST) in between can be seen. The former structures are surrounded by a thin basal lamina (BL) and the epithelial cells are tightly arranged. The stromal tissue is mainly formed by fibroblast-like cells.

Fig. 3. Part of a 10 day ovary showing a follicle similar to those found in normal ovaries of the same age. Some isolated steroid synthezising cells are seen (arrow). To the right of the follicle a steril epithelial cord can be seen . (light microscopy).

Fig. 4. Part of a two months old ovary in which several oocytes survived the busulphan treatment. Between two follicles of different size there is a large cluster of interstitial cells (IC). (Light microscopy).

Fig. 5. Ovary fixed on the second month after birth. One of the epithelial cords shows a well developed lumen (*). Among the stromal tissue collagen can be seen in abundance (Black stained).

Fig. 6. Electron micrograph of the same ovary shown in figure 5. On the left half of this figure part of an epithelial cord can be seen. The cells of this structure are tightly arranged with numerous "gap junctions". (arrows). To the basal lamina (BL) a layer of collagen fibers has been added. The right half shows a general view of the stromal tissue which surrounds the cords.

Fig. 7. Light microscope picture of an ovary taken in the third month. Epithelial cords somewhat reduced and a higher proportion of stromal tissue are seen.

Fig. 8. Electron micrograph showing the large vacuoles which appear around the third month in several epithelial cells.

Fig. 9. This light microscope picture shows an atretic oocyte surrounded by fully differentiated "luteal" tissue. Ovary fixed 30 days after birth.

Fig. 10. Two month ovary showing three "luteal" masses. (L). Such ovaries are always well vascularized (v).

Fig. 11. General view of luteinazed cells in which their fine morphology can be seen. The tubular crista of the mitochondria, lipid droplets, and abundant smooth endoplasmic reticulum are the typical features of steroid synthezising cells. Three month ovary in which some follicles are formed.

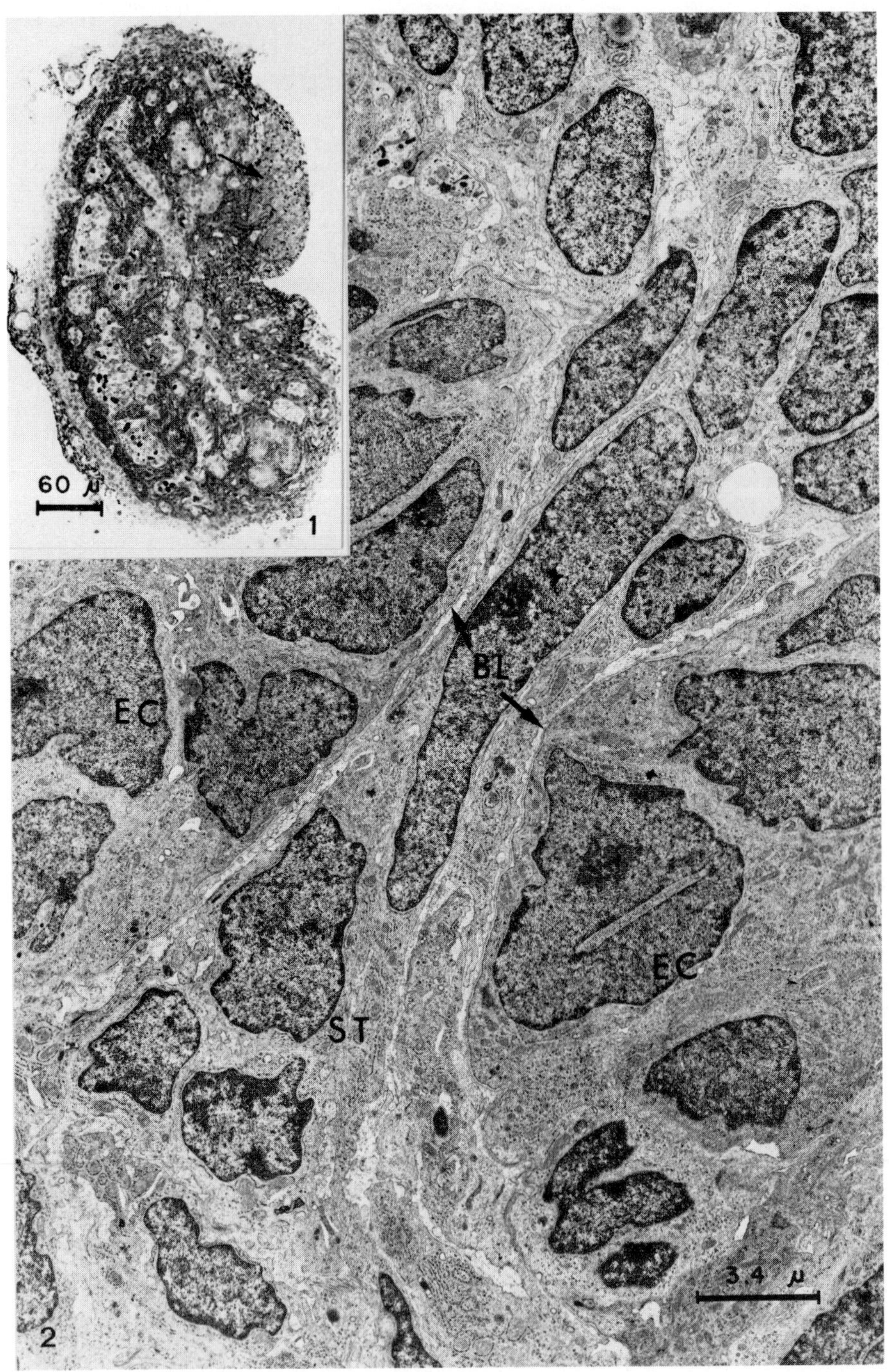
60 μ
1
BL
EC
EC
ST
3.4 μ
2

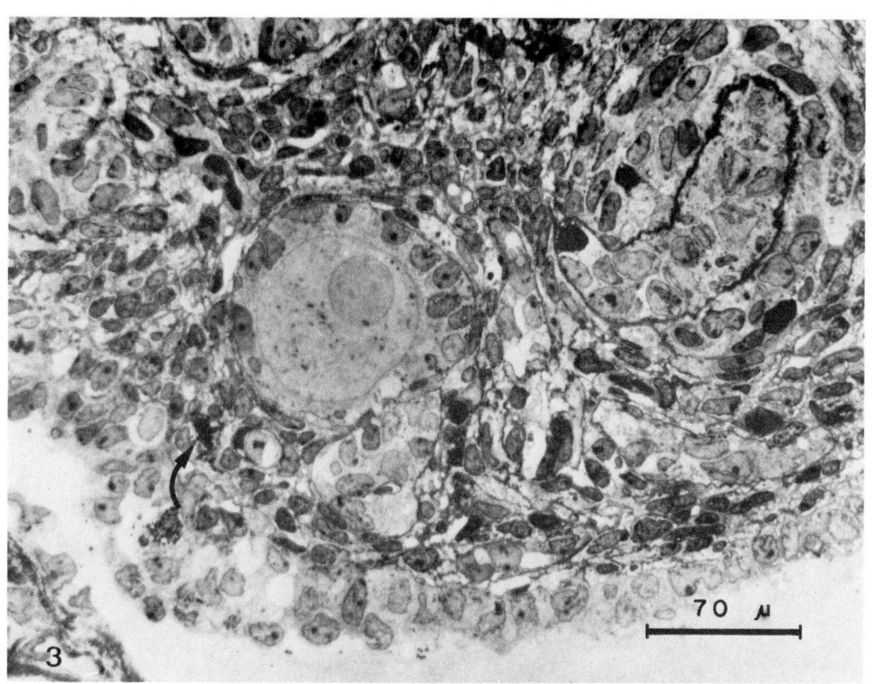
70 μ
3

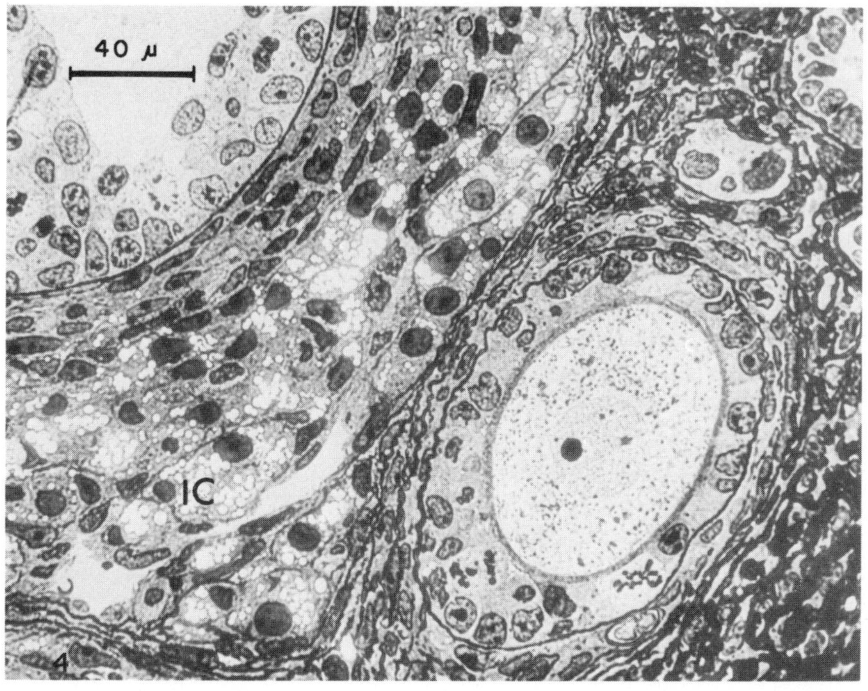
40 μ
IC
4

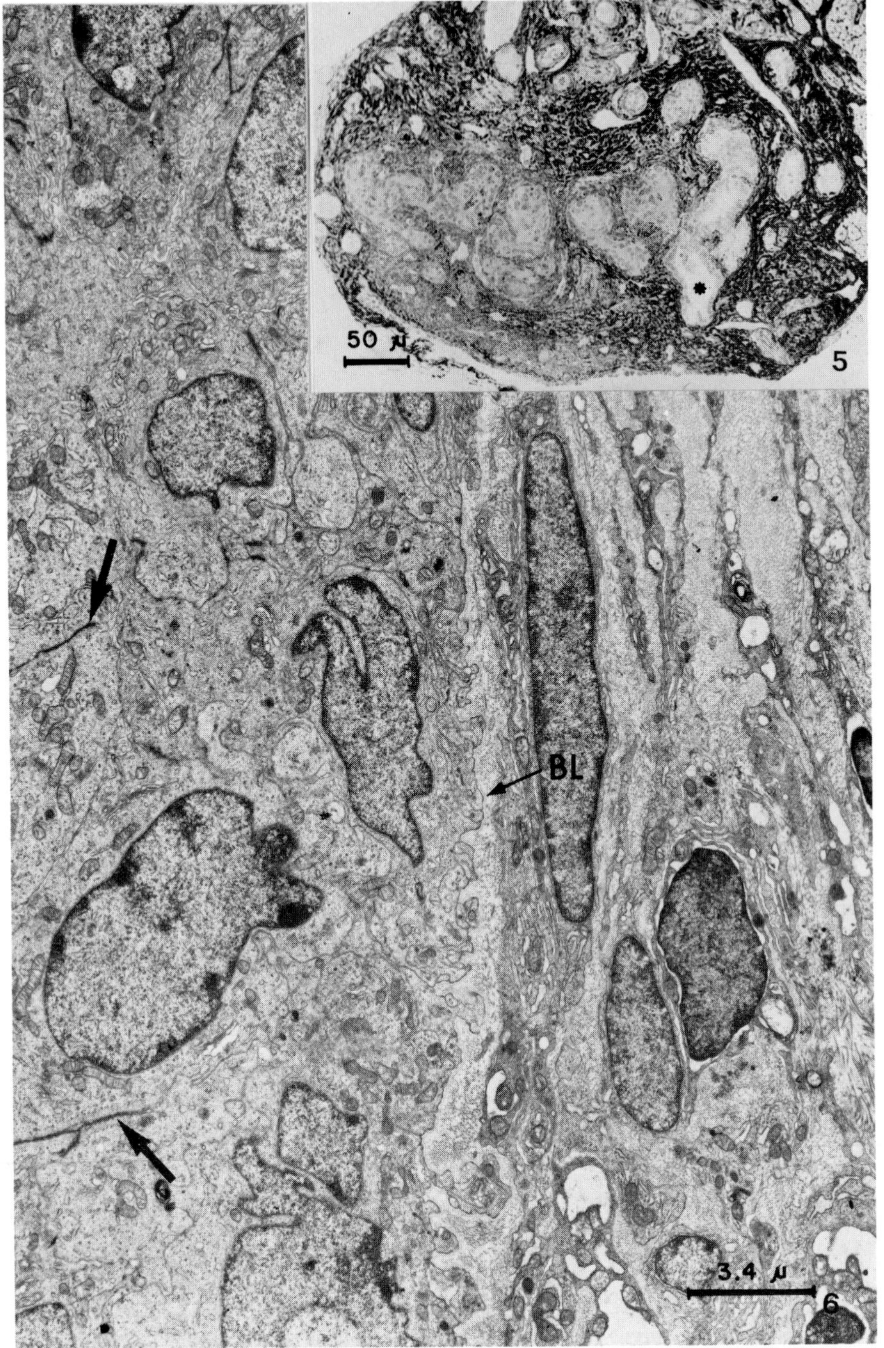
50 μ
5
BL
3.4 μ
6

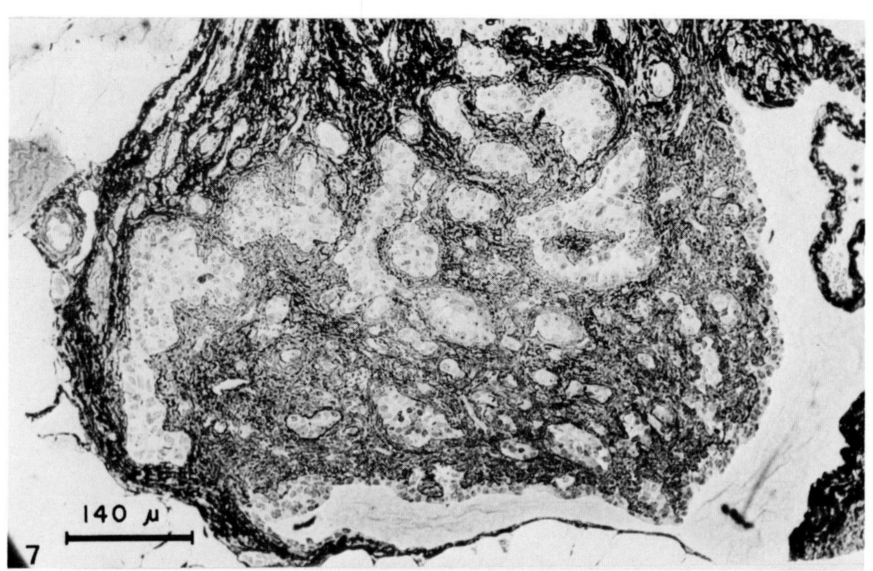
140 μ
7

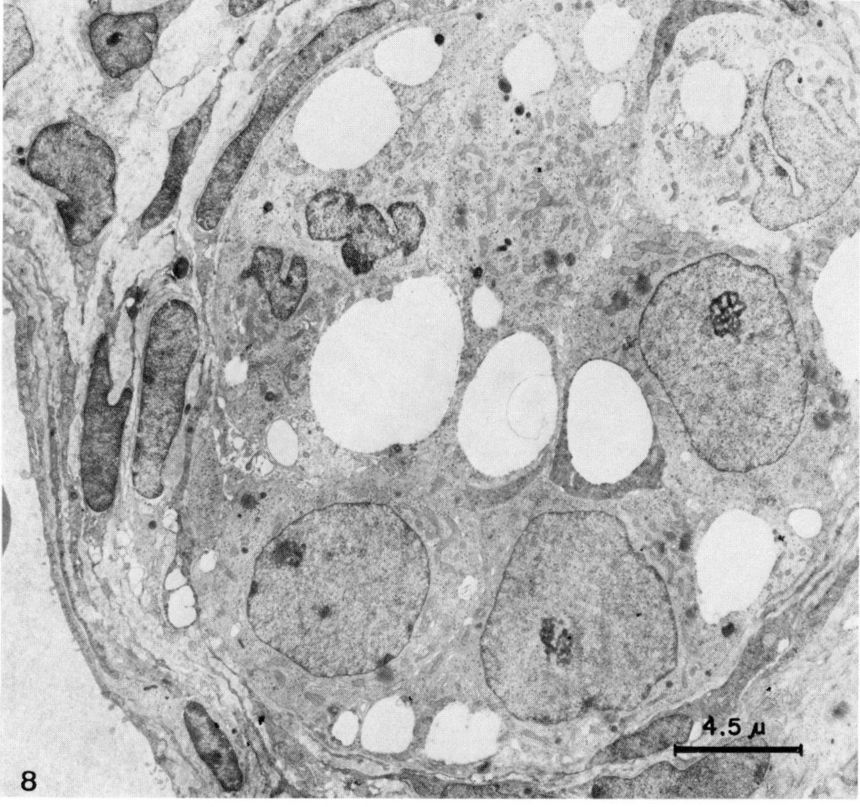
4.5 μ
8

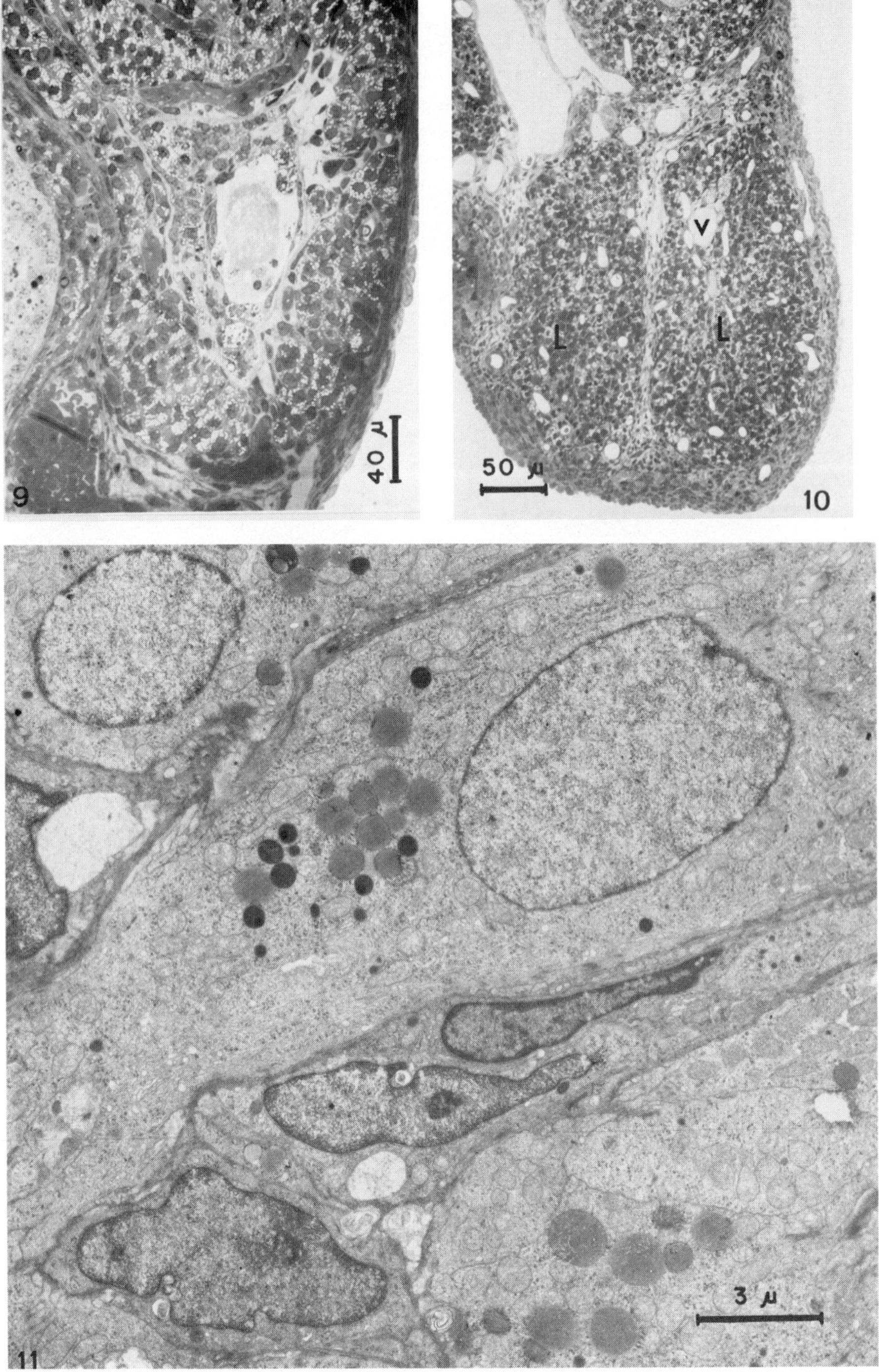
40 µ
9
V
L
L
50 µ
10
3 µ
11

## 4. Discussion

The present study suggests that the presence of germ cells in the epithelial cords at birth is a condition for the structural differentiation of somatic cells into steroid-synthezising cells in the rat ovary.

Although to my knowledge the problem of the postnatal morphogenesis and endocrine activity of the mammalian ovary has not been studied in terms of the role of the germ cells in such processes, there are in the literature several experimental reports which seem to support the present findings. Among these the reports dealing with the effects of x-ray irradiation on mammalian ovaries are the most numerous[5].

Beaumont[6], after irradiation of rats and mice _in utero_ stated "The presence of medium-sized and large follicles in the ovaries indicates that those oocytes which survived the treatment remained capable of growing and organizing around themselves investing layers of granulosa cells". Moreover, he observed that the granulosa cells surrounding the irradiated oocytes were clearly normal funcionally and responsive to gonadotrophic influences. Mandel[7] found that if the destruction of irradiated follicles during the first two weeks after birth was complete, no estrogenic activity remained. In contrast, if a few follicles persisted, oestros periods ocurred at irregular intervals and there was a clear response in all the estrogen-target tissues. Finally, when adult ovaries were irradiated it was always found that a certain amount of estrogenic activity remained detectable[8,9].

These data suggest that the presence of oocytes is necessary during a critical stage of development to induce the surrounding cells to respond to gonadotrophic stimulation; after that point, they remain responsive even in the total absence of oocytes.

The present ultrastructural study fully supports and extends a previous report given by Pugliatti and Allegra[10]. These workers, using a light microscope found a clear correlation between the amount of luteinized tissue left after busulphan treatment and the development of estrogen-target organs.

In studies of this sort it can always be argued that exposure of whole embryos to the action of a physical or chemical agent is very likely to alter other organs. However, there are some lines of evidence which suggest that in the present conditions, the busulphan treatment is quite specific.

The lack of response of the somatic cells in fully sterilized ovaries could be atributed to a poor secretion of gonadotrophins from an altered pituitary. If this were the case, exposure of the treated ovaries to an adequated concentration of gonadotrophins would induce the response. In this study, however, it was found that when the ovaries were fully sterile at birth, there was no response at all to the PMS, which always induces reactivation of the interstitial tissue[11]. Moreover, Vanhems and Bousquet[12] made grafts of busulphan sterilized ovaries on the kidney capsule of castrated female rats and found that they respond according to their tissular constitution. If they contained only "cords" they fail to respond at all, otherwise there is always stimulation of "theco-luteal" tissue.

The last point to be considered is in relation to the possible alteration of the ovaric somatic cells by the busulphan. In the present study, it was demonstrated that the surviving germ cells organize normal follicles and are surrounded by fully undifferentiated cords, it can therefore be concluded that it is the oocyte which in some way induces the formation of gonadotrophine receptors in the somatic cells. This process seems to occur only during a critical stage of the follicular development, and during the first month the stromal cells surrounding the follicular basal lamina become differentiated as an interstitial gland. After atresia of the oocyte, the granulosa cells could remain as steroid-synthezising cells, although in the present study this was not shown.

The way in which the oocyte induces the differentiation of the stromal tissue remain to be determined. Possibly they are necessary for the fragmentation of the epithelial cords, with the consequent formation of "follicular units" including epithelial and stromal tissue as well as blood vessels. Once formed, this anatomic unit will facilitate normal interaction between the different kinds of cells and the pituitary hormones. However, a more active role of the germ cell is possible. The germ cell may control the two processes (fragmentation and steroid-synthezising capacity) by means of "inductors" for each particular time of follicular development.

## References


1. Merchant, H., 1975, Develop. Biol. 44, 1.
2. El-Fouly, M.A., Cook, B. Nekola, M. and Nalvandov, A.B., 1970, Endocrinology 87, 288.
3. Nekola, M.V. and Nalvandov, A.V., 1971, Biol. Reprod. 4, 154.
4. Bernard, J., 1975, J. Reprod. Fert. 43, 453.
5. Lacassagne, A. Duplan, J.F., Marcovich, H. and Raynaud, A., 1962, in: The Ovary, vol. II eds. S. Zuckerman (Academic Press, New York and London) p. 463.
6. Beaumont, M.H. 1961 Int. J. Rad. Biol. 3, 59.
7. Mandel, J., 1935, Anat. Rec. 61, 295.
8. Mandl, A.M. and Zuckerman, S., 1956, J. Endocrin. 13, 243.
9. Mandl, A.M., 1959, J. Endocrin. 18, 426.
10. Pugliatti, V. and Allegra, P.S., 1966, Riv. Anat. Pat. Oncol. 29, 758.
11. Carithers, J.R. and Green, J., 1972, J. Ultrastructure Res. 39, 251.
12. Vanhems, E. and Bousquet, J., 1971, Ann. Endocrin. (Paris) 32, 753.

*Progress in Differentiation Research, ed. N. Müller-Bérat et al.*
*© 1976, North-Holland Publishing Company - Amsterdam, The Netherlands.* 

# DIETHYLSTILBESTROL AND PROGESTERONE REGULATION OF CULTURAL RABBIT ENDOMETRIAL CELLS *

L. E. Gerschenson and J.A. Berliner

Laboratory of Nuclear Medicine and Radiation Biology,
and Department of Pathology, University of California, Los Angeles,
California 90024, U.S.A.

## 1. Introduction

As our contribution to this conference, we wish to review our recent work (1,2,3) using a new experimental system of cultured rabbit endometrial cells that appears to be responsive to the addition of diethylstilbestrol and progesterone.

The endometrium is a target tissue for ovarian hormones, which have well known regulatory effects upon tissue metabolism, growth and differentiation (4,5,6). These hormones also play an important role in the development of endometrial adenocarcinoma, since estrogens have been described to increase the incidence of spontaneous endometrial adenocarcinoma and the induction of this neoplasm in rabbits by chemical carcinogens (5,7,8). On the other hand, progesterone has been described to inhibit cell division in endometrium and chick oviduct epithelia, to induce the secretion of specific proteins and to be necessary for decidualization (9,10 11,12,13). This hormone is also used for the treatment of endometrial adenocarcinoma (14).

## 2. Results and Discussion

We have described recently (1) a simple technique to isolate rabbit endometrial cells by everting the uterine horns and incubating them with an enzyme mixture. The cells can be cultured in a chemically defined medium for about two weeks. During these two weeks, the cells are responsive to the addition of either $10^{-7}$M diethylstilbestrol or $10^{-7}$M progesterone. Diethylstilbestrol was found to increase the percentage of DNA-replication cells, measured by radioautography after incubation with $^3$H-Thymidine. Progesterone had an opposite effect. The effect of both hormones upon the cells was found to be antagonistic and concentration dependent (1,2). However, both hormones induced a significant increase in the

---

* This investigation was supported by Grant CA-12136 from the NIH, Contract E (04-1) GEN 12 between ERDA and the University of California and Grants from the California Institute for Cancer Research and the UCLA Fund.

incorporation of labeled uridine and aminoacids into cold TCA precipitates, which are considered tentatively to mean increases in RNA and protein synthesis. These hormonal effects were rapidly reversed upon removal of the hormones from the culture medium and were also inhibited by Actinomycin D or cycloheximide.

The administration of estradiol to ovariectomized animals is known to rapidly induce inflammatory-like changes in uterus and these hormonal manifestations have been considered to be related to growth as causative factors and to result, among other things, from 3', 5'-cyclic AMP increases which may have lysosomal labilization effects. The anti-estrogenic effects of glucocorticoids were then related to their lysosomal stabilizing or anti-inflammatory effects (15,16). Acid and alkaline phosphatase, and lactic dehydrogenase activities, as well as the transport of amino acids and carbohydrates have been found to increase under the above described experimental conditions (17,18,19,4). Our studies using cultured cells (2) demonstrated that neither the addition of dibutyril 3',5'-cyclic AMP nor hydrocortisone had any effect on the diethylstilbestrol-mediated stimulation of DNA replication. The levels of the three enzymes and the rate of transport of the substrates mentioned above were also found to be unchanged by hormonal additions to our cultured cells (2). All the described results are supported by the electronmicroscopy studies (3) which show that cells treated with diethylstilbestrol had the appearance of a rapidly dividing culture having large euchromatic nuclei and prominent nucleoli with a cytoplasm containing many free ribosomes. Progesterone appeared to convert the cells to a more secretory type with larger amounts of rough endoplasmic reticulum and smaller nucleoli. Primary and secondary lysosome appearance was found to be unaltered by estrogenic stimulus.

## 3. Conclusions

Our findings suggest: a) It is possible to obtain estrogenic and progestogenic effects without metabolic and membrane transport changes, b) The inflammatory estrogenic effects may not be a primary cause for the hormonal manifestations but a secondary one, which has possibly been acquired during evolution as an advantageous change to amplify the hormonal stimulus, by increasing the availability of nutrients and c) Progesterone causes the cultured rabbit endometrial cells to differentiate into non-dividing, secretory cells. Such a process may be related to the decidualization phenomenon.

Work in progress, on the hormonal effects of the cell cycle, transcription, secretory proteins as well as the use of established cell lines will clarify further our preliminary data and working hypotheses.

## References


1. Gerschenson, L.E., Berliner, J.A. and Yang, J., 1974, Cancer Res. 34, 2873.
2. Gerschenson, L.E. and Berliner, J.A., 1976, in press, J. Steroid Biochem.
3. Berliner, J.A. and Gerschenson, L.E., 1976, in press, J. Steroid Biochem.
4. Roberts, S. and Szego, C.M., 1953, Physiol. Rev. 33, 593.
5. Dallenbach-Hellweg, G., 1971, Histopathology of the endometrium, New York: Springer-Verlag, Inc., p. 153.
6. Whitson, G.L. and Murray, F.A., 1974, Science 183, 668.
7. Meissner, W.A., Sommers, S.C. and Sherman, C., 1957, Cancer 10, 500.
8. Baba, N. and Von Haam, E., 1967, Progr. Exptl. Tumor Res. 9, 192.
9. Noyes, R.W., Hertig, A.T. and Rock, J., 1950, Fertility & Sterility 1, 3.
10. Smith, J.A., Martin, L., King, R.J.B. and Vertes, M., 1970, Biochem. J. 119, 773.
11. Socher, S.H. and O'Malley, B.W., 1973, Develop. Biol. 30, 411.
12. Zhinkin, L.N. and Samoshkina, N.A., 1967, J. Embryol. Exptl. Morphol. 17, 598.
13. Loeb, L., 1908, Proc. Soc. Exptl. Biol. Med. 5, 102.
14. Kelley, R.M. and Baxter, W.H., 1961, New England J. Med. 264, 216.
15. Spaziani, E., 1963, Endocrinology 72, 180.
16. Szego, C.M., 1971, in: Sex Steroids (Ed. E.W. McKerns), New York: Appleton-Century, Crofts, p. 1.
17. Leathen, J.H., 1958, Ann. N.Y. Acad. Sci. 75, 463.
18. Bever, A.I., 1958, Ann. N.Y. Acad. Sci. 75, 472.
19. Riggs, T.R., Pan, M.W. and Feng, H.W., 1968, Biochem. Biophys. Acta 150, 92.

*Progress in Differentiation Research, ed. N. Müller-Bérat et al.*



# MECHANISMS OF HISTIDASE DEVELOPMENT IN RAT LIVER AND EPIDERMIS

Muriel Feigelson, Madhu M. Bhargava and Coral A. Lamartiniere

Departments of Biochemistry and of Obstetrics and Gynecology, Columbia University College of Physicians and Surgeons, Roosevelt Hospital, New York, N. Y., U.S.A.

## 1. Differential developmental courses of skin and liver histidase activities.

Histidase (histidine ammonia-lyase, EC 4.3.1.3) activity undergoes markedly different developmental courses in rat liver and epidermis [1,2] (Fig. 1), the only two tissues in which it has been detected. In liver, histidase activity initially appears shortly following parturition [1], rises linearly until three weeks of age, plateaus for several weeks in the male, then ascends again during puberty, more rapidly and to higher levels in the female than in the male [3-6]. In epidermis, histidase activity is initially detectable on the 19th day of fetal life, at a time when no measurable hepatic enzyme activity is found. Skin histidase activity rises steeply during the perinatal period, attaining a maximum at one week post-partum, at a time when liver enzyme activities are barely discernable, following which there is an abrupt decline in the epidermal histidase activities, at a time when liver activities are rapidly mounting [1,7]. Skin histidase activity further decreases in males during puberty, achieving adult enzyme levels of approximately one-half that of females [2-8]. Thus, after the first postnatal week, histidase activities rise in liver postnatally and simultaneously descend in the epidermis of the same animals.

## 2. Differential hormonal regulation of skin and liver histidase activity.

Three hormonal inducers of hepatic histidase activity have been identified: estrogen [4], glucocorticoid [9] and glucagon [9] (Fig. 2). Induction of liver histidase activity by all three hormonal inducers is blocked by administration of cyclo-

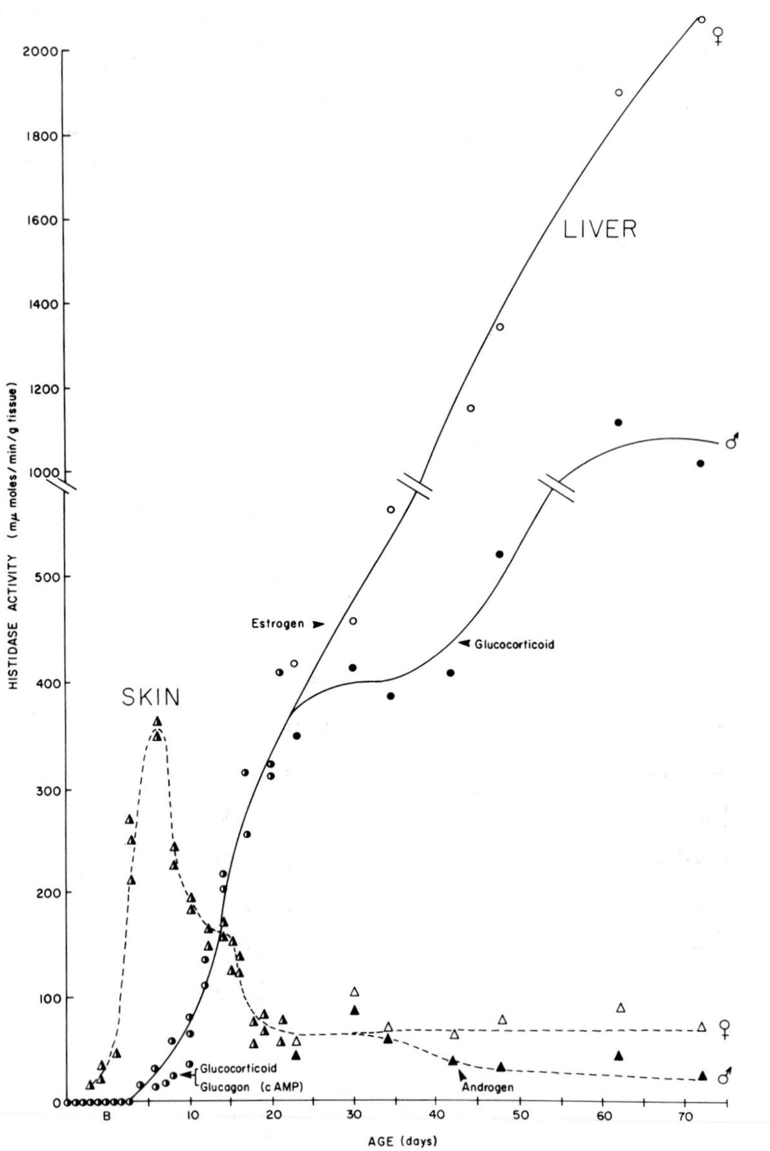


Fig. 1. Developmental courses of skin and liver histidase activities. Triangles= skin histidase; circles=liver histidase; shaded symbols=males; open symbols=females; half shaded-half open symbols=both sexes. Arrows denote specific stages and tissues where various hormones act to induce development of histidase activity.

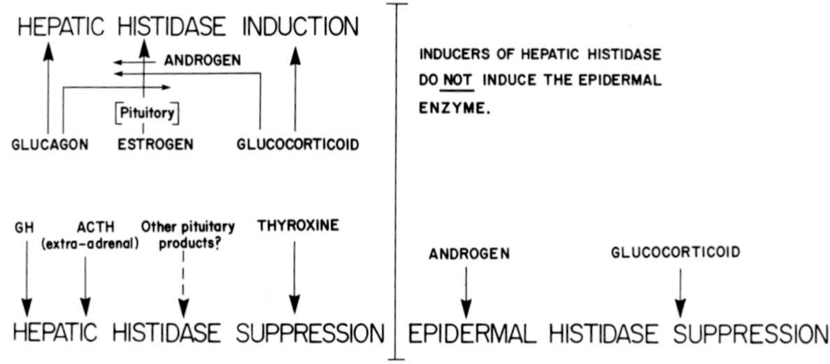


Fig. 2. Hormonal regulation of skin and liver histidase activity.

heximide, ethionine and actinomycin-D[4,6], inferring that both translational and transcriptional processes are involved in all these hormonal actions on this enzyme. The presence of an intact pituitary is mandatory for estrogenic induction of hepatic histidase. It has not been possible to restore estrogenic induction of hepatic histidase in the hypophysectomized animal with the administration of pituitary preparations or components[10]. Thus, estrogen may act on the pituitary or hypothalamus to inhibit liberation of some as yet unknown pituitary suppressor of hepatic histidase. Unlike estrogen, glucocorticoid and glucagon are capable of inducing liver histidase in hypophysectomized rats. Since dibutyryl cyclic AMP mimicks the action of glucagon *in vivo*[6], as in other systems, this pancreatic hormone is presumed to act via cyclic AMP in the regulation of histidase.

Not only is histidase activity itself altered during development, but its responsiveness to endocrine regulation undergoes hormone-specific developmental changes, as well[9]. Glucagon is capable of inducing liver histidase at all stages, from birth through adulthood. However, the liver is refractory to estrogenic induction during the first postnatal month, after which estrogen is a potent inducer of liver histidase. Glucocorticoid, on the other hand, is an effective inducer of this enzyme only during the first two postnatal months; at maturity, liver histidase activity fails to increase in response to this hormone.

The three inducing hormones of hepatic histidase have been shown to participate in effecting elevations in enzyme activity in liver at specific developmental stages (see Fig. 1). Loading neonatal animals with high doses of glucose, which is known to suppress glucagon secretion and cyclic AMP elaboration, inhibits the neonatal rise in liver histidase [6]; glucagon administration reverses this block. Similarly, blockage of glucocorticoid synthesis with the administration of cyanoketone or aminoglutethimide, likewise, curtails the neonatal rise in hepatic histidase [6]; glucocorticoid administration reverses this inhibition. Thus, both glucagon and glucocorticoid are believed to play a role in the neonatal emergence of this enzyme in liver. The above steroidogenic inhibitors or adrenalectomy significantly reduce the adolescent rise in hepatic histidase in the male [6], which is restored by hydrocortisone administration. Ovariectomy of prepubertal females results in the male pattern of adolescent development of hepatic histidase; administration of estradiol-17β to such ovariectomized rats restores the female pattern of enhanced pubertal liver histidase development [4]. Therefore, glucocorticoids participate in effecting the adolescent developmental rise in histidase in the male liver and estrogen secretion in the pubertal female is responsible for effectuating the more marked developmental rise in this enzyme characteristic of the female. Thus, each of the three hepatic histidase inducing hormones participate in implementing elevations of liver enzyme activity at specific developmental stages in each sex, as illustrated by the arrows in Fig. 1. These stage-specific actions on hepatic histidase development are compatible with the responsiveness of the liver to each hormone during development, as discussed above.

Although they are inducers of hepatic histidase, in the male or immature or ovariectomized female, glucocorticoid and glucagon block the estrogenic induction of this enzyme (Fig. 2) and thereby are actually suppressors of liver enzyme activity in mature females, which are fully estrogen induced [6]. Androgen, a non-inducer, also blocks estrogenic induction of this enzyme [11] (Fig. 2). Various pituitary hormones [12] (e.g., growth hormone and adrenocorticotrophin) and thyroxine [13-15] are *in vivo* suppressors of hepatic histidase in both sexes (Fig. 2).

Endocrine regulation of the epidermal histidase markedly differs from that of the hepatic enzyme. Although inducing the hepatic enzyme activity, neither estrogen, glucagon nor glucocorticoid administration is capable of inducing the epidermal enzyme[2] (Fig. 2). Androgen[2] and glucocorticoid are effective suppressors of the skin enzyme in the absence of estrogenization (Fig. 2). Orchiectomy of the prepubertal male results in the female pattern of adolescent skin histidase development (i.e., no significant change) thus, androgen secretion seems to be responsible for the decline in epidermal histidase in the pubertal male (Fig. 1).

### 3. Identity of epidermal and hepatic histidases.

The markedly different developmental courses and hormonal regulation of histidase in rat liver and epidermis suggests the possibility of differing enzymatic variants in these two tissues. However, a variety of evidence, physicochemical, kinetic and immunologic, indicates that this enzyme at these two tissue sites are identical proteins.

The enzyme from both tissue sources have the same isoelectric points, as established by isoelectric focusing on acrylamide gels, the pI's of hepatic and epidermal histidase both being in the range of 5.4-5.6. Furthermore, Lineweaver-Burk plots have indicated identical Michaelis constants for the enzyme from both tissues, viz., $K_m = 1.6 \times 10^{-3}$[1].

Employing a goat mono-specific antibody, prepared against purified female rat liver histidase (Lamartiniere, C.A. and Feigelson, M., unpublished results), histidases from both skin and liver have been shown to be immunologically identical by three criteria[1]: precipitin lines of identity in the Ouchterlony double diffusion system; identical mobilities of single precipitin arcs upon immunoelectrophoresis; and identical equivalence points upon immunotitration (indicating the same ratio of antigenic sites:catalytic sites in the enzymes from both tissues). No immunologically crossreacting materials are detectable in tissues devoid of histidase catalytic activity, e.g., spleen, kidney, brain, implying that gene products immunologically related to histidase are not formed in these tissues[1].

4. Identity of histidase during development in each tissue and sex.

Despite the marked developmental fluxes in histidase catalytic activities, which differ in magnitude in the livers of the two sexes during adolescence and which differ temporally and are indeed divergent in liver and epidermis (see Fig. 1), the immunologic identity of histidase in the livers of males and females[16] and in both tissues during postnatal development[1,17] has been established, employing the above described criteria in immunodiffusion, immunophoretic and immunotitration systems. Thus, no antigenic variants of histidase are elaborated during puberty in livers of either sex, nor during postnatal development of either tissue. Furthermore, using these test systems, no immunologically crossreacting materials are detectable in skin or liver, prior to the emergence of histidase catalytic activity[1], suggesting a block at some as yet unknown step in the expression of the histidase gene in skin and liver, which is lifted at specific perinatal stages of development in each tissue.

5. Developmental alterations in amounts of histidase protein in each tissue and sex.

Since immunotitration experiments have indicated identical equivalence points in both tissues and sexes during development, *i.e.*, the same amounts of antibody neutralize the same level of histidase activity in developing male and female liver and skin, it follows that the marked tissue and sex-specific developmental fluxes in catalytic activity are accompanied by equivalent alterations in *amounts* of enzyme protein. These observations are illustrated in the cases of adolescent development of hepatic histidase in males and females (Fig. 3) and postnatal development of the hepatic and epidermal enzyme in the male (Fig. 4). Both the level of catalytic activity and amount of hepatic histidase, as measured by immunotitration, rise proportionately in both sexes during the adolescent period, to higher levels in the female than in the male (Fig. 3). Similarly, it may be seen in

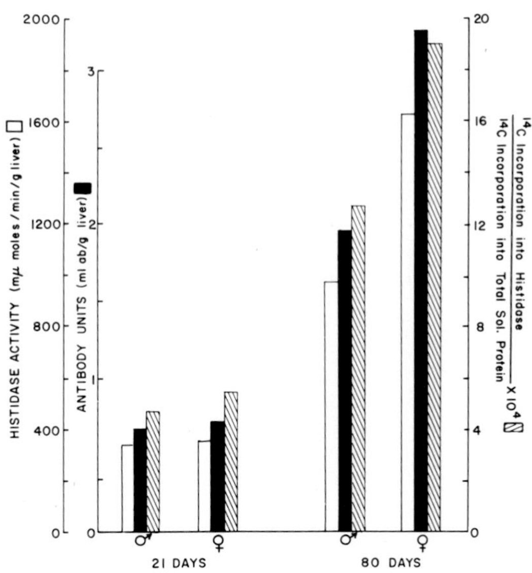


Fig. 3. Catalytic activity, quantity and relative synthetic rates of hepatic histi-in males and females during adolescent development. Rats of indicated ages received a 45 minute pulse *in vivo* of $^{14}$C-leucine[16]. High speed supernatants of liver homogenates were prepared, on which histidase activities were measured[18]; histidase levels were quantitated by immunotitration (*i.e.*, volume of antibody solution required to titrate the histidase activity per g tissue); and $^{14}$C-leucine incorporated into immunoprecipitated hepatic histidase, relative to that incorporated into total soluble protein, was determined[16], as a measure of enzyme synthetic rate.

Fig. 4, that in male liver both histidase catalytic activity and enzyme amount, as expressed in antibody units and in amounts of histidase immunoprecipitable protein, rise proportionately between 10 and 60 days of age. During the same developmental period, in the same animals, epidermal histidase catalytic activity and quantity (as determined by the same two measurements) fall precipitously and in parallel[17]. It may thus be inferred that the alterations in histidase catalytic activities, which differ quantitatively in the livers of the two sexes and are divergent in the two tissues, are a result of proportionate alterations in amounts of the same enzyme.

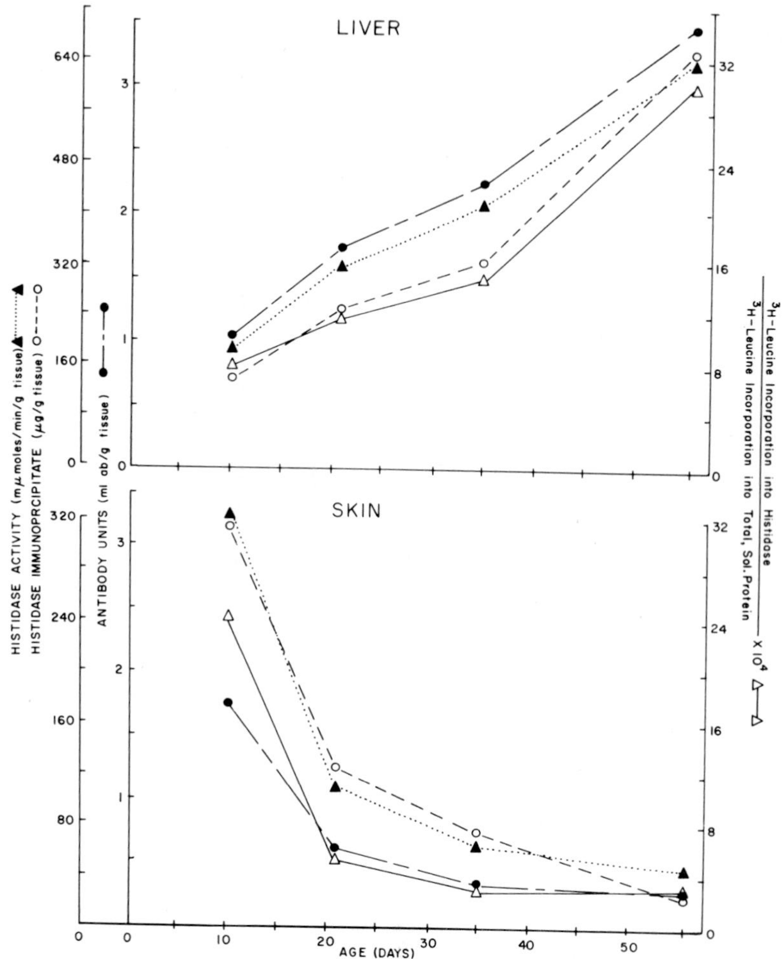


Fig. 4. Catalytic activity, quantity and relative synthetic rates of male liver and skin histidase during postnatal development. Rats of indicated ages received a 40 minute pulse *in vivo* of $^{3}$H-leucine[17]. High speed supernatants of liver and epidermal homogenates were prepared, on which histidase activities were measured[18]; levels of histidase were quantitated by immunotitration (see legend to Fig. 3) and by measurement of immunoprecipitated protein[17]; and histidase synthetic rates were measured by $^{3}$H-leucine incorporation into immunoprecipitated skin and liver histidase, relative to incorporation into total tissue soluble protein[17].

6. Developmental alterations in synthetic rates of histidase in each tissue and sex.

It may be seen from Figs. 3 and 4 that rates of pulse incorporation *in vivo* of $^{14}$C- or $^{3}$H-leucine into histidase immunoprecipitates, relative to that into total tissue soluble protein, vary directly with enzyme amount and catalytic activity during development in both sexes and tissues. Thus, histidase synthetic rate increases in liver proportionately to enzyme amount and activity in the male and significantly more in the female[16] (Fig. 3). In the developing male, the progressive elevation in liver histidase synthetic rate, which is proportionate to the elevations in enzyme amount and activity, is accompanied by a marked decline in histidase synthesis in the epidermis of same animals, which is likewise parallel to enzyme amount and activity[17] (Fig. 4). The validity of these pulse labelling experiments as a measure enzyme synthesis was established: precautions were taken to ensure complete immunoprecipitation; the identity and specificity of histidase immunoprecipitates, as assessed by SDS-acrylamide gel electrophoresis, indicated that the major radioactive protein band co-migrates with purified histidase subunits; the radioactivity in the minor contaminating bands corresponded to protein precipitated by non-immune serum and was applied as corrections to the radioactive data[17]. Thus, histidase synthetic rates are selectively altered proportionately to enzyme quantity and catalytic activity during development, increasing in male liver and more markedly in female liver and simultaneously declining in the skin of the same animals. Since glucocorticoid[9] and estrogen[4] have been implicated in the pubertal elevations of hepatic histidase activities of males and females, respectively, and androgen is responsible for the decline in epidermal enzyme activity[2], it follows that these hormones may act to increase, in the case of the liver, and decrease, in the case of the skin, histidase synthetic rates. This inference is compatible with the findings of inhibition of hormonal induction of hepatic histidase by protein synthesis inhibitors[4,6].

## 7. Conclusions.

It is inferred from these studies that tissue and sex specific differential development of histidase activity is due to alterations in amounts of the same enzyme protein, which in turn, result from hormonal modulation of histidase synthetic rates. In the case of the liver, the sites of regulation may be at transcriptional, mRNA processing or transport, or translational steps; in the case of skin, cell populations may be altered, as well.

Acknowledgements - Supported by NIH Grant No. HD-01951. The excellent technical assistance of Emily Scott is gratefully acknowledged.

References

1. Bhargava, M.M. and Feigelson, M., I, In press, 1976, Develop. Biol.

2. Feigelson, M., 1974, Abstr., J. Steroid Biochem. 5, 355.

3. Auerbach, V.H. and Waisman, H.A., 1959, J. Biol. Chem. 234, 304.

4. Feigelson, M., 1968, J. Biol. Chem. 243, 5088.

5. Makoff, R. and Baldridge, R.C., 1969, Biochim. Biophys. Acta 90, 282.

6. Feigelson, M., 1973, Enzyme 15, 199.

7. Baden, H.P., Sviokla, S., Mittler, B. and Pathak, A.M., 1968, Cancer Research 28, 1463.

8. Sahib, M.K. and Krishna Murti, C.R., 1969, J. Biol. Chem. 244, 4730.

9. Feigelson, M., 1973, Biochim. Biophys. Acta, 304, 669.

10. Feigelson, M., 1971, Biochim. Biophys. Acta, 230, 309.

11. Feigelson, M., 1971, Endocrinology, 89, 625.

12. Feigelson, M., 1971, Biochim. Biophys. Acta, 230, 296.

13. Freedland, R.A., Avery, E.H. and Taylor, A.R., 1968, Can. J. Biochem. 46, 141.

14. Noda, K. and Yoshita, A., 1969, Agric. Biol. Chem., 33, 31.

15. Neufeld, E., Harell, A. and Chayen, R., 1971, Biochim. Biophys. Acta, 237, 465.

16. Lamartiniere, C.A. and Feigelson, M., In press. DEVELOPMENTAL BIOLOGY-Pattern Formation-Gene Regulation, Vol. 3, eds., D. McMahon and C.F. Fox, W.A. Benjamin, Inc.

17. Bhargava, M.M. and Feigelson, M., II, In press, 1976, Develop. Biol.

18. Tabor, H. and Mehler, A.H., 1955, Methods in Enzymology 2, 228.

# Section 6.
# NORMAL AND MALIGNANT HEMOPOIESIS AS A MODEL OF DIFFERENTIATION

*Progress in Differentiation Research, ed. N. Müller-Bérat et al.*
*© 1976, North-Holland Publishing Company - Amsterdam, The Netherlands.*



# THE INTERRELATIONSHIP BETWEEN PROLIFERATION AND DIFFERENTIATION CONTROL

L.G. Lajtha

Paterson Laboratories,
Christie Hospital and Holt Radium Institute, Manchester

The haemopoietic system is ideally suited for studies of the interrationships between proliferation and differentiation and their control mechanisms. There are several reasons for this, one being the nature of the haemopoietic system exhibiting as it does high potential for proliferation and of course for differentiation, the other reason is that in recent years a number of assay methods, quantitative assays, have been developed for the analysis of the properties of the various sub-populations of cells which constitute the haemopoietic tissues.

1. The system is basically a chain of three inter-linked cell populations: first the stem cells which are pluripotent in their capacity to differentiate, and which have considerable proliferative capacity with a control mechanism (at least part local in nature) which can "shut down" proliferation. This can be shown by serial transplantation experiments, which also indicate that the self-duplicating capacity of the stem cells - qua stem cells - is not inexhaustible. Recent studies (using the alkylating agent isopropyl methane sulphonate) indicate that "stem cells" can exist in a pluripotent state - i.e. producing either erythroid or granulocytic or thrombocytic lines - with sufficient proliferative capacity to produce splenic colonies in excess of $10^5$ cells, but not being able to produce new stem cells! Whether this state of the pluripotent stem cell is an indication of an already (partially) differentiating state, future experiments will have to decide. From colony size analysis, such "differentiating" state does not affect the overall proliferative capacity of the stem cells and its descendants.

The "link" between proliferation and differentiation in the pluripotent stem cells is in a sense a negative one: under conditions of maximal demand for both, the capacity to differentiate is limited. This is shown by the colony growth kinetics of grafted marrow in the lethally irradiated recipient, in which, in the exponentially recovering (re-growing) stem cells "cell loss" for differentiation is limited to a maximum of 40 per cent cell cycle; This results in a population doubling time of about 24 hours, in spite of a cell cycle time of about 6 hours - its evolutionary value is preventing the "running down" of the stem cell population even under maximal demands for differentiated end product. The differentiation of the pluripotent stem cell is the "first step" differentiation event - the controlling factors of which are as yet unknown - which results in a "second set" of cell populations.

2. The second link in the "three tier" system is that of the "committed precursor cells". Two such populations have been identified so far - on the operational basis of being able to form colonies in suitable systems in vitro: (a) the in vitro granulocytic colony forming cells "$CFU_C$", and (b) the in vitro erythroid "burst" forming cells "BFU". For the latter an in vivo demonstration is the response to the humoral factor erythropoietin, hence this population is also broadly termed "ERC" (erythropoietin responsive cells).

a) The $CFU_C$ appears to be less "monophyletic" than its erythroid counterpart the BFU(ERC), since under varying culture conditions it can produce - as a "second" step differentiation - pure neutrophil granulocytic cells, or eosinophilic granulocytes, or even mononuclear cell colonies of the monocyte-macrophage type. The main factor in the "second" differentiation event appears to be the amount (or nature?) of the "colony stimulating factor(s)" supplied in the cultures. Various tissue can act as sources of colony stimulating activity, including a special population of cells in the bone marrow (which can be separated by their property of adhering to glass surfaces). The role of the CSA-s (colony stimulating activities) appears to be twofold: they certainly determine the "second step" differentiation pathways, but they equally play a role in the proliferation of the committed precursor cell (the $CFU_C$) itself. At least in vitro, no $CFU_C$ proliferation occurs without the presence of CSA. Of course the term CSA may hide a number of molecular species and full characterisation of its components is required before a truly "dual action" can be attributed to such factors.

b) The other "committed" precursor population, the erythroid BFU(ERC) differs from its granulocytic counterpart in a number of important aspects. Firstly, as far as is known the erythroid BFU is strictly monophyletic - it can only produce erythroblastic colonies as its "second step" differentiation. Secondly, the factor inducing its differentiation step - erythropoietin - is better characterised than CSA and it does exhibit a dual action: it causes the differentiation of ERC into the erythron (i.e. into pronormoblasts) but it also clearly stimulates proliferation of the BFU(ERC) itself. The third difference is a quantitative one: the proliferative capacity of the BFU/ERC population is considerably greater and is much more elastic than that of the $CFU_C$. While in the in vitro systems studied a growth potential of some 3-4 cell cycles may be attributed to $CFU_C$ (and in vivo, indirect information is compatible with this) the in vivo growth potential of the erythroid precursor (early ERC) ranges from 4-5 cell cycles (under normal steady state) to 9-10 cell cycles on maximal demand. Hence while both these committed precursors are amplifying transit populations, the ERC has a very considerably greater amplifying capacity.

As far as the capacity for the "second step" differentiation the BFU (ERC) population appears to have an "age structure" insofar as the early BFU are unable to turn into pronormoblasts, this effect of erythropoietin is only manifest in the later forms (the "mature ERC") of the population.

This means that the first step differentiation from the pluripotent stem cell to the early BFU results in a subsequent "maturation" of the BFU - irrespective of its degree of amplification, i.e. proliferation. No such age structure (or maturation process) has been noted in the granulocytic committed $CFU_c$ - unless the limited self-renewal capacity of the $CFU_c$ is considered a "maturation" phenomenon.

3. The third links in the three-tier system are the maturing, morphologically recognisable, end lines: the erythroid (pronormoblast → reticulocyte), and the granulocytic (myeloblast / promyelocyte → polymorphonuclear). These are the results of the respective "second step" differentiation processes, and, like in their respective "committed" precursors, the amplification through 4-6 cell cycles is accompanied by a maturation process. This maturation process could be called a "suicide maturation" since it results in cells incapable of further proliferation (reticulocytes and polymorphonuclear granulocytes) and with a limited life span.

The interaction between rates of maturation and rates of proliferation (i.e. cell cycle times) does act as a powerful regulator of cell output. Slowing down cell cycle times (or speeding up maturation rates) results in earlier (i.e. after fewer cell divisions) reaching a maturation stage at which no further proliferation is possible. This means decreased cell output. Conversely, shortening cell cycle times or slowing down maturation rates enables an increased number of cell divisions to occur before reaching the "nondividable" state, thus resulting in increased amplification during "transit" - increased cell output. Parellel speeding or slowing both processes leaves amplification unaltered while affecting transit times. The controlling effects of erythropoietin - almost certainly not the only factor in the processes are just beginning to be understood.

Cell line specific proliferation control has been demonstrated in two of the "third link" populations - most likely acting as cell cycle length modulators of the "recognisable" granulocytic and normoblastic lines, without apparently affecting maturation processes.

The megacaryocytic differentiation is different in many respects from the other two "recognisable" cell lines. Operationally no identification of its "committed" precursor has been made as yet, and amplification is connected with endomitosis.

The interrelationships between proliferation and differentiation control are only now beginning to emerge in the haemopoietic system - the examples indicated above only describe phenomena and some of the properties of the system rather than explaining the operating processes. The phenomenology is however slowly yielding to more and more mechanistic explanations.

*Progress in Differentiation Research, ed. N. Müller-Bérat et al.*



# REGULATION OF GRANULOPOIESIS

Malcolm A.S. Moore and Jeffrey I. Kurland
Sloan-Kettering Institute for Cancer Research, New York, New York 10021
This work was supported by grants NCI 08748-11A, ACS-CH-3, CA-17085

## Introduction

Considerable progress has been made in understanding control of granulopoiesis since the advent of the agar culture technique for cloning early granulocyte progenitor cells (CFU-c) present in bone marrow, spleen and blood[1,2,3]. Cell separation techniques have identified the CFU-c as a transitional mononuclear cell, distinct from the multipotential stem cells and capable of differentiating only into mature granulocytes, monocytes and macrophages[3,4]. The absolute requirement of a glycoprotein colony stimulating factor (CSF) for in vitro granulocytic differentiation has led to suggestions that this factor may be a granulopoietin[3,5]. Support for this has come from observations on correlations between fluctuations in neutrophil production and levels of serum and urine CSF in a variety of clinical and experimental situations[3,5-7]. Although in the mouse many tissues are found to contain or produce high levels of CSF[8], the distribution of this humoral factor in humans is more limited. The principal cell responsible for elaboration of CSF is the monocyte [6,9] or tissue macrophage[10] and this observation has led to the proposal that granulocyte-monocyte production is controlled, in part, by a positive feedback of CSF produced by the monocyte in which low levels of CSF direct the CFU-c into a monocyte pathway of differentiation and high levels switch differentiation into the granulocyte pathway with consequent decline in CSF production[9]. It has however become clear that control of steady-state granulopoiesis and the response of the system to perturbation cannot be accounted for solely on the basis of variations in CSF levels. Additional regulation of granulopoiesis by granulocytes via a negative feedback loop has been indicated, and evidence has accumulated which shows that granulocytes inhibit granulopoiesis in vitro[11-14]. The experimental data in this communication indicate that still further complexity of regulation may exist than hitherto considered and a model of regulation of granulopoiesis will be presented.

## Materials and Methods

The following assay systems were used:

(a) CFU-c assay: C57 BL/6 marrow cells were cultured at varying concentrations in McCoy's medium plus 15% fetal calf serum and 0.3% agar in the presence of varying concentrations of colony stimulating factor obtained from WEHI-3 conditioned media (a murine myelomonocytic leukemic cell line). Colonies of greater than 40 cells were scored at 7 days of incubation. The technique has been reported in detail elsewhere [2,3].

(b) Peritoneal macrophage colony assay: Thioglycollate stimulated C57BL/6 peritoneal exudate cells were cultured in McCoy's-agar medium in the presence of a concentrated source of WEHI-3 conditioned media using a modification of the technique of Lin and Stewart[15]. Cultures were scored at varying intervals and macrophage colonies of > 40 cells were counted after 3-4 weeks.

(c) B lymphocyte colony assay: C57BL/6 spleen and mesenteric lymph node cells were cultured in McCoy's-agar medium containing 5 x $10^{-5}$M 2-mercaptoethanol using a modification of the technique of Metcalf et al [16]. Colonies of Ig-bearing plasmacytoid cells were scored after 5 days of culture.

(d) Leukemic CFU-c assay: WEHI-3 developed as a spontaneous myelomonocytic leukemia in a Balb/c mouse and was shown to form colonies in agar in the presence of CSF[3]. It was subsequently adapted to culture as a continuous cell line which was predominantly monoblast-macrophage in morphology and continuously produced CSF. Its clonogenic capacity in agar (20-50% plating efficiency) has become autonomous with respect to CSF even when very low numbers of cells were cultured.

Production of Conditioned Media and Assay for CSF

Conditioned media were prepared from 1 x $10^6$ normal or thioglycollate stimulated mouse peritoneal exudate cells incubated for varying periods of time in 1ml of serum-containing McCoy's medium. WEHI-3 conditioned medium was prepared from continuous cell lines of WEHI-3 leukemic cells maintained in T-flasks in either serum-containing or serum-free McCoy's medium

Colony stimulating activity was measured in either fresh conditioned media or following 3 day dialysis against water. CSF was measured as previously described, using a standard target cell population of 75,000 C57BL/6 mouse bone marrow cells[3].

Prostaglandins of the F2$\alpha$, $E_1$ and $E_2$ series were kindly provided by Dr. John Hadden of the Sloan-Kettering Institute.

Results

A linear relationship between the number of colonies formed and the number of cells plated was seen using the various assay systems (Fig. 1).

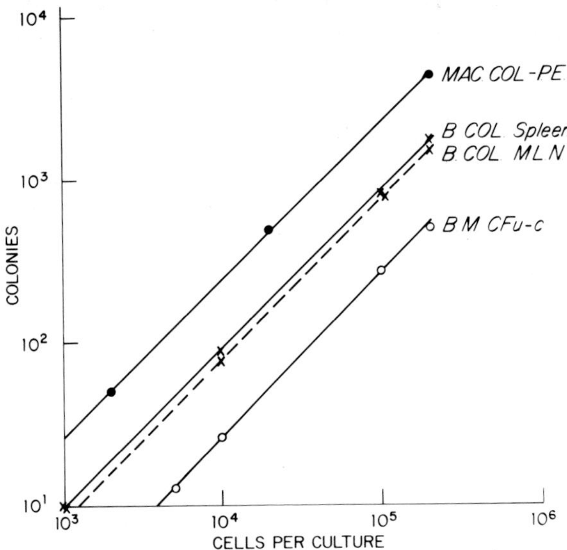


Fig. 1. The relationship between the number of colonies formed and the number of cells per culture. MacCol-PE macrophage colonies developing in cultures of activated mouse peritoneal exudate in the presence of CSF. B Col-B lymphocyte colonies developing in 2ME stimulated cultures of mouse spleen or mesenteric lymph node cells (MLN). CFU-c granulocyte-macrophage colonies developing in CSF stimulated mouse bone marrow culture.

Marrow CFU-c numbers in the presence of high concentrations of CSF were linear over the range of $10^2$ - 2 x $10^5$ cells per ml with an average incidence of 250 colonies/$10^5$. Thioglycollate-induced peritoneal macrophages formed an average of 3-5,000 colonies/$10^5$ in the presence of high concentrations of WEHI-3 conditioned media. Partial purification of the conditioned media showed that the factor stimulating CFU-c copurified with the factor stimulating macrophage proliferation and both activities appeared to reside in a common molecule (CSF) but that a considerably higher concentration was required to stimulate macrophage colony formation. 1 - 2% of spleen and mesenteric lymph node cells formed colonies of B-lymphocytes and Ig-bearing plasmacytoid cells and the relationship between the number of cells plated and colony incidence was linear over the range of $10^2$ - 2 x $10^5$ cells per ml. The cloning efficiency of B lymphocytes and WEHI-3 leukemic cells was not influenced by addition of CSF.

The various clonal assay systems measure a combination of proliferation and differentiation which it is reasonable to assume can be influenced by fluctuations in cyclic nucleotide levels within the cells. Addition of dBcAMP to the cultures inhibited colony formation over a wide range of concentrations ($10^{-4}$ - $10^{-12}$M). Of more physiological relevance was the influence of prostaglandins of the E series on colony formation. Fig. 2 shows that addition of $PGE_1$ to mouse bone marrow

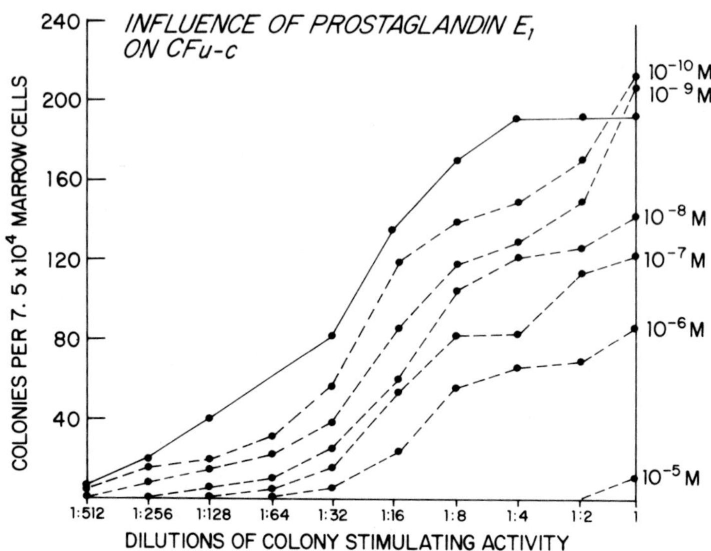


Fig. 2. The influence of varying concentrations of Prostaglandin $E_1$ on granulocytic colony formation by mouse bone marrow cells in the presence of varying dilutions of CSF. Note the sigmoid dose-response relationship of control cultures in the absence of $PGE_1$ (●——●).

cultures incubated with varying concentrations of CSF markedly suppressed CFU-c proliferation at concentrations of $10^{-5}$ - $10^{-8}$M. Significant inhibition was also seen at even lower concentrations of $PGE_1$ ($10^{-9}$ - $10^{-10}$M) when the CSF concentration was reduced; however, at the highest concentration of CSF significant potentiation of colony formation was found in the presence of very low concentrations of $PGE_1$. $PGE_2$ showed a similar interaction with CSF mediated colony formation but was somewhat less effective in inhibiting CFU-c proliferation. Prostaglandins of the F2α series were only inhibitory at highly unphysiological concentrations ($10^{-4}$M) and actually enhanced colony formation when present in culture at concentrations of $10^{-5}$ - $10^{-9}$M.

In the second CSF-dependent colony assay, the temporal course of development of colonies of macrophages from activated peritoneal exudate cell populations was similar to that reported previously[15]. Individual macrophages persisted in agar culture for 7-10 days prior to the onset of proliferation and then a subpopulation commenced active proliferation forming colonies of 40-500 macrophages between the 2nd and 4th week of culture (Fig. 3). This proliferation was CSF dependent and was blocked by addition of $10^{-6}$M $PGE_1$ added at day 0 of culture or as late as 16 days after the cultures were established. As in the CFU-c system, $PGE_1$ inhibition could be partially overcome by increasing the concentration of CSF in the cultures.

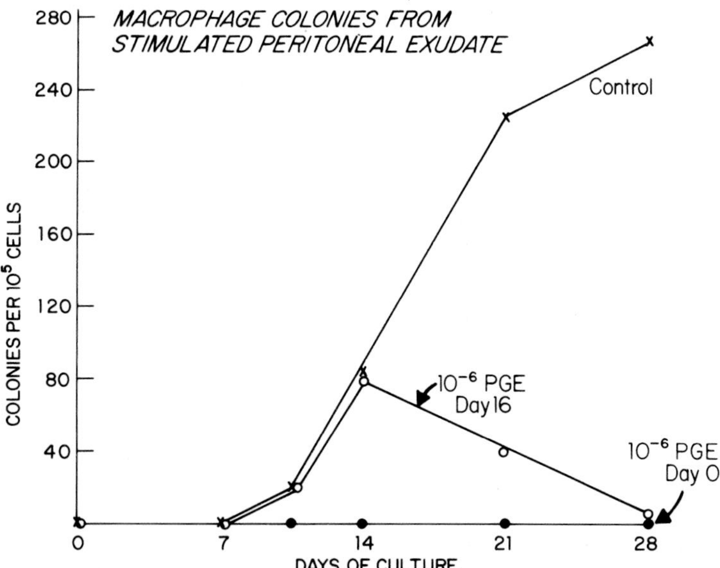


Fig. 3. The time course of production of macrophage colonies in agar cultures of 1 x $10^5$ thioglycollate activated Peritoneal exudate cells stimulated by CSF. Note the inhibition of colony formation when $10^{-6}$M $PGE_1$ was added at day zero (●——●) and at day 16 (o——o).

The two CSF-independent clonal assays for B lymphocytes and WEHI-3 leukemic cells showed marked sensitivity to $PGE_1$ and $PGE_2$ but not to F2α. 50% inhibition of spleen B lymphocyte colony formation was seen with concentrations of $10^{-8}PGE_1$ and significant inhibition of leukemic colony formation (20 ± 5%) was seen with

$10^{-12}$M $PGE_1$. In neither case was inhibition influenced by addition of varying concentrations of CSF.

## Kinetics of Macrophage CSF Production

We have observed that macrophage colony formation in activated peritoneal exudate cell cultures occurred spontaneously if the cells were plated at high cell concentrations ($2 \times 10^5$ - $1 \times 10^6$ cells per ml). This observation was not surprising in view of the known capacity of macrophages to elaborate CSF. In order to quantitate more directly macrophage CSF production, $1 \times 10^6$ normal or thioglycollate activated macrophages were cultured for 21 days with cultures sampled at daily intervals and assayed for CSF. Alternatively, the media was changed every 24 hours and assayed for cumulative CSF production. In Fig. 4, it can be seen

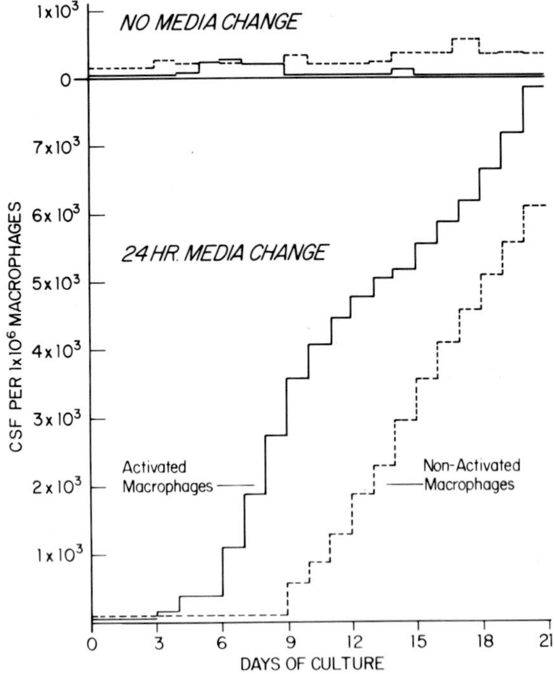


Fig. 4. CSF production by $1 \times 10^6$ normal or thioglycollate stimulated mouse peritoneal exudate cells maintained in 1ml of medium with either 24 hour complete media change or with no media change. In both groups conditioned media from triplicate plates was assayed at daily intervals. CSF is expressed in units of activity[3] and the histogram shows the cumulative production of CSF throughout the period of study.

that in the absence of daily media change, a base line concentration of CSF was present in the media but progressive accumulation of CSF did not occur. Non-activated macrophages were slightly more effective in elaborating CSF than were the activated in this system. In contrast, daily complete media change, after a lag period of 3 days in the case of activated macrophages and 9 days with non-activated cells, promoted a continuing and relatively constant daily production of CSF throughout the period of investigation. Preliminary observations with human macrophages maintained under similar conditions have shown that continuous CSF production with these cells occur over many months provided frequent media changes were made.

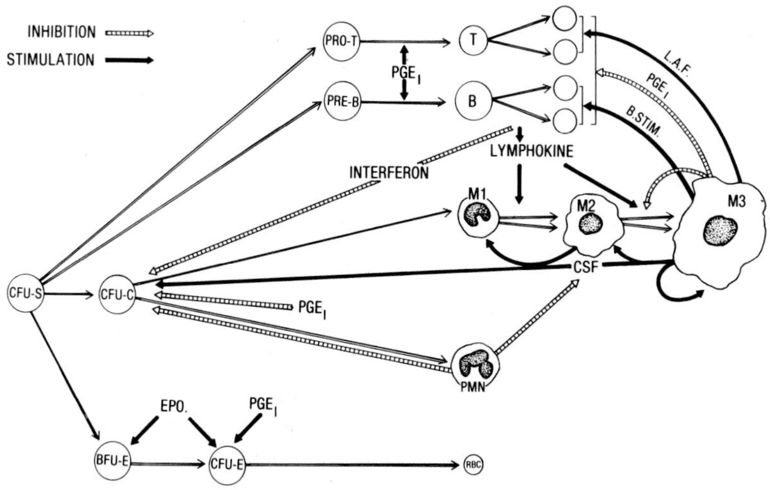


Fig. 5. A speculative model of regulation of granulopoiesis indicating the possible interaction between stimulatory and inhibitory factors and the interrelationship of granulopoiesis, lymphopoiesis and erythropoiesis. The various stem cell compartments are indicated. CFU-S - multipotential stem cell; CFU-C-committed granulocyte-macrophage stem cells; BFU-E-burst forming unit in methylcellulose proliferating in response to erythropoietin (EPO); CFU-E - a more differentiated erythropoietin responsive cell; Pro-T-prothymocyte precursor of T lymphocytes; Pre-B - precursor of B lymphocytes; M1-monocytes; M2-tissue macrophage; M3 is the activated macrophage; PMN-the mature granulocyte; L.A.F. -T lymphocyte activating factor[25]; B-stem - macrophage derived factor stimulating B cell proliferation[26].

## Discussion

The central role of phagocytic mononuclear cells in the control of hematopoiesis is shown in Fig. 5 which is a speculative model of regulatory interactions which may occur in vivo and influence the production of granulocytes. In this model, baseline production of CSF by circulating monocytes (M1) and by fixed tissue macrophages (M2) in hemopoietic tissues, lung, liver and peritoneal cavity provides a constant stimulus for CFU-c proliferation and differentiation into the mature granulocyte and monocyte-macrophage compartments. This positive feedback is in turn balanced by inhibitory factors released by mature granulocytes which (a) suppress CSF production by phagocytic mononuclear cells[14] and (b) act directly as CSF-antagonists to inhibit the proliferation of early myeloid precursor cells[12,13]. From the data presented on CSF production by macrophages following daily media replacement it is also possible that phagocytic mononuclear cells may control synthesis and/or release of CSF by sensing the CSF concentration in their external milieu. This possibility is made more plausible by the fact that macrophages not only produce CSF but can exhibit a proliferative response to exogenous CSF, indicating that they possess receptors for this molecule. The balance between the concentration of stimulatory and inhibitory regulators could thus control the normal steady state production of granulocytes, monocytes and macrophages. Not included in this model, but presumably of biological relevance, is the documented heterogeneity of CFU-c to CSF stimulation[3] and the functional heterogeneity of the different molecular forms of CSF[3].

The response of the host to bacterial infection causes an increase in granulocyte production and release by the bone marrow as well as activation and proliferation of phagocytic mononuclear cells. In the proposed model of regulation, the macrophage functions as a surveillance cell able to increase the production of CSF in response to a variety of stimuli of which endotoxin may be of particular importance since it is known to activate macrophages and stimulate the production and release of CSF[17]. The activated macrophage (M3) population can thus produce a higher circulating level of CSF which in turn promotes increased granulopoiesis. It should also be noted that in addition to recruitment of granulocytes from the committed stem cell compartment, high CSF concentrations promote proliferation of a subpopulation of activated macrophages in the peritoneal cavity and presumably elsewhere. In this context, activated macrophages can condition their own environment and promote active macrophage proliferation within foci of infection and inflammation. Endotoxin-induced granulopoiesis and macrophage proliferation would eventually be limited by negative feedback from the increased numbers of mature granulocytes and by the increased production of prostaglandins of the E series by activated macrophages[18]. The latter inhibition, due to the lability of the molecule, would be more localized than the systemic influence of inhibitors released by the granulocytes.

Increased granulopoiesis and elevated serum levels of CSF are also associated with a variety of immune responses including the host response to antigen administration, allograft or tumor rejection, and GVH reactions; thus immunocompetent cells may participate in regulation of granulopoiesis[3,19]. It has been demonstrated that lymphoid cell populations responding to antigenic or mitogenic stimuli produce large quantities of CSF[20,21]. The possibility that activated lymphocytes produce CSF is still questionable since it is not yet clear whether lymphokine production can occur in the total absence of macrophages or their precursors. It is however proven that mitogen-stimulated lymphocyte populations, extensively depleted of adherent cells, can produce large quantities of CSF. In Fig. 5, the activation of lymphocytes in the presence of macrophages and an appropriate antigenic or mitogenic stimulus leads to the production of lymphokines capable of activating macrophages[22] which in turn synthesise and release increased quantities of CSF. This macrophage mediated proliferative stimulus for granulopoiesis is again self-limiting since lymphokine activation of macrophages has been shown to increase prostaglandin synthesis[18] and, as shown in this study, $PGE_1$ is capable of suppressing CFU-c, macrophage and B lymphocyte proliferation at concentrations within the physiological range. A further consequence of lymphocyte activation is the macrophage-dependent production of interferon by both T- and B-lymphocytes[23]. As has been shown by McNeil[24], interferon has, in addition to its anti-viral role, the capacity to suppress granulopoiesis in vitro and to function as a CSF antagonist.

The speculative model of regulation of granulopoiesis presented in Fig. 5 attempts to integrate a number of disparate observations on humoral control of hematopoiesis, based mainly on in vitro studies. It has the virtue of explaining many responses of the granulopoietic system in terms of synergistic or antagonistic interactions of humoral factors, both cell line specific (CSF, erythropoietin, PMN inhibitors) and non-specific ($PGE_1$, interferon, lymphokines) at concentrations within the physiological range. It allows for in vivo situations where increased granulopoiesis may occur without increased CSF production (e.g., reduction in inhibitors which function specifically on granulopoiesis or which are not cell line specific such as $PGE_1$ and interferon) and for situations where increased granulopoiesis is associated with increased production of CSF with no change in inhibitor levels. The model also indicates the numerous potential control points that exist within the hemopoietic system each balanced by positive and negative humoral regulators. It also serves to explain systemic regulation of granulopoeisis since the regulatory macromolecules are present in the serum and presumably are active throughout the body whereas the more localized proliferative events may be explained on the basis of cell interaction and involvement of more labile humoral agents such as the prostaglandins.

References


1. Bradley, T.R. and Metcalf, D., 1966, Aust. J. Exp. Biol. and Med. Sci. 44, 287.
2. Pike, B.L. and Robinson, W.A., 1970, J. Cell. Comp. Physiol., 76, 77.
3. Metcalf, D. and Moore, M.A.S., 1971, in: Haemopoietic Cells, North Holland, Amsterdam.
4. Moore, M.A.S., Williams, N. and Metcalf, D., 1972, J. Cell. Physiol., 79, 283.
5. Weiner, H.L. and Robinson, W.A., 1971, Proc. Soc. Exp. Biol. and Med., 136, 29.
6. Moore, M.A.S., Spitzer, G., Metcalf, D. and Penington, D.G., 1974, Brit. J. Haemat. 27, 47.
7. Dale, D.C., Brown, C.H., Carbone, P. and Wolff, S.M., 1971. Science 173, 152.
8. Sheridan, J.W. and Stanley, E.R., 1971, J. Cell Physiol. 78, 451.
9. Moore, M.A.S. and Williams, N., 1972, J. Cell Physiol. 80, 195.
10. Golde, D.W., Finley, T.N. and Cline, M.J., 1972, Lancet i, 1397.
11. Haskill, J.S., McKnight, R.D. and Galbraith, P.G., 1972, Blood, 40, 394.
12. Rytoma, T. and Kiviniemi, K., 1968, Cell Tissue Kinet., 1, 34.
13. Lord, B.I., Cercek, B., Shah, G.P., Dexter, T.M. and Lajtha, L.G., 1974, Brit. J. Cancer, 29, 168.
14. Broxmeyer, H., Baker, F.L. and Galbraith, P.G., Blood, (In Press).
15. Lin, H-S, and Stewart, C.C., 1974, J. Cell Physiol., 84, 369.
16. Metcalf, D., Warner, N.L., Nossal, G.J.V., Miller, J.F.A.P., Shortman, K., and Rabellino, E., 1975, Nature, 255, 630.
17. Cline, M.J., Rothman, B., Golde, D.W., 1974, J. Cell. Physiol. 84, 193.
18. Bray, M.A., Gordon, D. and Morley, J., 1975, Brit. J. Pharmacol. (In Press).
19. McNeill, T.A., 1970, Immunology, 18, 61.
20. Parker, J.W. and Metcalf, D., 1974, J. Immunol. 112, 502.
21. Ruscetti, F.W. and Chervenick, P.A., 1975,
22. Morley, J., 1974, Prostaglandins, 8, 35.
23. Epstein, L.B., Kreth, H.W. and Herzenberg, L.A., 1974, Cell Immunol. 12, 407.
24. McNeill, T.A. and Gresser, I., 1973, Nature New Biol., 244, 173.
25. Gery, I. and Handschumacher, R.E., 1974, Cell Immunol., 11, 162.
26. Namba, Y. and Hanaoka, M., 1974, Cell Immunol. 12, 74.

*Progress in Differentiation Research, ed. N. Müller-Bérat et al.*



# REGULATION OF MURINE GRANULOPOIESIS BY BACTERIAL ENDOTOXINS

Ron N. Apte and Dov H. Pluznik

Department of Life Sciences
Bar-Ilan University, Ramat-Gan, Israel

## 1. Introduction

A desirable experimental model for studying differentiation is that in which three different components of cellular communication can be demonstrated: the mediators of communication, the producer cell which releases them and the target cell which responds to them. One of these models exhibiting all the three components is the model of in vitro granulopoiesis and macrophage formation. Committed progenitor cells for granulocyte and macrophage differentiation (colony forming cells - CFC) can proliferate in soft agar cultures to form colonies of mature granulocytes and/or macrophages[1,2]. Colony formation is wholly dependent upon the continuous presence of a colony stimulating factor (CSF). CSF could be derived from a number of sources, including various cell feeder layers, medium conditioned by cells in culture, mouse sera, tissue extracts and human urine (for a review see reference No. 3).

Endotoxins (ET), products of gram negative bacteria, are macromolecules composed of a specific polysaccharide (PS) covalently linked to a common lipid termed lipia A [4]. An injection of ET to mice causes a rise in tissue and serum CSF followed by a subsequent rise in marrow and spleen CFC levels [5,6].

The effects of certain normal and mutant genes upon granulopoiesis in the mouse may provide a suitable system for studying the control of this process. A strain of mice, $C_3H$/HeJ, defective in some ET induced responses, has been recently described. These mice are resistant to the lethal toxic effects of ET and increase their mononuclear cells in the peritoneal cavity after an injection of ET. These two responses are polygenically controlled[7]. In addition, these mice are also low responders to the mitogenic and immunogenic effects of ET, which are monogenically controlled [8].

## 2 Materials and Methods

Mice: Inbred mice of the $C_3H$/eB and $C_3H$/HeJ strains 9-12 weeks old were used. Reciprocal breeding was conducted to produce $F_1$, $F_2$ and backcross ($F_1$ x $C_3H$/HeJ) progeny.

Endotoxins: Lipopolysaccharide (LPS) from S. abortus equi was purchased from Difco Laboratories, Detroit, Mich., USA. LPS breakdown products (lipid A and PS), as well as glycolipids from rough (R) mutants of S. minnesota, were obtained from Dr. C. Galanos, Max Planck Institut für Immunbiologie, Freiburg, Germany.

Cloning of bone marrow and spleen cells in soft agar cultures: was performed as previously described [1]. Mouse embryo fibroblasts, conditioned medium or post-ET serum were used as sources of CSF.

## 3. Results and Discussion

The active part of the ET molecule in inducing granulopoietic responses: Lipid A obtained by acid hydrolysis of the ET and complexed to bovine serum albumin (BSA) (lipid A-BSA) was shown to be active in generating serum CSF and in increasing the splenic CFC levels, although it was less active than the parent ET, probably because of insufficient solubility or degradation of the lipid A by the acid hydrolysis during the isolation procedure. The PS showed no significant activity at the concentrations used (Table I).

TABLE I.

Effect of ET, PS and Lipid A on serum CSF and splenic CFC responses.

| | μg of the derivate injected/mouse | CFC response [(a)] (colonies/$10^6$ spleen cells) | CSF response [(b)] (colonies/$10^5$ bone marrow cells) |
|---|---|---|---|
| | none | 43 | 0 [(c)] |
| LPS | 10 | 870 | 129 |
| | 25 | NT [(d)] | 192 |
| | 50 | NT | 185 |
| | 100 | NT | 197 |
| PS | 10 | 38 | 0 |
| | 25 | 52 | 0 |
| | 50 | 56 | 0 |
| | 100 | 46 | 11 |
| Lipid A | 10 | 266 | 82 |
| | 25 | 314 | 100 |
| | 50 | 514 | 140 |
| | 100 | 786 | 160 |

(a) Splenic CFC levels were evaluated on the 6th day following the injection of ET. $10^6$ spleen cells were cloned in soft agar cultures supplemented with 33% of mouse embryo fibroblasts conditioned medium.

(b) Serum CSF was obtained 6 hours after the injection of ET. Samples of sera (1:4 dilution) were added to soft agar cultures at a final concentration of 10%. $10^5$ bone marrow cells were cloned in each plate.

(c) Each value represents average results obtained from 10 mice.

(d) NT = Not tested. ET was toxic and mice did not survive doses over 10μg/mouse.

ET (glycolipids) from R mutant of S. minnesota were active to the same extent as the complete ET (Table II). The fact that even the most defective ET from the R mutant R595, which contains only lipid A linked to a trisaccharide of 2-keto-3-deoxyoctanoate (KDO), is a potent ET, points to lipid A as the active principle of the ET molecule in inducing granulopoietic responses.

TABLE II.

Splenic CFC and serum CSF responses to LPS and glycolipids.

| Preparation | CFC responses (colonies/$10^6$ spleen cells) | CSF responses (colonies/$10^5$ bone marrow cells) |
|---|---|---|
| None | 90 | 0 |
| LPS (a) | 790 | 106 |
| mR 345 (b) | 728 | 117 |
| mR 7 (c) | 772 | 109 |
| mR 595 (d) | 718 | 99 |

Responses to an injection of 10μg of each preparation/mouse.

Serum CSF and splenic CFC responses were evaluated as described in the footnote to Table I.

(a) Lipid A-$(KDO)_3$-Hep-Hep-Glc-Gal-Glc-GlcNac-O antigen.
Gal

(b) Lipid A-$(KDO)_3$-Hep-Hep-Glc-Gal-Glc.
Gal

(c) Lipid A-$(KDO)_3$-Hep-Hep.

(d) Lipid A-$(KDO)_3$.

Abbreviations: Glc - Glucose; Gal-Galactose; GlcNac - N-acetylglucosamine; Hep - Heptose; KDO - 2-keto-3-deoxyoctanoate.

The genetic experimental system: Two inbred strains of mice differing in their response to ET were used: a high responder strain ($C_3H$/eB) reacting to ET (10μg/mouse) by an increase in serum CSF levels and by an elevation of splenic CFC levels, and a low responder strain ($C_3H$/HeJ) failing to show these responses (Table III).

The genetic control of ET induced granulopoietic responses: All mice of the $F_1$($C_3H$/HeJ x $C_3H$/eB) population, 45% of the backcross ($F_1$ x $C_3H$/HeJ) population and 81% of the $F_2$ population generated high levels of serum CSF (Fig. 1a). No difference in the CSF response between males and females was observed. The results suggest that the ability to generate serum CSF in response to ET is controlled by a single autosomal dominant gene. Splenic CFC levels of the $F_1$, the backcross ($F_1$ x $C_3H$/HeJ) and the $F_2$ hybrids were intermediate between the respective parental types. CFC values of the $F_2$ generation were more scattered than those of the $F_1$ and, in addition, a shift of the CFC response of the backcross ($F_1$ x $C_3H$/HeJ) mice towards lower levels was observed (Fig. 1b).

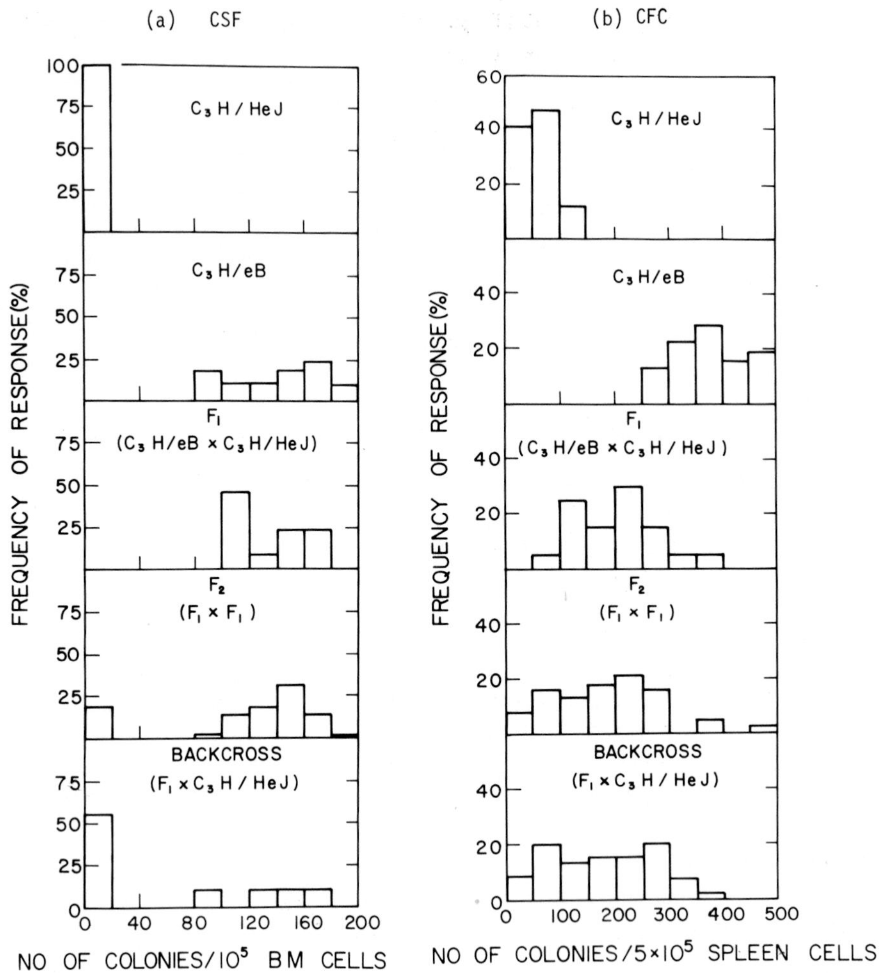


Fig. 1. Serum colony stimulating factor (CSF) (a) and splenic colony forming cells (CFC) (b) responses to LPS in the parental strains, the $F_1$ hybrid, the $F_2$ generation and the backcross ($F_1$ x $C_3H$/HeJ). Results obtained from 30 mice of each parental strain, 30 mice of the $F_1$ hybrid and 40 mice of $F_2$ and the backcross ($F_1$ x $C_3H$/HeJ) generations.

TABLE III.

Differential response of $C_3H$/eB and $C_3H$/HeJ mice to injection of ET

| ET µg/mouse | CSF colonies/$10^5$ BM cells | | CFC colonies/$10^6$ spleen cells | |
|---|---|---|---|---|
| | $C_3H$/eB | $C_3H$/HeJ | $C_3H$/eB | $C_3H$/HeJ |
| 0 | 0 | 0 | 57 | 34 |
| 5 | 155 | 0 | 291 | 63 |
| 10 | 166 | 0 | 374 | 51 |
| 25 | 244 | 0 | 334 | 48 |
| 50 | 240 | 0 | NT | 57 |
| 100 | 255 | 33 | NT | 48 |

Serum CSF tested 6 hours after injection of increasing doses of ET. Response of splenic CFC 6 days after injection of increasing amounts of ET.BM=bone marrow

The splenic CFC response to ET seems to follow the characteristic patterns of a polygenic inheritance control. The linkage relationships of CSF and CFC responsiveness was investigated in backcross ($F_1$ x $C_3H$/HeJ) and $F_2$ mice. Most mice which generated high levels of CSF showed a high or intermediate CFC response and most mice which did not generate any detectable levels of serum CSF showed a low CFC response (Table IV).

TABLE IV.

Summary of linkage analysis between CFC and CSF in $F_2$ and backcross ($F_1$ x $C_3H$/HeJ) mice injected with ET.

| CFC response (a) | | | CSF response (a) |
|---|---|---|---|
| low | intermediate | high | |
| 2 | 8 | 14 | high |
| 10 | 2 | 0 | low |

(a) No. of mice. Results obtained from 36 mice tested individually.

Each mouse was injected I.V. with 10µg of ET and 6 hours later bled from the orbital sinus; sera were collected and tested for the presence of CSF. Mice were allowed to remain alive 6 days later and then splenic CFC levels were evaluated.

The CSF and CFC responses were categorized according to appropriate controls of mice from the parental strains ($C_3H$/eB and $C_3H$/HeJ).

It seems from these results that CSF may play _in vivo_ a physiological regulative role as a granulopoetin. However, $F_1$ mice showed a high CSF response and an intermediate CFC response, indicating that _in vivo_ CSF may not be the sole regulator of granulopoiesis as observed _in vitro_, and other leukopoietic and endocrine factors may play a regulative role in this process.

The source of unresponsiveness to ET of $C_3H$/HeJ mice: It was found that the unresponsiveness of the low responder strain is in generating CSF, since CFC from these mice proliferate _in vitro_ in cultures supplmeneted with post ET-serum (CSF) obtained from the high responder strain (Table V).

TABLE V.

Stimulatory activity in sera from high and low responder strains of mice to ET on bone-marrow CFC from both strains

| Source of serum CSF | | No. of colonies/$10^5$ bone marrow cells (a) | |
|---|---|---|---|
| | | $C_3H$/eB CFC | $C_3H$/HeJ CFC |
| $C_3H$/eB | control | 12 | 4 |
| (high responder) | post-ET | 169 | 173 |
| $C_3H$/HeJ | control | 8 | 2 |
| (low responder) | post-ET | 16 | 5 |

(a) Plates were supplemented with mouse serum (CSF) (dilution 1:4) at a final concentration of 10%.

This unresponsiveness is not due to rapid clearance of ET from circulation, to inactivation of the ET by opsonizing antibodies (Table VI) or to the presence of high concentrations of serum inhibitors extractable by chloroform (Table VII). To identify the cellular population responsible for ET induced serum CSF, low responding mice ($C_3H$/HeJ) were lethally irradiated and hemopoietically reconstituted with bone marrow cells from $C_3H$/eB, $C_3H$/eB genetically thymus deficient (nude) or $C_3H$/HeJ mice. Only mice reconstituted with $C_3H$/eB or $C_3H$/eB nude cells restored the CSF response to ET (Table VIII). It can be therefore concluded that the cells generating serum CSF in response to ET are of bone marrow origin, and they are probably not thymus derived (T) lymphocytes. Experiments are under investigation to identify accurately which type of cell(s) generate serum CSF.

TABLE VI.

Effects of opsonizing antibodies to ET in the sera of both strains on the CFC response.

| Material injected | CFC/$10^6$ spleen cells |
|---|---|
| ET pretreated with $C_3H$/eB serum | 346 |
| ET pretreated with $C_3H$/HeJ serum | 315 |
| ET without serum | 453 |
| none | 48 |

$C_3H$/eB mice were injected with 10μg of ET preincubated with sera from both strains for 24 h. at 37°C or with untreated ET, and CFC levels were evaluated after 6 days.

TABLE VII.

Effect of chloroform treatment on stimulatory activity of sera from control and ET injected mice.

| mice | serum | No. of colonies/$10^5$ B.M. cells [(a)] | |
|---|---|---|---|
| | | untreated | chloroform-treated |
| $C_3H$/eB (high responder) | control | 2 | 4 |
| | postendotoxin | 156 | 197 |
| $C_3H$/HeJ (low responder) | control | 3 | 5 |
| | postendotoxin | 10 | 13 |

(a) $10^5$ bone-marrow cells from ICR mice were cloned in soft agar cultures supplemented with mouse serum (CSF) (dilution 1:4) at a final concentration of 10%.

TABLE VIII.

Serum CSF response of $C_3H$/HeJ mice which were lethally irradiated and hemopoietically reconstituted with different bone-marrow cells.

| Source of bone-marrow cells injected | CSF response (colonies/$10^5$ bone marrow cells) |
|---|---|
| $C_3H$/HeJ | 0 |
| $C_3H$/eB | 139 |
| $C_3H$/eB nude | 137 |

$C_3H$/HeJ mice were lethally irradiated (900R) and immediately injected with $10^7$ bone marrow cells. Mice were injected with 10μg of ET approximately 3 months after the irradiation and serum CSF activity was evaluated as previously mentioned (footnote to Table I).

References

1. Pluznik, D.H. and Sachs, L. 1965, J. Cell. Comp. Physiol. 66: 319.
2. Bradley, T.R. and Metcalf, D., 1966, Aust. J. Exp. Biol. Med. Sci.44: 287.
3. Robinson, W.A. and Mangalik, A., 1975, Sem. Hematol. 12: 7.
4. Lüderitz, O., Westphal, O., Staub, A.M. and Nikaido, H., 1971. Microbial Toxins. G. Weinbaum, S. Kadis and S.J. Ajl, editors. (Academic Press, New York) 4: 145.
5. Metcalf, D., 1971, Immunology 21: 427.
6. Quesenberry, P.J., Morley, A.A., Rickard, K.A., Garrity, M., Howard, E. and Stohlman, Jr., F., 1972, N. Eng. J. Med. 286: 1291.
7. Sultzer, B.M., 1972, Infect. Immun. 5: 107.
8. Watson, J. and Riblet, R., 1974, J. Exp. Med. 140: 1147.

*Progress in Differentiation Research, ed. N. Müller-Bérat et al.*



# REGULATION OF ERYTHROID DIFFERENTIATION IN NORMAL AND LEUKAEMIC CELLS

I. B. Pragnell[1], W. Ostertag[2], P. R. Harrison[1], R. Williamson[1] and J. Paul[1].

1. Beatson Institute for Cancer Research, Glasgow, Scotland.
2. Max-Planck Institut fur Experimentelle Medizin, Gottingen, W. Germany.

## 1. Introduction

Erythropoiesis in foetal liver, spleen or bone marrow has been studied extensively as a model system for investigating the regulation of differentiation by cytological, cell-kinetic and biochemical techniques. These approaches have clarified the way in which the multi-potential stem cell (CFU) becomes committed to erythroid differentiation in the micro-environment of certain organs[1], thus producing a series of committed cells (BFU-E→CFU-E) whose proliferation and differentiation into the morphologically recognisable erythroid series (proerythroblast→erythrocyte) is regulated by the hormone erythropoietin[2]. Our own work has shown that the large increase in haemoglobin synthesis induced by erythropoietin in foetal liver cultures is produced mainly by increasing the proliferation and differentiation of immature erythroblasts, rather than by increasing the rate of haemoglobin synthesis in each individual mature erythroid cell[3]. Nevertheless, high concentrations of erythropoietin can increase the rate of haemoglobin formation relative to the rate of morphological differentiation so that a series of atypical macro-erythroblasts is produced[3,4]. Using a novel technique for detecting globin messenger RNA in cytological preparations of erythroid cells[5], we have demonstrated that globin messenger RNA accumulates rapidly during the pro-erythroblast→basophilic erythroblast transition in early foetal liver development[6]. Other studies support this conclusion[7-9]. More recently, we have demonstrated that both in foetal liver <u>in vitro</u> or <u>in vivo</u> or in anaemic spleen <u>in vivo</u>, high levels of erythropoietin cause proerythroblasts both to proliferate and to accumulate globin messenger RNA[10]. The proerythroblast therefore seems to be an important stage at which significant regulatory events concerning erythroid differentiation maturation take place.

In recent years, the opportunity has arisen to compare the regulation of erythropoiesis in normal proerythroblasts with that in erythroleukaemic cells. Some years ago, Friend and co-workers discovered that an erythroleukaemia was induced in susceptible[11] mice by inoculation with the Friend virus complex (FV)[12]. This disease is characterised by an erythroblastosis in the spleen, which, depending on the strain of Friend virus may result in polycythemia (FV-P)[13] or anaemia (FV-A)[12] since the enhanced erythropoiesis seems in some way to be inefficient or defective[14]. Despite claims to the contrary[15], present evidence suggests that the target cells for Friend virus are the stem cells which are committed to the erythroid line (i.e. CFU-E's)[16]. In the case of FV-P infection, this gives rise to foci in the spleen which proliferate and undergo erythroid maturation both *in vivo* and *in vitro* in the absence of the hormone erythropoietin [17, 18], unlike normal CFU-E's. The extent of maturation of these erythropoietin-independent CFU-E's seems to be normal, at least in plasma clot culture *in vitro*[17]. Thus, Friend erythroleukaemia seems to be abnormal in the regulation both of the production of CFU-E's and of their dependence on erythropoietin, rather than in erythroid differentiation/maturation *per se*.

This conclusion is supported by studies with tissue culture lines (Friend cells) derived from spleen fragments from mice infected with FV-A or FV-P. These cells seem to be transformed CFU-E's which may have differentiated as far as the proerythroblast stage of development and as such can be compared with normal proerythroblasts. Although they do not respond to erythropoietin, Friend cells can be induced (for example, by DMSO[18]) to exhibit a whole series of functions characteristic of normal erythroid cell development, for example, haem[19] and delta amino laevulinic acid synthetase[20], globins[21, 22] and globin messenger RNAs[23, 24], specific membrane proteins, such as spectrin[25], carbonic anhydrase[26] and morphological changes resulting in the non-dividing orthochromatic cell[19]. Friend erythroleukaemia therefore represents a good model for studying the relationship between transformation and abnormal regulation of differentiation in a situation where both the virus and the programme of differentiation are known and can be characterised in detail. This report concerns in particular the interdependence of transformation and the induction of differentiation by DMSO in Friend cells.

## 2. Methods

Methods used for cell culture, assay of Friend virus components, large scale preparation of virus, induction of erythroid differentiation with DMSO, and preparation of DNA complementary to purified globin messenger RNA have all been described elsewhere[27-30]. Highly radioactive DNA complementary to viral RNA (viral cDNA) was prepared from lysed virions[31] isolated from Friend cells (clone F4/6) harvested after treatment with DMSO for 36 hours and 12 hours after renewal of medium. The concentration of viral sequences in RNA preparations was determined by titration with viral cDNA according to the method developed for globin cDNA[30] except that the hybridisation was carried out to a $D_0t$ of 2 (see reference 30) to allow for the greater base-sequence complexity and therefore slower rate of hybridisation of viral cDNA. The reiteration frequencies of viral sequences in DNA preparations were determined by annealing viral cDNA in vast DNA excess[29].

Characterisation of viral cDNA - A DNA copy of Friend virus RNA (viral cDNA) was prepared by transcribing with endogenous reverse transcriptase the viral RNA present in virus isolated from untreated (viral cDNA) or DMSO-treated (viral $cDNA^D$)

Friend cells (clone F4-6). These viral cDNAs were shown to be virus-specific since 90% of the cDNAs were hybridised by excess viral RNA of the type used as template, whereas less than 5% hybridised to E. coli RNA or to mouse embryo ribosomal RNA (Table 1).

Table 1

Extents of hybridisation of viral cDNAs to viral and cellular RNAs.

| Viral cDNA transcribed from virus extracted from cells | % cDNA hybridised by 200X excess: E. coli RNA | Mouse ribosomal RNA | 70S viral RNA from cells: untreated | 70S viral RNA from cells: DMSO-treated |
|---|---|---|---|---|
| untreated | 5% | - | 90% | 92% |
| DMSO-treated | 5% | 5% | 86% | 88% |

The extent of hybridisation was assayed by resistance to S1 nuclease[23]. Cells were treated with 1.5% DMSO for 2d.

In fact, the cross-hybridisation experiments summarised in Table 1 also show that sequences present in both viral cDNAs show extensive homologies to viral RNAs isolated from both untreated and DMSO-treated Friend cells. However, since these hybridisation experiments were carried out in vast RNA excess, they do not show whether or not the relative distributions of sequences in both viral RNAs are identical.

In this context it is clearly of great importance to ascertain how much of the viral RNA genome is represented in the viral cDNA transcript. This can be achieved by titrating viral cDNA with increasing ratios of viral RNA (Figure 1). It is clear that the viral RNA/viral cDNA$^D$ titration curve is very similar to that for globin messenger RNA/globin cDNA: in both cases, about 70% of the cDNA sequences are hybridised at an RNA/cDNA ratio of about 1.5-2.

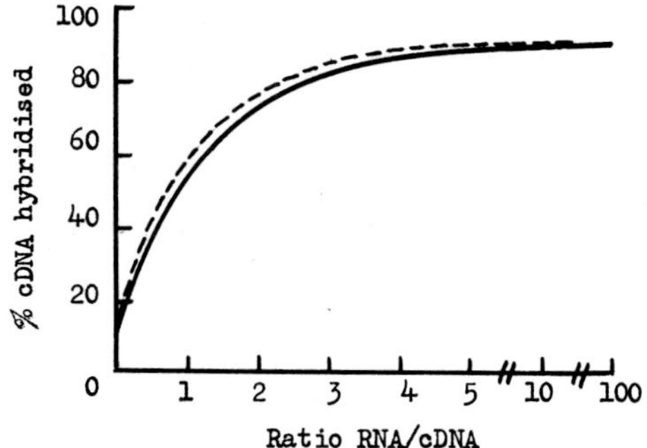


Fig. 1. Titration of globin (----) cDNA or viral (——) cDNA$^D$ from DMSO-treated cells with increasing ratios of globin messenger RNA and 70S viral RNA$^D$ respectively. See refs. 29, 30 for method.

This indicates[30] that more than half of the viral $RNA^D$ sequences have been transcribed into viral $cDNA^D$. Moreover, since the curvatures of the two titration curves are very similar, it can be concluded that the sequences in viral $cDNA^D$ are represented in approximately the same frequencies as they occur in the viral RNA released from DMSO-treated cells.

## 3. Results

a. Induction of viral sequences by DMSO - The virus complex released from untreated Friend cells (clone F4) contains two components, the defective spleen focus-forming virus (SFFV) and a helper virus, murine leukaemia virus (MLV-F). After treatment of Friend cells with DMSO for two days, release of C-type particles is increased 8-fold and spleen focus formation is increased 10-100 fold. (The differences in these values is due to the increased stability of intact viral RNA in DMSO-treated cells). This increase in virus release is transient: after treatment with DMSO for five to six days, Friend cells release much less virus than before treatment. Thus, the release of virus seems to correlate temporally with the period when globin messenger RNA accumulates[23,24].

Further studies have shown that the MLV component of Friend virus is not induced after DMSO treatment[27]. Whereas the virus released from untreated Friend cells forms spleen foci in both N- and B-type mice with similar efficiencies, the virus released from DMSO-treated cells forms spleen foci preferentially in N-type mice (Table 2).

Table 2

Spleen focus formation in N- and B-type mice

| Treatment of cells | Addition to cell supernatant | Specificity of host | |
|---|---|---|---|
| | | N-type | B-type |
| Nil | - | 1300 | 1300 |
| DMSO | - | 12000 | 300 |
| Nil | MLV | 1200 | 1100 |
| DMSO | MLV | 9000 | 9000 |

Values given are summarised from Fig. 2 of ref. 27 and give the numbers of spleen foci after injection of the supernatant from $10^6$ cells. In the second experiment, $10^7$ XC units of NB-tropic MLV-F were also injected simultaneously.

Furthermore, injection of NB-tropic MLV together with virus released from DMSO-treated cells increases the efficiency of formation of spleen foci in B-type mice to that in N-type mice (Table 2); but it does not increase the number of foci in N-type mice. This result shows that MLV is not limiting during DMSO induction and that SFFV is induced as well as endogenous virus. However, it is not certain what the relative amounts of SFFV and endogenous virus are in absolute terms.

b. Integration of viral sequences - Integration of viral sequences was studied by annealing viral $cDNA^D$ to excess DNA[29]. Whereas most of the viral $cDNA^D$ sequences are integrated into Friend cell DNA, only about 50% are integrated into mouse embryo DNA (Figure 2). By comparing the rate of annealing of viral $cDNA^D$ to carrier DNA with the rate of annealing of the carrier DNA itself, the extent of reiteration of the viral sequences can be calculated[29] to be about 5-10-fold both for the sequences integrated into the mouse embryo DNA and into Friend cell DNA.

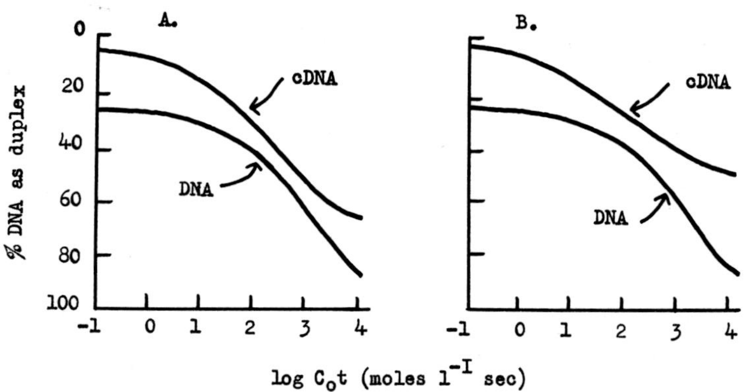


Fig. 2. Annealing of viral $cDNA^D$ to (250,000X) excess (w/w) mouse embryo (A) or Friend (B) DNA. cDNA and cellular DNAs were mixed, denatured and annealed at 60° in 0.12M phosphate to various $C_ot$ values (concentration of cellular DNA X time). The percentage of the cDNAs or cellular DNA in duplex form was determined by chromatography on hydroxyapatite as described previously[29].

This experiment therefore allows a distinction to be made between those sequences integrated in normal mouse DNA (in view of section 3a above, presumably endogenous virus sequences) and those integrated after Friend virus infection (SFFV and/or MLV sequences). As discussed above (Table 1), virus released from DMSO-treated cells (SFFV and endogenous virus) shows complete homology under our hybridisation conditions with virus released from untreated cells (SFFV and MLV). Therefore endogenous virus and MLV must have considerable sequence similarities. This is consistent with the fact that exogenous and endogenous type-C viruses can undergo genetic recombination[38]. These combined results therefore suggest that SFFV sequences (and also possibly some MLV sequences) are integrated into Friend cell DNA in addition to the endogenous virus sequences, which are present in normal mouse DNA.

c. Viral release and differentiation - By treating Friend cells in various ways, it has proved possible to prevent induction of viral release by DMSO without affecting the ability of the cells to differentiate (Table 3).

Therefore, induction of erythroid differentiation is not functionally dependent on induction of viral release. However, certain cell lines which cannot differentiate in response to DMSO do not show induction on virus release (Table 3).

Table 3

Virus release and differentiation in Friend cells

| Cell | Treatment | Stimulation by DMSO | |
|---|---|---|---|
| | | $Hb^+$ cells | Virus release |
| Inducible | - | 25 X | 10-100 X |
| Inducible | Interferon | 25 X | Nil |
| Inducible | Azidothymidine | 26 X | Nil |
| BUdR-resistant | - | 15 X | Nil |
| Inducible | BUdR | Nil | Nil |
| DMSO-resistant | - | Nil | Nil |
| Non-inducible | - | 1.5 X | Nil |

Data from references 27, 28, 32, 33.

This suggests that, conversely, some function induced during differentiation is required for induction of virus release. It is not certain what this function might be. However, a correlation has been established between induction of N-tropic endogenous virus (together with SFFV) and induction of thymidine kinase activity in clones of the BUdR-resistant line[28].

d. Virus-specific sequences in the polysomes of Friend cells - An attempt has been made to determine whether the types of viral sequences present on the polysomes of Friend cells change during induction of differentiation by DMSO. It is obviously important in this respect to distinguish viral sequences present in complete virions from those actually being translated on the polysomes. Polysomes were therefore prepared by centrifugation through 2M sucrose, through which the less dense virus particles cannot pass. The amounts of globin- or virus-specific sequences in polysomal RNA were then determined by titration with globin or viral cDNAs respectively. As a control, total cytoplasmic RNA from mouse embryos was also tested for the presence of viral sequences. The results indicate that viral sequences represent at most less than one part per million of mouse embryo RNA, which may not be significant.

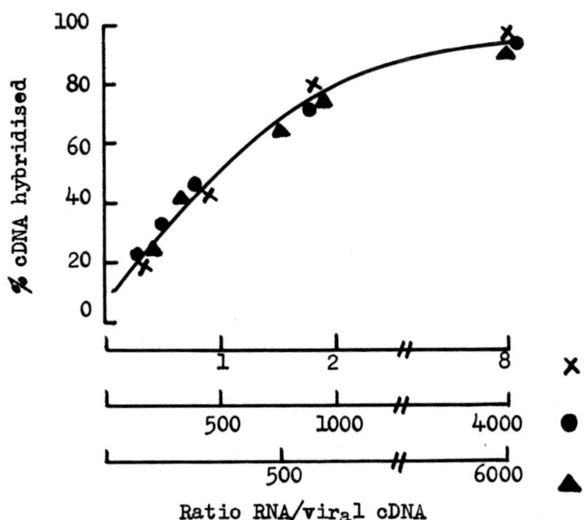


Fig. 3. Titration of viral $cDNA^{D}$ with 70S viral RNA from DMSO-treated cells (X) or polysomal RNA from untreated (●) on DMSO-treated (▲) cells. Details as Fig. 1.

In contrast, viral sequences are present at high levels in polysomal RNA from both untreated and DMSO-treated Friend cells (Fig. 3) and from a BUdR-resistant line (B8) which is induced to differentiate on DMSO-treatment but which fails to induce virus. These results show that there are no overall qualitative differences in the types of virus sequences present on the polysomes in these various cell types. Moreover, the shapes of the titration curves suggest that the ratios of SFFV sequences to helper virus sequences in the polysomes from both uninduced and induced cells are very similar. However, this statement has to be interpreted in the context of our previous conclusion that our present hybridisation methods do not distinguish between endogenous virus and MLV-F sequences.

From titration curves such as those given in Figure 3, the total amounts of virus-specific sequences on the polysomes from the two cell types can be quantitated (Table 4). For comparison, the amounts of globin messenger RNAs were also determined (Table 4). In F4-6, FLD/3 and B8-3 cell lines, the level of globin messenger RNAs associated with polysomes is increased very considerably after DMSO treatment. In fact, in DMSO-treated cells, globin messenger RNAs represent about 0.5 - 1.0% of the total polyA-containing (presumably total messenger RNA in the polysomes. By comparison, the level of viral sequences in untreated Friend cells (F4-6 or FLD/3) is considerable, representing up to 10% of the total polyA-containing RNA in the polysomes.

Table 4

Virus- and globin-specific sequences associated with polysomes in Friend cell variants

| Cell | Status | globin-specific | | virus-specific | |
|---|---|---|---|---|---|
| | | -DMSO | +DMSO | -DMSO | +DMSO |
| F4-6 | $Hb^{D+}$ $virus^{D+}$ | 3 | 200 | 2000 | 3000 |
| FLD/3 | $Hb^{D+}$ $virus^{D+}$ | 15 | 350 | 640 | 2100 |
| B8-3 | $Hb^{D+}$ $virus^{D-}$ | 30 | 400 | 1500 | 1600 |
| F4-12 | A $particle^{-}$ | - | - | 300 | - |

Values give the proportion of the total polysomal RNA which is globin- or virus-specific (in p.p.m.) at a treatment of cells with DMSO for the optimum period (2 or 4d respectively).

Treatment of these cells with DMSO increases the **yield** of virus particles 4-8 fold, and increases polysome-associated viral sequences 1.5-3 fold (Table 4). However, the variant line B8-3 which does not release virus on DMSO treatment, also has a high level of viral sequences associated with the polysomes; but this level is not increased during DMSO treatment. Perhaps significantly, in a Friend cell line (F4-12) which contains few A-type particles (as judged by electron microscopy), the level of Friend virus sequences associated with polysomes is lower than in the other cell lines.

## 4. Discussion

The nature of the events initiated by transforming agents which render normal cells malignant and unable to differentiate normally is generally acknowledged to be a subject of outstanding importance. It has been recognised for some time that studies with leukaemic cells offer many possibilities of experimental approach both at cellular and virological levels. As argued in detail above, the discovery of Friend erythroleukaemia represents a very useful model system in which to investigate the relationship between cell transformation and abnormal differentiation. In particular it offers the opportunity to investigate how the integration and expression of specific viral sequences correlates with changes in the pattern of differentiation, especially with respect to the activity of the globin genes.

Current evidence suggests that Friend cells are derived from stem cells transformed by the Friend virus complex after commitment to the erythroid line. These cells do not differentiate in culture, unless treated with DMSO[19] or a range of other agents[34-37]. The reason for this failure to differentiate is not clear: it could be a consequence of selecting for cells which fail to differentiate terminally during culture <u>in vitro</u>; alternatively, it may be related more specifically to the state of integration or expression of virus sequences in Friend cells. It is in this context that our present results may be of some significance. Friend cells in culture normally release MLV and SFFV; whereas on treatment with DMSO, the cells differentiate and release increased amounts of SFFV plus an endogenous virus. This fact raises the interesting possibility that differentiation/maturation of transformed CFU-E's may be prevented by transformation with MLV/SFFV; whereas DMSO treatment releases the block to differentiation and concomitantly activates the release of endogenous virus/SFFV. Further work has shown that differentiation and release of endogenous virus/SFFV are not uniquely interdependent: under conditions in which erythroid maturation proceeds normally in response to DMSO, Friend cells treated with various agents fail to release virus in the normal manner. However, under the same conditions, an increase in A-type particles occurs intracisternally; so that it may be these virus particles which are associated with induction of differentiation. As a corollary of this, it has been found that cells which cannot be induced to differentiate are also not inducible for release of SFFV/endogenous virus. This result shows that the release of virus is dependent on differentiation and is not an affect of DMSO itself.

Further analysis of the relationship between transformation and differentiation clearly requires more detailed elucidation of the extent of integration, transcription and translation of specific viral sequences. A preliminary approach along these lines has been made by preparing viral cDNA transcripts of viral RNAs released from untreated or DMSO-treated cells (i.e. MLV/SFFV and endogenous virus/SFFV respectively). By hybridising viral cDNA released from DMSO-treated cells to excess DNA, evidence has been obtained that about half the sequences present in this viral cDNA (presumably those of the endogenous virus) are integrated about five to ten times in mouse embryo DNA; whereas all the viral sequences (endogenous virus plus SFFV?) are integrated about ten times in DNA from Friend cells.

The results of cross-hybridisation experiments between viral RNAs and cDNAs show that homology exists between MLV and the endogenous virus induced by DMSO. Further work will be necessary to determine whether viral cDNA transcripts from specific viruses can be purified by cross-hybridisation under greater conditions of stringency. Until this is achieved, our hybridisation experiments can measure only quantitative differences in viral sequences and cannot distinguish qualitative changes in virus components. Nevertheless, our present results do show that a large pool of virus-specific sequences is present on the polysomes of Friend cells (representing up to 10% of the total poly(A)-containing messenger RNA in the polysomes) and that the numbers of such sequences increase in differentiating Friend cells but not in cells incapable of releasing virus in response to DMSO. However, it is significant that the amounts of viral sequences being translated on the polysomes are much lower in cells which do not show evidence of intracisternal A-type particles.

## 5. References


1. Trentin, J.J., 1970, In: Regulation of Haematopoiesis, vol, 1, ed. A. S. Gordon (Appleton-Century-Crofts, New York), p.159.

2. Lajtha, L.G. and Schofield, R., 1974, Differentiation, 2, 313.

3. Paul, J., Conkie, D. and Burgos, H., 1973, J. Embryol. exp. Morph., 29, 453.

4. Harrison, P.R., Conkie, D. and Paul, J. 1973, In: Brit. Soc. Dev. Biol. Symp. "The Cell cycle in development and differentiation (Cambridge University Press), p.341.

5. Harrison, P.R., Conkie, D., Paul, J. and Jones, K. 1973, FEBS Letters, 32, 109.

6. Harrison, P.R., Conkie, D., Affara, N. and Paul, J. 1974, J. Cell Biol., 63, 402.

7. Terada, M., Cantor, L., Metafora, S., Rifkind, R.A., Bank, A. and Marks, P. 1972, Proc. Natn. Acad. Sci. U.S.A., 69, 3575.

8. Ramirez, F., Gambino, R., Maniatis, G.M., Rifkind, R.A., Marks, P.A. and Bank, A. 1975, Trans assoc. Am. phys., **in press.**

9. Ramirez, F., Gambino R., Maniatis, G.M., Rifkind, R.A., Marks, P.A. and Bank, A. 1975, J. Biol. Chem, in press.

10. Conkie, D., Kleiman, L., Harrison, P.R. and Paul, J. 1976, Exp. Cell Res. in press.

11. Lilly, F., 1972, J. Natn. Cancer Inst., 49, 927.

12. Friend, C., 1957, J. exp. Med., 105, 307.

13. Mirand, E.A., Prentice, T.C., Hoffman, J.C. and Grace, J.T. 1961, Proc. Soc. exp. Biol. Med., 106, 423.

14. Tambourin, P.E., Gallien-Lartigue, D., Wendling, F. and Huaulme, D., 1973, Br. J. Haematol., 24, 511.

15. Seidel, H.J., 1973, In: 'Unifying Concepts of Leukaemia' (S. Korger, Basel), 935.

16. Tambourin, P. and Wendling, F., 1975, Nature, 256, 320.

17. Mirand, E.A., Steeves, R.A., Lange, R.D. and Grace, J.T., 1968, Proc. Soc. exp. Biol. Med., 128, 844.

18. Liao, S.K. and Axelrad, A.A., 1975, Int. J. Cancer, 15, 467.

19. Friend, C., Scher, W., Holland, J.G. and Sato, T., 1971, Proc. Natn. Acad. Sci. U.S. 68, 378.

20. Ebert, P.S. and Ikawa, Y., 1974, Proc. Soc. exp. Biol. Med., 146, 601.

21. Ostertag, W., Melderis, H., Steinheider, G., Kluge, W. and Dube, S., Nature New Biol., 239, 231.

22. Boyer, S.H., Wuu, K.D., Noyes, A.W., Young, R., Scher, W., Friend, C., Preisler, H.D. and Bank, A., 1972, Blood, 40, 823.

23. Gilmour, R.S., Harrison, P.R., Windass, J., Affara, N. and Paul, J. 1974, Cell Differentiation, 3, 9.

24. Harrison, P.R., Gilmour, R.S., Affara, N., Conkie, D. and Paul, J., 1974, Cell Differentiation, 3, 23.

25. Eisen,H., personal communication.

26. Kabat, D., Sherton, C.C., Evans, L.H., Bigley, R. and Koler, R.D., 1975, Cell 5, 331.

27. Dube, S.K., Pragnell, I.B., Kluge, N., Gaedicke, G., Steinheider, G., and Ostertag, W., 1975, Proc. Natn. Acad. Sci., U.S., 72, 1863.

28. Ostertag, W., Roesler, G., Krieg, C.J., Kind, J., Cole, T., Crozier, T., Gaedicke, G., Steinheider, G., Kluge, W. and Dube, S., 1974, Proc. Natn. Acad. Sci., U.S. 71, 4980.

29. Harrison, P.R., Birnie, G.D., Hell, A., Humphries, S., Young, B.D., and Paul, J., 1974, J. Mol. Biol. 84, 539.

30. Young, B.D., Harrison, P.R., Gilmour, R.S., Birnie, G.D., Hell, A., Humphries, S. and Paul, J. 1974, J. Mol. Biol., 84, 555.

31. Sveda, M., Fields, B.N. and Soeiro, R., 1974, Cell, 2, 271.

32. Swetly, P. and Ostertag, W., 1974, Nature, 251, 342.

33. Ostertag, W., Cole, T., Crozier, T., Gaedicke, G., Kind, J., Kluge, W., Krieg, J.C., Roesler, G., Steinheider, G., Weissmann, B.J. and Dube, S.K. 1974, in: Proc. Fourth Symp. Princess Takamatsu Res. Fund 'Differentiation and Control of Malignancy in Tumour Cells' (University of Tokyo Press) p.493.

34. Scher, W., Preisler, H.D. and Friend, C., 1973, J. Cell Physiol. 81, 63.

35. Tanaka, Lev J., Terada, M., Breslow, R., Rifkind, R.A. and Marks, P., 1975, Proc. Natl. Acad. Sci. U.S., 72, 1003.

36. Preisler, H.D. and Lyman, G., 1975, Cell Differentiation, 4, 179.

37. Leder, A. and Leder, P., 1975, Cell, 5, 319.

38. Stephenson, J.R., Anderson, G.R., Tronick, S.R. and Aaronson, S.A. 1974, Cell, 2, 87.

*Progress in Differentiation Research, ed. N. Müller-Bérat et al.*



# INDUCTION OF LEUKOPOIETIC STEM CELL PROLIFERATION AND DIFFERENTIATION IN SPLEEN CULTURES OF NORMAL AND FRIEND LEUKEMIA VIRUS-INFECTED MICE

D.W. Golde and C. Friend

Division of Hematology-Oncology, Department of Medicine
UCLA School of Medicine, Los Angeles, California 90024 and
Center for Experimental Cell Biology, Mount Sinai School of Medicine,
New York, New York 10029, U.S.A.

## Summary

The effect of colony-stimulating activity (CSA) on granulopoiesis and monocytopoiesis was studied in cultures of spleen cells from normal and leukemic DBA-2 mice. The cells were cultured in liquid suspension using the Marbrook chamber method and in soft gel with the double-layer agar technique (CFU-C). Gravid mouse uterus extract was the source of CSA. Liquid cultures were maintained for up to 14 days, with viable and differential cell counts performed at intervals. Pulse $^{3}$H-thymidine labeling indices were determined and cellular differentiation was assessed by peroxidase and α-naphthyl butyrase histochemistry and tests of leukocyte function. Normal spleen cells did not proliferate actively in cultures without CSA whereas viable cell counts were three- to five-fold greater in CSA-stimulated cultures. With CSA stimulation, the labeling index rose from less than 2% at day 0 to 30-40% by 5-8 days. By day 5, individual cells had the morphology of blasts, promyelocytes and promonocytes. Differentiation then proceeded normally and synchronously with mature cells assuming their characteristic histochemical and functional markers by day 8-10. The total number of CFU-C was low at 0 time but increased by up to 70-fold after 7 days in liquid culture. Leukemic spleen cells also were dependent on CSA in vitro. Cloning efficiency (CFU-C) of Friend leukemic spleen cells was about 10 times that observed in uninfected cells, but differentiation in liquid and semi-solid culture approximated that of normal cells. Total CFU-C per leukemic spleen was approximately 300 times normal. In Friend leukemia, splenic CFU-C are present in increased numbers but have normal capacity for granulocytic differentiation in vitro. These results indicate that stem cells present in normal and

---

Grant support: USPHS CA 15688, 15619, 10000, and 13047.

leukemic spleens may be stimulated by CSA to synthesize DNA and proceed to mature to the level of morphologically recognizable granulocytic and monocytic precursors. The model system described provides a means to study factors regulating the in vitro generation and differentiation of committed granulopoietic stem cells in normal and leukemic conditions.

## 1. Introduction

Although the murine spleen is predominantly composed of lymphoid cells concerned with immune responses, it is also an important hematopoietic organ containing pluripotent stem cells (CFU-S)[1] and committed stem cells of the erythrocytic (CFU-E)[2] and granulocytic series (CFU-C).[3,4] The concentration of pluripotent and committed stem cells in the spleen is substantially lower than in the bone marrow[4] and splenic CFU-C are reported to be largely in a resting state and not in active cycle.[5]

The spleen is prominently involved in Friend leukemia and the target cell for Friend virus is believed to be a committed erythroid precursor -- the erythropoietin-responsive cell.[6] The Friend leukemia cell is thought to be erythroid in nature because derived cell lines clearly exhibit erythrocytic differentiation.[7]

We undertook a series of experiments to examine the characteristics of normal and leukemic splenic granulopoiesis in vitro. Colony-stimulating activity (CSA) stimulated the proliferation of CFU-C in suspension culture with subsequent cellular differentiation along the granulocytic and mononuclear phagocyte lines. Leukemic spleens contained a markedly increased number of CFU-C which appeared to differentiate normally in semi-solid and liquid culture.

## 2. Materials and Methods

Young adult female DBA-2 mice were used for all experiments. The leukemia was initiated by intravenous injection of a crude splenic extract[8] and subsequently passed by transferring living leukemic cells intraperitoneally. Studies were performed on leukemic mice in the terminal phase of the disease. Spleens were removed aseptically and single-cell suspensions prepared. The cells were cultured in liquid suspension using the Marbrook diffusion chamber as described previously[9,10] and pregnant mouse uterus extract served as the source of colony-stimulating activity.[11]

The suspension cultures were harvested at intervals for $^{3}$H-thymidine radioautography and for viable and differential cell counts. Before culture in liquid suspension, and at intervals thereafter, spleen cells were plated in agar culture to assay for colony-forming cells (CFU-C). The double-layer agar culture technique was used as previously described and McCoy's 5A medium with 15% fetal calf serum was used in all experiments.[12] Cellular differentiation in culture was assessed morphologically on Giemsa-stained preparations and histochemically with peroxidase and α-naphthyl butyrase stains.[13] Leukocyte phagocytic capacity was assessed with opsonized heat-killed Candida albicans.

3. Results

Normal and leukemic spleen cells did not grow well in suspension culture without added CSA. In CSA-stimulated cultures there was a fall in cell numbers until day 3-4, and thereafter a prominent rise in viable cell counts was observed (table 1). Labeling studies with $^{3}$H-thymidine showed that CSA stimulated DNA synthesis in both normal and leukemic cell cultures (table 2). Before culture there was little DNA synthetic activity in normal spleen cells (labeling index 2%) but leukemic cells showed a labeling index of 40%. Cellular differentiation proceeded along the granulocytic and mononuclear phagocyte lines in both leukemic and normal cell cultures (table 3). Immature cells were prominent in suspension culture up to one week but thereafter mature granulocytes and macrophages predominated (table 3; fig. 1&2). These mature cells appeared morphologically and histochemically normal and exhibited active phagocytic capacity. The mature cells in leukemic cultures were indistinguishable from those seen in the normal spleen cultures.

The generation of CFU-C in liquid culture was also largely dependent on the presence of CSA (fig. 3). CFU-C in normal spleen cell cultures increased by a mean of 70-fold after 1 week in culture. Leukemic cell cultures showed a roughly similar pattern of CFU-C generation.

The cloning efficiency of freshly isolated leukemic spleen cells in agar was substantially greater than normal spleen cells (table 4). No colonies were seen in agar plates that did not contain CSA and the CSA-stimulated colonies were composed of normally differentiated granulocytes and monocytes. Since the leukemic spleens were more than 30 times normal size, the total content of CFU-C in the leukemic spleens was 100-300 times greater than normal.

Table 1

Viable cell counts in suspension cultures of normal and leukemic spleen cells (X $10^5 \pm$ SE)

| Days in Culture | CSA Normal | CSA Leukemic | NO CSA Normal | NO CSA Leukemic |
|---|---|---|---|---|
| 0 | 30 | 30 | 30 | 30 |
| 3 - 4 | 4 ± 1 | 7 ± 4 | 3 ± 1 | 5 |
| 5 - 6 | 7 ± 1 | 16 ± 4 | 3 ± 1 | 2 ± 1 |
| 7 - 8 | 26 ± 5 | 18 ± 6 | 8 ± 3 | 2 ± 1 |
| 9 -10 | 30 ± 4 | 12 ± 5 | 7 ± 3 | 2 ± 1 |
| 11 -12 | 21 ± 3 | 6 ± 2 | 9 ± 5 | 1 |
| 13 -15 | 15 ± 7 | 3 | 3 ± 1 | --- |

Table 2

$^3$H-thymidine labeling in suspension cultures of normal and leukemic spleen cells

(Total cells labeled X $10^3$ - labeling index in parentheses)

| Days in Culture | CSA Normal | CSA Leukemic | NO CSA Normal | NO CSA Leukemic |
|---|---|---|---|---|
| 0 | 60 (2%) | 1,200 (40%) | 60 (2%) | 1,200 (40%) |
| 1-2 | 50 (5%) | 750 (25%) | < 50 (3%) | 500 (23%) |
| 5-6 | 300 (47%) | 700 (40%) | 58 (18%) | < 50 (13%) |
| 7-8 | 650 (25%) | 450 (26%) | 152 (19%) | < 50 (12%) |
| 9-10 | 290 (10%) | 100 (6%) | < 50 (4%) | < 50 (1%) |

Table 3

Differential cell counts of CSA-stimulated suspension cultures (%)

| Days in Culture | | Immature cells | Mature granulocytes | Mature macrophages | Lymphoid cells |
|---|---|---|---|---|---|
| 0 | normal | 4 | 4 | 2 | 90 |
| | leukemic | 98 | 1 | 0 | 1 |
| 5 | normal | 70 | 10 | 6 | 14 |
| | leukemic | 55 | 13 | 30 | 2 |
| 7-8 | normal | 34 | 29 | 27 | 10 |
| | leukemic | 15 | 20 | 65 | 0 |
| 9-10 | normal | 17 | 31 | 47 | 5 |
| | leukemic | 18 | 17 | 65 | 0 |

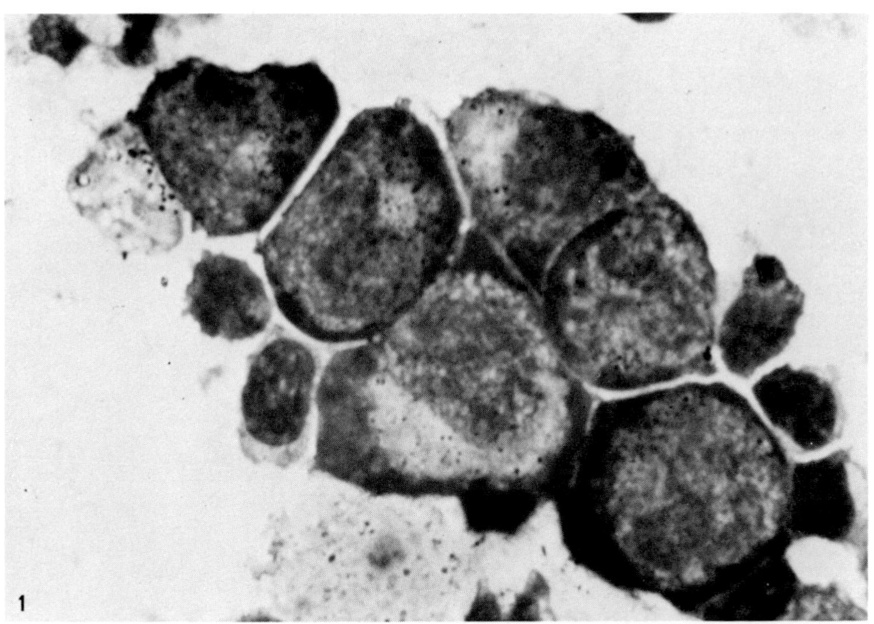


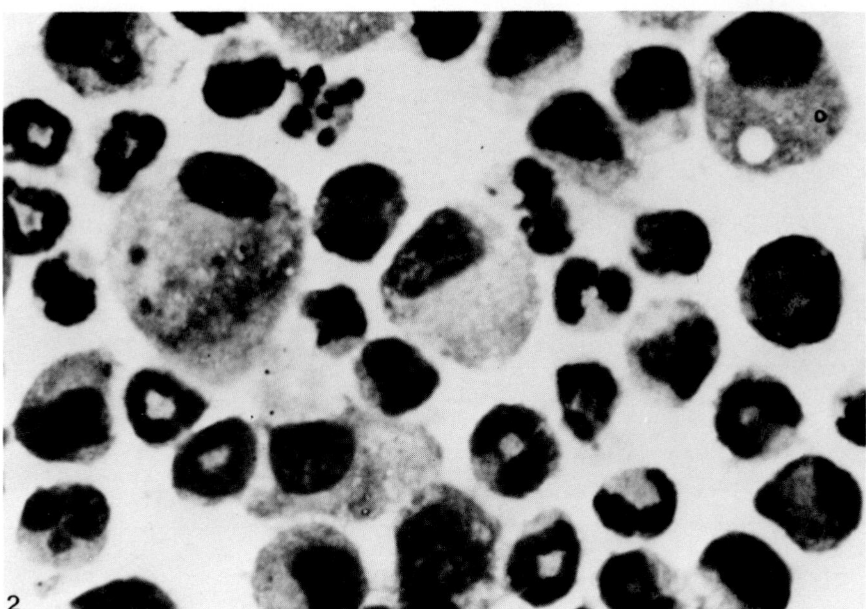


Fig. 1. Immature cells in CSA-stimulated suspension culture of normal spleen cells at 5 days.

Fig. 2. Mature granulocytes and macrophages in 10-day suspension culture of normal spleen cells.

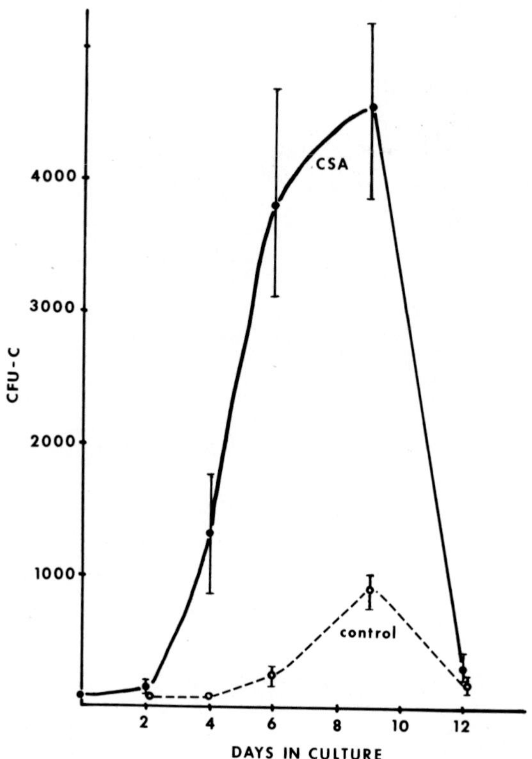


Fig. 3. Pattern of CFU-C generation in liquid culture. Control cultures did not have added CSA.

Table 4
Spleen cell colony formation in agar

| | $CFU-C/2X10^5$ (mean and range |
|---|---|
| normal spleen | 4 ( 2 - 10) |
| leukemic spleen | 39 (30 - 49) |

## 4. Discussion

The hematopoietic potential of the murine spleen is most apparent in fetal and neonatal life and during times of stress such as hemorrhage or infection.[4,14] Whereas erythropoietin is the important humoral factor governing the induction of splenic erythropoiesis, relatively little is known about the factors initiating granulopoiesis in the spleen. Committed granulopoietic stem cells are present in the spleen in relatively low numbers as compared to bone marrow and the spleen CFU-C are believed to be largely in a resting state.[5] The present in vitro studies suggest that splenic CFU-C can be induced by CSA to enter the proliferative cycle, generate more CFU-C, and to differentiate into mature granulocytes and macrophages. The CSA-stimulated cells demonstrate an orderly pattern of $^{3}$H-thymidine incorporation, CFU-C proliferation, and cellular maturation, suggesting that the cultures are partially synchronized.

Leukemia induced by the anemic strain of Friend virus is characterized by the proliferation of undifferentiated cells in the spleen, liver, and bone marrow.[8,15] There is considerable evidence that these leukemic cells are erythroid precursors.[7,16-18] In this study, the in vitro growth of leukopoietic stem cells from leukemic spleens was dependent on CSA and the leukemic spleen contained 100-300 times the normal number of CFU-C. However, the granulopoietic stem cells from leukemic mice formed morphologically normal colonies in semisolid gel culture and differentiated normally to mature granulocytes and macrophages in suspension culture. The leukemic spleen, therefore, appears to contain increased numbers of CFU-C with the potential for normal differentiation.

These results indicate that committed stem cells present in normal and leukemic spleen may be induced by CSA in vitro to synthesize DNA and thereafter then develop into mature leukocytes. The system described may be a useful model for studying normal and leukemic granulopoietic stem cell activation and differentiation in vitro.

References

1. Blaži-Poljak, M. and Boranić, M., 1975, Exp. Hematol. 3, 85.
2. Gregory, C.J., McCulloch, E.A. and Till, J.E., 1973, J. Cell. Physiol. 81, 411.
3. Metcalf, D. and Stevens, S., 1972, Cell Tissue Kinet. 5, 433.
4. Metcalf, D. and Moore, M.A.S., 1971, in: Frontiers of Biology, Vol. 24, eds. A. Neuberger and E.L. Tatum (North-Holland, Amsterdam).
5. Metcalf, D., 1972, Proc. Soc. Exp. Biol. Med. 139, 511.
6. Tambourin, P. and Wendling, F., 1971, Nature New Biol. 234, 230.

7. Friend, C., Scher, W., Holland, J.G. and Sato, T., 1971, Proc. Natl. Acad. Sci. U.S.A. 68, 378.

8. Friend, C., 1957, J. Exptl. Med. 105, 307.

9. Sumner, M.A., Bradley, T.R., Hodgson, G.S., Cline, M.J., Fry, P.A. and Sutherland, L., 1972, Br. J. Haematol. 23, 221.

10. Golde, D.W. and Cline, M.J., 1973, Blood 41, 45.

11. Bradley, T.R., Stanley, E.R. and Sumner, M.A., 1971, Aust. J. Exp. Biol. Med. Sci. 49, 595.

12. Golde, D.W. and Cline, M.J., 1972, J. Clin. Invest. 51, 2981.

13. Golde, D.W., Byers, L.A. and Cline, M.J., 1974, Cancer Res. 34, 419.

14. McCulloch, E.A., Thompson, M.W., Siminovitch, L. and Till, J.E., 1970, Cell Tissue Kinet. 3, 47.

15. Metcalf, D., Furth, J. and Buffett, R.F., 1959, Cancer Res. 19, 52.

16. Rossi, G.B. and Friend, C., 1970, J. Cell. Physiol. 76, 159.

17. Horoszewicz, J.S., Leong, S.S. and Carter, W.A., 1975, J. Natl. Cancer Inst. 54, 265.

18. Liao, S-K. and Axelrad, A.A., 1975, Intern. J. Cancer 15, 467.

*Progress in Differentiation Research, ed. N. Müller-Bérat et al.* 


# ERYTHROID COLONY GROWTH IN CULTURE: EFFECTS OF DESIALATED ERYTHROPOIETIN, NEURAMINIDASE, DIMETHYL SULFOXIDE, AND AMPHOTERICIN B

Fritz Sieber

Friedrich Miescher-Institut, P.O. Box 273, CH-4002 Basel, and Institut für Zellbiologie, ETH, CH-8049 Zürich, Switzerland

## 1. Introduction

Erythroid and granulocyte/macrophage progenitors can be detected in freshly explanted mouse bone marrow by their capacity to proliferate and form morphologically recognizable colonies in culture. "Erythropoietin-dependent burst forming units" (BFU-E), apparently primitive members of the erythroid pathway, give rise to colonies ("bursts") which reach macroscopic dimensions after 10 days in culture[1]. "Erythropoietin-dependent colony forming units" (CFU-E) are more mature cells which form small erythroid clusters after 36 hours in culture[2]. In the same culture system[2], granulocyte/macrophage progenitors ("colony forming units-culture" or "CFU-C") give rise to colonies in the presence of granulocyte colony stimulating factor (CSF) which reach maximum size after 7 days in culture.

The cell surface and in particular complex carbohydrates on the cell surface have been implicated in mechanisms controlling growth and development. It is quite possible that the glycoprotein hormone erythropoietin also exerts its action at the surface of its target cells[3]. The effects of desialated erythropoietin and neuraminidase added to the culture medium were accordingly examined. In addition, the interaction of erythropoietin with its target cells in the presence of dimethyl sulfoxide (DMSO) was studied. This agent induces differentiation of Friend virus-transformed mouse erythroleukemia cells[4], and there is evidence that it may also exert its action at the cell surface[5].

## 2. Materials and Methods

Human erythropoietin was extracted from pooled urine from severely anemic patients by adsorption to benzoic acid[6] and further purified by ion exchange chromatography on DEAE-cellulose[1] and gel filtration on Sephadex G-100[2]. This preparation (266 units/mg protein) contained CSF as an impurity. Murine plasma erythropoietin from severely anemic mice (courtesy of Dr. O. Gallien-Lartigue, INSERM, Orsay) was partially purified by gel filtration on Sephadex G-100.

Erythropoietin was desialated by mild acid hydrolysis[7]. Equal volumes of the same erythropoietin solution heated in distilled water instead of 0.03 M HCl served as controls. The neuraminidase added to the culture medium was a commercial preparation (Serva) from Vibrio cholerae with a specific activity of ~32,000 units/mg protein (1 unit defined as the amount of enzyme required to liberate 1 µg of N-acetylneuraminic acid from a human serum glycoprotein substrate in 15' at 37 °C). The enzyme solution was dialyzed for 4 hours against culture medium immediately before addition to the culture. No protease activity was detected in this neuraminidase preparation using Azocoll (Calbiochem) as a substrate.

Amphotericin B was either obtained as a prepared solution (GIBCO) or in powder form (Calbiochem). The latter was dissolved in 0.1 ml of DMSO/mg of dry powder.

Bone marrow cells were taken from the femurs of adult male C57BL or BDF1 mice. $10^5$ nucleated cells were plated in 35 mm Petri dishes containing 1 ml of culture medium. Culture methods, scoring of colonies, and standardization of erythropoietin have been described previously[1,2].

## 3. Results

### Effect of desialated erythropoietin

Fig. 1 shows that CFU-E and BFU-E responded better to non-saturating doses of chemically desialated human urinary erythropoietin than to respective doses of the untreated control. The number of colonies obtained at saturating hormone concentrations remained unchanged.

### Effect of neuraminidase in culture

When neuraminidase was continuously present in the culture medium, thus exposing both hormone and cell surfaces (as well as sub-

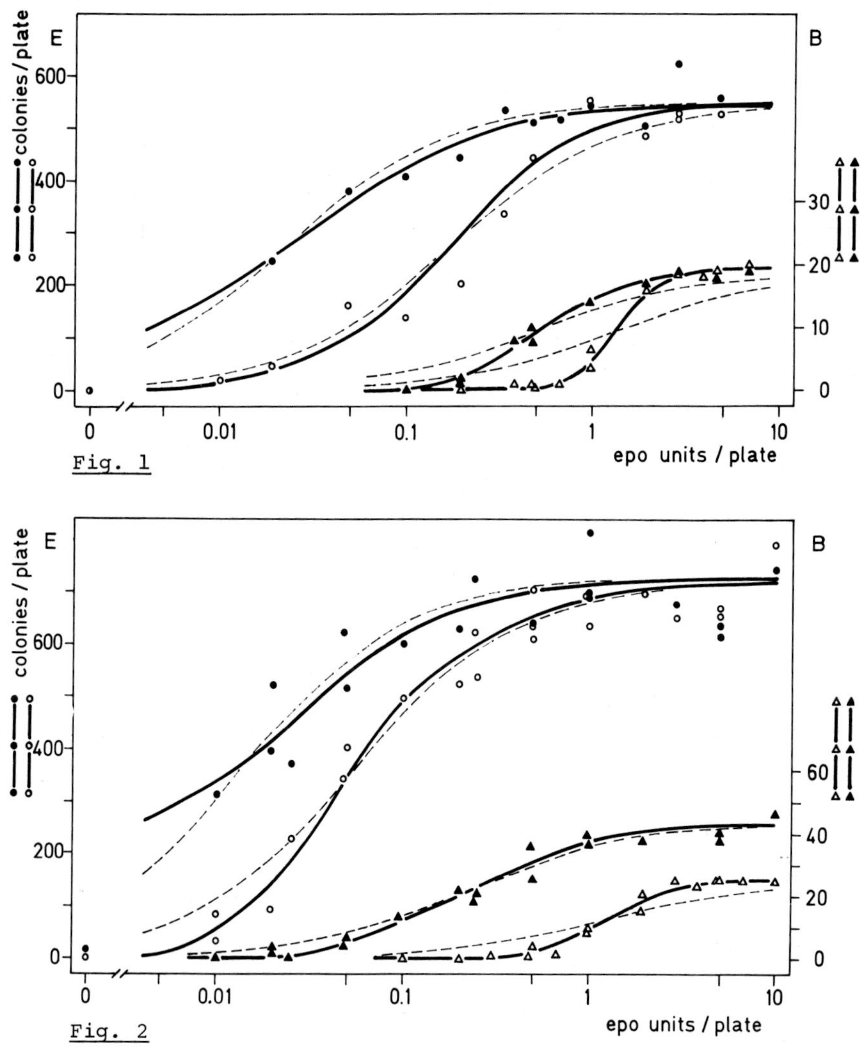


Fig. 1 : Relationship between dose of control (●▲) or chemically desialated (●▲) erythropoietin and number of erythroid clusters (E) and bursts (B) formed per $10^5$ nucleated bone marrow cells.

Fig. 2 : Relationship between dose of erythropoietin and number of erythroid clusters (E) and bursts (B) formed per $10^5$ nucleated bone marrow cells in the presence (●▲) and absence (○△) of 40 units of neuraminidase per culture plate.

(-----) : Langmuir isotherms

strates in the serum) to desialation, the plating efficiency of BFU-E and CFU-E was again higher at non-saturating doses of erythropoietin (Fig. 2). At saturating erythropoietin concentrations the number of erythroid colonies at 2 days remained unchanged, but in contrast to the experiment described in Fig. 1, the plating efficiency of BFU-E at saturating erythropoietin concentrations was increased (Fig. 2). The results of Fig. 2 were reproduced in a separate experiment, when human urinary erythropoietin was replaced by mouse erythropoietin. Addition of neuraminidase to a culture containing chemically desialated human erythropoietin additionally enhanced the plating efficiency of BFU-E, but not of CFU-E. The effect of neuraminidase on the dose-response curve for CFU-E plateaued at 10 neuraminidase units/ml. The highest number of bursts (2.2-fold enhancement) was obtained in the presence of saturating amounts of erythropoietin and 800 units of neuraminidase/ml (toxic for CFU-E). The routinely used enzyme concentration of 40 units/ml was not toxic for CFU-E and increased the number of bursts 1.7-fold at saturating erythropoietin levels. Preincubation of bone marrow cells with neuraminidase for 15'-60' did not enhance their sensitivity to erythropoietin.

Effects of dimethyl sulfoxide and Amphotericin B

The sensitivity of CFU-E to non-saturating doses of erythropoietin was enhanced in the presence of DMSO in the culture medium, but the number of colonies obtained at saturating hormone concentrations remained unchanged (Fig. 3). In contrast, growth of granulocyte/macrophage colonies and bursts was inhibited by DMSO (Fig. 3,4).

Etiocholanolone and allyl isopropyl acetamide are, like DMSO, able to induce hemoglobin synthesis in Friend virus-transformed cells[8]. However, in the present system they failed to stimulate erythropoietin-independent colony growth, and only etiocholanolone had a slight (~10 %) enhancing effect on added erythropoietin.

Amphotericin B is a synthetic polyene antibiotic which increases membrane permeability by the formation of complexes with cholesterol, leading to the formation of distinct pores[9,10]. Amphotericin B antagonized DMSO by reducing to normal the response of CFU-E observed in the presence of 1 % DMSO (Table I). Size and appearance of these colonies were unchanged. Amphotericin B only partially antagonized

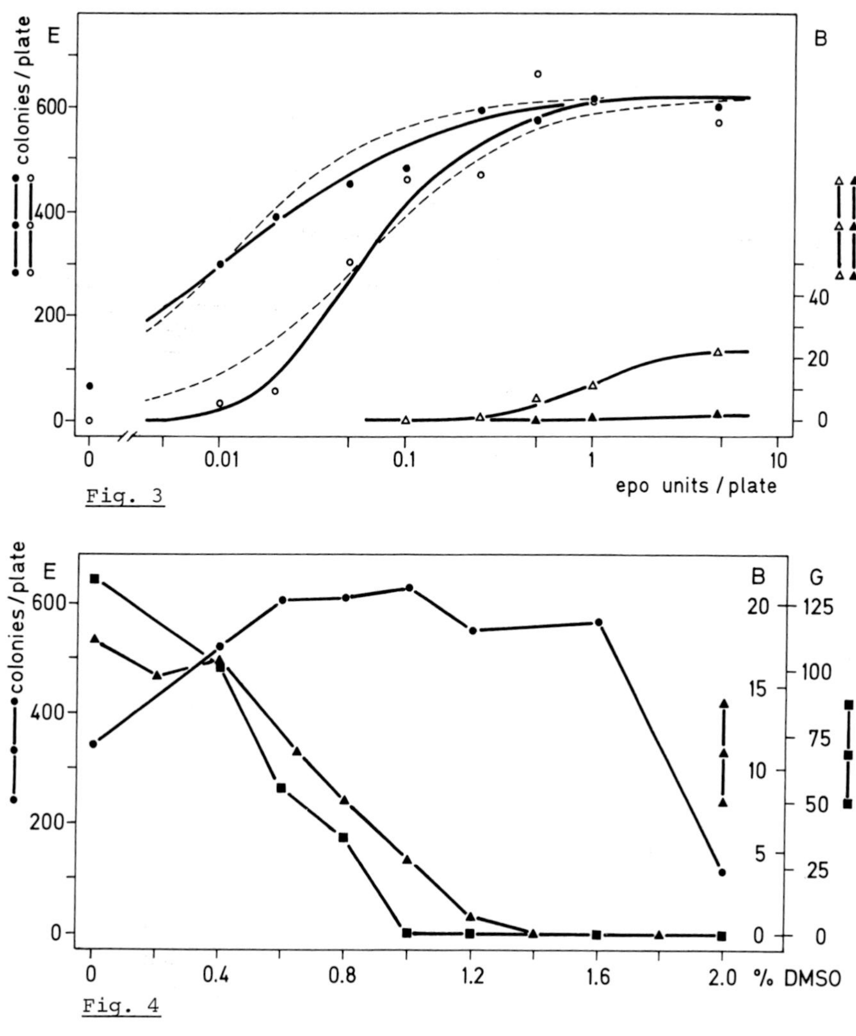


Fig. 3 : Relationship between dose of erythropoietin and number of erythroid clusters (E) and bursts (B) formed per $10^5$ nucleated bone marrow cells in the presence (●▲) and absence (o△) of 1 % DMSO. (– – – –) : Langmuir isotherms

Fig. 4 : Relationship between DMSO concentration and number of erythroid clusters (E), bursts (B), and granulocyte/macrophage colonies (G) formed per $10^5$ nucleated bone marrow cells, in the presence of a constant amount of erythropoietin (0.04 units/ml for E; 3.9 units/ml for B) and CSF.

Table I : Effect of Amphotericin B on growth of CFU-E, BFU-E, and CFU-C

| | colonies/$10^5$ | % |
|---|---|---|
| | CFU-E | |
| erythropoietin (0.05 units/ml) | 78 | 100 |
| erythropoietin + Amphotericin B (4 µg/ml) | 69 | 88 |
| erythropoietin + DMSO (1 %) | 171 | 219 |
| erythropoietin + DMSO + Amphotericin B | 93 | 119 |
| | BFU-E | |
| erythropoietin (5 units/ml) | 16 | 100 |
| erythropoietin + Amphotericin B (4 µg/ml) | 28 | 175 |
| erythropoietin + DMSO (1 %) | 1 | 6 |
| erythropoietin + DMSO + Amphotericin B | 5 | 31 |
| | CFU-C | |
| CSF | 204 | 100 |
| CSF + Amphotericin B (4 µg/ml) | 227 | 111 |
| CSF + DMSO (1 %) | 0 | 0 |
| CSF + DMSO + Amphotericin B | 139 | 68 |

the inhibitory effect of DMSO on the growth of bursts, but it enhanced the response of BFU-E to a saturating amount of erythropoietin by 75 % (Table I). Bursts grown in the presence of DMSO were considerably smaller than normal. The dose of Amphotericin B required to restore growth of granulocyte/macrophage colonies in the presence of DMSO (Table I) was in the toxic range; colony size was reduced and with certain lots of Amphotericin B the effect could not be demonstrated.

## 4. Discussion

Incorporation of radioiron into heme of bone marrow cells is enhanced in the presence of neuraminidase, and it has been proposed that this may be due to a direct effect of the enzyme on the cells[11] or alternatively to enhanced potency of desialated erythropoietin[7]. The present results show that mouse CFU-E and BFU-E are more sensitive to desialated human erythropoietin than to the native form.

This might be explained on the basis of enhanced binding of the hormone to a receptor (due to reduced negative charge, altered steric configuration or less rigid structure) or of increased efficacy of the desialated hormone. This observation may have little relevance in the intact animal, where desialated erythropoietin is rapidly removed from the circulation[7]. The additional bursts that grew in the presence of neuraminidase at saturating erythropoietin levels are more difficult to explain. One possibility is that some primitive erythroid progenitors respond to erythropoietin only after their surface has been desialated (unmasking of receptors). The present state of the experimental system does not permit clear experimental distinction between this suggestion and other possibilities such as activation or inactivation of serum components or cell products or effects of impurities in the enzyme preparation.

DMSO failed to make proliferation of normal mouse bone marrow cells erythropoietin-independent, but it enhanced the response of CFU-E to non-saturating doses of erythropoietin. Similar results have been obtained with normal mouse fetal liver cells[12]. Amphotericin B antagonized the effect of DMSO on proliferation of CFU-E, BFU-E and CFU-C and enhanced the sensitivity of BFU-E to erythropoietin. These observations suggest that control of proliferation and differentiation of hemopoietic cells may be a membrane-mediated process. They lend support to the suggestion that DMSO induces differentiation of Friend erythroleukemia cells by decreasing the permeability of the cell membrane[5].

If erythropoietin reacted with identical, independent receptors on a homogeneous cell population, and if the biological effect (here, colony formation) were a linear function of the number of occupied sites, dose-response curves should follow simple Langmuir isotherms. Dose-response curves for CFU-E deviated onls slightly from a Langmuir isotherm (Fig. 1-3), but dose-response curves for BFU-E were distinctly "cooperative" (Fig. 1,2). The inherent scatter of the bioassay made it impossible to quantify this "cooperative" response to erythropoietin by accurate Hill-coefficients. Despite this scatter it can be seen that neuraminidase and desialated hormone reduced the apparent cooperativity in the dose-response curves. The observations that desialation of the hormone and addition of

DMSO to the culture medium caused the same shift in the dose-response curve of CFU-E, and that these effects were not additive, suggest that the two agents may exert their potentiating effect through a common underlying mechanism.

References

1. Iscove, N.N. and Sieber, F., 1975, Exp. Hemat. 3, 32.
2. Iscove, N.N., Sieber, F. and Winterhalter, K.H., 1974, J. Cell. Physiol. 83, 309.
3. Chang, S.C.-S., Sikkema, D. and Goldwasser, E., 1974, Biochem. Biophys. Res. Commun. 57, 399.
4. Friend, C., Scher, W., Holland, J.G. and Sato, T., 1971, Proc. Nat. Acad. Sci. USA 68, 378.
5. Dube, S.K., Gaedicke, G., Kluge, N., Weimann, B.J., Melderis,H., Steinheider, G., Crozier, T., Beckmann, H. and Ostertag, W., 1973, in: Differentiation and control of malignancy of tumor cells, eds. W. Nakahara, T. Ono, T. Sugimura and H. Sugano (University Park Press, Baltimore, London, Tokyo) p. 99.
6. Espada, J. and Gutnisky, A., 1970, Biochem. Med. 3, 475.
7. Goldwasser, E., Kung, C.K.-H. and Eliason, J., 1974, J. Biol. Chem. 249, 4202.
8. Ebert, P.S. and Ikawa, Y., 1974, Proc. Soc. Exp. Biol. Med. 146, 601.
9. De Kruijff, B., Gerritsen, W.J., Oerlemans, A., Demel, R.A. and Van Deenen, L.L.M., 1974, Biochim. Biophys. Acta 339, 30.
10. De Kruijff, B. and Demel, R.A., 1974, Biochim. Biophys. Acta 339, 57.
11. Lukowsky, W.A. and Painter, R.H., 1972, Can. J. Biochem. 50,909.
12. Conkie, D. and Harrison, P.R., personal communication.

Acknowledgement

This communication is part of a Ph. D. thesis at the Swiss Federal Institute of Technology in Zürich. I am grateful to my supervisor Prof. H. Ursprung, and to Drs. H. Eppenberger, N. Iscove and L. Guilbert for their interest and help. I am particularly indebted to Dr. D. Turner, whose suggestion it was to apply current concepts of drug action to the analysis of interactions between erythropoietin and its target cells.

*Progress in Differentiation Research, ed. N. Müller-Bérat et al.*



# MACROMOLECULES CODED FOR BY THE MAJOR HISTOCOMPATIBILITY COMPLEX

Per A. Peterson, Lars Östberg, and Lars Rask

Department of Medical and Physiological Chemistry, Biomedical Center, University of Uppsala, Uppsala, Sweden

For a long period of time, it has been well recognized that organ transplantation between non-syngeneic individuals will result in rejection of the transplant. As long ago as in 1937 Gorer (1) demonstrated that a cell surface antigen was the most important single contributing factor for the rejection. Later studies showed, that the cause of rejection of the transplant was more complex than just the difference in a single cell surface antigen between the donor and the recipient. We know now that Gorer had discovered one antigen of many derived from a narrow genetic region which encompasses several distinct loci, all of which seem to participate in various immunobiological events.

This genetic region which constitutes the main barrier against tissue transplantation has been identified in almost all species examined. The most extensive studies have been performed in the mouse; the main reason being the existence of a number of congenic strains differing from each other only at the chromosomal segment bearing the major histocompatibility complex (MHC). In the mouse this genetic region is located on chromosome 17 (2) and is called H-2. In man its counterpart is termed HLA. Several excellent reviews have recently been published dealing with the MHC of man (3) and mouse (4,5).

Due to the availability of the H-2 congenic lines and recombinants derived thereof, it has been possible to construct quite a detailed gene map for the H-2 region. Fig. 1 shows that the H-2 complex contains four distinct loci called K, I, S, and D. The recombination frequency between the K- and D-ends is only about 0.5 %. Outside the D-end of the H-2 complex another locus is located, called Tla. Its distance to the D-end is about 1 map unit. This series of loci controls a variety of immune phenomena (see 3-5).

The serologically detected, classical alloantigens are controled by the K and D regions. The common belief is that these loci contain the structural genes for the classical transplantation antigens. The mixed leucocyte reaction (MLR) is governed by genes located in the I-region whereas the related cell-mediated lympholysis (CML) reaction is greatly dependent on the K and D loci. In fact, the target antigens of the latter reaction appears identical to the classical transplantation antigens. The mixed leucocyte reaction is commonly believed to be an *in vitro* correlate to the graft-versus-host reaction (GvHR) which accordingly is

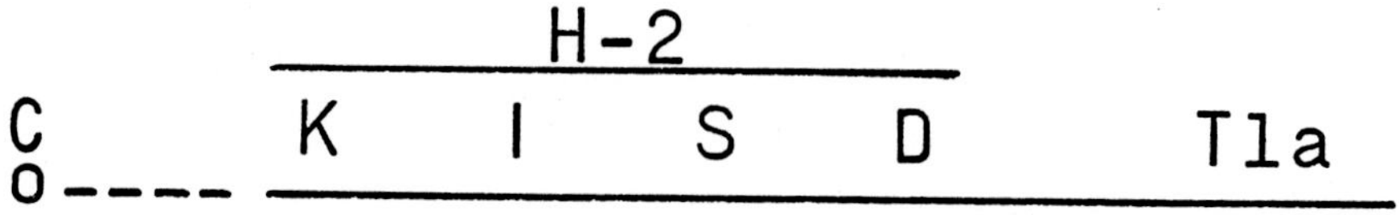

H-2

C K I S D Tla

o----

Fig. 1. Loci present on the 17th mouse chromosome related to the major histocompatibility complex. C denotes the centromere.

controled by the I-region with contributions from the K and D loci.

Since in MLR and GvHR cell surfaces seem to be engaged, it is apparent that the I-region controls the expression of membrane molecules. These have now been identified and are called Ia antigens (see 5). The I-region is not only involved in the GvHR but controls the immune response to most thymus-dependent antigens as well. The existence of such immune response genes has led Benacerraf and McDevitt to suggest that the antigen receptor on T-lymphocytes is the product of the I-region (6). Also another aspect of the immune response is under I-region control. Normally, T- and B-lymphocyte collaboration is needed to elicit antibody production. This collaboration is greatly hindered unless there is I-region identity (7).

The I-region has also been implicated in susceptibility to Gross virus leukaemogenesis as well as to tumorigenesis due to other RNA tumor viruses (8). The responsible factor seems to influence the host's immune response to the antigens associated with virus infection.

The S-region controls the quantitative expression of a plasma protein, the Ss-protein, which exhibits testosterone-regulated allotypic variation (9). In addition, the serum complement level is dependent on the S-region genotype (10).

The Tla locus which really is situated outside the H-2 region proper controls the expression of a series of allelic cell surface proteins (11), normally present only on thymocytes during a certain stage of their differentiation. Quite frequently leukemia cells also express these antigens. Therefore, they are tumor antigens as well as differentiation antigens.

Structural features of MHC-coded gene products. It has become increasingly apparent that the MHC contains structural genes and a considerable amount of work has been performed to elucidate the structure of these molecules. Scarcity of material, the high degree of polymorphism, and difficulties in solubilizing the cell membrane-bound molecules have been major obstacles in defining even the gross anatomy of these antigens. However, in a series of experiments Nathenson and co-workers (12) have devised reproducible methods to solubilize the MHC-coded proteins. In addition, the availability of congenic mouse strains and recombinants derived thereof has allowed the production of seemingly monospecific antisera directed against single molecular species derived from the MHC. With use of such antisera and the solubilization procedures devised it is now relatively simple to obtain small amounts of endogeneously or exogeneously labelled, highly purified MHC coded antigens by immunoprecipitation techniques. However, since analyses of immune precipitates usually have to be performed under denaturing conditions to abolish the antigen-antibody interaction, tests for conformation or even biological activity of the isolated molecules are not readily achieved. From the foregoing it is obvious that major technical obstacles remain to be solved before a full understanding of the structure-function relationship for the MHC antigens can be reached. The emerging picture of the biochemistry of the MHC molecules (see below) is exciting and although much tedious work is needed, the biological importance of the major histocompatibility complex promises to unravel new and fundamental structural findings.

The following discussion will focus upon certain structural aspects of the MHC-coded macromolecules. The mouse H-2 system has provided most of the information, but where knowledge is available the human HLA-region coded antigens will be mentioned.

The H-2K and D gene products. The classical transplantation antigens are cell surface glycoproteins. The earliest employed, reproducible method to solubilize these molecules consists of controlled proteolytic digestion of crude membrane fractions (see 12). This method yields water-soluble polypeptide fragments which carry all or most of the particular alloantigenic determinants. Radioactive labelling of the solubilized macromolecules followed by specific sandwich-immunoprecipitation of the H-2K and D-antigens with use of specific antisera provide a simple, one-step isolation procedure. Analysis of the precipitated transplantation antigens on SDS-polyacrylamide gel electrophoresis usually reveals the presence of two polypeptide chains. As can be seen in Fig. 2 the immuno-precipitated polypeptide chains are of different sizes. The small subunit is invariant regardless of whether H-2K or D-antigens have been isolated, whereas the larger polypeptide chain carries the alloantigenic markers (13-17). Accordingly, it is the larger chain which is coded for by the H-2K and D-loci and whose specificity seemingly is involved in the CML reaction. The small chain has attracted attention for quite another reason. It was established that this chain is identical to $\beta_2$-microglobulin (13-17). Following the determination of the amino acid sequence of $\beta_2$-microglobulin (18-20) it was realized that its primary structure is as similar to anyone of the immunoglobulin G domains as they are to one another. Based on this finding it was suggested that $\beta_2$-microglobulin had evolved from the same primordial gene as that giving rise to regular immunoglobulin light and heavy chains (19).

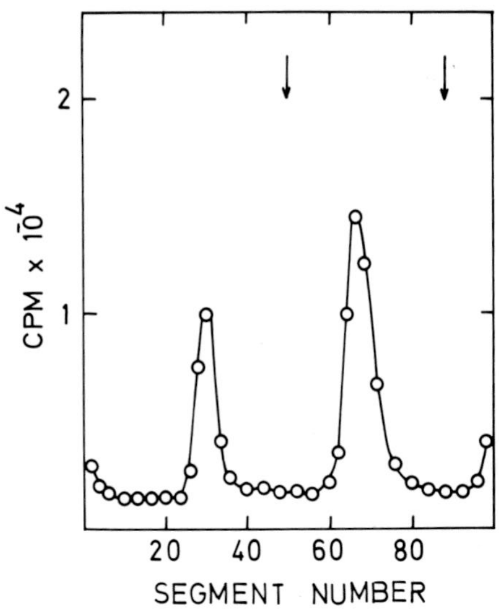


Fig. 2. SDS-polyacrylamide gel electrophoresis of $^{125}$I-labelled H-2K antigen solubilized by papain treatment and isolated by indirect immunoprecipitation. The anode is to the left.

Solubilizing transplantation antigens with use of proteolytic digestion is supposed to yield less than the intact molecule. Therefore, it appeared of interest to examine the detergent solubilized antigens, since it seemed possible that such molecules should be more representative of the whole structure. With various techniques and under physiological as well as denaturing conditions, such analyses were performed. Molecular weight estimations, summarized in Table I, suggest that the intact H-2 antigens under physiological conditions display an apparent weight of somewhat less than 130,000. On denaturation of this molecule two main polypeptide species are apparent. The small subunit, with an apparent molecular weight of about 12,000, is identical to $\beta_2$-microglobulin, whereas the larger component has a molecular weight of about 100,000. In several analyses a minor component with a molecular weight of about 50,000 is present. Similar findings have also been recorded in Nathenson's laboratory (21). On reduction of the 100,000-dalton

Table I

Molecular weight determinations of H-2 alloantigens

| Solubilization procedure | Method | Treatment | Molecular weight |
|---|---|---|---|
| Detergent | Sedimentation-Diffusion | - | 122,000 |
| " | SDS-PAGE | - | 96,000 and 12,000 |
| " | SDS-PAGE | Reduction | 50,000 and 12,000 |
| Proteolysis | Sedimentation-Diffusion | - | 51,000 |
| " | SDS-PAGE | - | 36,000 and 12,000 |
| " | SDS-PAGE | Reduction | 37,000 and 12,000 |

component under denaturing conditions, the molecular weight is reduced to half, suggesting that two very similar polypeptide chains are held together by one or more disulfide bridges. It is at present impossible to clearly establish whether the inter-chain disulfide bonds are created fortuitously due to the solubilization conditions, leaving a small portion of the larger subunit in its native form (i.e. the 50,000-dalton species noted in the absence of reducing agents), or if the free 50,000-dalton component is an artifact due to the presence of a very labile inter-chain disulfide bridge. One may speculate that the transplantation antigens are indeed tetrameric proteins on the cell surface since the mixed type of dimers, K- and D-coded or even K- and K-coded from different haplotypes in the heterozygous state, have not yet been found. Further structural analyses are, however, needed to resolve this question.

With the above mentioned reservations in mind it appears that the classical transplantation antigens coded for by the K- and D-loci may be composed of two molecules of $\beta_2$-microglobulin and two molecules of the larger, alloantigenic polypeptide chain. On limited proteolytic digestion of this molecule, fragments displaying all or most of the alloantigenic determinants are recovered. Each fragment has an apparent molecular weight of about 50,000 under physiological conditions and is composed of two types of subunits. Intact $\beta_2$-microglobulin is part of the structure and the remaining portion comprises two-thirds of the heavy chain (see Table I). It is important to note that this 37,000-dalton fragment carries all the alloantigenic determinants, suggesting that it represents the outer part of the molecule. Since no interchain disulfide bonds exist in the fragments recovered after limited proteolysis it seems reasonable to conclude that the reactive half-cystines are present in the part of the heavy, alloantigenic chain which lies in closest proximity to the cell membrane.

Also for the human HLA antigens data have been provided which suggest a tetrameric structure (22-25). In the human system three structural loci coding for classical transplantation antigens have been discovered (see 3). The subunit composition of these antigens is very similar (22,26). Since $\beta_2$-microglobulin is the small subunit for all molecules, at least the part of the structure of the various heavy chains which is engaged in binding the small subunit has to be very similar if not identical. In a recent study we have obtained data suggesting that the similarity may be more extensive than so. With use of a heteroantiserum directed against HLA antigens it could be shown that molecules derived from the three subloci to a great extent shared antigenic determinants. Furthermore, cyanogen bromide fragmentation of the three different gene products gave rise to very similar peptide patterns. This study, therefore, provides evidence for the view that the three HLA antigen subloci have arisen by gene duplications from a common ancestral gene.

Since $\beta_2$-microglobulin generally appears to be expressed concomitantly with the alloantigenetic HLA polypeptide chain, it appeared of interest to examine if the two subunits are under separate genetic control. A unique opportunity to study this question is provided by the Daudi cell line which does not manufacture $\beta_2$-microglobulin (27). In fact, the Daudi cells display a deletion in chromosome 15, which appears to carry the structural gene for $\beta_2$-microglobulin (28). With use of heteroantibodies directed against HLA antigens it could be demonstrated that Daudi cells express normal or slightly less than normal amounts of the heavy, alloantigenic chain on their surfaces (29). The isolated HLA polypeptide chain did not contain $\beta_2$-microglobulin, as expected, nor any $\beta_2$-microglobulin-like subunit. In a number of laboratories, as well as in ours, it has been impossible to assess the HLA antigen phenotype of Daudi cells. Since all such analyses are dependent on alloantisera, the protein isolated with use of the heteroantiserum may represent a molecule exhibiting immunological crossreactivity with regular HLA antigens, rather than being a true HLA antigen. However, $\beta_2$-microglobulin may have to be bound to the heavy HLA antigen subunit to allow a proper configuration of the alloantigenic determinants, which recently has been suggested (30). Since the HLA antigen-like molecules isolated from Daudi cells bind $\beta_2$-microglobulin, it seems reasonable to conclude that $\beta_2$-microglobulin is not under the same genetic control system which governs the expression of the HLA antigen region genes (see below).

A tetrameric structure for the classical transplantation antigens and $\beta_2$-microglobulin displaying an immunoglobulin domain-like structure provide suggestive evidence for the existence of structural similarities between immunoglobulins and transplantation antigens. To obtain further evidence for this possibility it appeared mandatory to examine the heavy, alloantigenic HLA chain in some detail. There are two outstanding features of the immunoglobulins which possibly could be shared with the transplantation antigens. First, the regular domain structure of the immunoglobulin heavy and light chains (31) could have its counterpart in the transplantation antigens especially since $\beta_2$-microglobulin, which represents a free domain, interacts with the heavy transplantation antigen chain. It should be noted that all immunoglobulin domains occur in pairs. Furthermore, limited proteolytic digestion of the immunoglobulins results in cleavage of the polypeptide chains in those extended stretches connecting the domains. In this context it is worth recalling that the detergent solubilized transplantation antigen heavy chain is reduced in weight by approximately 10,000 to 15,000 on limited proteolysis, i.e. the loss equals the weight of an immunoglobulin domain. Secondly, in each immunoglobulin domain there is a single disulfide bridge encompassing about 65 amino acid residues (31). $\beta_2$-Microglobulin displays the same type of disulfide loop.

To examine these features papain-solubilized transplantation antigens were isolated. Such molecules, as pointed out above, do not represent the intact heavy chain. Limited proteolysis of the heavy chain fragment yielded two distinct products irrespective of whether trypsin, chymotrypsin, pepsin, papain, or thermolysin had been used. The molecular weights of the two proteolytically derived polypeptide chains were about 13,000 and 20,000, respectively. The 20,000-dalton component interacted with $\beta_2$-microglobulin and carried most if not all of the carbohydrate. By indirect techniques suggestive evidence were obtained that the 13,000-dalton component contained an immunoglobulin-like disulfide loop. Due to the presence of large amounts of carbohydrate in the 20,000-dalton fragment the data for the existence of a disulfide bond in this part of the molecule is as yet less convincing. Since the carbohydrate portion may represent as much as half to one-third of the weight of the 20,000-dalton component its polypeptide portion seems, however, to be of a size similar to the immunoglobulin domains.

Based on these and similar analyses we have suggested that the transplantation antigen molecule may be very similar in structure to the constant part of the immunoglobulin G molecule (25). Fig. 3 depicts this model. Much detailed work is needed to establish the structural features of the transplantation antigens. If a structure similar to that in Fig. 3 will turn out to be the correct one, it appears reasonable to suggest that there has been an interrelated evolution for transplantation antigens and immunoglobulins. The degree of similarity will undoubtedly be revealed by amino acid sequence determinations. However, regardless of the outcome of such studies it appears highly likely that at least the $\beta_2$-microglobulin site of the 20,000-dalton domain will be similar to the site making contact between the heavy and light immunoglobulin chains. This anticipation is based upon the knowledge that amino acid residues creating this site have been conserved in the $\beta_2$-microglobulin structure (32).

The TL antigens. In close association to the H-2 region, a compound locus coding

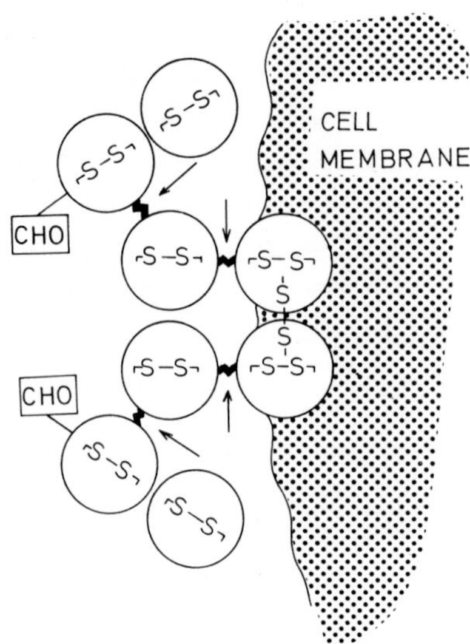


Fig. 3. Schematic representation of the proposed structure for H-2 and HLA alloantigens.

for the thymus leukemia antigens is located. These antigens are normally expressed on the surface of thymocytes of some strains of mice. At least three alleles coding for certain antigenic specificities are known for the Tla locus. The antigenic specificities are also sometimes expressed on leukemia cells, also in strains whose thymocytes are TLa-negative. The expression of TL antigens on the cell membrane can be modulated in the presence of specific anti-TLa antibodies such that TLa-positive cells may become phenotypically TLa-negative. Excellent reviews on the expression and genetics of the TL antigens have been published (33,34).

Previous work had suggested that TL antigens, solubilized by proteolysis, are very similar to H-2 antigens (36,37). To extend this information to the intact molecule we have recently isolated and partially characterized detergent-solubilized TL antigens (36). Under physiological conditions such antigens consist mainly of molecules of a size similar to immunoglobulin G. A single component of TLa was apparent on sucrose density gradient ultracentrifugation of solubilized thymocyte membrane macromolecules as monitored by indirect immunoprecipitation. The sedimentation constant for the TL antigens, 5.8 S, is considerably less than that for immunoglobulin G. The combined gel chromatography and ultracentrifugation data suggest an apparent molecular weight for TL antigens of about 120,000. The TL antigens are composed of two types of polypeptide chains. The smaller subunit has been identified as $\beta_2$-microglobulin (36-38). The larger chain, which carries the alloantigenic determinants, displays a molecular weight of slightly less than 50,000 after reduction. Without reduction most of the larger chain material exhibits an apparent molecular weight of about 100,000, suggesting that most of the molecules are pairwise linked by one or more disulfide bonds. Consequently, the TL antigen molecule is composed of two disulfide-linked heavy chains and two $\beta_2$-microglobulin subunits.

Limited proteolytic digestion of the detergent solubilized 120,000-dalton component yields fragments with an apparent molecular weight of about 50,000. Such fragments contain part of the heavy chain but all of the alloantigenic determinants and $\beta_2$-microglobulin.

Current knowledge is consistent with the view that H-2 alloantigens and TL antigens have similar molecular characteristics. Both are tetrameric molecules and contain $\beta_2$-microglobulin as one of the two types of subunits. Although the H-2 and TL alloantigenic polypeptide chains are immunologically distinct they display similar molecular weights. Furthermore, both polypeptide chains are digested by papain similarly and, upon such treatment, yields proteolytic fragments of indistinguishable size. It is commonly believed that the H-2 complex arose by events of gene duplications (39). The present observations would suggest that also the Tla locus may have arisen by the same process.

The Ss protein. Among the serveral distinct loci of the major histocompatibility complex only one seems to control a molecule normally not integrated into the cell surface. This exception is provided by the S region. This genetic segment controls the level of a plasma protein, the Ss-protein (40), but regulates the hemolytic complement level as well (10). This has prompted the suggestion that the Ss-protein may be a complement component (10,41).

We have recently isolated the Ss-protein and raised a monospecific rabbit antiserum against this component (42). A radioimmunoassay was developped with use of this antiserum and it was shown that male mice contained about twice as much of the Ss-protein as did female mice regardless of strain. Furthermore, testosterone-treated females increased their plasma concentration of the Ss-protein to the male level. These data strongly support the view that the antiserum is indeed directed against the Ss-protein, since Shreffler and Passmore have noted similar findings earlier(9).

Preliminary investigations of the isolated Ss-protein suggested that it did not represent the entire molecule. Consequently attempts were made to isolate the protein as rapidly as possible to minimize break-down during the various fractionation steps. Therefore, purification was achieved in a single step with use of an immunosorbent column. A single molecular species with an apparent molecular weight of slightly less than 200,000 was obtained from such an anti-Ss protein column. On reduction and alkylation under denaturing conditions, the Ss-protein resolved into three distinct polypeptide chains. Two of the chains appeared in the 70,000 to 80,000-dalton region on gel filtration in 6M guanidine hydrochloride. The third chain displayed an apparent molecular weight of about 23,000. All chains are held together by disulfide bonds. It seems reasonable to conclude that this molecule probably represents the intact Ss-protein. The Ss-protein isolated by conventional techniques most likely represents proteolytically degraded forms of this molecule.

The structure proposed for the Ss-protein is indeed very similar to the human complement component C4 (43). The only obvious difference between the Ss-protein and human C4, noted in our studies, is in the size of the small polypeptide chain, which is considerably larger for the human molecule. This discrepancy in the polypeptide chain structure may be genuine but could also have arisen due to uncontroled proteolysis since serum contains a C4b inactivator (44).

Further proof for the similarity between human C4 and the murine Ss-protein was achieved by the finding that antiserum against human C4 crossreacted extensively with the Ss-protein and vice versa. It was also shown that mouse serum depleted of the Ss-protein had lost all its C4-complement activity. From these and similar analyses it is apparent that the Ss-protein is the murine equivalent of complement component C4 (42). Similar results have independently been obtained in other laboratories (45-47).

The present data reinforces the importance of the H-2 region in various immunobiological systems. In addition to the long recognized role of the major histocompatibility complex in cell mediated immunity, it now appears evident that it is involved in an effector function of the humoral immunity system as well. Not only C4 but C2 (48), C3 (49), and factor B of the alternate pathway (50) are regulated by this genetic region. Other components of the complement system may also be controled by this region, although as yet unknown.

The I-region defined antigens. The most complex locus of the H-2 system is the I-region. From a number of informative recombinant mice it has been possible to subdivide the I-region into several parts denoted A,B,C,E, and F. The MLR and the GvHR are controled by the entire I-region. It is particularly the stimulatory parts of these reactions which are governed by this region. Immune response regulating genes are present at least in the A and B parts.

Considering the immunobiological effects of the I-region, it was believed for some time that the I-region had to control the expression of cell surface antigens. Recently, this has been shown to be the case (see ref. 5), and there is intensive research in several laboratories to define the structure and biological role of such Immune response associated (Ia) antigens. These molecules are, however, hard to come by since they seem to be present in very low amounts. In addition, most antisera available seem to recognize antigens derived from several parts of the I-region. With an antiserum recognizing the A,C, and E-parts of the I-region, we obtained a mixture of molecules, as can be seen in Fig. 4. Although separation was achieved with regard to size only, at least three distinct components are resolved. The number of Ia-antigens reacting with this antiserum may, however, be much great-

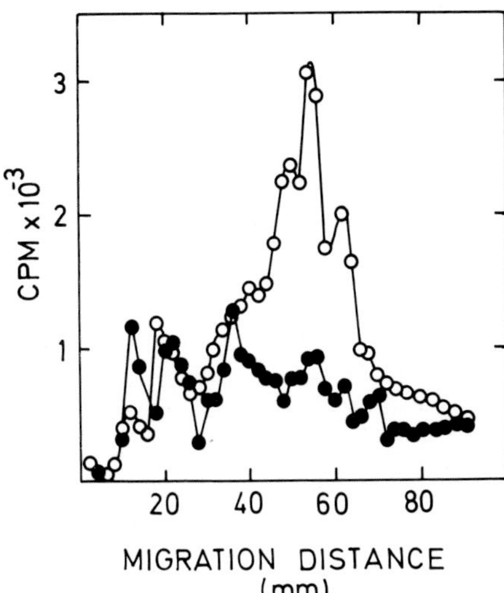


Fig. 4. SDS-polyacrylamide gel electrophoretic analysis of $^{125}$I-labelled, detergent-solubilized $Ia^{K}$-antigens isolated with use of a specific anti-Ia serum (o). In a control experiment the anti-Ia serum was replaced by normal mouse serum (•).

er than that.

Neither in our laboratory nor elsewhere has any evidence for the association of Ia-antigens and $\beta_2$-microglobulin been recorded with use of the above described type of antisera. However, indirect evidence suggests that there may exist I-region defined molecules which contain $\beta_2$-microglobulin. This may be inferred from studies of the MLR reaction. Usually, B-lymphocytes are the best stimulator cells in this reaction, whereas T-cells always are the responder cells. If B-lymphocytes are coated with antibodies against $\beta_2$-microglobulin a perfectly normal MLR will be obtained. On the other hand, if antibodies against Ia-antigens react with the stimulator cell, no or at most a very weak MLR response is recorded. These data are consistent with the view that the MLR-stimulating macromolecules are Ia-antigens which do not contain $\beta_2$-microglobulin. However, on treating the responder T-cells with antibodies or antibody fragments against $\beta_2$-microglobulin the MLR response will be completely quenched (51). The kinetics of this effect makes it likely that the antibodies are really interfering with the recognition phase of the reaction (52). We do not know much about the putative T-cell MLR-receptor, but from our data it appears reasonable to conclude that this structure contains $\beta_2$-microglobulin.

Recently, Katz and his collaborators have provided direct evidence for $\beta_2$-microglobulin being associated with a T-lymphocyte-produced molecule containing Ia-antigenic determinants (53). It is not unlikely that in the near future additional Ia-antigens containing $\beta_2$-microglobulin will be defined.

Recently an important finding was reported by Dickler and Sachs (54). They discovered that antibodies against Ia antigens will block the binding of aggregated IgG to the Fc-receptor present on B-lymphocytes. To obtain a defined I-region controled antigen, we have recently isolated a murine spleen cell Fc-receptor (55). A series of distinct molecules were obtained with the largest component exhibiting an apparent molecular weight of about 65,000. This molecule is very sensitive to proteolysis and various tests suggested that the proteins displaying lower molecular weights may represent, at least partly, fragments derived from the 65,000-dalton component. So far all the antibody activity in anti-Ia antigen sera is defined by their cytotoxic activity. By various tests it could be ruled out that the isolated Fc-receptor carried any of the cytotoxically defined antigenic specificities (55). However, recently we have found that at least some anti-Ia sera contain antibodies that are not cytotoxic and preliminary evidence suggests that such antibodies will bind to the isolated Fc-receptor. It is, therefore, reasonable to conclude that the Fc-receptor probably is controled by the I-region, but it does not carry any of the "regular" Ia-antigenic specificities.

A most intriguing aspect of the I-region is its control of the immune response. In a series of studies Munro and Taussig (see 56) have shown that the I-region manufactures a specific antigen-binding molecule which may well be the elusive T-cell receptor. This factor interacts with specific structures on the B-lymphocyte, which thereby is triggered to become an antibody-producing cell. The structure on the B-cell surface which recognize the T-cell antigen-binding factor appears to be controled by the I-region (56). This provides the important conclusion that there exists distinct gene products of the I-region which display molecular complementarity. It would not be surprising if it turns out that much of the I-region (including adjacent loci) codes for a series of "lock-and-key" molecules, some of which are restricted in their expression only to certain types of cells.

Ontogeny of MHC antigens. $\beta_2$-Microglobulin is apparently associated with MHC gene products derived from several loci (K,D,I, and Tla). Since it is so ubiquitous, one may ask if there are other loci whose gene products are also physically linked with $\beta_2$-microglobulin. To answer this question we have examined early stages of the mouse embryo since it is well documented that the H-2 antigens do not appear until about the 7th day of gestation (for review, see 57). However, fluorescent-labelled antibodies against $\beta_2$-microglobulin stain the 4 1/2 -day embryo, *i.e.* the blastocyst. We have not found any detectable H-2 gene products on the blastocyst, so it seems reasonable to conclude that $\beta_2$-microglobulin is indeed associated with molecules which are derived from loci not belonging to the H-2-Tla complex. Preliminary evidence suggests that the blastocyst molecules containing $\beta_2$-microglobulin share several structural features with H-2 and TL alloantigens.

Vitetta and her colleagues (58) have recently shown that one antigen controled by the T-locus contain a $\beta_2$-microglobulin-like subunit as well as a heavy chain similar to that of H-2 antigens. The T-locus, which is located on the same chromosome as the H-2 complex (see 57), controls well defined stages of the ectodermal differentiation. It apparently codes for a series of cell surface molecules whose temporal expression is restricted to early stages of the embryo. It is therefore quite possible that the $\beta_2$-microglobulin present on the blastocyst is associated with one or more T-locus products.

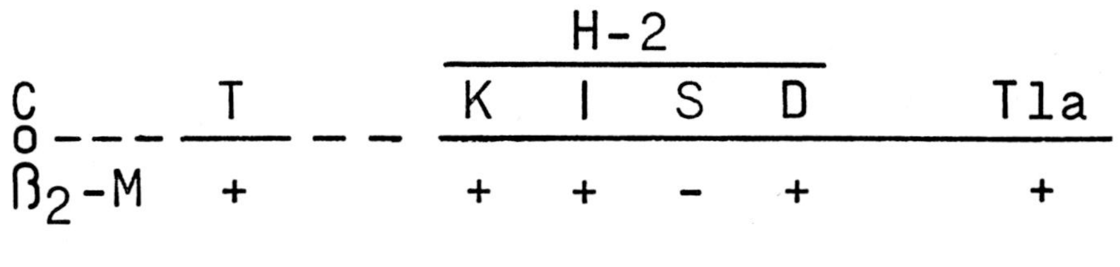

H-2

C T K I S D Tla
o
$\beta_2$-M + + + - + +

Fig. 5. Loci on the 17th mouse chromosome whose gene products may associate with $\beta_2$-microglobulin.

Evolutionary aspects of the MHC region. The present data demonstrates that $\beta_2$-microglobulin on the cell surface is associated with gene products coded for by several loci on the 17th mouse chromosome, as outlined in Fig. 5. It is reasonable to postulate that at least the $\beta_2$-microglobulin-binding portions of these molecules are very similar, since $\beta_2$-microglobulin is a small protein which probably can interact with other proteins only in a limited number of ways. Indeed, the common, gross structure of the K-, D-, I-, Tla-, and T-locus gene products suggest that even larger parts of the molecules may be similar. This similarity lends support to the view that all these loci may have evolved from a common ancestral gene, because not only the structure but also the function of these gene products may be related. Although the function of the K- and D-alloantigens is far from understood it may be suggested that they participate in a final stage of a series of differentiation events. This notion may gain some support from the fact that the TL and T antigens, which are true differentiation antigens, display reciprocity with regard to H-2 antigens in their cell surface expression.

The conspicuous similarity in structure between MHC-coded antigens and regular immunoglobulin suggests that these molecules may have had an interrelated evolution. A highly speculative scheme for such a relationship is depicted in Fig. 6. It is possible that already at the unicellular stage of evolution, a primordial gene coding for a polypeptide of about 100 amino acid residues was present. As soon as evolution took the step from unicellular to colony forming, multicellular organisms a cell sorting device had to appear. After gene duplications and subsequent diversification a higher level if discrimination may have been obtained so

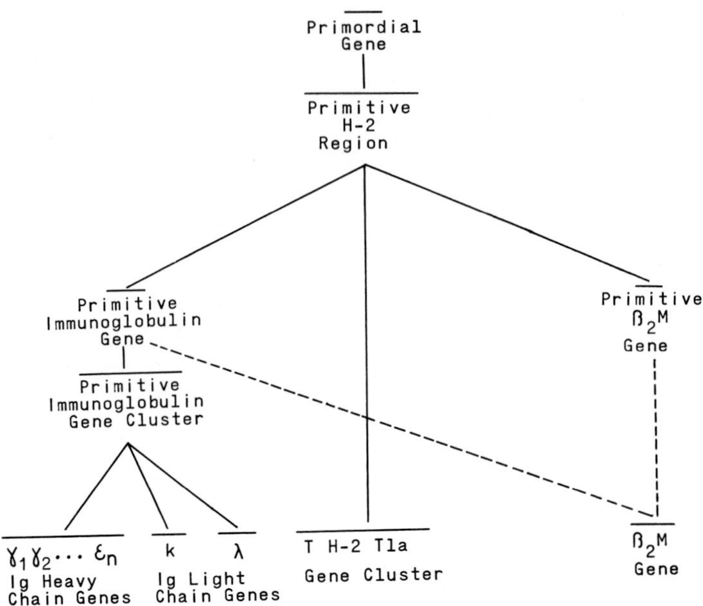


Fig. 6. Tentative schema for the evolution of the major histocompatibility complex and the immunoglobulins.

that like could be distinguished from unlike. Consequently, the appearance of a primitive cell mediated immunity system as well as the occurrence of some limited sort of differentiation antigen system may go all the way back to this stage. It is also possible that genetic material from this gene cluster has been translocated to other chromosomes to ultimately give rise to heavy and light immunoglobulin chains and $\beta_2$-microglobulin. A consequence of this scheme is that evolution of the proposed morphogenetic unit, part of the present-day 17th chromosome, has been under the guidance of various types of molecular interactions. The interaction of the antigen with the antigen-binding receptor on B- and T-lymphocytes could, thus, be but a specialized version of a more general type of recognition machinery in which the various H-2-Tla-T-loci products are engaged. That such an interacting system may exist is evident from findings that T-cell produced I-region molecules interact with B-lymphocytes (53,56). At least in one case it has been documented that the B-cell receptor is also under I-region control (56). It would therefore be of importance to extend these observations and examine if other MHC-gene products may also bind to receptors distinct from the regular antigen receptors present on immunocytes.


## ACKNOWLEDGMENTS

The excellent secretarial assistance of Ms C. Sjöholm is gratefully acknowledged. Work from the authors laboratory has been supported by grants from the Swedish

Cancer Society, the Swedish Medical Research Council, the Harald and Greta Jeanssons Fund, the Marcus Borgströms Fund, and Konung Gustaf V 80-års Fund.

## REFERENCES


1. Gorer, P.A. (1937) J. Path. Bact., 44, 691
2. Miller, D.A., Kouri, R.E., Dev, V.G., Grewal, M.S., Hutton, J.J., and Miller, O.J. (1971) Proc. Natl. Acad. Sci. USA, 68, 1530
3. Thorsby, E. (1974) Transpl. Rev., 18, 75
4. Démant, P. (1973) Transpl. Rev., 15, 162
5. Shreffler, D.C., and David, C.S. (1975) Advan. Immunol. 20, 125
6. Benacerraf, B., and McDevitt, H.O. (1972) Science, 175, 273
7. Katz, D.H., and Benacerraf, B. (1975) Transpl. Rev., 22, 175
8. Lilly, F., and Pincus, T. (1973) Advan. Cancer Res., 17, 231
9. Shreffler, D.C., and Passmore, H.C. (1971) In: Immunogenetics of the H-2 system. p. 56, Karger, Basel
10. Démant, P., Capková, J., Hinzová, E., and Vorácová, B.(1973) Proc. Natl. Acad. Sci. USA, 70, 863
11. Old, L.J., Boyse, E.A., and Stockert, E. (1963) J. Natl. Cancer Inst., 31, 977
12. Nathenson, S.G. (1970) Annu. Rev. Genet., 4, 69
13. Nakamuro, K., Tanigaki, N., and Pressman, D. (1973) Proc. Natl. Acad. Sci. USA, 70, 2863
14. Peterson, P.A., Rask, L., and Lindblom, J.B. (1974) Proc. Natl. Acad. Sci. USA, 71, 35
15. Grey, H.M., Kubo, R.T., Colon, S.M., Poulik, M.D., Cresswell, P., Springer, T., Turner, M.J. and Strominger, J.L. (1973) J. Exp. Med. 138, 1608
16. Rask, L., Lindblom, J.B. and Peterson, P.A. (1974) Nature, 249, 833
17. Silver, J. and Hood, L. (1974) Nature, 249, 764
18. Smithies, O. and Poulik, M.D. (1972) Science, 175, 187
19. Peterson, P.A., Cunningham, B.A., Berggård, I. and Edelman, G.M. (1972) Proc. Natl. Acad. Sci. USA, 69, 1697
20. Cunningham, B.A., Wang, J.L., Berggård, I. and Peterson, P.A. (1973) Biochemistry, 12, 4811
21. Schwartz, B.D., Kato, K., Cullen, S.E. and Nathenson, S.G. (1973) Biochemistry, 12, 2157
22. Rask, L., Östberg, L., Lindblom, J.B., Fernstedt, Y. and Peterson, P.A. (1974) Transpl. Rev., 21, 85
23. Strominger, J.L., Cresswell, P., Grey, H., Humphreys, R.E., Mann, D., McCune, J., Parham, P., Robb, R., Sanderson, A.R., Springer, T.A., Terhorst, C. and Turner, M.J. (1974) Transpl. Rev., 21, 126
24. Cresswell, P., and Dawson, J.R. (1975) J. Immunol. 114, 523
25. Peterson, P.A., Rask, L., Sege, K., Klareskog, L., Anundi, H. and Östberg, L. (1975) Proc. Natl. Acad. Sci. USA, 74, 1612
26. Rask, L., Lindblom, J.B., and Peterson, P.A. (1975) Eur. J. Immunol. (in press)
27. Nilsson, K., Evrin, P.-E. and Welsh, K.I. (1974) Transpl. Rev., 21, 53
28. Goodfellow, P.N., Jones, E.A., van Heyningen, V., Solomon, E., Bobrow, M., Miggiano, V. and Bodmer, W.F. (1975) Nature, 254, 267
29. Östberg, L., Rask, L., Nilsson, K., and Peterson, P.A. (1975) Eur. J. Immunol. 5, 462
30. Cresswell, P., Robb, R.J., Turner, M.J., and Strominger, J.L. (1974) J. Biol. Chem. 249, 2828
31. Gally, J.A., and Edelman, G.M. (1972) Annu. Rev. Genet., 6, 1
32. Poljak, R.J. (1975) Nature, 256, 373
33. Boyse, E.A., Old, L.J. and Stockert, E. (1966) IV International Symp. Immunopathology (Ed. P. Grabar and P.A. Mieseher), Monte Carlo, p. 23
34. Boyse, E.A. and Old, L.J. (1969) Annu. Rev. Genet. 3, 269
35. Muramatsu, T., Nathenson, S.G., Boyse, E.A. and Old, L.J. (1973) J. Exp. Med. 137, 1256
36. Anundi, H., Rask, L., Östberg, L., and Peterson, P.A. (1975) Biochemistry (in press)

37. Vitetta, E., Uhr, J.W. and Boyse, E.A. (1975) J. Immunol. 114, 252
38. Östberg, L., Rask, L., Wigzell, H. and Peterson, P.A. (1975) Nature, 253, 735
39. Klein, J., and Shreffler, D.C. (1971) Transpl. Rev. 6, 3
40. Shreffler, D.C., and Owen, R.D. (1963) Genetics, 48, 9
41. Hansen, T.H., Shin, H.S. and Shreffler, D.C. (1975) J. Exp. Med. 141, 1216
42. Curman, B., Östberg, L., Sandberg, L., Malmheden-Eriksson, I., Stålenheim, G., Rask, L., and Peterson, P.A. (Submitted)
43. Schreiber, R.D., and Muller-Eberhard, H.J. (1974) J. Exp. Med. 140, 1324
44. Cooper, N.R. (1975) J. Exp. Med. 141, 890
45. Meo, T., Krasteff, T., and Shreffler, D.C. (1975) Proc. Natl. Acad. Sci. USA (in press)
46. Lachmann, P.J., Grennan, D., Martin, A., and Démant, P. (1975) (Submitted)
47. Capra, J.D. (personal communication)
48. Fu, S.M., Kunkel, H.G., Brusman, H.P., Allen, F.H., and Fotino, M. (1974) J. Exp. Med. 140, 1464
49. Ferreira, A., and Nussenzweig, V. (1975) J. Exp. Med. 140, 513
50. Allen, F.H. (1974) Vox Sanguinis, 27, 173
51. Lindblom, J.B., Östberg, L., and Peterson, P.A. (1974) Tissue Antigens, 4, 186
52. Östberg, L., Lindblom, J.B., and Peterson, P.A. (Submitted)
53. Armeding, D., Kubo, R.T., Grey, H.M. and Katz, D.H. (1975) Proc. Natl. Acad. Sci. USA (in press)
54. Dickler, H.B., and Sachs, D. (1974) J. Exp. Med. 140, 779
55. Rask, L., Klareskog, L., Östberg, L., and Peterson, P.A. (1975) Nature, 257, 231
56. Munro, A.J., and Taussig, M.J. (1975) Nature, 256, 103
57. Klein, J. (1975) Biology of the mouse histocompatibility-2 complex, Springer, New York
58. Vitetta, E., Artzt, K., Bennett, D., Boyse, E.A., and Jacob, F. (1975) Proc. Natl. Acad. Sci. USA (in press)

*Progress in Differentiation Research, ed. N. Müller-Bérat et al.*



# SEQUENTIAL APPEARANCE OF SURFACE MARKERS IN THE ONTOGENY OF B LYMPHOCYTES

J. Abbott, M.K. Hoffmann, A.F. Chin and U. Hämmerling
Memorial Sloan-Kettering Cancer Center,
New York, N.Y. 10021

## 1. Introduction

Lymphocytes consist of two independent lineages, one thymus-dependent ("T") and the other thymus-independent ("B"), both of which originate in bone marrow and mature through a number of defined ontogenetic stages to form immunologically competent lymphocytes. The exact number of compartments through which B and T cells must pass from stem cell status to immunocompetent cells is not known although a minimal number of compartments can be inferred from the requirement for residence in specialized organs (e.g. thymus, bursa).

Years of immunogenetic research have paved the way for investigating lymphocyte ontogeny, furnishing a growing number of surface markers which distinguish B from T cells, as well as characterizing subpopulations within each lineage (reviewed in 1,2). The observation that such markers can be induced in vitro in these subpopulations[3] has lent support to the hypothesis that for each surface antigen there is a given timepoint at which it is phenotypically expressed for the first time and retained in some subsequent stages in the cell lineage. Thus, lymphocyte induction can be understood as the transition of cells from one compartment to the next.

In studying the induction of specific surface phenotypes (Ig, Ia and CR) in precursor populations of B cells, a preliminary understanding of their sequential ontogenetic relationship has been achieved.

## 2. Methods and Results

The basic induction system in lymphocytes, as first devised by Komuro and Boyse[3], consists of an initial enrichment of inducible cells from spleen and bone marrow populations by centrifugation through a discontinuous BSA density gradient. Cells from the 23 to 26% interface ("C layer cells") are cultured for a short time in the presence and absence of an inducer. The inducing agent can be either a

---

List of abbreviations: Ig: surface associated immunoglobulin of B cells; Ia: immune response gene associated antigen. CR: receptors of B cells for the third complement component. BSA: bovine serum albumin. LPS: lipopolysaccharide of E. coli. poly A:U: polyadenylic uridylic acid. C: rabbit complement. EAC: sheep erythrocytes coated with 19S anti-erythrocyte antibody and mouse complement.

specific hormone (e.g. thymopoietin[4]) or an agent with a broad range of ligand specificities (e.g. LPS, poly A:U, etc.). The conversion of phenotypically negative to phenotypically positive cells ("induction") is assessed by determining the relative proportion of cells carrying specific surface markers after culture with and without inducer. The enumeration of positive cells is performed serologically by i) determining the number of cells sensitive to cytotoxic antibody and complement (C) (e.g. Ig, Ia, Thy-1[5]); ii) by binding of fluorescent antibody (Ig and Ia); iii) by binding of sensitized erythrocytes (EAC) for detecting CR-bearing lymphocytes. For the Ia marker the increase in the proportion of cells lysed by cytotoxic antibody has been correlated with an increase in antigen quantity[6], as assessed by quantitative absorption. Such experiments demonstrate that the increase in cytotoxicity is due to new antigen being expressed, rather than merely an increase in cell fragility caused by the experimental procedure.

The principles of the induction experiments are illustrated in this paper for the B cell marker Ig. Spleen and bone marrow contain a subpopulation of Ig cells which are committed to the B lineage and which can be induced *in vitro* by LPS to express Ig. This is shown by experiments (see Table 1) in which bone marrow cells are fractionated on a discontinuous BSA gradient, and the cells accumulating at the interfaces then cultured for 2.5 hr in the absence or presence of 30$\mu$g/ml LPS. We have observed an approximately three-fold relative increase in the number of $Ig^+$ cells in cultures with LPS as compared to those without. The enumeration of $Ig^+$ cells is based on cytotoxicity assays with anti-immunoglobulin sera and C. The typing antibody for $Ig^+$ cells contains equal proportions of anti-$\mu$ and anti-kappa. Anti-$\gamma$ is used as control antibody, which in all cases gives negative results. Since anti-Ig antibodies are purified by adsorption of rabbit anti-immunoglobulin sera on immunoadsorbants and elution at pH 2.0[6] heteroantibodies do not constitute a problem.

Table 1

Induction of Surface Ig in Four Fractions of Bone Marrow Cells Separated in a Discontinuous BSA Gradient

| Layer | BSA concentration (%) | Cell recovery (%) | Cytotoxicity index and standard deviation<br>Without LPS | <br>With LPS | <br>Net induction |
|---|---|---|---|---|---|
| B | 15/23 | 3 | 10 ± 4 | 27 ± 8 | 17 ± 9 |
| C | 23/26 | 8 | 2 ± 9 | 25 ± 4 | 23 ± 7 |
| D | 26/29 | 25 | 10 ± 13 | 10 ± 10 | 0 ± 6 |
| E | 29/35 | 28 | 1 ± 1 | 2 ± 3 | 1 ± 2 |

Mean values of 3 independent experiments. Induction of Ig is obtained in the B and notably the C layer cells.

Recent work has demonstrated that Ia and CR can also be induced in B lymphocytes[6], with kinetic and biochemical principles similar to those for induction of Ig and Thy-1 from C layer cells. Three alternatives can be considered as to the relationship of inducible B cells to one another:

1) The three B cell markers appear simultaneously on one and the same subpopulation, 2) two or more unrelated subpopulations express a single marker, or a combination of two markers, when induced, and 3) a temporal ontogenetic relationship exists between subpopulations, whereby the induction of the markers is sequential (i.e. inducible cells are derived from one another). A subpopulation acquiring one surface antigen, on further maturation would switch on a second marker.

Since the knowledge of the phenotype of a cell would allow its assignment to a particular compartment, our experimental approach to these alternatives was to determine whether any surface markers were already present on each cell population prior to induction. The experiments were done indirectly by first eliminating cells bearing specific surface antigens (Ig, Ia or CR) from inducible precursor populations and secondarily inducing with LPS. Cytotoxic antibody and C were used to remove Ig and Ia bearing cells, and exhaustive depletion of EAC rosettes to remove $CR^+$ cells.

Phenotype of Ig inducible cells

Bone marrow (see Table 1) and spleen contain a subpopulation of Ig precursor cells, which can be induced to express Ig. Induction is accomplished under the conditions of the Komuro-Boyse assay[3]. It has been found that Ig induction is insensitive to pretreatment with anti-Ig and C (by definition), or with anti-Ia and C, and is also insensitive to CR depletion (see Table 2). Thus, cells undergoing conversion from $Ig^-$ to $Ig^+$ can be assigned the phenotype $Ig^-Ia^-CR^-$.

Phenotype of Ia inducible cells

These are found equally in bone marrow and spleen populations, from which they can be enriched by density gradient centrifugation. Ia inducible cells are completely eliminated by anti-Ig and C, and hence have the $Ig^+$ phenotype. Anti-Ia and C predictably has no effect on Ia induction. Similarly, removal of $CR^+$ cells does not deplete Ia inducible cells (see Table 2). Therefore, it can be concluded that Ia inducible cells are $Ig^+Ia^-CR^-$.

Phenotype of CR inducible cells

The CR induction system is similar to Ig and Ia induction systems[7]. Inducible $CR^-$ precursor cells are more numerous in spleen than in bone marrow. CR induction is completely abolished after pretreatment with either anti-Ig and C or anti-Ia and C, while CR depletion has no effect (Table 2). Thus CR inducible cells have the phenotype $Ig^+Ia^+CR^-$.

Prothymocyte induction[3], as characterized by the prototype marker Thy-1, is unaffected by anti-Ig, anti-Ia, or EAC treatment.

Table 2

| Pretreatment of precursor cells | Net Induction for Markers and Reduction (%) Ig | Ia | CR | Thy-1 |
|---|---|---|---|---|
| I | | | | |
| anti IgM + anti $F(ab)_2$+C | 16 ± 10 | 0 ± 5 | -2 ± 3 | 33 ± 8 |
| anti IgG + C (control) | 17 ± 10 | 26 ± 6 | 8 ± 4 | 27 ± 4 |
| % reduction | 6 | 100 | 125 | -22 |
| II | | | | |
| anti $Ia^k$ + C | 12 ± 8 | 24 ± 13 | -2 ± 4 | 29 ± 10 |
| anti $Ia^s$ + C (control) | 18 ± 8 | 25 ± 10 | 8 ± 4 | 24 ± 11 |
| % reduction | 33 | 4 | 125 | -17 |
| III | | | | |
| EAC | 22 ± 14 | 36 ± 15 | 3 ± 8 | 26 ± 9 |
| EA (control) | 20 ± 5 | 28 ± 9 | 5 ± 3 | 33 ± 16 |
| % reduction | -10 | -29 | 40 | 21 |

100 x $10^6$ spleen cells of 4 week old (B6 x A)$F_1$ mice were treated in 1.5 ml volume with antibody and rabbit complement (final dilution = 1:30), for 45 min at 37°C. Group I: anti-IgM (μ κ), or control anti-IgG/Fc (γ) antibody[6]; Group II: A.TH anti A.TL antiserum ("anti-$Ia^k$") or control: A.TL anti A.TH ("anti-$Ia^s$")[8]. Group III: EAC, or control: EA. Treated spleen cells were fractionated on discontinuous BSA gradients, and C layer cells (see Table 1) cultured in medium RPMI-1640 containing 1% FBS for 2.5 hr in a $CO_2$ incubator. Induced cultures contained 30μg/ml LPS, and control cultures were without LPS. Cytotoxicity indices[3] for Ig and Ia, and percentage of EAC rosette forming cells in the presence of 0.1mM EDTA[9] were determined in triplicate. Reduction of inducible cells was calculated using the formula:

$$\frac{\text{Net induction in treated cells - Net induction in control cells}}{\text{Net induction in control cells}} \times 100$$

Tabulated data are means of 3 to 5 independent experiments and standard deviation. Similar induction data were obtained with bone marrow populations, with the exception of CR induction which is low in bone marrow populations due to a lack of precursor cells. Negative inhibition signifies enrichment for certain precursor cells (see e.g. increase in prothymocyte induction after anti-Ig treatment).

## 3. Discussion

It is becoming abundantly clear that cells acquire their specialized properties only after progression through a number of tightly regulated steps during which they undergo cell division and restricted genetic programming. By definition, each of these steps, or compartments, contain cell populations with unique properties. The B lymphocyte cell lineage in mice is a particularly useful model with which to analyze cell differentiation since several cell surface markers are now known by which precursor subpopulations can be distinguished.

Recent studies of lymphopoiesis in developing embryos and neonatal mice show that $Ig^+$ cells appear earlier than $CR^+$ cells[10,11]. Although double labeling techniques reveal Ig on virtually all $CR^+$ cells there are substantial numbers of $Ig^+$ $CR^-$ cells in both immature and mature lymphoid organs[12] (i.e. fetal liver, spleen and bone marrow). Thus during normal development, it is likely that $Ig^+CR^-$ cells precede $Ig^+CR^+$ cells.

$Ig^+$ cells in fetal liver also lack Ia and CR, and they are functionally different from mature B cells, being incapable of responding to antigen stimulation by antibody production. In addition, $Ig^+$ $Ia^+CR^-$ B cells are immunologically incompetent. This follows from the observation that B cell populations in the spleen depleted of $CR^+$ cells cannot respond to sheep erythrocytes in an *in vitro* primary immune response (M. Hoffmann *et al.*, unpublished results). This defect is not permanent, as inducing factors such as LPS can restore immune responsiveness, presumably by converting $Ig^+$ $Ia^+$ $CR^-$ to $Ig^+$ $Ia^+$ $CR^+$ cells. Therefore, at least in mice, presence of the CR marker characterizes a subpopulation of immunocompetent B cells.

From our serological studies of the induction of B cell surface markers *in vitro* it is clear that there are at least four subpopulations of cells in the murine B cell lineage that exist together in the spleen and bone marrow. The evidence that a restricted array of surface markers is required before a cell can acquire a particular marker lends strong support to the hypothesis that B cell differentiation proceeds by the orderly progression of precursor cells through strictly defined compartments. We propose that the four compartments described thus far contain cells having the following surface phenotypes, which appear sequentially during ontogeny:

$$\underline{Ig^- \ Ia^- \ CR^-} \text{ to } \underline{Ig^+ \ Ia^- \ CR^-} \text{ to } \underline{Ig^+ \ Ia^+ \ CR^-} \text{ to } \underline{Ig^+ \ Ia^+ \ CR^+}$$

Our model does not limit the number of compartments which may yet be discovered, nor does it rule out the possibility of one compartment giving rise to more than one cell type. For instance, recent work has shown that exposure of early $Ig^+$ cells to antigen or to anti-Ig leads to loss of Ig expression[13] and possibly to tolerance[14] (programmed cell death on contact with antigen?).

It is clear from our studies that newly induced cells are not immediately inducible for the next surface marker. We assume that in order to acquire the property of inducibility one or more metabolic and genetic events must take place. From studies of other differentiating cell lineages it is quite possible that DNA synthesis and cell proliferation may be required for programming these cells for progression to the next compartments (see Holtzer, these Proceedings). This would be in contrast to unicellular organisms such as Acetabularia where an orderly sequence of morphogenetic events occurs without DNA synthesis[15]. It is of particular interest, therefore, to analyze the regulation of this crucial step.

Our present investigation has focused on the induction phase of the phenotypic conversion. By developing culture conditions which will allow long term viability of lymphocytes we hope to clarify all of the progenitor-successor relationships in the B cell lineage and identify the regulatory mechanisms which control this progression.

References


1. Boyse, E.A. and Old, L.J., 1969, Ann. Rev. Genetics 3, 269.
2. Schlesinger, M., 1972, Progr. Allergy 16, 214.
3. Komuro, K. and Boyse, E.A., 1973, The Lancet, April 7, 740.
4. Goldstein, G., 1974, Nature (London) 247, 11.
5. Reif, A.E. and Allen, J.M.V., 1964, J. Exp. Med. 120, 413.
6. Hämmerling, U., Chin, A.F., Abbott, J. and Scheid, M.P., 1975, J. Immunol., in press.
7. Scheid, M.P., Goldstein, G., Hämmerling, U. and Boyse, E.A., 1975, Ann. N.Y. Acad. Sci. 249, 531.
8. Hämmerling, G., Deak, B.D., Mauve, G., Hämmerling, U., and McDevitt, H.O., 1974, Immunogenetics 1, 68.
9. Bianco, C., Patrick, R. and Nussenzweig, V., 1970, J. Exp. Med. 132, 702.
10. Gelfand, M., Elfenbein, G.J., Frank, M.M., and Paul, W.E., 1974, J. Exp. Med. 139, 1125.
11. Kincade, P.W. and Moore, M.A.S., 1975, in The Lymphocyte: Structure and Function. (J.F. Marchalonis, ed.)., Marcel Dekker, New York, in press.
12. Parish, C.R. and Hayward, J.A., 1974, Proc. R. Soc. Lond. B. 187, 65.
13. Owen, J.J.T., Cooper, M.D. and Raff, M.C., 1974, Nature (London), 249, 361.
14. Nossal, G.J.V. and Pike, B., 1975, J. Exp. Med. 141, 904.
15. Hämmerling, J., 1934, Naturwissenschaften 50, 55.


Supported in part by NCI grants CA-08748, CA-16889, CA-17085 and NIH grant HD-08415.

Progress in Differentiation Research, ed. N. Müller-Bérat et al.



# DIFFERENTIATION OF MURINE T AND B LYMPHOCYTES FROM RESTING TO EFFECTOR CELLS, AS CHARACTERIZED BY ALA-1 (ACTIVATED LYMPHOCYTE ANTIGEN-1)

A.J. Feeney and U. Hämmerling

Sloan-Kettering Institute for Cancer Research
New York, New York

## I. Introduction

In Burnet's clonal selection theory[1], antigen initiates an immune response by selecting and expanding a pre-existing lymphocyte clone of the corresponding specificity. In the work presented here, we address the question of whether clonal proliferation is a sufficient condition for establishing a measurable immune response, or whether additional differentiative events are involved. In other words: are the effector lymphocytes phenotypically identical to their progenitor cells, or have they undergone a differentiative step or steps in the course of expansion? In the B cell lineage, the available evidence abundantly supports the latter view, because plasma cells are both morphologically and antigenically different from resting B lymphocytes[2]. However, there is to date no evidence that the same is true of effector T cells (activated helper, killer and suppressor cells). Here we furnish evidence that effector cells of both T and B origin in the mouse express an alloantigen, Ala-1 (activated lymphocyte antigen-1), which is absent from unstimulated, quiescent lymphocyte precursors. On this basis, we propose that differentiation is a prerequisite for the production of effector T and B cells.

## II. Methods and Results

Procedures will be briefly outlined below; further technical details will be published later.

### A. Serology of Ala-1

#### 1. Production of antisera

Anti Ala-1.2. Since stimulation of a lymphocyte population by any given antigen can trigger clones present in only a very low frequency, we have employed T cell mitogens which cause most of the T cell population to undergo blast transformation[3]. C58 spleen and lymph node cells were cultured for 24 hrs in the presence of 2μg/ml of phytohemagglutinin (PHA). C3H/An mice were immunized with increasing numbers of PHA-stimulated lymphocytes (from 5 x $10^6$ to 100 x $10^6$) at

Abbreviations used: C, complement; PHA, phytohemagglutinin; Con A, concanavalin A; LPS, lipopolysaccharide; B6, C57BL/6 mice; TNP, trinitrophenyl

monthly intervals, and bled one week after the third and subsequent inoculations.

C58 and C3H/An differ with regard to at least three lymphocyte alloantigens. Anti Ly-1[4], and anti Ly-3[5], and other thymic antibodies were removed by exhaustive absorption with C58 thymocytes. Autoantibodies were frequently encountered and were removed by absorption with C3H/An lymphoid tissues.

In the course of the serological analysis, another antibody specificity, distinct from anti Ala-1, was discovered. The corresponding antigen is distinguished from Ala-1 by a different strain distribution, and is expressed only on subsets of lymph node and spleen cells, but not on mitogen-stimulated cells. This activity was removed by absorption with spleen and lymph node cells of HTG mice which express this antigen and also carry the Ala-1 allele.

Anti-Ala-1.1. The reciprocal immunization, C58 mice with PHA-stimulated C3H/An lymphocytes, yielded Ala-1.1 antiserum of low titer. After a similar regimen of absorptions with donor strain (C3H/An) thymocytes and recipient strain (C58) lymphocytes, the antiserum was Ala-1.1 specific.

2. Expression of Ala-1 on lymphoid subpopulations

In direct cytotoxicity tests, thymocytes, lymph node and spleen lymphocytes were negative for Ala-1, but > 90% of lymph node cells cultured for 48 hrs with concanavalin A (Con A) (2μg/ml) were lysed. The T cell mitogen Con A was chosen because it produced virtually no cell agglutination and yielded over 80% blast cells. When tested with anti Thy-1[6] serum, more than 80% of the stimulated lymph node cells were identified as T cells. Spleen cells cultured for 48 hrs with the B cell mitogen lipopolysaccharide[7] (LPS) at 5μg/ml, and treated with anti Thy-1 and complement (C) to eliminate T cells, were also lysed (>90%). Thus, both activated T and B lymphocytes express Ala-1.

Quantitative absorptions were also performed on unstimulated lymphocyte populations. These studies showed that the number of normal lymph node and spleen cells required to absorb out anti Ala-1 activity was 10-20 times greater than the number of Con A-stimulated T cells. We attribute this weak absorption to the small number of activated lymphocytes resident in unprimed lymph node and spleen populations, although we have not yet formally excluded the possibility that a low concentration of Ala-1 may be present on the unstimulated population. Absorption with thymocytes did not remove Ala-1 activity, even when 20 times more thymocytes than the number of stimulated T cells required to remove Ala-1 activity were used.

To verify that both activated T and B cells carry the same antigen, quantitative cross-absorptions were performed. As shown in Fig. 1, LPS-stimulated B cells removed Ala-1 activity to the same extent as Con A-activated T cells, indicating that both cell types express Ala-1, and in similar quantities.

3. Ala-1 is not a blast cell antigen

Thymocytes were cultured with Con A, yielding a large percentage of blast

Figure 1

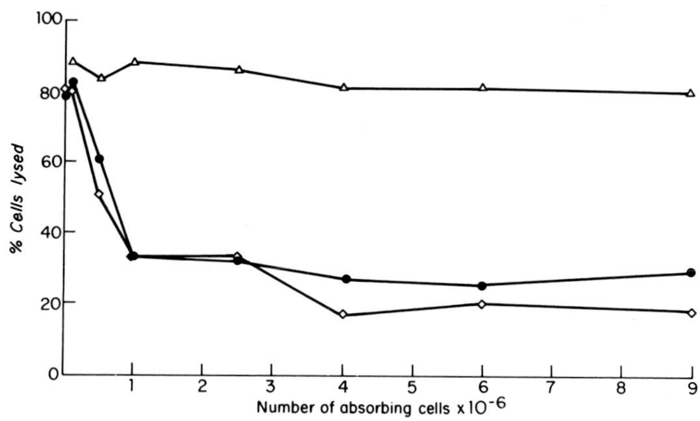


Anti Ala-1.2 serum, prediluted to two doubling-dilutions before the 50% cytotoxic endpoint, was absorbed with graded numbers of LPS-stimulated B cells (◇), Con A-stimulated lymph node cells (●), and thymocytes (△), all from B6 mice. The absorbed antisera were then tested for residual activity on Con A-stimulated lymph node cells. Maximal kill in this experiment was 94%; the C control level was 28%.

---

cells. These stimulated thymocytes were Ala-1$^-$ in a direct cytotoxicity test. Hence, Ala-1 does not appear to be an antigen of blast cells. Rather, it is expressed at a later time point in the differentiation scheme of T and B cells.

4. Tissue distribution

Erythrocytes, sperm, and dissociated cells from brain, skin, liver, kidney, thymus lymph node, and spleen were all found to be negative by qualitative absorption tests.

5. Strain distribution

Lymph node cells from different strains of mice were cultured with Con A and then tested with both anti Ala-1.1 and anti Ala-1.2 in cytotoxicity tests. The two antisera were mutually exclusive, indicating allelism of Ala-1.1 and Ala-1.2. The strain distribution is shown in Table 1.

B. Expression of Ala-1 on functional subsets of lymphocytes

The general strategy in these experiments involving antigen-primed T and B cells was to investigate whether removal of Ala-1$^+$ cells by treatment with Ala-1 antiserum and C would affect the relevant function: (1) antibody secretion by B cells, (2) ability of primed T cells to cooperate with unprimed B cells in initiating a humoral immune response (helper function), (3) T cell mediated lysis

Table 1

Strain Distribution of Ala-1

| Strain | Ala-1.1 | Ala-1.2 |
|---|---|---|
| C3H/An | + | - |
| BALB/c | + | - |
| CBA/T6 | + | - |
| A.By | + | - |
| A.SW | + | - |
| A | + | - |
| HTG | + | - |
| CE/J | + | - |
| 129 | + | - |
| H-2H | + | - |
| B6 | - | + |
| B6/Ly-1.1 | - | + |
| B6/Ly-2.1 | - | + |
| B6/Ly-2.1 Ly-3.1 | - | + |
| AKR | - | + |
| C58 | - | + |
| DBA/2 | - | + |
| GR | - | + |
| C57BR/J | - | + |
| SJL/J | - | + |
| Swiss HSFS/N | - | + |
| I/HaBoy | - | + |

of target cells (killer function). A reduction in any of the functions would indicate the presence of Ala-1 on the responsible lymphocyte subpopulation.

1. IgM plaque-forming cells are Ala-1$^+$

B6 mice were immunized intravenously with $10^8$ sheep erythrocytes. Four days later their spleens were removed, and the splenocytes were treated with anti Ala-1.2 and C. Controls included treatment with normal mouse serum and C, or C alone. The cells were then assayed in the Jerne hemolytic plaque assay[8], as modified by Mishell and Dutton[9], with sheep erythrocytes as the antigen, to enumerate IgM plaque-forming cells surviving after mass cytolysis. In five such experiments, the number of plaque-forming cells was reduced on the average by 76% (range 52-93%) after treatment with anti Ala-1 and C. However, by the criterion of trypan blue uptake, only a very low proportion of cells were killed with anti Ala-1 and C, not exceeding the proportion of dead cells obtained with control sera or C alone. This indicates a low frequency of plaque-forming cells in immune spleen populations.

2. Helper cells are Ala-1$^+$

Spleen cells from B6 mice primed with $10^6$ sheep erythrocytes were used as a source of helper T cells. The titration analysis was performed according to Hoffmann and Kappler[10]. Briefly, the primed cells, after treatment with various antisera and C, were titrated with normal spleen cells, keeping the total cell number constant, and were cultured with sheep erythrocytes conjugated with the hapten trinitrophenyl (TNP) in Mishell-Dutton cultures[9]. After four days, the cells were assayed for TNP-specific plaque-forming cells on TNP-conjugated horse erythrocytes[11]. Pretreatment of the carrier-primed spleen cells with anti Ala-1.2 and C reduced helper activity by an average of 79% (range 59-94%) (see Fig. 2) although, as in the case of the plaque-forming cells, low numbers of cells were lysed. Anti Thy-1.2 pretreatment resulted in almost complete elimination of helper cell activity. In control assays, pretreatment of the primed spleen cells with C alone, or with normal mouse serum and C, did not affect the helper activity.

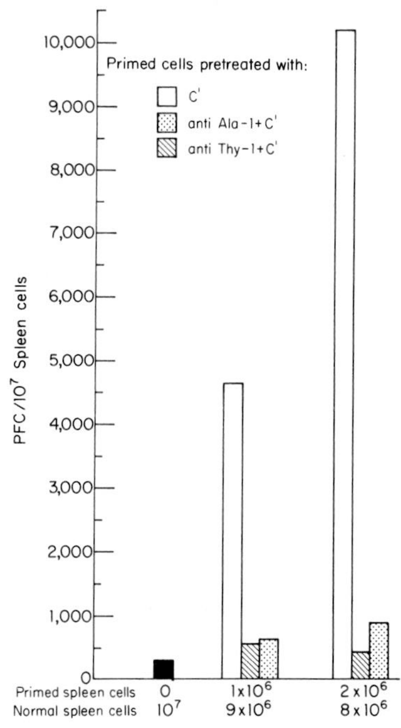


Fig. 2. Effect of Ala-1 and C pretreatment on helper cell activity.

3. Precursors of helper cells and of IgM plaque-forming cells are Ala-1$^-$

Normal spleen cells of B6 mice were treated with anti Ala-1.2 and C, anti Thy-1.2 and C, or C alone, and were then cultured with sheep erythrocytes in Mishell-Dutton cultures. Four days later, the cells were assayed to determine the number of sheep erythrocyte-specific plaque-forming cells generated. A reduction in the number of plaque-forming cells would indicate that either the precursors of helper cells or precursors of IgM plaque-forming cells were depleted. Preliminary evidence shows that treatment with anti Ala-1.2 and C did not reduce the number of plaque-forming cells, as compared with the sample treated with C alone. Anti Thy-1.2 pretreatment gave the expected reduction of plaque-forming cells, indicating that precursors of helper cells are Thy-1$^+$. These results are consistent with the absence of Ala-1 from unstimulated lymphocyte populations.

4. Killer T cells are Ala-1$^+$

Killer T cells were obtained from the peritoneal cavity of B6 mice immunized with P815 tumor cells, a histoincompatible mastocytoma. After purification by passage through a nylon fiber column[12], the killer T cells were added to

$^{51}Cr$-labeled P815 cells and co-cultured for 2 hrs. $^{51}Cr$ was released into the medium as the target cells were lysed; hence, the amount of $^{51}Cr$ in the supernatant of the cultures, as measured in a gamma counter, was taken to be proportional to the percent lysis[13]. The ratio of untreated killer cells to target cells was adjusted to give approximately 80% lysis within the 2 hr culture period.

Treatment of killer T cells with normal mouse serum and C, or C alone did not affect lymphocytotoxicity. However, mass cytolysis of the killer cell population with anti Ala-1.2, or anti Thy-1.2, both of which lysed over 80% of the population, eliminated > 95% of the killer cell activity.

## III. Discussion

Evidence has been put forth by Dutton[14] and Schimpl and Wecker[15] that the generation of IgM plaque-forming cells from antigen-responsive, unstimulated B cells involves both proliferation and differentiation. In response to antigen-triggering, B cells proliferate, regardless of whether or not helper T cells are present. B cells can then become plaque-forming cells, but only if a differentiative signal is provided by helper T cells. These findings, coupled with the fact that plasma cells express the alloantigens Pc-1 and Ala-1, both of which are not present on resting B cells, strongly suggest that effector B cells are distinct from unstimulated B cells, and that they have undergone at least one differentiative step to acquire their new function and surface phenotype.

From the observed Ala-$1^-$ to Ala-$1^+$ conversion, we infer that T cells must also undergo a differentiative process to become T effector cells. It follows that they are expressing a genetic program not manifest in their progenitors.

Previous work by other investigators has revealed at least three compartments in the T cell lineage. (1) Prothymocytes are a subpopulation of lymphocytes found in bone marrow and spleens and are negatively defined by the absence of any of the known thymocyte markers. They are positively defined by their ability to be induced *in vitro* into cells of the thymocyte phenotype[16]. (2) Several phenotypically distinct subgroups exist within the thymus. The vast majority (cortical thymocytes) are characterized by Thy-$1^+$ $TL^+$[17] Ly-$1^+$ Ly-$2/3^+$ Ala-$1^-$. A smaller population (medullary thymocytes) have a lower density of Thy-1 and no TL[18]. Within medullary thymocytes, cells can be assigned to three groups with regard to the expression of Ly antigens: Ly-$1^+$ Ly-$2/3^+$; Ly-$1^-$ Ly-$2/3^+$; and Ly-$1^+$ Ly-$2/3^-$ [19]. The precise relationship of the latter cells to each other, as well as to the cortical thymocytes is unclear, as several diverging pathways may operate in the thymus[20]. (3) In the post-thymic population, two sublines can be distinguished according to the following phenotypes: Thy-$1^+$ $TL^-$ Ly-$1^+$ Ly-$2/3^-$ (helper cell line)[21], and Thy-$1^+$ $TL^-$ Ly-$1^-$ Ly-$2/3^+$ (killer cell line)[22]. The above phenotypes are displayed by the antigen-activated helper and killer cells as well as their immediate precursors[19]. A third functionally defined subline, antigen-specific suppressor cells, express the same phenotype as killer cells (M. Feldman, personal communication), but information on the Ly phenotype of suppressor cell precursors is lacking.

On the basis of Ala-1 expression, we can subdivide the killer cell and helper cell lines into a pre-antigen activated compartment (Ala-$1^-$) and a compartment comprised of activated cells (Ala-$1^+$). While most of the resting lymphocytes are not called upon by antigen in their life cycle, the few clones that are activated appear to undergo differentiation and are distinguished by Ala-1 expression. Thus, Ala-1 has both theoretical interest as a marker for T cell differentiation and practical interest for the identification of effector cells.

(Supported in part by NCI grants CA-08748, CA 16889 and NIH grant HD 08515.)

References

1. Burnet, F.M., 1959, The clonal selection theory of acquired immunity (Vanderbilt University Press, Nashville).
2. Takahashi, T., Old, L.J., and Boyse, E.A., 1970, J. Exp. Med. 131, 1325.
3. Janossy, G. and Greaves, M.F., 1972, Clin. Exp. Immunol. 10, 525.
4. Boyse, E.A., Miyazawa, M., Aoki, T., and Old, L.J., 1968, Proc. Roy. Soc. B. 170, 175.
5. Boyse, E.A., Itakura, K., Stockert, E., Iritani, C.A., and Miura, M., 1971, Transplantation 11, 351.
6. Reif, A.E., and Allen, J.M.L., 1964, J. Exp. Med. 120, 413.
7. Andersson, J., Möller, G., Sjoberg, O., 1972, Cell. Immunol. 4, 381.
8. Jerne, N.K., and Nordin, A.A., 1963, Science 140, 405.
9. Mishell, R.I., and Dutton, R.W., 1967, J. Exp. Med. 126, 423.
10. Hoffmann, M., and Kappler, J.W., 1973, J. Exp. Med. 137, 721.
11. Kettman, J., and Dutton, R.W., 1970, J. Immunol. 104, 1558.
12. Julius, M.H., Simpson, E., and Herzenberg, L.A., 1973, Europ. J. Immunol. 3, 645.
13. Brunner, K.T., Mauel, J., Cerottini, J.-C., and Chapuis, B., 1968, Immunology 14, 181.
14. Dutton, R.W., 1975, Transplant. Rev. 23, 66.
15. Schimpl, A.L., and Wecker, E., 1975, Transplant. Rev. 23, 176.
16. Komuro, K., and Boyse, E.A., 1973, Lancet April 7, 740.
17. Boyse, E.A., Old, L.J., and Stockert, E., 1965, in: Immunopathology IVth International Symposium, Monaco, eds. P. Grabar and P.A. Miescher (Schwabe and Co., Basel) p. 23.
18. Leckband, E., and Boyse, E.A., 1971, Science 172, 1258.
19. Cantor, H., and Boyse, E.A., 1975, J. Exp. Med. 141, 1376.
20. Shortman, K., and Jackson, H., 1974, Cell. Immunol. 12, 230.
21. Kisielow, P., Hirst, J.A., Shiku, H., Beverley, P.C.L., Hoffmann, M.K., Boyse, E.A., and Oettgen, H.F., 1975, Nature 253, 219.
22. Shiku, H., Kisielow, P., Bean, M.A., Takahashi, T., Boyse, E.A., Oettgen, H.F., and Old, L.J., 1975, J. Exp. Med. 141, 227.

*Progress in Differentiation Research, ed. N. Müller-Bérat et al.*
*© 1976, North-Holland Publishing Company - Amsterdam, The Netherlands.*


# THE DIFFERENTIATION OF THE T LYMPHOID SYSTEM. THE ATHYMIC (NUDE) MOUSE AS A TOOL.

F. Loor*, G.E. Roelants, B. Kindred, K.S. Mayor and L.-B. Hägg.

Basel Institute for Immunology, Grenzacherstrasse 487, Postfach 4005 Basel 5, Switzerland.

In the mouse species, the nude mutant (symbol nu, autosomal recessive, location in chromosome 11) is almost completely hairless, and this gene mutation has pleiotropic effects (1,2). The homozygous nu/nu mice show, besides many abnormalities, a marked deficiency of their immune system, especially of its T cell compartment (a T cell being defined as a thymus processed lymphocyte): they are unable to reject foreign tissues or cells and they cannot mount normal antibody responses to T cell dependent antigens (2). It is yet unclear if the T cell deficiency has only a cellular basis or if a hormonal defect is also involved: a) there is no macroscopic thymus during the embryonic life nor after birth (3-5); b) in the secondary lymphoid organs the areas where T cells selectively home (6) either contain very reduced numbers of lymphocytes, or lymphocytes simply cannot be found (7), but germinal centers, though typically B cell homing area (8), are not organized either (9); c) the pool of recirculating lymphocytes is smaller in nudes than in phenotypically normal mice (3,10) and it seems to be contributed only by the B cell pool (11,12), whose size is comparable to that found in normal mice (13,14); d) the lymphoid population is strongly deficient in cells bearing easily detectable amounts of the Thy-1 determined antigen (Θ-antigen), a mouse T cell membrane marker (15), both by the criteria of cytolysis by α-Θ alloantibody plus complement and of staining by fluorochrome labelled α-Θ alloantibodies (15). Only very low percentages of Θ-bearing lymphocytes (less than 1.5%) can be identified by immunofluorescence in nude spleen or lymph nodes (16,17), instead of the 30-80% found in normal mice (15,18). Slightly higher values (5-6%) have been obtained by cytotoxicity (19), and also the capacity of nude lymph node cells to absorb the cytotoxic activity of a titrated α-Θ antiserum was unequivocal (15). The existence of more frequent cells having low density of the Θ antigen was however suggested (18) and has recently found supporting evidence (20,21).

The origin of this obvious lack of T cells seems to reside essentially in an abnormal development of the thymus anlage (4,5,10) rather than in a lack or defect

---

*Fellow of the Belgian National Funds for Scientific Research, on leave of absence from the Department of Molecular Biology of the Free University of Brussels.

at the stem cell level. Indeed in the day 14 fetus, some thymic rudiment does actually appear, but it never becomes lymphoid (3-5); when examined after birth the very small nude thymus has a cystic appearance (4) and lacks, besides the lymphocytes, many of the structures which are characteristic for a normal thymus (5). This is especially the case for the epithelial cells of cortical type, which are said to be needed for the T lymphocyte differentiation and multiplication (5,10,22-24), and for the medullary cystic cells to which the secretory, hormonal functions of the thymus (25) have been attributed (5,24,26). The most likely explanation given for the nu/nu mouse thymus malformation (5,27) is a primary defect in the ectoblastic component of the thymus anlage which originates from the third branchial slit (28) with a consequent defect in the organisation of the endodermic part of the thymus derived from the third pharyngeal pouch (28).

In contrast to a deficient thymus anlage, it is not clear if the stem cell potential of the adult nu/nu mouse is affected by the mutation. Indeed the spleen colony forming capacity of nu/nu, nu/+ and +/+ adult bone marrow in lethally irradiated +/+ recipients was reported to show similar values (29); precursors able to repopulate the thymus and thymus dependent areas of lymphoid organs of lethally irradiated +/+ recipients were found in the nu/nu bone marrow and spleen (10,18, 29). However, it was also reported that, when compared to +/+ bone marrow cells, nu/nu bone marrow showed a defective capacity to restore hemopoiesis, permanent chimerism and immunological responsiveness (30). The stem cell potential of neonatallly thymectomized +/+ mice also shows a reduced colony forming capacity and qualitative changes in the intrinsic nature of the colony forming unit population (more of erythroid type and less of granuloid type, in absence of thymus [31,32]). The defective capacity of nu/nu bone marrow cells, though potentially being another limiting factor of the response of nude mice (33) is thus less likely due to a direct effect of the pleiotropic nu gene rather than to the consequence of a lack of hormonal influence of the thymus gland on the quality, the activity and/or the differentiation of the bone marrow stem cell potential.

The actual existence of potential precursors for the T cell lineage was proven by experiments involving the implantation of a +/+ mouse thymus to a nude mouse (10,18,29) and looking for thymus and host derived T cells by use of chromosome (T6) markers or of membrane markers. When a neonatal allogeneic thymus is grafted to a nude, host derived lymphocytes progressively repopulate it (18,29). This seems to work with a xenogeneic thymus too (34). These lymphocytes show large amounts of the θ antigen (18) characteristic of the T cell membrane (15), and, in $TL^+$ strains, of the TL (thymus leukemia) antigen (18) characteristic of the thymocyte membrane (35). It should be recalled here that in $TL^-$ mouse strains, TL antigens can be found on the surface of leukemic lymphocytes, thus showing that the gene is not lacking, but simply not expressed. Since a $TL^-$ mouse strain thymus epithelium is, however, able to support the expression of TL antigens on $TL^+$ mouse strain host derived thymus cells, (18), it appears that TL expression is not or not exclusively dependent on the thymus epithelium but rather on a humoral or cellular inducer of extra thymic origin which is lacking in $TL^-$ mouse strains and present in $TL^+$ mouse strains.

While the thymus itself becomes repopulated by host derived cells, thymus graft derived lymphocytes transiently appear in the peripheral lymphoid organs; later host derived T cells (presumably derived from the regenerated thymus) appear there as well in increasing numbers (18,29). The secondary lymphoid organs become repopulated and reorganized (36) and there is a persistent restoration of T cell dependent immunity (10,29,37-39)). This regeneration by the neonatal thymus graft is specific in that grafting a spleen does not allow the differentiation of detectable amounts of host derived T cells. A humoral effect alone is not enough (41); a traffic of precursors through the thymus (18,22,23,40) is needed since such a differentiation is not obtained when the thymus is included in a cell impermeable, diffusion chamber (42,43). The differentiation of host derived T cells is not mediated by the lymphocytes of the grafted thymus, since suspensions of allogeneic thymus lymphocytes fail to induce it and to reconstitute a persistent T cell dependent immunity (38,44). The differentiation of host derived T cells is thus likely to depend on the interaction of precursor cells with the thymus epithelium, though it is not yet clear if the thymus structure has to be preserved (45). Curiously enough if a thymus epithelium devoid of lymphocytes is grafted to a nude mouse, it never becomes repopulated by large numbers of lymphocytes, even though host derived T cells appear in the peripheral lymphoid organs (46); this implies that these cells can be differentiated from T precursors during a rather short transit through the thymus epithelium. Finally the ability of the grafted thymus to regenerate the T lymphoid system of the host nude was found to be markedly dependent on the age of the thymus donor (adult thymus is inefficient) (46) and on the TL characteristics of the host and of the thymus donor (39).

A series of T cell characteristic reactivities that nu/nu mice completely lack and that thymus grafted nudes show were found to be and to persistently remain below the levels found in +/+ mice. Thus the antibody responses to two thymus dependent antigens (T4 phage and sheep red blood cells) were intermediate between those of +/+ and those of nu/nu mice (37). The _in vitro_ mitogenic responses (of spleen and lymph node suspensions, to phytolectins known to stimulate T cells or to depend on their presence for the stimulation of B cells [47]) were much lower than in +/+ mice (37,39). Even 6-12 months after transplantation of a neonatal thymus these mitogenic responses were, compared to the responses of +/+ mice, 5-10% for concanavalin A, 15% for Phaseolus vulgaris phytohaemagglutinin and 25% for the pokeweed mitogen. The shapes of the dose response curves and the lectin doses giving the maximal responses were, however, similar to those of +/+ mice. It is remarkable that though poorly responsive these animals showed in their spleen and lymph nodes mean absolute and relative numbers of $\theta^+$ lymphocytes which were as high as 90-95% of the mean numbers found in +/+ animals (39). Thymus grafted nudes also show a reconstitution of the response to oxazalone (29) and they also recover the ability to reject skin grafts (10,29,37-39), but again they remain inferior to +/+ mice in that more time is required to achieve a full rejection of the skin graft, and very abnormal patterns of rejection are occasionally found (37-39). Moreover, it has been so far impossible (39,48) to correlate the fate of skin grafts syngeneic or semiallogeneic to the thymus graft (in some strain combinations, more frequently rejection, in others more frequently acceptance) with 1) the fate (acceptance or rejection) of the thymus graft, 2) the persistence of thymus donor type lymphocytes within the thymus (chimerism), 3) the presence of circulating antibodies directed against the skin and thymus grafts type θ antigens, 4) the Tla locus differences of the thymus donor and of the host (though Tla (associated) loci seem to be important for the fate of the thymus graft itself), 5) the in vitro responses of thymus grafted nude lymphocytes to allogeneic H-2 recognition (mixed lymphocyte culture). It is thus likely that several of these factors may play a role.

The actual reasons for the lower responsiveness of the T cells which differentiate in thymus grafted nudes are not yet known. Several non-mutually exclusive reasons can be put forward: 1) there is some other unknown effect of the pleiotropic nu gene; 2) not all T cell subsets are generated; 3) the lack of thymus during embryonic and early postnatal life irreversibly modifies the stem cell potential of the adult mouse bone marrow; 4) the chimerism found in the majority of old thymus grafted nudes is a direct cause for a general low (suppressed?) responsiveness of the T cells; etc...

A likely candidate for a precommitted T cell precursor function was recently identified (20,21,49-50). This T cell lineage precursor was present in all lymphoid organs of athymic mice (spleen, lymph nodes, bone marrow, Peyer's patches) (49). It was found not only in congenitally athymic mice but also in surgically thymus deprived mice (50). It lacks membrane immunoglobulins (20,21) and it has a low density of the θ antigen (20,21,49); in the spleen (but not in the bone marrow) of nu/nu mice (but not in +/+ mice) on a $TL^+$ mouse strain background (but not in $TL^-$ ones), that cell type has also some TL antigens (21,50). When analysed by cell electrophoresis, it shows a rather low overall negative charge, its electrophoretic mobility being much inferior to that of peripheral T cells and very much like that of thymus lymphocytes of +/+ mice (49). The T cell lineage precursor does not recirculate in the lymph but is found in the peripheral blood and is especially abundant in the spleen where it may represent some 20% of all the lymphocytes (49). *In vivo* it incorporates $^{3}$H-thymidine at a high rate compatible with lifespan as short as 1-2 days (50), which is markedly different from the lifespan of +/+ peripheral T cells which is of the order of weeks and months.. This cell type is normally produced in the bone marrow and is continuously entering the spleen as shown by its rapid disappearance from the spleen when the bone marrow is selectively destroyed by *in vivo* $^{89}$Sr irradiation (50). This cell type appears in the spleen within a few days following a surgical thymus deprivation, and it disappears from it within a few days after thymus (epithelium) grafting, even when the thymus is included in a diffusion chamber, but not after injection of thymus lymphocytes or spleen grafting (50). Therefore the production and/or release of the T cell lineage precursors by the bone marrow might be under humoral control of the thymus.

The role of the cells with low θ density as prethymic precursors is further supported by their presence in the thymus of +/+ day 13 embryos. Thus they are found at early stages of thymus differentiation (50) and in thymus grafted on adult nu/nu mice at early stages of repopulation by host cells (Loor *et al.*, unpublished).

Finally this cell type was found in all nudes (51,52) studied so far including newborn nudes born from homozygous parents which could in no possible way have been under the influence of a thymus product of mother, littermate or foster mother origin (Loor, Roelants and Dukor, unpublished). Therefore the differentiation step from the multipotential stem cells towards the T cell lineage precursors occurs in the complete absence of thymus influence i.e. contact with a thymus epithelium or thymus hormones.

In conclusion, as schematically shown below,

1. The differentiation towards the T lineage does not require the thymus
2. The thymus controls the size of the T cell lineage precursor pool through a humoral pathway.
3. Further differentiation of the T cell lineage precursors towards the more mature T cells requires their actual traffic through the thymus presumably for a close interaction with its epithelial component.

The diversification into various subsets of T lymphocytes and the post-thymic maturation of the T cells are not considered here (for a review, see 53).

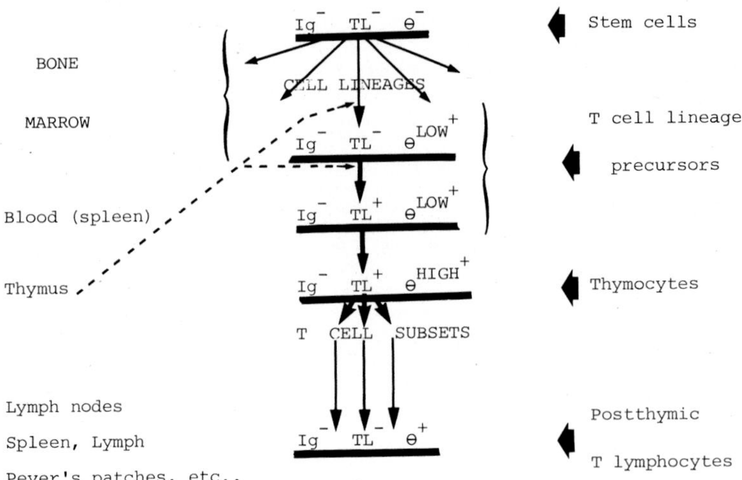


The pathway of T cell differentiation is shown schematically (see text for details). The thymus may exert its homeostatic control as indicated by the broken arrows. T cell lineage precursors have a lower Θ density and thymocytes a higher Θ density than postthymic T lymphocytes.

REFERENCES


1. Flanagan, S.P. 1966. Genet. Res. Cam. 8:295-309.
2. Pantelouris, E.M. 1973. Differentiation 1:437-450.
3. Pantelouris, E.M. 1968. Nature 217:370-371.
4. Pantelouris, E.M. and J. Hair. 1970. J. Embryol. exp. Morph. 24:615-623.
5. Cordier, A.C. 1974. J. Ultrastructure Res. 47:26-40.
6. Parrott, D.M.V., M.A.B. de Sousa and J. East. 1966. J. exp. Med. 123:191-204.
7. Sousa, M.A.B. de, D.M.V. Parrot and E.M. Pantelouris. 1969. Clin exp. Immunol. 4:637-644.
8. Guttman, G.A. and I.L. Weissman. 1972. Immunology 23:465-479.
9. Mitchell, J., J. Pye, M.C. Holmes and G.J.V. Nossal. 1972. Aust. J. exp. Med. Sci. 50:637-650.
10. Wortis, H.H., S. Nehlson and J.J. Owen. 1971. J. exp. Med. 134:681-692.
11. Basten, A., J.F.A.P. Miller, J. Sprent and J. Pye. 1972. J. exp. Med. 135:610-626.
12. Sprent, J. and J.F.A.P. Miller. 1972. Eur. J. Immunol. 2:384-387.

13. Miller, J.F.A.P., A. Basten, J. Sprent and C. Cheers. 1971. Cell. Immunol. 2:469-495.

14. Raff, M.C. and J.J.T. Owen. 1971. Eur. J. Immunol. 1:27-30.

15. Raff, M.C. 1971. Transplant. Rev. 6:52-80.

16. Lamelin, J.P., B. Lisowska-Bernstein, A. Matter, J.E. Ryser and P. Vassalli. 1972. J. exp. Med. 136:984-1007.

17. Raff, M.C. 1973. Nature 246:350-351.

18. Loor, F. and B. Kindred. 1973. J. exp. Med. 138:1044-1055.

19. Raff, M.C. and H.H. Wortis. 1970. Immunology 18:931-942.

20. Loor, F. and G.E. Roelants. 1974. Nature 251:229-230.

21. Loor, F. and G.E. Roelants. 1975. Ann. N.Y. Acad. Sci. 254:226-242.

22. Mandel, T. 1969. Aust. J. exp. Biol. Med. Sci. 47:153-155.

23. Owen, J.J.T. and M.A. Ritter. 1969. J. exp. Med. 129:431-437.

24. Mandel, T., P.J. Russell and W. Byrd. 1972. in: Cell Interactions Proceedings of the Third Lepetit Colloquium. L.G. Silvestri ed: North Holland. Amsterdam. 183-191.

25. Trainin, N. 1974. Physiol. Rev. 54:272-315.

26. Mandel, T. 1968. Aust. J. exp. Biol. Med. Sci. 46:755-767.

27. Cordier, A.C. and J.F. Heremans. 1975. 4:193-196.

28. Crisan, C. 1935. Z. Anat. Entwickl.-Gesch. 104:327-358.

29. Pritchard, H. and H.S. Micklem. 1973. Clin. exp. Immunol. 14:597-607.

30. Zipori, D. and N. Trainin. 1973. Blood 42:671-678.

31. Trainin, N. and P. Resnitzky. 1969. Nature 221:1154-1155.

32. Resnitzky, P., D. Zipori and N. Trainin. 1971. Blood 37:634-646.

33. Holub, M., I. Hajdu, L. Jaroskova and I. Trebichavsky. 1974. Z. Immun.-Forsch. 146:322-333.

34. Loor, F. and B. Kindred. 1974. in: Proc. 1st Intern. Workshop on Nude mice. J. Rygaard and C.D. Povlsen eds. Gustav Fisher Verlag. Stuttgart. 141-147.

35. Loor, F., N. Block and J.R. Little. 1975. Cell. Immunol. 17:351-365.

36. de Sousa, M. and H. Pritchard. 1974. Immunology 26:769-776.

37. Kindred, B. and F. Loor. 1974. J. exp. Med. 139:1215-1227.

38. Kindred, B. and F. Loor. 1974. in: Proc. 1st Intern. workshop on Nude mice. J. Rygaard and C.O. Povlsen. eds. Gustav Fisher Verlag. Stuttgart. 149-154.

39. Loor, F., B. Kindred and L.-B. Hägg. 1974. submitted for publication.

40. Mosier, D.E. and C.W. Pierce. 1972. J. exp. Med. 136:1484-1500.

41. Schlesinger, M. and I. Yron. 1970. J. Immunol. 104:798-804.

42. Isaak, D. 1973. Thesis. Montana State Univ. Bozeman.

43. Stutman, O. 1974. Fed. Proc. abstracts. 33:736.

44. Kindred, B. and F. Loor. 1975. Cell. Immunol. 16:432-438.

45. Kindred, B. and F. Loor. 1975. Am. Zool. 15:175-179.

46. Loor, F. and L.-B. Hägg. 1974. submitted for publication.

47. Loor, F. 1974. Eur. J. Immunol. 4:210-220.

48. Kindred, B. 1974. submitted for publication.

49. Roelants, G.E., F. Loor, H. von Boehmer, J. Sprent, L.-B. Hägg, K.S. Mayor and A. Rydén. 1975. Eur. J. Immunol. 5:127-131.

50. Roelants, G.E., K.S. Mayor, L.-B. Hägg and F. Loor. 1974. Eur. J. Immunol. in press.

51. Loor, F., L.-B. Hägg, K.S. Mayor, and G.E. roelants. 1975. Nature 255:657-658.

52. Roelants, G.E., K.S. Mayor, L.-B. Hägg and F. Loor. 1975. in: Proc. of the fifth Intern. Conf. on Lymphatic Tissue and Germinal Centers in Immune Reaction. M. Feldman. ed. Plenum Publish. Corp. New York, in press.

53. Cantor, H., and E.A. Boyse, 1975. J. Reticuloendoth. Soc. 17:115-118.

*Progress in Differentiation Research, ed. N. Müller-Bérat et al.*



# ANALYSIS OF THE TIME OF CELL DETERMINATION IN THE IMMUNE SYSTEM

J.P. Johnson

Basel Institute for Immunology
Basel 4058, Switzerland

## 1. Introduction

The mammalian immune system is a useful model for the study of determination and differentiation of specific cell populations. An almost infinite number of different antibody forming cell (AFC) populations, distinguishable by the specificity of their antibody (Ab) products, can be detected, quantitated and induced to express their differentiated function by presentation of the specific antigen (Ag). In addition, the existence of allelic exclusion results in a natural mosaicism within the AFC populations of heterozygous animals. Individual AFCs produce unique Ab proteins. This Ab specifity is a clonal trait and the progeny of individual AFCs produce the same Ab molecule. The restriction of a precursor cell to the production of a single Ab molecule, and the establishment of its clonal inheritance is the determination (or commitment) of the cell. Differentiation is the process by which committed precursors become populations of active AFCs.

The binomial sampling model can be applied to mosaic animals to calculate "primordial precursor pool sizes" for various differentiated cell populations[1]. These precursor numbers reflect the number of cells which have contributed progeny to the population under investigation, i.e. the number of clones comprising the population[2]. With respect to individual AFC populations, these primordial precursor pools should reflect the number of clones of AFC precursors which are committed to the production of a specific Ab whose progeny make up the observed AFC population. The bionomial sampling model was applied to the natural AFC mosaicism existing in heterozygous mice and to the artificial AFC mosaicism existing in tetraparental mice to generate information about the determination of AFC precursors and its relationship to Ag exposure and to allelic exclusion.

## 2. Materials and Methods

The mosaic development model has been reviewed[1] and the assumptions involved in its application of the development of AFCs discussed[3]. Briefly this model envisions the determination of AFC precursors as a binomial sampling situation in which cells are withdrawn from the undetermined precursor pool randomly with respect to mosaic marker (immunoglobulin allotype) and set aside to give rise through proliferation and differentiation to specific AFC populations. When a particular AFC population is examined in a series of mosaic individuals, the mosaic composition of the samples will follow a binomial distribution $(p+q)^N$. Using the binomial nature of the situation, it is possible to calculate the number of cells sampled from the precursor population to form the specific AFC populations. The sample size N is given by the equation $N=pq/$ variance $P_{AFC}$ where p and q represent the mosaic composition of the precursor pool $(p+q=1)$, and $P_{AFC}$ is the mosaic composition of the specific AFC population.

Mice heterozygous (C3H x C57Bl/6 $F_1$, Ig-1a/Ig-1b) or chimeric (C3H↔C57Bl/6, Ig-1a↔Ig-1b) for the immunoglobulin allotypic marker of the Ig-1 locus were used. The Ig-1a and Ig-1b in serum Ab and in total immunoglobulin was determined using a quantitative hemagglutination inhibition assay[4]. This parameter was used to characterize the mosaic composition of specific AFC populations and the AFC precursor population (estimated from the mosaic composition of the total serum immunoglobulin). The mosaic composition of a given AFC population was estimated from the allotype proportion of its secreted Ab, and was expressed as its $P_a$ where $P_a$ = (Ig-a)/(Ig-1a + Ig-1b). Antigens were prepared using standard procedures.

3. The Relationship of Antigen Exposure to AFC Precursor Determination

One of the great advantages of the immune system for developmental studies lies in the ability of the investigators to induce the expression of selected populations of cells simply by presenting the Ag. However, it is not clear whether the Ag acts simply to select those precursors already committed to the production of Ab of the proper specificity, or whether the Ag may in fact act to commit the multi-potential cell to the production of a given specificity[5,6]. The inability to easily control environmental Ag exposure has made this question difficult to approach. However if the Ag is able to trigger the determination of cells rather than simply acting to trigger the differentiation of committed precursors, it should be possible to alter the number of clonal precursors calculated using the binomial sampling model by variations in the dose and presentation of Ag. If AFC precursors have undergone determination prior to Ag encounter, the number of clonal precursors calculated for a given AFC population should be a finite number which varies only with the number of different determinants on the Ag and is unalterable by changes in the amount of Ag presented.

230 Ig-1a/Ig-1b heterozygotes and 17 Ig-1a ↔ Ig-1b chimeras were immunized with 100 ug of a hapten-protein complex (DNPLS-papain, DNP-HSA, or Xg-BGG) in incomplete adjuvant. The mice were bled 14 to 28 days later and the amount of Ig-1a and Ig-1b in the specific Ab determined. The mean mosaic composition of the total serum immunoglobulin in these mice was p = .465, q = .535 for heterozygotes and p = .579, q = .421 for chimeras. From the variance of the mosaic composition ($P_a$), precursor numbers were calculated for each of the specific AFC populations (table 1).

Table 1

Precursor pool sizes for different AFC populations

| AFC population | number of mice | mean Pa | $s^2$ | N |
|---|---|---|---|---|
| Ig-1a/Ig-1b mice | | | | |
| DNPLS | 85 | .499 | .051 | 5+ |
| Xp | 14 | .499 | .050 | 5 |
| Papain | 96 | .532 | .018 | 14**++ |
| BGG | 14 | .492 | .007 | 36** |
| DNP | 24 | .454 | .005 | 50** |
| HSA | 18 | .484 | .003 | 83** |
| Total Ig | 230 | .465 | .018 | 14** |
| Ig-1a ↔ Ig-1b mice | | | | |
| DNPLS | 12 | .466 | .089 | 3+ |
| Papain | 12 | .552 | .028 | 9**++ |
| Total Ig | 17 | .526 | | 4 |

** $P < 0.01$ N values different from N=5 by F test
+ n.s. by F test
++ n.s. by F test

Superficially it would appear that Ags with a larger number of different determinants (HSA, papain, BGG, DNP) trigger the differentiation of cells derived from a larger number of precursor clones than do Ags with a smaller number of different determinants (DNPL, Xp). The observed difference between the anti DNP AFC populations responding to DNP-HSA and DNPLS-papain is consistent with known differences in the number of different DNP determinants in these two Ags. DNP-papain is prepared by the covalent coupling of a molecule of DNPL to the single SH group on the papain molecule. Since the DNPL group can therefore only exist in a single molecular environment, the DNP determinant should approach a single Ag. In contrast, the hapten protein DNP-HSA is a highly heterogeneous Ag with DNP groups existing in a wide variety of molecular environments, a situation which would be expected to lead to the presence of a large number of different DNP determinants in this Ag.

The observed differences in calculated precursor pool sizes between AFC populations responding to different Ags could be a result of differences in the total number of determinants presented (i.e. dose), rather than a reflection of the actual differences in the absolute number of precursors of different specificities. To investigate this point, animals were immunized with doses of DNPLS papain covering a 10 fold range. In addition, the number of haptenic groups per molecule of protein was varied over a 20 fold range. As can be seen from Table 2, these manipulations failed to significantly alter the calculated precursor clone sizes for any of the AFC populations. Further investigation revealed that the calculated precursor pool sizes for the various AFC populations were remarkably constant and could not be altered by the manner of Ag presentation by time after immunization, or by repeated Ag exposure.

Table 2

Effect of antigen dose on precursor pool sizes in heterozygous animals

| Antigen | Dose (μg) | AFC population | $S^2$ | No. of mice | N |
|---|---|---|---|---|---|
| $DNPLS_{0.3}$papain | 10 | DNPLS | .035 | 13 | 7* |
| | 100 | " | .052 | 18 | 5* |
| $DNPLS_{1.3}$papain | 100 | " | .051 | 16 | 5* |
| $DNPLS_{3.0}$papain | 100 | " | .064 | 19 | 4* |
| $DNPLS_{7.0}$papain | 100 | " | .038 | 7 | 6* |
| $DNP_{48}$ HSA | 100 | DNP | .005 | 24 | 50** |
| $DNP_{3}$ HSA | 100 | " | .004 | 17 | 62** |

* values not different from N = 5 by F test
** values not different from N = 50 by F test

4. The Time of Allelic Exclusion relative to the Time of Precursor determination

As presented in Table 1, the precursor clone sizes calculated for a given AFC population are not significantly different in allelic exclusion mosaics and tetraparental chimeras. If allelic exclusion mosaicism occurred significantly later than AFC precursor commitment, the clone numbers in Ig-1a/Ig-1b mice would be much larger than those in Ig-1a↔Ig-1b mice, as the generation of mosaicism in a large population of specific AFCs would look like a large sample from a mosaic precursor population. Therefore allelic exclusion must occur at least as early as AFC determination. The variance of the mosaic composition of the total serum immunoglobulin can be used in an analagous manner to estimate the number of primordial precursors of the entire AFC precursor pool, which is generated from hematopoietic stem cells. This comparison leads to the conclusion that the latest allelic exclusion can occur is between first and second cell division of the primordial AFC precursor cells.

References

1. Nesbitt, M.N. and Gartler, S.M. 1971. Ann.Rev.Gen. 5:143.
2. McLaren, A. 1972. Nature 239:274.
3. Johnson, J.P. and Coons, A.H. 1975, in preparation.
4. Johnson, J.P. and Coons, A.H. 1975, submitted, J. Immunol.
5. Kim, Y.B. 1975, in Immunodeficiency in Man and Animals, ed. D. Borgsma (Sinauer Associates, Inc., Sunderland, Mass.) p.549.
6. Cooper, M.D., Lawton, A.R. and Kincade, P.W. 1972. Contemporary Topics in Immunology, 1:33.

*Progress in Differentiation Research, ed. N. Müller-Bérat et al.*
*© 1976, North-Holland Publishing Company - Amsterdam, The Netherlands.*



# DIFFERENTIATION OF B CELLS INTO SPECIFIC ANTIBODY FORMING CELLS INDUCED BY RNA EXTRACTED FROM ACTIVATED T CELLS

B. Kayibanda and C. Rosenfeld

Département de Culture de Cellules Humaines
I.C.I.G. - Hôpital Paul Brousse - Villejuif

## ABSTRACT

Cytoplasmic RNA extracted from T cells activated by a thymus-dependent antigen (informational RNA : iRNA) induced a great increase in DNA synthesis in B cells of syngenic mice which had been thymectomized at birth. This DNA synthesis was followed by a parallel increase in the synthesis of total cellular RNA and proteins, finally resulting in proliferation and differentiation of specific antibody-forming B cells.

The biological activity of iRNA was associated with an 8-12 S single-stranded RNA and was abolished by RNase but not by DNase or pronase.

The preparations of iRNA were free of any protein and DNA, as determined by appropriate tests.

RNA from non-activated T cells (niRNA) of syngenic animals had no effect on differentiation of B cells.

## INTRODUCTION

Differentiation of specific antibody-forming B cells for most complex antigens requires "T cell help"[1]. Certain experiments have indicated that activated T cells release soluble factors which stimulate the differentiation of B cells[2]. Some of these factors seem to be specific[3] and appear to be under control of T cell genes other than those for immunoglobulin[4]. These immune response genes (iR) are linked to the histocompatibility locus (H-2 complex)[4].

The intermediate between T and B cells must be an informed and informational substance (factor) synthesized by activated T cells and transferable to B cells[5,6].

The messenger type of RNA is both informed and informational, contains RNA which can be transferred and translated[7].

Furthermore, the transfer of specific information by extracts containing RNA from activated T cells has been described by different authors[5,6,8]; therefore, we believed that RNA would be a good candidate for study. In this paper, we will describe experiments supporting this hypothesis.

## MATERIALS AND METHODS

### 1. Animals, antigens and cells

CBA mice were used for experiments. Sheep red blood cells(SRBC), human red blood cells(HRBC) and encephalomyocarditis virus (EMC) labelled with (U-$^{14}$C)-amino acids were used as antigens.

Pure population of B cells were obtained *in vivo* by neonatal thymectomy and recovered from spleen[10].

T cells activated *in vivo* by antigens were obtained by injection of thymus cells into heavily irradiated CBA mice and subjected to a strong immunogenic stimulus. As a result, the cells become specifically activated and proliferate to produce a progeny of T cells which can be recovered from the spleen[11].

Non-activated T cells were obtained in the same way as the activated T cells except for the omission of the immunogenic stimuls.

Labelling of activated T cells with either $^3$H-uridine, $^3$H-thymidine or $^{14}$C-amino acids was carried out as previously described[12,13].

### 2. Isolation and purification of informational RNA (iRNA) and non-informational RNA (niRNA)

Cytoplasmic ribonucleoprotein particles (cRNP) were isolated from activated and non-activated T cells according to KAYIBANDA[14]. iRNA was extracted from cRNP isolated from activated T cells and niRNA was extracted from cRNP isolated from non-activated T cells;extraction and analysis of RNA was carried out as previously described[15,16]. The preparations of RNA were purified by $Cs_2SO_4$ density gradient centrifugation[17] and only fractions pooled from the 1.68 g/cm$^3$ density region (Fig. 3) were used after removal of excess salt by gel filtration and dilution in DEAE-Dextran phosphate buffer solution.

### 3. Measurement of biological activity of RNA and antigens

The biological activity of RNA and antigens were measured by estimation of plaque forming cells (PFC) according to Jerne et al.[18]

## RESULTS

### The evidence of specific differentiation of B cells induced by iRNA

The protocol and data from a representative experiment using CBA mice thymectomized at birth as source of B cells demonstrating the involvement of iRNA in the physiological differentiation of B cells is shown in Table I.

The thymectomized CBA mice were divided into 5 groups: group 1 received intraperitoneally, an injection of $10^9$ sheep red blood cells (SRBC) as antigen, group 2 received intraperitoneally injection of 200 µg of iRNA extracted from T cells of SRBC-sensitized mice. The iRNA was dissolved in sodium phosphate buffer, pH 7.2, containing 100 µg/ml of DEAE-Dextran. Group 3 received intraperitoneally an injection of 200 µg of niRNA extracted from non-activated T cells dissolved in the same buffer as in group 2. Group 4 received only the buffer used in groups 2 and 3. Group 5 : untreated mice. Differentiated B cells were recovered from the spleens of these mice and measured 4 days after stimulation by estimation of anti-SRBC direct plaque forming cells. As shown in Table I, only the mice which had been treated with iRNA produced differentiated B cells synthesizing anti-SRBC antibody (Table I).

TABLE I

Effect of iRNA on differentiation of B cells to synthesize specific hemolysin anti-SRBC

B cells synthesizing anti-SRBC hemolysin were measured 4 days after injection. The comparison of means between the group of thymectomized mice + SRBC and the group of thymectomized mice + iRNA was made according to the Student Fisher Test and was significant at 0.1%.

| Substance injected | B cells differentiated into anti-SRBC PFC/spleen |
|---|---|
| SRBC (antigen) | 43 |
| iRNA extracted from SRBC activated T cells | 1 281 |
| niRNA extracted from non-activated T cells | 88 |
| Buffer of RNA | 57 |
| Untreated control | 32 |

RNA extracted from spleen cells containing a mixture of B cells and macrophages of antigen-stimulated mice thymectomized at birth did not induce differentiation of B cells, indicating that the iRNA cannot be of macrophage or B cell origin, but is dependent on the presence of T cells.

There is a dose dependent relationship between injected iRNA and the number of differentiated B cells obtained. It has to be remarked that a plateau is observed: at high concentration of iRNA, indicating that the number of target B cells may be limited (Table II).

The kinetics of appearance of differentiated B cells in the spleen of recipient mice following injection with iRNA show an exponential increase of differentiated B cells between day 2 and day 4. The injection of normal non-thymectomized mice with SRBC is followed by a similar pattern (Fig. 1 A and B).

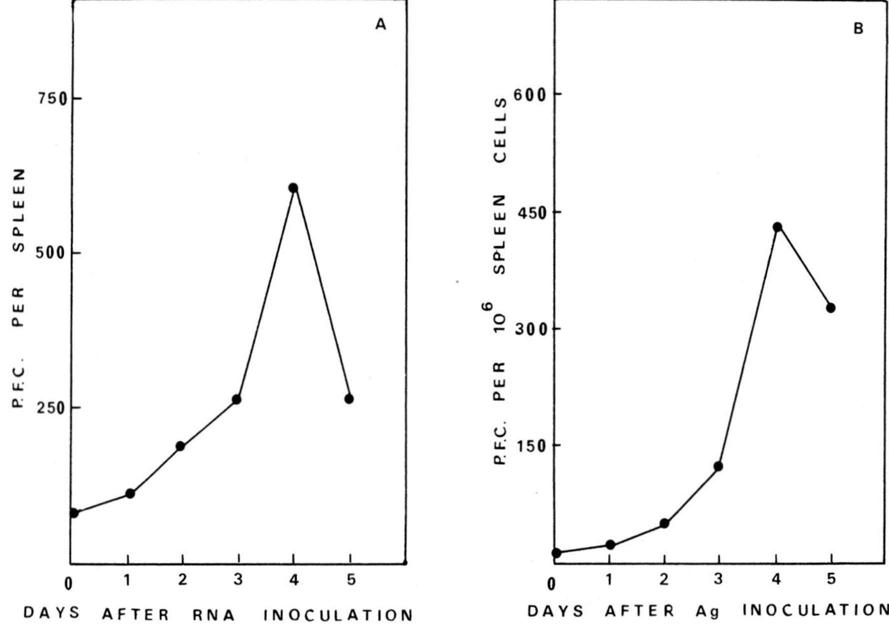


Fig. 1 : Kinetics of the rate of appearance of B cells differentiated to synthesize anti-SRBC hemolysin after injection of iRNA or SRBC in thymectomized mice. Kinetics of appearance of B cells differentiated into anti-SRBC-PFC spleen in thymectomized CBA mice after injection of 200 μg of iRNA per mouse(A) and in normal non-thymectomized mice after injecting $10^9$ SRBC per mouse (B). Each dot represents the geometric mean of values obtained from 10 mice.

TABLE II

Effect of iRNA dose on the number of B cells differentiated to synthesize specific hemolysin anti-SRBC

B cells synthesizing anti-SRBC-hemolysin were measured 4 days after RNA injection. Each figure represents the geometric mean of values obtained from 10 mice.

| Dose injected of iRNA extracted from SRBC-activated T cells/ mice | Differentiated B cells producing anti-SRBC-PFC/spleen |
|---|---|
| 0 | 51 |
| 50 μg of iRNA | 328 |
| 100 μg of iRNA | 500 |
| 200 μg of iRNA | 910 |
| 300 μg of iRNA | 1098 |

## Specificity of differentiated B cells induced by iRNA

The specificity of differentiation of B cells was verified by the following experiments : iRNA extracted from SRBC activated T cells induced B cell production of anti-SRBC hemolysin synthesis, but not anti-HRBC. In corresponding experiments, it was found that iRNA extracted from HRBC-activated T cells induce specifically formation of B cells synthesizing hemolysin but not anti-SRBC (Table III).

TABLE III

Specificity of differentiated B cells induced by iRNA

B cells synthesizing specific hemolysin were measured 4 days after RNA injection. Each figure represents the geometric mean of values obtained from 10 mice.

| Substance injected | Differentiated B cells Anti-SRBC PFC/spleen | Anti-HRBC PFC/spleen |
|---|---|---|
| iRNA extracted from T cells educated by SRBC | 971 | 0 |
| iRNA extracted from T cells educated by HRBC | 0 | 788 |

The specificity of response provides strong evidence against the concept that iRNA serves simply as adjuvant.

The iRNA induced a large increase in DNA, RNA and protein synthesis in B cells. The kinetic studies have revealed a direct correlation between this DNA synthesis and proliferation of differentiated B cells (results to be published).

## Properties of the informational RNA

In the experiments described above total cytoplasmic RNA extracted from iRNA of activated T cells were used; it was therefore necessary to characterise which RNA species had the differentiation capacity.

a) Preparation of cytoplasmic iRNA from activated and $^3$H-uridine labelled T cells were fractionated on a sucrose gradient. Different regions of the gradient were pooled and assayed for their ability to induce specific differentiation of B cells. Under all experimental conditions tested the ability to induce differentiation of B cells was associated with material exhibiting a sedimentation coefficient of 8-9 S which indicates the molecular weight to be about 1.14 x $10^5$ to 1.76 x $10^5$ daltons (Fig. 2 and Table IV).

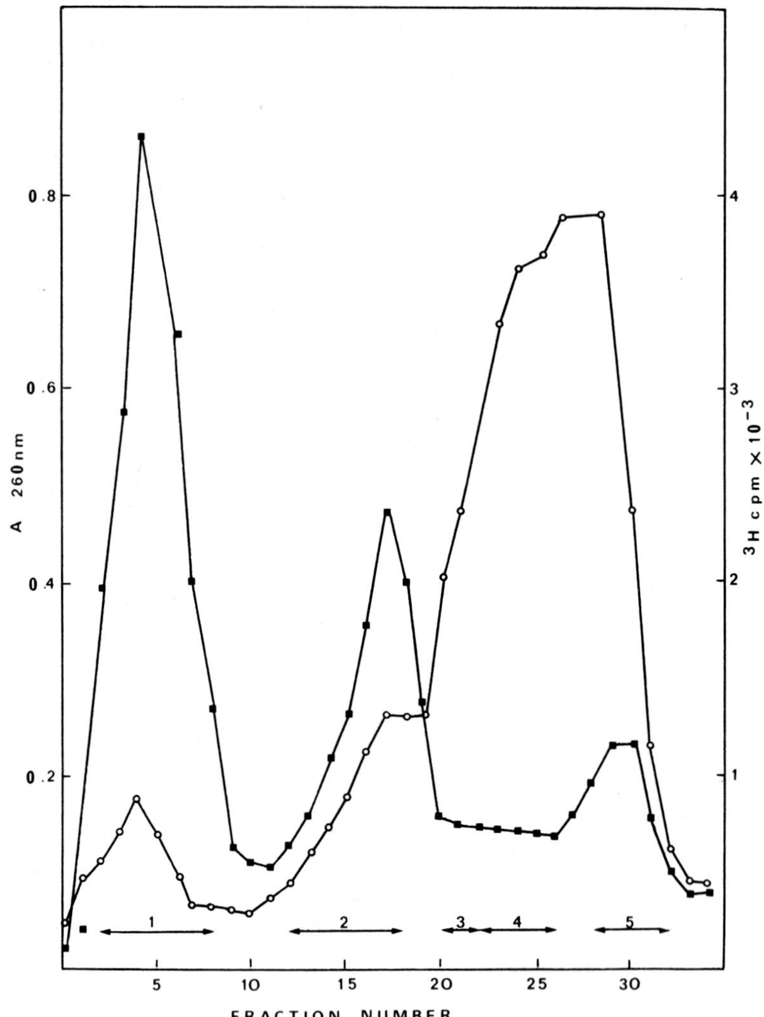


Fig. 2 : Sucrose gradient analyais of iRNA.
Sucrose gradients (5-20%) containing $10^{-2}$M NaCl, $10^{-2}$M Triethanolamine-HCl (pH 7.4) and $10^{-3}$M EDTA were prepared. The cytoplasmic RNA extracted from SRBC-activated and $^{3}$H-uridine labelled T cells were layered on top of the gradients. The gradients were centrifuged in the SW 25 rotor at 24 000 rpm for 24h at 4°C. The gradients were collected and the $A_{260}$ of every other fraction was determined (■—■). An aliquot of each fraction was then precipitated with trichloroacetic acid and assayed for radioactivity (o—o). Samples from RNA gradients were pooled into five groups (as indicated) and assayed for ability to induce anti-SRBC hemolysin synthesis by B cells ( see Table IV).

TABLE IV

Differentiation of B cells into specific antibody forming cells by fractionated cytoplasmic iRNA extracted from T cells educated by SRBC (see Fig.2, pool 4)

Different pools of iRNA fractionated by sucrose gradient (see Fig. 2) were incubated with $10^8$ syngeneic spleen B cells and injected in thymectomized mice. 4 days later the animals were killed; their spleens were removed and single suspensions were prepared for quantification of B cells differentiated into anti-SRBC-PFC/spleen.

| RNA fractions | Quantity inoculated | B cells differentiated into anti-SRBC PFC+ |
|---|---|---|
| None | - | 50 |
| Pool 1 : 28 S | 200 µg | 68 |
| Pool 2 : 18 S | 200 µg | 113 |
| Pool 3 : RNA heavy region of gradient | 20 µg | 307 |
| Pool 4 : RNA light region of gradient | 20 µg | 988 |
| Pool 5 : 4 S | 200 µg | 144 |

+ The number of B cell anti-SRBC-PFC represents the mean from 10 mice.

b) On a $Cs_2SO_4$ gradient, the active iRNA displayed a density of 1.68 g/cm$^3$, corresponding to that of pure single-stranded RNA (Fig. 3).

c) When activated T cells used for cytoplasmic RNA preparation were labelled with $^3$H-thymidine or with $^{14}$C-amino acids no radioactivity was found in the RNA density range, indicating the absence of DNA, DNA-RNA hybrids and protein contamination. Attempts to detect significant quantities of DNA and proteins in the most purified RNA fractions by colorimetric methods have been unsuccessful

d) Furthermore, when (U-$^{14}$C)-amino acid labelled EMC was used as antigen to activate T cells, no radioactivity could be detected in the RNA density zone, indicating that the persistence of antigens is most unlikely.

In addition, concentration of antigen small enough to escape detection are also too small to stimulate antibody production; for example, a 1% contamination of the iRNA by protein antigens would be equivalent to only 0.2 ug of iRNA injected (see Table IV, pool 4). Whereas $10^9$ SRBC did not elicit a response.

e) $^3$H-uridine labelled iRNA was completely precipated with 2 M lithium chloride and could be eluted from hydroxyapatite column by a low (0.20 M) concentration of phosphate buffer (Fig. 4), strongly suggesting that iRNA is a single-stranded molecule.

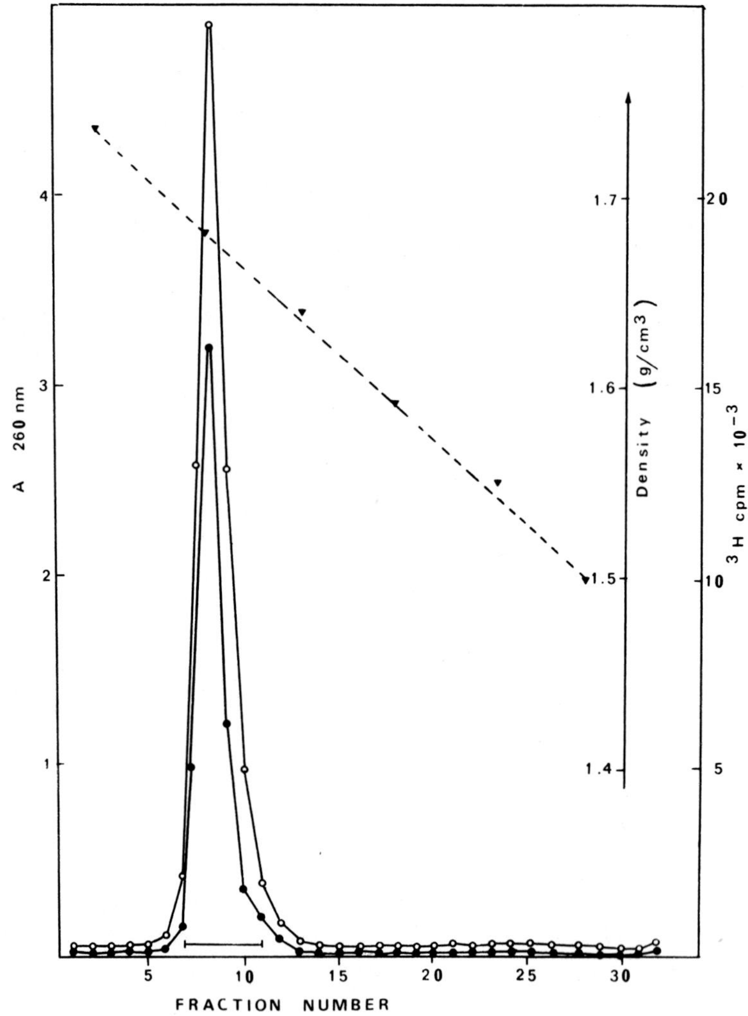


Fig. 3 : Bouyant density profile of iRNA by $Cs_2SO_4$ equilibrium density. Pool 4 of iRNA (see Fig. 2 and Table IV) was precipitated with ethanol, dissolved in buffer containing 10mM NaCl-10mM Tris-HCl (pH 7.4)-3mM EDTA, and an equal volume of saturated $Cs_2SO_4$ was added. Centrifugation was performed in the Spinco 50.1 Ti rotor at 33 000 rpm at 15°C for 72h, after which 10 drops per fraction were collected. The density (▲---▲) and the absorbance of the fractions at 260nm (●—●) were determined. Aliquots of each gradient for radioactivity (o—o).

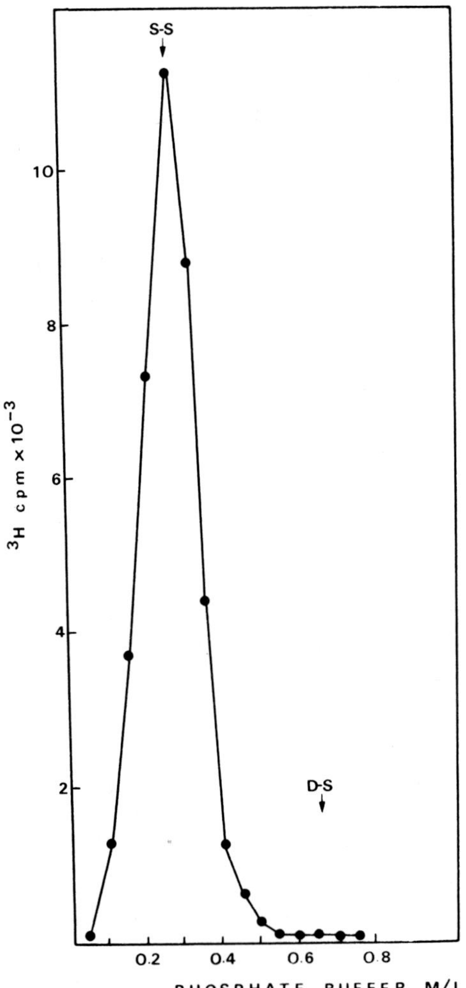


Fig. 4: Hydroxylapatite chromatography of iRNA. $^3$H-uridine-labelled RNA from pool 4 of Fig. 2 was loaded on hydroxylapatite and eluted at different phosphate buffer (pH7.8) molarities, processed for trichloroacetic acid precipation, and assayed for radioactivity (●——●). The arrows indicate the molarities of elution of single-stranded (SS) and double-stranded (DS) DNA molecules as marker.

f) Chromatography on poly-U-sepharose reveal that 73% of labelled iRNA contains a polyadenylic acid sequence.

The results of these experiments taken together, indicate that a single-stranded RNA from activated T cells can induce differentiation of B cells into specific antibody-forming cells without the participation of T cells and antigen. The antigen itself (SRBC or HRBC) was unable to induce differentiation of B cells in T cells deprived animals.

## DISCUSSION

### How does the iRNA act to produce differentiation of B cells?

1) One may hypothesize that iRNA is the immunoglobulin messenger, but it can only be translated into protein by B cells. This hypothesis however requires several assumptions which are not adequately supported by the experimental results.
a) The fact that exponential kinetics over several days precede maximal differentiation of B cells (Fig. 1) indicates that the iRNA cannot be directly translated (as, *e.g.*, when mRNA is injected in oocytes[6]) but that additional events must precede the expression of iRNA-contained information.
b) The iRNA sediments not as immunoglobulin mRNA but at approximately 8-9 S, which indicates a molecular size hardly sufficient to code for the full immunoglobulin chains.
c) Genes coding H and L chains have been shown to be localized in B cells, probably on the same chromosome and their expression has been shown to be under the same regulation[19].

2) An alternative hypothesis is that iRNA acts on B cells as a regulating agent activating specific information which pre-existed in B cells. However, this is unlikely in view of the results of Bell *et al.*[20] which showed that the antibodies produced after injection of iRNA are those of the donor alleles and not those of the recipient animal.

3) Finally, it is possible that iRNA represents the specific mRNA for the variable part of the immunoglobulin. This RNA, once introduced into the B cells may be transcribed into DNA by a "reverse transcriptase", and the DNA product integrated into the B cell DNA to form the complete antibody genes. The hypothesis that RNA-directed DNA synthesis may be involved in immunological processes has been formulated by different authors[21, 22].

The DNA transcribed from iRNA might code for variable regions of immunoglobulin after integration in the chromosome of B cells; the DNA of B cells would then code for constant regions and transcription might occur then in the form of single molecule of immunoglobulin mRNA after integration and recombination of variable and constant genes.

Alternatively, this portion of DNA transcribed on the iRNA's could remain as extrachromosomal genetic information in B cells (episome) and would be the actual repository of immunological memory.

Suggestive in this respect are the demonstration that reverse transcriptase may be induced in immunocompetent cells after contact with antigens[23] and our preliminary results (to be published: B. Kayibanda, N. Fossar and J. Huppert) which reveal that iRNA is used as template by AMV reverse transcriptase in the absence of added primer.

This indicates that iRNA contains natural endogenous primer and is transcribed into DNA in vivo.

The data and observations presented in this paper give a possible model for differentiation of B cells into specific antibody synthesizing cells, and provide the basis for understanding the diversity of antibody by the mechanism of the reverse transcription, integration and recombination of immunoglobulin genes. But it remains to be explained why only a part of B cells is sensible to iRNA.

ACKNOWLEDGMENTS

We wish to thank Drs : J. HUPPERT and L. PRITCHARD for critical reading of this manuscript; and for gifts of encephalomyocarditis virus labelled with $^{14}C$-amino acids. This investigation was supported by a grant from I.N.S.E.R.M. (ATP No. 71.5.440.2).

REFERENCES

1 CLAMAN H.N. et al. (1966) J. Immunol., 97, 828

2 FELDMANN R.E. et al. (1973) Cell Immunol., 9, 1

3 KATZ D.H. et al. (1975) in Molecular Approaches to Immunol., SMITH E.E. and RIBBONS, Academic Press, New York, p.221

4 KAYIBANDA B. and ROSENFELD C. (1974) in Interaction Cellulaire, INSERM Paris (Ed. 13)

5 KAYIBANDA B. and ROSENFELD C. (1974) C.R. Acad. Sci., Paris, 279, 1645

6 LANE C.D. et al. (1971) J. Mol. Biol., 61, 73

7 COHEN E.P. et al. (1973) Annals of N.Y. Acad. Sci., 207, 83

8 KAYIBANDA B. and ROSENFELD C. (1975) in Abstracts FEBS Meeting in Paris, (1975), p. 1592

9 KAYIBANDA B. and ROSENFELD C. (1975) in Abstracts Second International Conference on Differentiation, p. 109

10 MILLER J.F.A.P. and MITCHELL G.F. (1969) Transplant. Rev. 1, 3

11 MILLER J.F.A.P. and MITCHELL G.F. in Control Processes in Multicellular Organisms, G.E.W. WOLT STENHOLME and J. KNIGHT, eds., Ciba Found. Symp., Churchill, London, p. 238, 1970

12 KAYIBANDA B., AMIEL J.L. et al. (1968) C.R. Acad. Sci. Paris, 267, 977

13 KAYIBANDA B., AMIEL J.L. et al. (1970) Europ J. Clin. Biol. Res. 15, 437

14 KAYIBANDA B. (1973) These de Doctorat des Sciences, Université de Paris VI

15 MOREL C., KAYIBANDA B. et al. (1971) FEBS Letters 18, 84

16 SPOHR G. and KAYIBANDA B. (1972) Europ. J. Biochem., 31, 194

17 ERICKSON R.L. (1969) in Fundamental Techniques in Virology, HABEL and SALZAMAN eds., Acad. Press, New York, p. 406

18 JERNE N.K. and NORDIN A.A. (1963) Science, 140, 505

19 MAGE R. (1971) Ann. N.Y. Acad. Sci., 180, 203

20 BELL C. and DRAY J. (1970) J. Immunol., 105, 541

21 BALTIMORE D. (1970) Cold Spring Harbor Symp. Quant. Biol., 35, 843

22 GRIPPA M. and TOCCHINI VALENTINI G.P. (1971) Proc. Natl. Acad. Sci., U.S., 68, 2769

23 JACHERTZ D. (1973) Ann. N.Y. Acad. Sci., 207, 122

# Author index

# Subject index